THE

THEORY OF

FUNCTIONS

THE THEORY OF FUNCTIONS

BY

E. C. TITCHMARSH, M.A., F.R.S.
Late Savilian Professor of Geometry in the University of Oxford

SECOND EDITION

OXFORD UNIVERSITY PRESS

Oxford University Press, Walton Street, Oxford OX2 6DP

Oxford New York Toronto
Delhi Bombay Calcutta Madras Karachi
Petaling Jaya Singapore Hong Kong Tokyo
Nairobi Dar es Salaam Cape Town
Melbourne Auckland

and associated companies in
Beirut Berlin Ibadan Nicosia

Oxford is a trade mark of Oxford University Press

Published in the United States
by Oxford University Press, New York

ISBN 0 19 853349 7

First edition 1932
Second edition 1939
Reprinted 1944 (corrected), 1947 (corrected), 1949, 1950
1952 (corrected), 1958, 1960, 1964, 1968 (corrected), 1975, 1978, 1979, 1983, 1985, 1986

Printed in Great Britain by
Antony Rowe Ltd, Chippenham

PREFACE TO SECOND EDITION

A CONSIDERABLE number of minor corrections and improvements have been made in this reprint of the first edition. I have to thank a large number of colleagues who have helped me in the revision. The only major change is that I have inserted in Chapter VIII a short introduction to the theory of meromorphic functions. This has been made possible by compressing some comparatively unimportant sections, and transferring the theory of the gamma-function to Chapter IV, where it now includes a more complete discussion of Stirling's formula.

E. C. T.

PREFACE TO FIRST EDITION

THIS volume is a development of the notes from which I have lectured in recent years to students at University College, London, and Liverpool University. It consists of some rather disconnected introductions to various branches of the theory of functions, both real and complex. I think the average student finds the existing literature on these subjects rather formidable, and I hope that these chapters will do something to bridge the gap between the elementary text-books and the systematic treatises on the theory of functions.

A knowledge of elementary analysis is assumed. By elementary analysis we mean, roughly, what is contained in Hardy's *Course of Pure Mathematics*. Apart from this the work is self-contained. The order in which the chapters occur is to a certain extent arbitrary. The last four chapters might well come after Chapter I. Apart from occasional references forward, the earlier part of the book is independent of these chapters; but what they contain is part of the necessary equipment of the analyst of to-day, just as much as the older theory of analytic functions.

A number of miscellaneous examples are given at the ends of the chapters. Some of them are more or less immediate applications of the book-work. Others are more difficult theorems which have not found a place in the text; these are accompanied by indications of the solution, and references to the sources.

When I first proposed to put my notes into the form of a book, Professor Hardy very generously offered to work through them in connexion with his lectures at Oxford, and they have been revised with the help of the notes which he made during this process. I have adopted a very large number of improvements from Professor Hardy's notes, and I wish to express my very deep gratitude for the assistance which he has given.

I have also to thank Mr. U. S. Haslam-Jones and Dr. B. M. Wilson, who have read the proofs and made a large number of useful suggestions.

E. C. T.

REFERENCES TO HARDY'S *PURE MATHEMATICS*

This book refers to the sixth edition of the above but the corresponding references in the seventh edition may be found from the following table:

Sixth Edition	*Seventh Edition*	*Sixth Edition*	*Seventh Edition*	*Sixth Edition*	*Seventh Edition*
99	100	160, 161	165, 166	193–4	200–1
101–2	102–3	167, 168	173, 175	206	213
105–6	106–7	175	181	208	215
125	126	177–8	184–5	213–14	220–1
146	149	180–1	187–8	222	229
153, 154	157, 159	184–5	191–2	224	231
156–64	161–9	189–90	196–7	233	240

The following alterations to references should also be noted:

Page 10, footnote, *for* Ex. 27 *read* Ex. 16
19, line 20, *for* ex. 32 *read* ex. 36
20, line 2 from foot, *for* § 184 *read* § 203
31, line 3, *for* ex. (xv) *read* ex. (xvii)
43, lines 2–3, *for* § 181, exs. LXXVI, 9–10 *read* § 188, exs. LXXVI, 8–9
51, footnote †, *for* ex. 20 *read* ex. 15
305, footnote *, *for* ex. 42 *read* ex. 44

CONTENTS

CHAPTER I

INFINITE SERIES, PRODUCTS, AND INTEGRALS

CHAPTER II

ANALYTIC FUNCTIONS

CHAPTER III

RESIDUES, CONTOUR INTEGRATION, ZEROS

CHAPTER IV

ANALYTIC CONTINUATION

CHAPTER V

THE MAXIMUM-MODULUS THEOREM

CHAPTER VI

CONFORMAL REPRESENTATION

CHAPTER VII

POWER SERIES WITH A FINITE RADIUS OF CONVERGENCE

CHAPTER VIII

INTEGRAL FUNCTIONS

CHAPTER IX

DIRICHLET SERIES

CHAPTER X

THE THEORY OF MEASURE AND THE LEBESGUE INTEGRAL

CHAPTER XI

DIFFERENTIATION AND INTEGRATION

CHAPTER XII

FURTHER THEOREMS ON LEBESGUE INTEGRATION

CHAPTER XIII

FOURIER SERIES

CHAPTER I

INFINITE SERIES, PRODUCTS, AND INTEGRALS

1. Introduction. In this opening chapter we supplement the knowledge of elementary analysis which the reader is supposed to have at his disposal. We deal particularly with series, each term of which is a function of a variable; with integrals involving variable parameters; and with a variety of those double-limit problems which are so common in all branches of analysis. As we have explained in the preface, we take Hardy's *Pure Mathematics* (to which we refer as *P.M.*) as a starting-point, and refer to it whenever possible.

We shall use the following notation. In any argument, a number independent of the main variables is called a constant. A number not depending on any variable is called an absolute constant. We use A to denote an absolute positive constant, not necessarily the same one each time it occurs. The reader may find statements such as '$f(x) < A$, hence $2f(x) < A$' a little disconcerting at first, but he will soon get used to them. A constant depending on one or more parameters is usually denoted by K.

By $f(x) = O\{\phi(x)\}$ we mean generally that $|f(x)| < A\phi(x)$ if x is sufficiently near to some given limit. In particular, $O(1)$ means a bounded function. Thus

$$\sin x = O(|x|), \qquad (x+1)^2 = O(1)$$

as $x \to 0$; and

$$\sin x = O(1), \qquad (x+1)^2 = O(x^2)$$

as $x \to \infty$.

Sometimes, however, $f(x) = O\{\phi(x)\}$ is used to mean

$$|f(x)| < K\phi(x),$$

but it is usually sufficiently obvious what parameters are involved.

By $f(x) = o\{\phi(x)\}$ we mean that $f(x)/\phi(x) \to 0$ as x tends to a given limit. Thus

$$\sin x = o\,(x^2), \qquad (x+1)^2 = o\,(x^3)$$

as $x \to \infty$. In particular, $o(1)$ means a function which tends to zero.

By $f(x) \sim \phi(x)$ we mean that $f(x)/\phi(x) \to 1$ as x tends to a given limit.

We use ϵ to denote a variable which is to be given arbitrarily small values, and so may be thought of as small.

By $\max(a, b, ...)$ we mean the greatest of a, b,..., and by $\min(a, b, ...)$ the least.

1.1. Uniform convergence. The reader should be familiar with the idea of a *convergent series*.* Our standard notation for an infinite series is

$$u_1 + u_2 + u_3 + ... = \sum_{n=1}^{\infty} u_n = \sum u_n,$$

the limits of summation being $(1, \infty)$, unless other limits are definitely assigned. The nth partial sum of the series is

$$s_n = u_1 + u_2 + ... + u_n.$$

We begin by recalling the definition of convergence. The series is said to be convergent to the sum s if, given any positive number ϵ, however small, we can find a number n_0, depending on ϵ, such that

$$|s - s_n| < \epsilon \qquad (n > n_0).$$

In other words, s_n tends to the limit s as n tends to infinity.

Suppose now that each term of the series is a function of a real variable x. This variable is usually supposed to range over a closed interval, $a \leqslant x \leqslant b$, say; but the range of variation may equally well be an open interval, $a < x < b$; or indeed any set of points. We now write the series

$$u_1(x) + u_2(x) + ... = \sum u_n(x),$$

and its nth partial sum is $s_n(x)$. The series may, of course, be convergent for some values of x and divergent for others. If it is convergent for all the values of x considered, its sum is a function of x, defined for these values of x. We denote it by $s(x)$.

DEFINITION. *The series $\sum u_n(x)$ is said to be uniformly convergent over the interval (a, b) if, given any positive number ϵ, however small, we can find a number n_0, depending on ϵ* **but not on** *x, such that*

$$|s(x) - s_n(x)| < \epsilon$$

for $n > n_0$, and for every value of x in the interval (a, b).

It is clear that uniform convergence implies convergence for every value of x in the interval; but a series may (as we shall

* *P.M.* § 76.

show by examples) be convergent for every value of x in an interval without being uniformly convergent. It may be true that to every pair of values of x and ϵ corresponds a number n_0 such that $|s(x)-s_n(x)|<\epsilon$ for $n>n_0$; but at the same time it may happen that, as x approaches some point of the interval, the number n_0 may become indefinitely large. The series would then not be uniformly convergent.

Notice that uniform convergence is a property associated with *an interval* (or set of points), not with a single point.

1.11. Tests for uniform convergence. Just as there are tests for the convergence of a series of constants, so there are tests for the uniform convergence of a series of functions. The simplest and most useful test, due to Weierstrass, is as follows:

The series $\sum u_n(x)$ *is uniformly convergent over the interval* (a, b) *if there is a convergent series of positive constant terms,* $\sum a_n$ *say, such that*

$$|u_n(x)| \leqslant a_n$$

for all values of n and x.

In the first place, the series $\sum u_n(x)$ is convergent for every value of x, by the ordinary comparison theorem (*P.M.* §§ 167, 184). It therefore has a sum $s(x)$ for every value of x. Also

$$|s(x)-s_n(x)| = |u_{n+1}(x)+u_{n+2}(x)+\ldots| \leqslant a_{n+1}+a_{n+2}+\ldots,$$

which can be made less than any given ϵ by taking n greater than a certain number n_0. Since the a_n series is independent of x, the number n_0 is independent of x. This proves the theorem.

Notice that the result still holds if $|u_n(x)| \leqslant a_n$, not necessarily for *all* values of n, but for all sufficiently large values of n

A more general test of the same type, which is sometimes useful, is that $\sum u_n(x)$ is uniformly convergent if $|u_n(x)| \leqslant v_n(x)$, and $\sum v_n(x)$ is uniformly convergent. We leave the proof of this to the reader.

Examples. (i) The power series $\sum_{n=0}^{\infty} x^n$ is uniformly convergent for $a \leqslant x \leqslant b$, if $-1<a<b<1$. [Take $a_n = |a|^n$ or $|b|^n$, whichever is the greater.]

(ii) The trigonometrical series

$$\sum_{n=1}^{\infty} \frac{\cos nx}{n^2}$$

is uniformly convergent over any interval.

(iii) The Dirichlet series $\sum_{n=1}^{\infty} n^{-s}$ is uniformly convergent for $a \leqslant s \leqslant b$, if $1 < a < b$.

[Take $a_n = n^{-s}$; see *P.M.* § 175. The sum of this important series is denoted by $\zeta(s)$.]

(iv) The series

$$\{1-(1-x^2)\}^{\frac{1}{2}} = \sum_{n=0}^{\infty} a_n(1-x^2)^n$$

is uniformly convergent for $-1 \leqslant x \leqslant 1$.

(v) A similar definition of uniform convergence may be framed for series such as

$$\sum_{n=0}^{\infty} r^n \cos n\theta,$$

where the general term is a function of two (or more) variables, here r and θ. This series is uniformly convergent for $0 \leqslant r \leqslant b < 1$ and any range of values of θ.

1.12. Other tests. In a general way, any test for convergence becomes a test for uniform convergence if its conditions are satisfied independently of x. For example (*P.M.* § 168), $\sum u_n(x)$ is convergent for a particular value of x if there is a number r, less than 1, such that

$$\left|\frac{u_{n+1}(x)}{u_n(x)}\right| \leqslant r$$

for all values of n. In general, the value of r for which this is true will depend on x. Suppose, however, that we can find a number r such that the condition is satisfied for all values of x with this same value of r. Then the series is uniformly convergent, provided that $u_1(x)$ is bounded. For repeated application of the above inequality gives

$$|u_n(x)| \leqslant r^{n-1}|u_1(x)| \leqslant Mr^{n-1},$$

if $|u_1(x)| \leqslant M$, and the result follows from the comparison test.

Other tests for convergence may be extended in the same way. Take, for example, Dirichlet's test (*P.M.* § 189). The analogous test for uniform convergence is as follows:

If ϕ_n is a positive function of n which tends steadily to zero as $n \to \infty$, and if there is a constant A such that

$$\left|\sum_{n=1}^{N} u_n(x)\right| \leqslant A$$

for all values of N and x, then the series

$$\sum \phi_n u_n(x)$$

is uniformly convergent.

The reader should have no difficulty in formulating the rigorous proof.*

Examples. (i) If the numbers a_n are positive and decrease steadily to zero, the series

$$\sum a_n \sin nx$$

is uniformly convergent in any closed interval not including a multiple of 2π. [Compare *P.M.* ex. LXXIX, 2. Use the identity

$$\sin x + \sin 2x + \ldots + \sin nx = \frac{\cos \tfrac{1}{2}x - \cos(n+\tfrac{1}{2})x}{2 \sin \tfrac{1}{2}x}.]$$

(ii) Under the same conditions, the series

$$\sum a_n x \sin nx$$

is uniformly convergent in an interval including $x = 0$.

1.13. A necessary and sufficient condition for uniform convergence. *The series $\sum u_n(x)$ is uniformly convergent if and only if the following condition is satisfied. Given any positive number ϵ, we can find n_0, depending on ϵ but not on x, such that*

$$|s_m(x) - s_n(x)| < \epsilon$$

for all values of m and n greater than n_0.

This corresponds to the 'general principle of convergence' for ordinary series (*P.M.* §§ 83, 84).

As in the case of ordinary series, the condition is easily seen to be necessary; for

$$|s_m(x) - s_n(x)| \leqslant |s(x) - s_m(x)| + |s(x) - s_n(x)|,$$

so that, if the series is uniformly convergent, the condition is satisfied. In the case of ordinary series, the proof of sufficiency is more difficult. But, once the difficulty has been overcome in the 'ordinary' case, there is no further difficulty in the 'variable' case. For suppose that the condition is satisfied. Then, by the theorem for ordinary series, the series $\sum u_n(x)$ is convergent for every x. Let its sum be $s(x)$. Given ϵ, choose n_0 so that

$$|s_m(x) - s_n(x)| < \epsilon \qquad (m > n_0,\ n > n_0).$$

Keeping m fixed, make $n \to \infty$. Then, since $s_n(x) \to s(x)$,

$$|s_m(x) - s(x)| \leqslant \epsilon$$

provided only that $m > n_0$. Hence the convergence is uniform.

1.131. The following theorem† on a class of trigonometrical series is an excellent example of the above principle.

* See Bromwich's *Infinite Series*, ed. 2, § 44.

† Chaundy and Jolliffe (1).

If the numbers b_n are positive and steadily decreasing, a necessary and sufficient condition that the series

$$\sum b_n \sin nx$$

should be uniformly convergent throughout any interval is that $nb_n \to 0$.

To show that the condition is necessary, observe that, if $x = \pi/(2p)$, and* $n = [\frac{1}{2}p+1]$,

$$b_n \sin nx + b_{n+1}\sin(n+1)x + \ldots + b_p \sin px$$
$$> b_p(\sin nx + \ldots + \sin px) > b_p(\tfrac{1}{2}p-1)\sin\tfrac{1}{4}\pi,$$

since there are at least $\frac{1}{2}p-1$ terms in the bracket, in each of which $mx > \frac{1}{4}\pi$. Since the given series is uniformly convergent in an interval including the origin, the left-hand side of the above inequality tends to zero as $p \to \infty$. Hence $pb_p \to 0$.

In proving the sufficiency of the condition, we require the following result, known as Abel's lemma:

If $$b_1 \geqslant b_2 \geqslant \ldots \geqslant b_n \geqslant 0,$$
and if $$m \leqslant a_1+a_2+\ldots+a_n \leqslant M$$
for all values of n, then

$$b_1 m \leqslant a_1 b_1 + a_2 b_2 + \ldots + a_n b_n \leqslant b_1 M$$

for all values of n.

Let $s_n = a_1+\ldots+a_n$. Then†

$$a_1 b_1+\ldots+a_n b_n = b_1 s_1 + b_2(s_2-s_1)+\ldots+b_n(s_n-s_{n-1})$$
$$= s_1(b_1-b_2)+s_2(b_2-b_3)+\ldots+s_{n-1}(b_{n-1}-b_n)+s_n b_n.$$

Since each bracket is positive or zero, the sum is not decreased if each s_m be replaced by M; and this gives

$$M(b_1-b_2)+M(b_2-b_3)+\ldots+Mb_n = Mb_1,$$

the required upper bound. Similarly we obtain the required lower bound. This proves the lemma.

In the series in question, it is sufficient to consider the interval $0 \leqslant x \leqslant \pi$, since each term is odd and has the period 2π. Consider the sum

$$s_{n,p} = b_n \sin nx + \ldots + b_p \sin px,$$

where now n and p are unconnected. Let $\mu_n = \max_{m \geqslant n}(mb_m)$, so

* $[x]$ means the integral part of x.

† Compare *P.M.* § 189.

that $\mu_n \to 0$. If $x \geqslant \pi/n$, we apply Abel's lemma. We have

$$|\sin nx + \ldots + \sin rx| = \left|\frac{\cos(n-\frac{1}{2})x - \cos(r+\frac{1}{2})x}{2\sin\frac{1}{2}x}\right| \leqslant \frac{1}{\sin\frac{1}{2}x}$$

for all values of n and r, and, since $\sin\theta/\theta$ is steadily decreasing for $0 < \theta < \frac{1}{2}\pi$,

$$\frac{1}{\sin\frac{1}{2}x} \leqslant \frac{\pi}{x},$$

and we deduce that

$$|s_{n,p}| \leqslant \frac{b_n\pi}{x} \leqslant nb_n \leqslant \mu_n.$$

If $x \leqslant \pi/p$, we have, since $\sin\theta < \theta$,

$$|s_{n,p}| \leqslant b_n nx + \ldots + b_p px \leqslant p\mu_n x \leqslant \pi\mu_n.$$

If $\pi/p < x < \pi/n$, we combine the two arguments. We have

$$|s_{n,p}| \leqslant |s_{n,k}| + |s_{k+1,p}|,$$

and, applying Abel's lemma to the second part, and the other method to the first part, obtain

$$\begin{aligned} |s_{n,p}| &\leqslant k\mu_n x + b_{k+1}\pi/x \\ &\leqslant \mu_n[kx + \pi/\{(k+1)x\}]. \end{aligned}$$

Taking $k = [\pi/x]$, we have

$$|s_{n,p}| \leqslant \mu_n(\pi+1).$$

Hence in any case $\qquad |s_{n,p}| < A\mu_n,$

and, since $\mu_n \to 0$, the result follows.

1.14. Uniform convergence and continuity. So far, of course, we have not suggested any reason for considering uniformly convergent series at all. They are important for many reasons, not all of which can be explained in this chapter. The first reason is the following theorem:

The sum of a uniformly convergent series of continuous functions is a continuous function.

We use the same notation as before, and write

$$s(x) = s_n(x) + r_n(x),$$

so that $r_n(x)$ is the remainder after n terms of the series. Then, if x and $x+h$ are any two points of the interval considered,

$$\begin{aligned} |s(x+h) - s(x)| &= |s_n(x+h) - s_n(x) + r_n(x+h) - r_n(x)| \\ &\leqslant |s_n(x+h) - s_n(x)| + |r_n(x+h)| + |r_n(x)|. \end{aligned}$$

Having given ϵ, we can choose n_0 so that

$$|r_n(x+h)| < \epsilon, \qquad |r_n(x)| < \epsilon \qquad (n > n_0),$$

for all values of h. We now fix on a definite value of n which satisfies this condition. Having fixed n, $s_n(x)$ is a continuous function of x, since it is the sum of n continuous functions. We can therefore choose δ so small that

$$|s_n(x+h)-s_n(x)| < \epsilon \qquad (|h| < \delta).$$

Hence, combining the above inequalities,

$$|s(x+h)-s(x)| < 3\epsilon \qquad (|h| < \delta),$$

which proves that $s(x)$ is continuous.

Notice that the result is true if the functions $s_n(x)$ are merely continuous at the single point x considered; for all we have used is that $s_n(x+h) \to s_n(x)$ as $h \to 0$, x being fixed. We can therefore state the result as follows:

The limit of the sum of a uniformly convergent series of functions, each of which tends to a limit, is the sum of the limits of the separate functions.

1.2. Series of complex terms.* The theory of uniform convergence may be extended to series of the form

$$u_1(z)+u_2(z)+\dots,$$

in which the general term $u_n(z)$ is a function of the complex variable z. Instead of uniform convergence in an interval, we shall now have uniform convergence throughout some region of the z-plane, such as the interior of a circle or a square. The reader should have no difficulty in extending the definitions and tests to this case. It should also be noticed that the theorem on the continuity of the sum of a uniformly convergent series can be extended at once to series of complex functions.

Example. The series $\sum_{n=1}^{\infty} n^{-s}$, where s is a complex variable, is uniformly convergent throughout any finite region in which $\mathbf{R}(s) \geqslant a > 1$.

The function $\zeta(s)$, defined as the sum of the series, is continuous at all points of the region $\mathbf{R}(s) > 1$.

[Compare ex. (iii), § 1.11.]

1.21. Power series. One of the simplest cases of uniform convergence of a series of complex terms is that of a power series. We know (*P.M.* § 193) that a power series

$$\sum_{n=0}^{\infty} a_n z^n$$

* *P.M.* § 190.

has a *radius of convergence* R (which may be zero or infinite), such that the series is convergent for $|z| < R$, and divergent for $|z| > R$.

The series is uniformly convergent for $|z| \leqslant R'$, where R' is any positive number less than R.

For let ρ be a number between R' and R. Since the series is convergent for $z = \rho$, there is a number K, independent of n, such that $|a_n \rho^n| < K$ for all values of n. Hence, for $|z| \leqslant R'$,

$$|a_n z^n| = \left| a_n \rho^n \cdot \left(\frac{z}{\rho}\right)^n \right| < K\left(\frac{R'}{\rho}\right)^n,$$

and the last term is independent of z, and is the general term of a convergent geometrical progression. Hence (by the analogue for complex functions of the test of § 1.11) the series is uniformly convergent.

We have thus shown that any circle interior to the circle of convergence is a region of uniform convergence. The circle of convergence itself is not necessarily a region of uniform convergence; in fact on the circle the series does not necessarily converge at all.

Example. For the series $\sum z^n/n^2$, the circle of convergence is a region of uniform convergence.

1.22. Abel's theorem. There is one interesting possibility which the above discussion so far leaves open. Suppose, to take the simplest case, that we have a real power series

$$\sum_{n=0}^{\infty} a_n x^n \tag{1}$$

with radius of convergence 1. Suppose further that the series

$$\sum_{n=0}^{\infty} a_n \tag{2}$$

is convergent. Does the interval of uniform convergence, in this case, extend right up to the point $x = 1$? The answer is in the affirmative.

If the series

$$\sum_{n=0}^{\infty} a_n$$

is convergent, and has the sum s, then the series

$$\sum_{n=0}^{\infty} a_n x^n$$

is uniformly convergent for $0 \leqslant x \leqslant 1$, *and*

$$\lim_{x\to 1} \sum_0^\infty a_n x^n = s.$$

The proof is an immediate consequence of Abel's lemma (see § 1.131). Let

$$s_{n,p} = a_n + a_{n+1} + \ldots + a_p.$$

Then, given ϵ, we can choose n_0 so large that $|s_{n,p}| \leqslant \epsilon$ $(n_0 \leqslant n < p)$. Since the numbers x^n are non-increasing if $x \leqslant 1$, Abel's lemma gives, for $n_0 < n < p$,

$$|a_n x^n + \ldots + a_p x^p| \leqslant \epsilon x^n \leqslant \epsilon \qquad (0 \leqslant x \leqslant 1),$$

and this is the condition for uniform convergence.

The second part of the theorem now follows from the continuity theorem of § 1.14.

Example. From the expansion (*P.M.* § 213)

$$\log(1+x) = x - \tfrac{1}{2}x^2 + \tfrac{1}{3}x^3 - \ldots \qquad (|x| < 1),$$

deduce that

$$\log 2 = 1 - \tfrac{1}{2} + \tfrac{1}{3} - \ldots.$$

1.23. Tauber's theorem. The direct converse of the 'continuity' part of Abel's theorem would be that if

$$f(x) = \sum_{n=0}^\infty a_n x^n \to s$$

as $x \to 1$, then $\sum_0^\infty a_n$ converges to the sum s. That this is false is shown by the simple example

$$f(x) = \sum_{n=0}^\infty (-1)^n x^n = \frac{1}{1+x},$$

in which $f(x) \to \frac{1}{2}$, but $\sum_0^\infty a_n$ is not convergent.

If, however, we impose on the coefficients a_n a restriction as to their order of magnitude, it is possible to prove a converse theorem.

If $a_n = o(1/n)$, *and* $f(x) \to s$ *as* $x \to 1$, *then* $\sum_0^\infty a_n$ *converges to the sum* s.

We first prove the following simple lemma:*

Lemma. *If* $b_n \to 0$ *as* $n \to \infty$, *then*

$$\frac{b_0 + b_1 + \ldots + b_n}{n+1} \to 0.$$

* *P.M.* Ch. IV, Misc. Ex. 27.

For, if $|b_n| < K$ for all values of n, and $|b_n| < \epsilon$ for $n > n_0$, then

$$\left|\frac{b_0+b_1+\ldots+b_n}{n+1}\right| \leqslant \left|\frac{b_0+\ldots+b_{n_0}}{n+1}\right| + \left|\frac{b_{n_0+1}+\ldots+b_n}{n+1}\right|$$
$$\leqslant \frac{(n_0+1)K}{n+1}+\frac{(n-n_0)\epsilon}{n+1} < 2\epsilon$$

if $n > (n_0+1)K/\epsilon$. This proves the lemma.

To prove Tauber's theorem, it is sufficient to prove that

$$\sum_0^\infty a_n x^n - \sum_0^N a_n \to 0$$

as $x \to 1$, where $N = [1/(1-x)]$. That is, we have to show that

$$\sum_{N+1}^\infty a_n x^n - \sum_0^N a_n(1-x^n) \to 0.$$

Call these two sums S_1 and S_2. Given ϵ, choose N so large that $|na_n| < \epsilon$ $(n > N)$. Then

$$|S_1| = \left|\sum_{N+1}^\infty na_n \cdot \frac{x^n}{n}\right| < \frac{\epsilon}{N+1}\sum_{N+1}^\infty x^n < \frac{\epsilon}{(N+1)(1-x)} < \epsilon.$$

Also $$1-x^n = (1-x)(1+x+\ldots+x^{n-1}) < n(1-x),$$

and so $$|S_2| < (1-x)\sum_0^N n|a_n| \leqslant \frac{1}{N}\sum_0^N n|a_n|,$$

which tends to zero, by the lemma. Hence $|S_2| < \epsilon$ if N is large enough, and so $|S_1+S_2| < 2\epsilon$. This proves the theorem.

1.3. Series which are not uniformly convergent. Up to this point, the reader may still suspect that convergence throughout an interval is the same thing as uniform convergence. We shall show by means of examples that this is not so.

Examples. (i) We can construct a series for which

$$s_n(x) = \frac{1}{1+nx} \qquad (0 \leqslant x \leqslant 1)$$

by taking $u_1(x) = 1/(1+x)$, and

$$u_n(x) = \frac{1}{1+nx} - \frac{1}{1+(n-1)x} = \frac{-x}{(1+nx)\{1+(n-1)x\}} \qquad (n > 1).$$

This function $s_n(x)$ is *a continuous function which tends to a discontinuous limit.* For, if $x > 0$, $s_n(x)$ obviously tends to zero as $n \to \infty$. But if $x = 0$, $s_n(x) = 1$ for all values of n, and so its limit is 1. The sum of the series is therefore discontinuous. Hence the series cannot be uniformly convergent.

(ii) Consider the series

$$\sum_{n=0}^\infty xe^{-nx}.$$

Here $s_n(0) = 0$, so that $s(0) = 0$. When $x > 0$,

$$s_n(x) = x\frac{1-e^{-nx}}{1-e^{-x}}, \qquad s(x) = \frac{x}{1-e^{-x}}.$$

As $x \to 0$, $s(x) \to 1$. Hence $s(x)$ is discontinuous, and, as before, the series is not uniformly convergent in any interval ending at $x = 0$.

In fact, if $x = 1/n$,

$$s\left(\frac{1}{n}\right) - s_n\left(\frac{1}{n}\right) = \frac{xe^{-1}}{1-e^{-x}} \to e^{-1},$$

so that $|s(x)-s_n(x)|$ is not 'uniformly small' near $x = 0$.

(iii) Consider similarly

$$\sum_{n=0}^{\infty} x^n(1-x).$$

(iv) As in example (i), we can construct a series for which

$$s_n(x) = nx(1-x)^n \qquad (0 \leqslant x \leqslant 1).$$

Obviously $s(0) = 0$. Also, if $x > 0$, $n(1-x)^n \to 0$ as $n \to \infty$ (*P.M.* § 206). Hence $s(x) = 0$ for all values of x. In this case, therefore, the sum of the series is continuous. But the series is not uniformly convergent. It is a simple exercise in differential calculus to find the maximum of $s_n(x)$; it is

$$\left(\frac{n}{1+n}\right)^{1+n},$$

and thus tends to the limit e^{-1} as $n \to \infty$ (*P.M.* §§ 73, 208). Hence, however large n may be, the function $s_n(x)-s(x)$ takes values nearly as large as e^{-1}. Thus the convergence is not uniform.

The reader should draw the graph of $s_n(x)$. It has a wave which approaches the origin, and diminishes indefinitely in breadth, but not in height.

Notice that uniformity of convergence may be altered by multiplying by a factor independent of n. For example, if

$$s_n(x) = \frac{x}{1+nx},$$

then $|s_n(x)| \leqslant 1/n$ $(0 \leqslant x \leqslant 1)$, so that the series converges uniformly to zero. But the series obtained by multiplying by $1/x$ is not uniformly convergent (ex. (i)).

On the other hand, if we multiply a uniformly convergent series by a *bounded* factor independent of n, the resulting series is also uniformly convergent. This is easily seen from the definition.

1.31. Uniform convergence of series of positive terms. It is clear from the above examples that uniform convergence is not a necessary condition for continuity, though it is a sufficient condition. There is, however, one interesting case in which uniform convergence and continuity are equivalent.*

If $\sum u_n(x) = s(x)$ *is a series of continuous positive terms in a given closed interval, a necessary and sufficient condition that* $s(x)$ *should be continuous is that the series should be uniformly convergent over the interval.*

We have to prove that the condition is necessary, i.e. that, if $s(x)$ is continuous, the series is uniformly convergent.

Employing our usual notation, the function $s(x)-s_n(x)$ is continuous, and so (*P.M.* § 102, Th. 2) has an upper bound, ϵ_n say, which is attained at some point x_n of the interval. It is sufficient to prove that $\epsilon_n \to 0$; for, since the terms are positive,

$$|s(x)-s_n(x)| \leqslant |s(x)-s_N(x)| \leqslant \epsilon_N$$

for $n \geqslant N$ and all x; and this implies uniform convergence if $\epsilon_N \to 0$ as $N \to \infty$.

Suppose on the contrary that ϵ_n does not tend to zero. Then, since it is steadily decreasing (because the terms are positive), it has a positive lower bound, δ say. Also the numbers x_n have a limit-point, ξ say, in the interval (*P.M.* § 19). Choose N so large that $s(\xi)-s_N(\xi) < \delta$. Then, if ξ is an interior point of the interval, there is an interval $(\xi-h, \xi+h)$ throughout which $s(x)-s_N(x) < \delta$ (*P.M.* § 101, Th. 1). If ξ is an end-point, the same is true of $(\xi, \xi+h)$ or $(\xi-h, \xi)$. Hence $\epsilon_n < \delta$ for those values of n for which $|x_n-\xi| < h$. This gives a contradiction, and the theorem is proved.

1.4. Infinite products. An infinite product is an expression of the form

$$(1+a_1)(1+a_2)(1+a_3)... \tag{1}$$

containing an infinity of factors. We denote it by

$$\prod_{n=1}^{\infty} (1+a_n).$$

We suppose that no a_n is equal to -1.

Writing p_n for the partial product

$$p_n = \prod_{m=1}^{n} (1+a_m),$$

we say that *the infinite product is convergent if, as* $n \to \infty$, p_n *tends to a limit other than zero.* We might, of course, admit the

* See Hardy (11) for a detailed discussion.

limit zero as well; but we shall see later that this would often be inconvenient.

If the product is not convergent, it is said to be divergent. If $p_n \to 0$, it is said to diverge to zero.

Examples. (i) The product

$$(1-\tfrac{1}{2})(1+\tfrac{1}{3})(1-\tfrac{1}{4})(1+\tfrac{1}{5})\dots$$

is convergent.

(ii) If the product (1) is convergent, $a_n \to 0$.

1.41. We begin by considering two simple cases.

If $a_n \geqslant 0$ for all values of n, the product $\prod(1+a_n)$ and the series $\sum a_n$ converge or diverge together.

Since, in this case, p_n is a non-decreasing function of n, it either converges or tends to positive infinity. Now

$$a_1+a_2+\dots+a_n \leqslant (1+a_1)\dots(1+a_n) \leqslant e^{a_1+a_2+\dots+a_n}.$$

The left-hand inequality is obvious on multiplying out the product; and the right-hand inequality follows from the fact that $1+a \leqslant e^a$ for every positive a. The two inequalities show that p_n and $a_1+\dots+a_n$ are bounded or unbounded together, and this gives the result.

If $a_n \leqslant 0$ for all values of n, write $a_n = -b_n$, and consider the product

$$\prod_{n=1}^{\infty}(1-b_n).$$

If $b_n \geqslant 0$, $b_n \neq 1$, for all values of n, and $\sum b_n$ is convergent, then $\prod(1-b_n)$ is convergent.

Since $\sum b_n$ is convergent, we can choose N so large that

$$b_N+b_{N+1}+\dots < \tfrac{1}{2},$$

and, in particular, $b_n < 1$ $(n \geqslant N)$. Then

$$(1-b_N)(1-b_{N+1}) \geqslant 1-b_N-b_{N+1},$$

$$\begin{aligned}(1-b_N)(1-b_{N+1})(1-b_{N+2}) &\geqslant (1-b_N-b_{N+1})(1-b_{N+2})\\ &\geqslant 1-b_N-b_{N+1}-b_{N+2},\end{aligned}$$

and so generally

$$(1-b_N)(1-b_{N+1})\dots(1-b_n) \geqslant 1-b_N-\dots-b_n > \tfrac{1}{2}.$$

Hence p_n/p_{N-1} is steadily decreasing for $n \geqslant N$, and has a positive lower bound. Hence it tends to a positive limit. Since p_{N-1} is not zero, the result follows.

If $0 \leqslant b_n < 1$ for every n, but $\sum b_n$ diverges, then $\prod(1-b_n)$ diverges to zero.

For $1-b \leqslant e^{-b}$ if $0 \leqslant b < 1$, so that

$$(1-b_1)(1-b_2)...(1-b_n) \leqslant e^{-b_1-b_2...-b_n}.$$

The right-hand side tends to zero, and the result follows.

In particular, *if* $0 \leqslant b_n < 1$, *the product* $\prod (1-b_n)$ *and the series* $\sum b_n$ *converge or diverge together.*

1.42. The general case. Now let the numbers a_n be any numbers, real or complex, other than -1.

DEFINITION. *The product* $\prod (1+a_n)$ *is said to be absolutely convergent if the product* $\prod (1+|a_n|)$ *is convergent.*

It is clear from the first result of § 1.41 that *a necessary and sufficient condition that the product should be absolutely convergent is that* $\sum |a_n|$ *should be convergent.*

We next show that *an absolutely convergent product is convergent.*

To prove this, let p_n denote the same partial product as before, and let

$$P_n = \prod_{m=1}^{n} (1+|a_m|).$$

Then

$$p_n - p_{n-1} = (1+a_1)...(1+a_{n-1})a_n,$$
$$P_n - P_{n-1} = (1+|a_1|)...(1+|a_{n-1}|)|a_n|,$$

and it is plain that

$$|p_n - p_{n-1}| \leqslant P_n - P_{n-1}.$$

Now, if $\prod (1+|a_n|)$ is convergent, P_n tends to a limit, and so $\sum (P_n - P_{n-1})$ is convergent. Hence, by the comparison theorem, $\sum (p_n - p_{n-1})$ is convergent, i.e. p_n tends to a limit.

This limit cannot be zero. For, since $\sum |a_n|$ is convergent and $1+a_n \to 1$, the series

$$\sum \left| \frac{a_n}{1+a_n} \right|$$

is also convergent. Hence, by what we have just proved, the product

$$\prod_{m=1}^{n} \left(1 - \frac{a_m}{1+a_m}\right)$$

tends to a limit. But this product is equal to $1/p_n$. Hence the limit of p_n is not zero.

Example. The factors of an absolutely convergent product may be taken in any order, without altering the value of the product. (Compare *P.M.* § 185.)

1.43. The logarithm of an infinite product. If

$$\prod_{n=1}^{\infty}(1+a_n)=p,$$

is it necessarily true that

$$\sum_{n=1}^{\infty}\log(1+a_n)=\log p\ ?$$

Here $\log z$ denotes the principal value of the logarithm of z, i.e. the value whose imaginary part lies between $-\pi$ and π (*P.M.* § 224).

The result is obviously true if all the numbers a_n are real and positive, for then all the logarithms have their ordinary arithmetical value. But, in the general case, the formula requires modification.

Let p_n denote the nth partial product, and let $p_n=\rho_n e^{i\phi_n}$, so that p_n and ρ_n tend to limits, and so does ϕ_n if its values are suitably chosen. Let $1+a_n=r_n e^{i\theta_n}$, where $-\pi<\theta_n\leqslant\pi$; then, since $a_n\to 0$, $\theta_n\to 0$ as $n\to\infty$.

Let
$$s_n=\sum_{m=1}^{n}\log(1+a_m).$$

Then
$$s_n=\log p_n+2k_n i\pi, \tag{1}$$

where k_n is an integer. Now

$$2k_n\pi=\theta_1+\ldots+\theta_n-\phi_n,$$

so that
$$2\pi(k_{n+1}-k_n)=\theta_{n+1}-(\phi_{n+1}-\phi_n),$$

and the right-hand side tends to zero. Hence, if n is sufficiently large,

$$|2\pi(k_{n+1}-k_n)|<2\pi,$$

and so $k_{n+1}=k_n$, since k_n is always an integer. Thus k_n has a constant value, k say, if n is sufficiently large; i.e.

$$s_n=\log p_n+2ki\pi \qquad (n>n_0),$$

and, making $n\to\infty$,

$$\sum\log(1+a_n)=\log p+2ki\pi.$$

The sum of the series is therefore a value, but not necessarily the principal value, of the logarithm of the product.

Notice also that it follows from the proof that

$$\sum_{N+1}^{\infty}\log(1+a_n)=\log p-\log p_N$$

for all sufficiently large values of N.

If we start with the series of logarithms, and assume that

$$\sum_{n=1}^{\infty} \log(1+a_n) = s,$$

we have, on taking the exponential of (1),

$$e^{s_n} = p_n,$$

and so $$p_n \to p = e^s,$$

i.e. the product converges to the exponential of the sum.

Examples. (i) If $\sum a_n$ and $\sum |a_n|^2$ are convergent, then $\prod (1+a_n)$ is convergent. [Use the equation $\log(1+a_n) = a_n + O(|a_n|^2)$.]

(ii) If $\sum a_n$, $\sum a_n^2, \ldots, \sum a_n^{k-1}$, $\sum |a_n|^k$, are all convergent, then $\prod (1+a_n)$ is convergent.

(iii) If a_n is real, and $\sum a_n$ is convergent, the product $\prod (1+a_n)$ converges, or diverges to zero, according as $\sum a_n^2$ converges or diverges.

(iv) The product

$$\left(1-\frac{1}{\sqrt{2}}\right)\left(1+\frac{1}{\sqrt{3}}\right)\left(1-\frac{1}{\sqrt{4}}\right) \cdots$$

is divergent.

(v) Show that, if

$$a_{2n-1} = -\frac{1}{\sqrt{(n+1)}}, \quad a_{2n} = \frac{1}{\sqrt{(n+1)}} + \frac{1}{n+1} + \frac{1}{(n+1)\sqrt{(n+1)}},$$

the product $\prod (1+a_n)$ converges, though both $\sum a_n$ and $\sum a_n^2$ diverge.

(vi) The product $\prod \left(1+\frac{i}{n}\right)$ is divergent, but $\prod \left|1+\frac{i}{n}\right|$ is convergent.

(vii) If $\sum |u_n^2|$ is convergent, so is $\prod (1-u_n)e^{u_n}$; and if $\sum |u_n|^3$ is convergent, so is $\prod (1-u_n)e^{u_n+\frac{1}{2}u_n^2}$. [As $u \to 0$, $(1-u)e^u = 1+O(u^{-2})$ and $(1-u)e^{u+\frac{1}{2}u^2} = 1+O(u^{-3})$; or we may consider the series of logarithms, as in (i). Products of this type are of great importance in Chapter VIII, and are discussed fully there.]

1.44. Uniform convergence of infinite products. The infinite product

$$\prod_{n=1}^{\infty} \{1+u_n(z)\},$$

where the factors are functions of a variable z, real or complex, is said to be uniformly convergent if the partial product

$$p_n(z) = \prod_{m=1}^{n} \{1+u_m(z)\}$$

converges uniformly in a certain region of values of z to a limit which is never zero.

The simplest test for the uniform convergence of a product is as follows:

The product $\prod \{1+u_n(z)\}$ *is uniformly convergent in any region where the series* $\sum |u_n(z)|$ *converges uniformly to a bounded sum.*

The proof consists of a re-examination of the convergence-argument of § 1.42 from the point of view of uniformity. Let M be the upper bound of the sum $\sum |u_n(z)|$ in the region considered. Then

$$\{1+|u_1(z)|\}...\{1+|u_n(z)|\} < e^{|u_1(z)|+...} \leqslant e^M.$$

Let
$$P_n(z) = \prod_{m=1}^{n} \{1+|u_m(z)|\}.$$

Then

$$P_n(z)-P_{n-1}(z) = \{1+|u_1(z)|\}...\{1+|u_{n-1}(z)|\}|u_n(z)| < e^M|u_n(z)|.$$

Hence $\sum \{P_n(z)-P_{n-1}(z)\}$ is uniformly convergent, and the result follows as in § 1.42.

Examples. (i) The product

$$\prod\left(1-\frac{1}{\varpi^s}\right),$$

where ϖ runs through the prime numbers 2, 3, 5,..., is uniformly convergent in any finite region throughout which $\mathbf{R}(s) \geqslant a > 1$; for the same thing is true of the series $\sum |\varpi^{-s}|$, which consists of some of the terms of the series $\sum |n^{-s}|$ (§ 1.2, example).

The value of the product is $1/\zeta(s)$. For

$$(1-2^{-s})\zeta(s) = 1+3^{-s}+5^{-s}+...,$$

all terms containing the factor 2 being omitted on the right. Next

$$(1-2^{-s})(1-3^{-s})\zeta(s) = 1+5^{-s}+7^{-s}+11^{-s}+...,$$

all terms containing the factors 2 or 3 being omitted. So generally, if ϖ_n is the nth prime,

$$(1-2^{-s})...(1-\varpi_n^{-s})\zeta(s) = 1+l^{-s}+...$$

where all numbers containing the factors 2, 3,..., ϖ_n are omitted. Since all the numbers up to ϖ_n are of this form,

$$|(1-2^{-s})...(1-\varpi_n^{-s})\zeta(s)-1| \leqslant |(\varpi_n+1)^{-s}|+|(\varpi_n+2)^{-s}|+...,$$

which tends to 0 as $\varpi_n \to \infty$. Hence

$$\lim_{n\to\infty} (1-2^{-s})...(1-\varpi_n^{-s})\zeta(s) = 1,$$

the result stated.

(ii) If $\mathbf{R}(s) \geqslant 2$

$$\log \zeta(s) = -\sum_{\varpi} \log(1-\varpi^{-s}),$$

all the logarithms having their principal values.

[We deduce from the above example and § 1.43 that

$$\log \zeta(s) = -\sum \log(1-\varpi^{-s})+2ki\pi,$$

where k is an integer, which depends *prima facie* on s. If s is real, k is obviously 0. Also, as long as $\mathbf{R}(s) > 1$, the real part of $1-\varpi^{-s}$ remains positive, and so its amplitude remains between $-\frac{1}{2}\pi$ and $\frac{1}{2}\pi$. Each term $\log(1-\varpi^{-s})$ is therefore continuous for $\mathbf{R}(s) > 1$. Hence the sum of the series is continuous.

Similarly $\log \zeta(s)$ is continuous, provided that $\mathbf{R}\{\zeta(s)\} > 0$. This is certainly true if $\mathbf{R}(s) \geqslant 2$, since, if $\mathbf{R}(s) = \sigma \geqslant 2$,

$$\mathbf{R}\{\zeta(s)\} \geqslant 1-2^{-\sigma}-3^{-\sigma}-\ldots \geqslant 1-2^{-2}-3^{-2}-\ldots$$

$$> 1-\frac{1}{1.2}-\frac{1}{2.3}-\ldots = 0.$$

It follows that k is continuous for $\mathbf{R}(s) \geqslant 2$, and so zero throughout this region.]

(iii) The convergence of the product $\prod (1+a_n)$ does not imply that of $\prod (1+a_n x)$, except for $x = 0$ and $x = 1$.*

(iv) The convergence of $\prod (1+a_n)$ does not imply that

$$\lim_{x\to 1} \prod (1+a_n x^n) = \prod (1+a_n).$$

[In fact, Hardy (5) gives an example in which

$$\lim \prod (1+a_n x^n) = 2 \prod (1+a_n).$$

The result is, of course, in striking contrast to Abel's theorem on the continuity of power series.]

(v) The product
$$\prod \left(1+\frac{e^{in\theta}}{\log n}\right)$$
is not convergent for any rational value of θ/π, but is convergent if θ/π is an algebraic number (*P.M.* Ch. I, ex. 32) which is not rational.

[The problem of the behaviour of this product, suggested by Hardy (5), was solved by Littlewood (2).]

1.5. Convergence of infinite integrals. We assume that the reader is familiar with the elementary properties of the Riemann integral of a continuous function (*P.M.* §§ 156–64). If $f(x)$ is continuous over a finite closed interval (a, b), the Riemann integral

$$\int_a^b f(x)\,dx$$

exists. Similarly the indefinite integral

$$F(x) = \int_a^x f(t)\,dt$$

exists for $a \leqslant x \leqslant b$; and $F(x)$ is continuous, and has a differential coefficient equal to $f(x)$. We assume a knowledge of the usual rules of integration by parts, integration by substitution, etc., and of the mean-value theorems (*P.M.* §§ 160–1).

We next extend the definition to a class of discontinuous

* Hardy (5).

functions. Suppose that the interval (a,b) can be divided up into a finite number of parts (a,x_1), (x_1,x_2),..., (x_n,b), such that $f(x)$ is continuous except at x_1,..., x_n, and such that the limits $f(x_1-0)$, $f(x_1+0)$,..., etc., exist (*P.M.* § 99). Then the integral of $f(x)$ over each partial interval exists, and the integral over the whole interval is defined as being the sum of the integrals over the partial intervals; i.e.

$$\int_a^b f(x)\,dx = \int_a^{x_1} f(x)\,dx + \int_{x_1}^{x_2} f(x)\,dx + \ldots + \int_{x_n}^b f(x)\,dx.$$

An infinite integral is defined in *P.M.* § 177. If $f(t)$ is integrable over (a, x) for all values of x, and

$$\lim_{x\to\infty} \int_a^x f(t)\,dt = l,$$

then we say that the infinite integral

$$\int_a^\infty f(t)\,dt$$

is convergent, and has the value l.

Similarly, if $f(t)$ tends to infinity, or oscillates, as $x \to c$, but

$$\lim_{x\to c} \int_a^x f(t)\,dt = l,$$

then we define the integral of $f(t)$ over (a,c) to be equal to l (*P.M.* § 180).

There is no difficulty in extending the rules for integration by parts and substitution to these cases.

A number of tests for convergence, such as the comparison test, for the case where $f(x)$ is positive, are given in *P.M.* § 178.

Suppose now that $f(x)$ is not necessarily positive. If $f(x)$ and $|f(x)|$ are both integrable in one of the senses already explained, and if the integral

$$\int_a^\infty |f(t)|\,dt$$

is convergent, then the integral

$$\int_a^\infty f(t)\,dt$$

is said to be *absolutely convergent* (cf. *P.M.* § 184).

An absolutely convergent integral is convergent. For, if the integral of $|f(t)|$ is convergent, so are the integrals of

$$\phi(t) = |f(t)| + f(t), \qquad \psi(t) = |f(t)| - f(t),$$

by the comparison test, ϕ and ψ both being positive. Hence the integral of $\frac{1}{2}\{\phi(t) - \psi(t)\} = f(t)$ is convergent

The result may be extended to the case where $f(t)$ is a continuous complex function, by considering separately its real and imaginary parts.

An integral which is convergent, but not absolutely convergent, is said to be *conditionally convergent*.

The most important tests for conditionally convergent integrals are the analogues of Dirichlet's and Abel's tests for series (*P.M.* § 189).

Analogue of Dirichlet's test. *If $\phi(x)$ has a continuous derivative, and decreases steadily to zero as $x \to \infty$, and*

$$F(x) = \int_a^x f(t)\, dt$$

is bounded, then the integral

$$\int_a^\infty \phi(x) f(x)\, dx$$

is convergent.

We integrate by parts, this being the process for integrals analogous to the 'partial summation' by which Dirichlet's test is proved. We have

$$\int_a^X \phi(x) f(x)\, dx = \phi(X)F(X) + \int_a^X \{-\phi'(x)\} F(x)\, dx.$$

The integrated term tends to zero as $X \to \infty$; and the last integral is absolutely convergent by the comparison test; for $|F(x)|$ is bounded and $-\phi'(x)$ is positive, and

$$\int_a^X \{-\phi'(x)\}\, dx = \phi(a) - \phi(X) \to \phi(a).$$

This proves the theorem.

Examples. (i) The integrals

$$\int_0^\infty \frac{\sin x}{x}\, dx, \qquad \int_1^\infty \frac{\cos x}{x}\, dx$$

are conditionally convergent.

(ii) State and prove the analogue for integrals of Abel's test.

We note finally a *necessary and sufficient condition* for the convergence of the integral

$$\int_a^\infty f(x)\,dx;$$

it is that, given ϵ, we can find X_0 such that

$$\left|\int_X^{X'} f(x)\,dx\right| < \epsilon$$

for $X' > X \geqslant X_0$. This may be proved in the same way as (or deduced from) the corresponding theorem for series (*P.M.* §§ 83–4).

Examples. (i) Use this principle to prove that an absolutely convergent integral is convergent.

(ii) Prove Dirichlet's test for convergence by means of this principle and the second mean-value theorem (*P.M.* § 161, exs. 11–12).

1.51 Uniform convergence of infinite integrals. We can now extend the idea of uniform convergence to infinite integrals. Let $f(x, y)$ be an integrable function of x over the interval $a \leqslant x \leqslant b$, for $\alpha \leqslant y \leqslant \beta$, and for all values of b. Suppose that the integral

$$\phi(y) = \int_a^\infty f(x, y)\,dx$$

is convergent for all values of y in the interval (α, β). *Then the integral is said to be uniformly convergent if, given ϵ, we can find a number X_0, depending on ϵ but not on y, such that*

$$\left|\phi(y) - \int_a^X f(x, y)\,dx\right| < \epsilon, \qquad (X \geqslant X_0).$$

A similar definition may be framed for integrals which are infinite by reason of the integrand becoming infinite in the range of integration.

The simplest test for uniform convergence is the analogue of the series-test of § 1.11. *The above integral is uniformly convergent if there is a positive function $g(x)$, independent of y, such that $|f(x, y)| \leqslant g(x)$ for all values of x and y, and such that the integral*

$$\int_a^\infty g(x)\,dx$$

is convergent.

This may be proved in the same way as the corresponding result for series.

Other tests may be extended in a similar way. For example, in Dirichlet's test, if f and ϕ are functions of x and y, we assume that $\partial\phi/\partial x$ is continuous, and $\phi(x)$ tends to zero steadily and uniformly with respect to y, and that $|F|$ is less than a constant independent of x and y. The integral of ϕf is then uniformly convergent.

Examples. (i) Consider the convergence of the integral

$$\Gamma(x) = \int_0^\infty t^{x-1}e^{-t}\,dt.$$

[Suppose first that x is real. The integral is convergent at the upper limit for all values of x, since $t^{x+1}e^{-t}$ is bounded for all x, and we can compare the integral with that of $1/t^2$; but for convergence at the lower limit we must have $x > 0$ (*P.M.* § 180).

The integral is uniformly convergent over any finite x-interval (a, b), where $a > 0$. To prove this, we divide it into integrals over $(0, 1)$ and $(1, \infty)$, and compare the two parts with

$$\int_0^1 t^{a-1}\,dt, \qquad \int_1^\infty t^{b-1}e^{-t}\,dt,$$

which are convergent and independent of x.

Similarly, if x is complex, the integral is uniformly convergent over any finite region throughout which $\mathbf{R}(x) \geqslant a > 0$; for if $x = \xi + i\eta$, then $|t^{x-1}| = t^{\xi-1}$, and the result can now be proved as before.]

(ii) The integral

$$\int_0^\infty \frac{\sin xy}{x^s}\,dx$$

is absolutely convergent for $1 < s < 2$, and any y. For a fixed s in this range, it is uniformly convergent for $0 < \alpha \leqslant y \leqslant \beta$, for any β.

It is conditionally convergent if $0 < s \leqslant 1$, $y > 0$, and uniformly convergent for s in this range and $0 < \alpha \leqslant y \leqslant \beta$.

For fixed $y > 0$, it is absolutely and uniformly convergent in $1 < s_1 \leqslant s \leqslant s_2 < 2$, and uniformly, but not absolutely, convergent in $0 < s_1 \leqslant s \leqslant 1$.

1.52. The continuity theorem. In this section we shall prove the analogue for integrals of the theorem that the sum of a uniformly convergent series of continuous functions is continuous.

We first require the following theorem on continuous functions of two variables, similar to the theorem for one variable proved in *P.M.* § 106.

Let $f(x,y)$ be a continuous function of x and y throughout the rectangle $a \leqslant x \leqslant b$, $\alpha \leqslant y \leqslant \beta$. Then, given ϵ, we can divide up the given rectangle into a finite number of sub-rectangles $x_\mu \leqslant x \leqslant x_{\mu+1}$, $y_\nu \leqslant y \leqslant y_{\nu+1}$, in such a way that

$$|f(x,y)-f(\xi,\eta)| < \epsilon$$

provided that (x,y) and (ξ,η) belong to the same sub-rectangle.

We prove this by the method of subdivision. Suppose that the given rectangle has not the required property. Then, if we divide it into quarters by the lines $x = \frac{1}{2}(a+b)$, $y = \frac{1}{2}(\alpha+\beta)$, at least one of the four quarter-rectangles has not the required property. Choose that one which has not; or, if more than one have not, choose one of them—to give a definite rule, choose one on the left-hand side if possible, and then, having fixed the side, choose the lower of two on the same side.

We next subdivide the chosen rectangle into quarters; and so the process of subdivision proceeds indefinitely, there being always at least one quarter which has not the required property. The left-hand sides of the chosen rectangles form an increasing sequence, and the right-hand sides form a decreasing sequence, and so each sequence has a limit; and the limits are the same, since the length of the side tends to zero. Call the limit X. Similarly the upper and lower sides tend to a limit Y.

We now use the fact that the function is continuous at (X, Y). Given ϵ, we can find δ so that

$$|f(x,y)-f(X,Y)| < \tfrac{1}{2}\epsilon \qquad (|x-X| < \delta,\ |y-Y| < \delta),$$

and so
$$|f(x,y)-f(\xi,\eta)| < \epsilon$$

if (x,y) and (ξ,η) both lie in the square with centre (X, Y) and side 2δ. Thus the rectangles chosen in the construction have the required property when they lie in this square, as they ultimately do. We have thus obtained a contradiction, and the theorem is proved.

We also deduce the following result: *Given ϵ, we can find δ such that*

$$|f(x,y)-f(\xi,\eta)| < \epsilon$$

provided that $|x-\xi| < \delta$ and $|y-\eta| < \delta$, δ depending on ϵ only, and not on x, y, ξ, or η.

For divide up the rectangle so that $|f(x,y)-f(\xi,\eta)| < \frac{1}{2}\epsilon$ if (x,y) and (ξ,η) belong to the same sub-rectangle. Let δ be the minimum of the sides of sub-rectangles. Then δ is the required number. For if $|x-\xi| < \delta$ and $|y-\eta| < \delta$, (x,y) and (ξ,η) belong to the same or to adjacent rectangles, and in either case the theorem follows.

The result may be expressed by saying that *a function of two variables which is continuous in a rectangle (boundary included) is uniformly continuous in the rectangle.*

We can now proceed with the properties of integrals.

If $f(x,y)$ is continuous in the rectangle $a \leqslant x \leqslant b$, $\alpha \leqslant y \leqslant \beta$, then

$$\phi(y) = \int_a^b f(x,y)\,dx$$

is a continuous function of y in (α,β).

For

$$\phi(y+k)-\phi(y) = \int_a^b \{f(x,y+k)-f(x,y)\}\,dx,$$

and, given ϵ, we can choose k_0 so that

$$|f(x,y+k)-f(x,y)| \leqslant \epsilon \qquad (|k| < k_0),$$

for all values of x and y. Hence

$$|\phi(y+k)-\phi(y)| \leqslant \epsilon(b-a) \qquad (|k| < k_0),$$

the required result.

If $f(x,y)$ is continuous in the rectangle $a \leqslant x \leqslant b$, $\alpha \leqslant y \leqslant \beta$, for all values of b, and the integral

$$\phi(y) = \int_a^\infty f(x,y)\,dx$$

converges uniformly with respect to y in the interval (α,β), then $\phi(y)$ is a continuous function of y in this interval.

We have

$$|\phi(y+k)-\phi(y)| = \left|\int_a^\infty \{f(x,y+k)-f(x,y)\}\,dx\right|$$

$$\leqslant \left|\int_a^X \{f(x,y+k)-f(x,y)\}\,dx\right| + \left|\int_X^\infty f(x,y+k)\,dx\right| + \left|\int_X^\infty f(x,y)\,dx\right|.$$

Given ϵ, we can choose X_0 so that each of the last two terms is less than ϵ for $X > X_0$, for all values of k. Having fixed X, the first term tends to zero with k, by the previous theorem. The result now follows.

In the above theorems, the continuity of $\phi(y)$ at the end-points α and β is one-sided, e.g. $\phi(y) \to \phi(\alpha)$ as $y \to \alpha$ by values greater than α.

Examples. (i) If the integral

$$\int_0^\infty f(x)\,dx$$

is convergent, then

$$\int_0^\infty e^{-xy}f(x)\,dx$$

is uniformly convergent in $0 \leqslant y \leqslant \beta$, and so continuous at $y = 0$.

[This is the analogue for integrals of Abel's theorem on power series. It may be proved as follows. Let

$$F(x) = \int_x^\infty f(t)\,dt,$$

so that $F(x) \to 0$ as $x \to \infty$. Suppose that $|F(x)| < \epsilon$ for $x > X_0$. Then

$$\left|\int_X^{X'} f(x)e^{-xy}\,dx\right| = \left|F(X)e^{-Xy} - F(X')e^{-X'y} - y\int_X^{X'} F(x)e^{-xy}\,dx\right|$$

$$\leqslant \epsilon + \epsilon + y\epsilon\int_X^{X'} e^{-xy}\,dx < 3\epsilon$$

for $X' > X > X_0$ and all $y \geqslant 0$; and the result follows.]

(ii) The integral

$$\int_0^\infty \frac{\sin xy}{x}\,dx$$

is convergent for every y; but it is not uniformly convergent in the neighbourhood of $y = 0$, since it is discontinuous at this point.

[To prove this, observe that it is (*a*) constant for $y > 0$ (put $x = u/y$), (*b*) positive for $y = 1$ (express it as

$$\sum \int_{(n-1)\pi}^{n\pi} \frac{\sin x}{x}\,dx,$$

i.e. as a series of decreasing terms of alternate signs), and (*c*) an odd function of y.

We may prove directly that it is not uniformly convergent by considering the 'remainder'

$$\int_X^\infty \frac{\sin xy}{x}\,dx = \int_{Xy}^\infty \frac{\sin u}{u}\,du,$$

and putting, e.g., $X = \pi/y$.

The value of the integral will be obtained later (§ 1.76).]

1.6. Double series. A double series consists of a double array of terms

$$\sum\sum a_{m,n},$$

where each of the suffixes m and n runs from 1 to infinity. There is no single method of summing the series, such as the '$\lim s_n$' method for single series, which obviously claims our attention. We can form partial sums of the series in a great variety of different ways, and each way gives rise to a method of summing the series. We may, for example, consider 'rectangular' sums

$$\sum_{m=1}^{M}\sum_{n=1}^{N} a_{m,n},$$

and then make M and N tend to infinity in various ways. Or we may consider sums such as

$$\sum_{m+n\leqslant N} a_{m,n},$$

taken over triangular regions. Or, finally, we may convert the double series into a 'repeated' series, first evaluating the sums

$$\sum_{n=1}^{\infty} a_{m,n},$$

and then finding the sum of their sums. We write this repeated series as

$$\sum_{m=1}^{\infty}\sum_{n=1}^{\infty} a_{m,n},$$

the inner sum being found first. We call this the 'sum by rows'. If we proceed in the opposite order, we obtain another repeated series

$$\sum_{n=1}^{\infty}\sum_{m=1}^{\infty} a_{m,n}.$$

We call this the 'sum by columns'.

1.61. Double series of positive terms. *If all the terms $a_{m,n}$ of the series are positive, all methods of summation are equivalent. Either we obtain a finite limit, the same in all cases; or, however we sum, the series diverges to positive infinity.*

To prove this we consider in turn the various possibilities.

We call a set of pairs of numbers (m,n) a region. Let Δ_p $(p = 1, 2,...)$ denote a sequence of finite regions, each of which includes the one before, and such that, however large N is, Δ_p includes the square $m \leqslant N$, $n \leqslant N$, if p is large enough.

There are now two possible cases. Suppose first that the finite sums

$$a_{m_1,n_1}+a_{m_2,n_2}+...+a_{m_k,n_k},$$

selected in any manner from the series, have an upper bound G. Then plainly

$$\sum_{\Delta_p} a_{m,n} \leqslant G$$

for all values of p. On the other hand, given ϵ, we can find one of these finite sums greater than $G-\epsilon$. But Δ_p includes every term of this finite sum, if p is large enough, and then

$$\sum_{\Delta_p} a_{m,n} > G-\epsilon.$$

Hence, since $\sum_{\Delta_p}$ is non-decreasing,

$$\lim_{p\to\infty} \sum_{\Delta_p} a_{m,n} = G;$$

that is to say, the series is convergent when summed in this particular way, and its sum is G. In this case the series is said to be convergent, it being unnecessary to specify the particular sequence of regions taken.

Suppose secondly that there is no such upper bound G. Then, having given any positive number H, there is a finite sum

$$a_{m_1,n_1}+\ldots+a_{m_k,n_k} > H.$$

Since we can find a number p such that Δ_p includes this sum, we have

$$\sum_{\Delta_p} a_{m,n} > H$$

for this value of p. Hence

$$\sum_{\Delta_p} a_{m,n} \to \infty.$$

In this case the series is said to be divergent.

These two cases are the only possibilities; and, since the results are independent of the particular regions Δ_p considered, we have proved the theorem, so far as finite partial sums are concerned.

Repeated series do not, so far, come under our analysis. To include them we have to replace our finite regions Δ_p by infinite regions.

Suppose first that the double series is convergent. Let D be any region, finite or infinite. Let $b_{m,n} = a_{m,n}$ if (m,n) is a point of D, and otherwise $b_{m,n} = 0$. Then clearly $\sum b_{m,n}$ converges if $\sum a_{m,n}$ does. We write

$$\sum_{D} a_{m,n} = \sum b_{m,n},$$

this being the definition of the left-hand side. It is clear that

$$\sum_{D} a_{m,n} \leqslant G.$$

Now let D_p $(p = 1, 2, \ldots)$ be a sequence of regions, finite or infinite, having the property characteristic of Δ_p. Then we have

$$\sum_{D_p} a_{m,n} \leqslant G,$$

and we can prove precisely as before that

$$\sum_{D_p} a_{m,n} > G - \epsilon \qquad (p \geqslant p_0).$$

Hence

$$\lim_{p \to \infty} \sum_{D_p} a_{m,n} = G.$$

In particular, if we take D_p to be the infinite region defined by $m \leqslant p$, we find that the sum by rows is equal to G. Similarly the sum by columns is G.

Secondly, suppose that the double series is divergent. Then it may happen that the series

$$\sum_{D_p} a_{m,n}$$

is divergent for a definite value of p. In this case the process comes to an end at this point. On the other hand, if $\sum_{D_p}$ is convergent for every p, we can, as before, show that

$$\sum_{D_p} a_{m,n} > H$$

for every H and $p \geqslant p_0(H)$. Hence

$$\sum_{D_p} a_{m,n} \to \infty.$$

In particular, if the double series is divergent, either some column is divergent, or every column is convergent, but their sums form a divergent series. The same thing is true of rows.

1.62. As the case of repeated series is particularly interesting, we give an alternative proof for this case.

If $a_{m,n} \geqslant 0$, then

$$\sum_{m=1}^{\infty} \sum_{n=1}^{\infty} a_{m,n} = \sum_{n=1}^{\infty} \sum_{m=1}^{\infty} a_{m,n}, \tag{1}$$

in the sense that, if either side converges, then so does the other, and to the same sum.

Suppose, for example, that the left-hand side is convergent. This means that all the series

$$\sum_{n=1}^{\infty} a_{m,n}$$

are convergent, to sums A_m, say, and that

$$\sum_{m=1}^{\infty} A_m$$

is convergent, to sum S, say.

Since $a_{m,n} \leq A_m$ for all values of m and n, it follows from the comparison test that all the series

$$\sum_{m=1}^{\infty} a_{m,n}$$

are convergent. Let their sums be $A^{(n)}$. Then

$$\sum_{n=1}^{N} A^{(n)} = \sum_{n=1}^{N} \sum_{m=1}^{\infty} a_{m,n} = \sum_{m=1}^{\infty} \sum_{n=1}^{N} a_{m,n} \leqslant \sum_{m=1}^{\infty} A_m = S.$$

Hence the series $\sum A^{(n)}$ is convergent, and, if its sum is S', then $S' \leqslant S$. But we can now reverse the whole argument, and, starting with the convergence of the right-hand side, prove that $S \leqslant S'$. Hence $S = S'$.

1.621. Still another method of proof is as follows. Suppose that the left-hand side of 1.62 (1) is convergent. Then

$$\sum_{n=1}^{N} \sum_{m=1}^{\infty} a_{m,n} = \sum_{m=1}^{\infty} \sum_{n=1}^{N} a_{m,n}, \tag{1}$$

this being merely the addition of a finite number of convergent series, whose convergence follows from that of $\sum A_m$. It is now sufficient to prove that, as $N \to \infty$,

$$\sum_{m=1}^{\infty} \sum_{n=N+1}^{\infty} a_{m,n} \to 0; \tag{2}$$

for this expression, together with the right-hand side of (1), is S; and it will then follow that the left-hand side of (1) tends to S, which is what is required.

This, however, follows from the uniform convergence theorem of § 1.14; for the series (2) is of the form

$$\sum_{m=1}^{\infty} u_m(N).$$

It converges uniformly with respect to N, since

$$|u_m(N)| \leqslant A_m,$$

and $\sum A_m$ is convergent; and, for each value of m, $u_m(N) \to 0$. This proves the theorem.

This method is of interest for the following reason. In less simple cases, where the numbers $a_{m,n}$ are not all positive, we can still start from (1), and so reduce the problem to the proof of (2). This can then be proved by some special method—see, for example, § 1.66, ex. (xvii).

1.63. The comparison test. Series of positive and negative terms. We first note the comparison test for the convergence of a double series of positive terms: *if* $a_{m,n} \leqslant b_{m,n}$, *and* $\sum b_{m,n}$ *is convergent, then* $\sum a_{m,n}$ *is convergent.* We leave the proof of this to the reader.

Suppose now that some of the numbers $a_{m,n}$ are positive, and some are negative. Then *the series* $\sum a_{m,n}$ *is said to be absolutely convergent if the series* $\sum |a_{m,n}|$ *is convergent.*

Let $\alpha_{m,n} = a_{m,n}$ if $a_{m,n} \geqslant 0$, and otherwise let $\alpha_{m,n} = 0$; let $\beta_{m,n} = -a_{m,n}$ if $a_{m,n} \leqslant 0$, and otherwise $\beta_{m,n} = 0$. Then $0 \leqslant \alpha_{m,n} \leqslant |a_{m,n}|$, $0 \leqslant \beta_{m,n} \leqslant |a_{m,n}|$. Hence, by the comparison test, the series

$$\sum \alpha_{m,n}, \qquad \sum \beta_{m,n}$$

are convergent if $\sum |a_{m,n}|$ is convergent. Let the sums of these series be α and β. Then, with our previous notation,

$$\sum_{\Delta_p} a_{m,n} = \sum_{\Delta_p} \alpha_{m,n} - \sum_{\Delta_p} \beta_{m,n} \to \alpha - \beta.$$

The same thing is true if the finite region Δ_p is replaced by an infinite region D_p, but now the above equation is taken as the definition of the left-hand side.

The 'sum' $\alpha - \beta$ is independent of the region Δ_p or D_p. We state the result simply by saying that *the series* $\sum a_{m,n}$ *is convergent; that is, an absolutely convergent double series is convergent.*

The comparison test can now be extended to series of this type.

1.64. Series of complex terms. Suppose now that $a_{m,n}$ is a complex number, say $a_{m,n} = b_{m,n} + ic_{m,n}$. Then the series $\sum a_{m,n}$ is said to be absolutely convergent if $\sum |a_{m,n}|$ is convergent. Since

$$|b_{m,n}| \leqslant |a_{m,n}|, \qquad |c_{m,n}| \leqslant |a_{m,n}|,$$

this involves the absolute convergence, and so the convergence, of $\sum b_{m,n}$ and $\sum c_{m,n}$. If b and c are the sums of these series,

$$\sum_{\Delta_p \,(\text{or } D_p)} a_{m,n} = \sum_{\Delta_p} b_{m,n} + i \sum_{\Delta_p} c_{m,n} \to b + ic,$$

and the series is said to be convergent; that is, *an absolutely convergent series of complex terms is convergent.*

All our conclusions about series of positive terms can now be extended to absolutely convergent series.

1.65. Multiplication of series. Let

$$a_0+a_1+a_2+\ldots, \qquad b_0+b_1+b_2+\ldots$$

be two absolutely convergent series. The series obtained by multiplying these series by Cauchy's rule is

$$c_0+c_1+c_2+\ldots$$

where $$c_n = a_0 b_n + a_1 b_{n-1} + \ldots + a_n b_0.$$

The rule has its origin in the case of power series, where $a_n = \alpha_n x^n$, $b_n = \beta_n x^n$, and where, in the multiplied series, we collect together terms involving the same power of x.

If the series $\sum_0^\infty a_n$, $\sum_0^\infty b_n$, are absolutely convergent, and their sums are a and b, then $\sum_0^\infty c_n$ is absolutely convergent, and its sum is ab.

This follows at once from the above theorems on double series. For the double series $\sum a_m b_n$ is absolutely convergent. Its sum by rows or columns is ab, and

$$\sum_{n=0}^{N} c_n = \sum_{m+n \leqslant N} a_m b_n,$$

which also tends to the limit ab.

If the series $\sum_0^\infty a_n$, $\sum_0^\infty b_n$, $\sum_0^\infty c_n$, are all convergent, to sums a, b, and c, then $c = ab$.

We apply the above theorem to the series $\sum_0^\infty a_n x^n$, $\sum_0^\infty b_n x^n$, $\sum_0^\infty c_n x^n$, which are absolutely convergent if $0 < x < 1$, and then make $x \to 1$ and use Abel's theorem (§ 1.22).

1.66. Miscellaneous examples on double and repeated series.

(i) If $|a_{m,n}| < Am^\alpha n^\beta$, where A, α, β are constants, and $|x| < 1$, $|y| < 1$, then $\sum a_{m,n} x^m y^n$ is absolutely convergent.

(ii) If $\sum a_{m,n} x_0^m y_0^n$ is absolutely convergent, then $\sum a_{m,n} x^m y^n$ is absolutely convergent for $|x| \leqslant |x_0|$, $|y| \leqslant |y_0|$.

(iii) The series $\sum m^{-\alpha} n^{-\beta}$ is convergent if $\alpha > 1$, $\beta > 1$.

(iv) The series $\sum (m^2+n^2)^{-\alpha}$ is convergent if $\alpha > 1$.

[Compare the terms for which $m \leqslant n$ with $\sum_{n=1}^{\infty} \sum_{m=1}^{n} n^{-2\alpha}$.]

The same result holds for the corresponding doubly infinite series, in which each variable ranges from $-\infty$ to ∞, omitting $m = n = 0$.

(v) The series $\sum (am^2+2bmn+cn^2)^{-\alpha}$ is convergent if $a > 0$, $b^2 < ac$, and $\alpha > 1$. [For

$$\frac{am^2+2bmn+cn^2}{m^2+n^2}$$

has a positive minimum value.]

(vi) If the ratio z/z' is complex, and a is not equal to any of the numbers $-mz-nz'$, then $\sum |a+mz+nz'|^{-\alpha}$ is convergent if $\alpha > 2$.

(vii) By expanding the function

$$\log(1-2x\cos\theta + x^2) = \log(1-xe^{i\theta})+\log(1-xe^{-i\theta})$$

in two different ways, show that

$$\cos n\theta = 2^{n-1}\cos^n\theta - \frac{n}{1!}2^{n-3}\cos^{n-2}\theta + \frac{n(n-3)}{2!}2^{n-5}\cos^{n-4}\theta - \ldots,$$

and also obtain a series of ascending powers of $\cos\theta$.

[The rearrangement may be justified by the double series theorem; we also use the theorem on the uniqueness of a power series, *P.M.* § 194.]

(viii) If $|a| < 1$ and $|x| < 1$,

$$\sum_{n=0}^{\infty} \frac{1}{n!}\frac{a^n}{1+x^2a^{2n}} = e^a - x^2e^{a^3} + x^4e^{a^5} - \ldots.$$

(ix) If x is not a negative integer,

$$\sum_{n=0}^{\infty} \frac{1}{x(x+1)\ldots(x+n)} = e\left\{\frac{1}{x} - \frac{1}{1!}\frac{1}{(x+1)} + \frac{1}{2!}\frac{1}{(x+2)} - \ldots\right\}.$$

(x) If $d(n)$ denotes the number of divisors of n, and $|x| < 1$,

$$\sum_{n=1}^{\infty} \frac{x^n}{1-x^n} = \sum_{n=1}^{\infty} d(n)x^n.$$

(xi) **Dirichlet multiplication.** If $\sum a_n n^{-s}$, $\sum b_n n^{-s}$ are absolutely convergent, and $c_n = \sum_{pq=n} a_p b_q$, then

$$\sum a_n n^{-s} . \sum b_n n^{-s} = \sum c_n n^{-s}.$$

In particular

$$\{\zeta(s)\}^2 = \sum d(n) n^{-s}.$$

(xii) If $a_{m,n} = 1$ $(m = n+1,\ n = 1, 2,\ldots)$, $a_{m,n} = -1$ $(m = n-1,\ n = 2, 3,\ldots)$, and otherwise $a_{m,n} = 0$, then

$$\sum_{m=1}^{\infty}\sum_{n=1}^{\infty} a_{m,n} \neq \sum_{n=1}^{\infty}\sum_{m=1}^{\infty} a_{m,n}.$$

(xiii) Prove a similar result if

$$a_{m,n} = \frac{1}{m^2-n^2}\ (m \neq n), \qquad a_{m,n} = 0\ (m = n).$$

[Here (terms $m = n$ being omitted)

$$\sum_{m=1}^{\infty}{}' \frac{1}{m^2-n^2} = \lim_{N\to\infty} \frac{1}{2n} \sum_{m=1}^{N}{}' \left(\frac{1}{m-n} - \frac{1}{m+n}\right)$$

$$= \lim_{N\to\infty} \frac{1}{2n}\left(-\sum_{\nu=1}^{n-1} \frac{1}{\nu} + \sum_{\nu=1}^{N-n} \frac{1}{\nu} - \sum_{\nu=n+1}^{N+n} \frac{1}{\nu} + \frac{1}{2n}\right)$$

$$= \frac{3}{4n^2},$$

and the sums by columns and rows are $\pi^2/8$ and $-\pi^2/8$. See Hardy (3).]

(xiv) If $\sum |u_n|$ is convergent, then $\prod (1+u_n z)$ is absolutely and uniformly convergent in any finite region, and it may be rearranged as a power series in z,

$$\prod (1+u_n z) = 1+z\sum u_n + z^2 \sum_{m\neq n}\sum u_m u_n + \dots .$$

[The first part has been proved already (§ 1.44), and it is simply a question of justifying the rearrangement. Let z be fixed. Let

$$P_N = \prod_1^N (1+|u_n||z|) = 1+C_1^{(N)}|z|+\dots+C_N^{(N)}|z|^N.$$

Then $P_N \to P$, and, for each m, $C_m^{(N)}$ is non-decreasing and $\leqslant P|z|^{-m}$. Hence $C_m^{(N)} \to C_m$, say. Plainly

$$1+\sum_{m=1}^{k} C_m^{(N)}|z|^m \leqslant P_N \leqslant 1+\sum_{m=1}^{N} C_m|z|^m \qquad (k \leqslant N).$$

Making $N \to \infty$, then $k \to \infty$, we deduce that

$$P = 1+\sum_{m=1}^{\infty} C_m|z|^m.$$

Let $$p_N = \prod_1^N (1+u_n z) = 1+c_1^{(N)}z+\dots+c_N^{(N)}z^N.$$

By an obvious extension of the theorem of absolute convergence to multiple series, $c_m^{(N)} \to c_m$, and plainly $|c_m - c_m^{(N)}| \leqslant C_m - C_m^{(N)} \leqslant C_m$. Hence if $k \leqslant N$,

$$\left|\sum_{m=1}^{\infty} c_m z^m - \sum_{m=1}^{N} c_m^{(N)} z^m\right| \leqslant \sum_{m=1}^{k} (C_m - C_m^{(N)})|z|^m + \sum_{k+1}^{\infty} C_m|z|^m,$$

which tends to 0 by choosing first k and then N. Hence the result.

(xv) Assuming that

$$\frac{\sin x}{x} = \prod_1^{\infty}\left(1-\frac{x^2}{n^2\pi^2}\right),$$

deduce that $$\sum_{n=1}^{\infty} \frac{1}{n^2} = \frac{\pi^2}{6}, \qquad \sum_{n=1}^{\infty} \frac{1}{n^4} = \frac{\pi^4}{90}.$$

(xvi) Let $s_{m,n} \to s$ mean that $|s_{m,n}-s| < \epsilon$ if m and n are both $> n_0 = n_0(\epsilon)$.

Show that, if $s_{m,n} \to s$, and $\lim_{n\to\infty} s_{m,n}$ exists for each m, then

$$\lim_{m\to\infty}\left(\lim_{n\to\infty} s_{m,n}\right) = s.$$

It has been proved by Pringsheim that, if the double series $\sum\sum a_{m,n}$ converges to sum s in the sense that

$$\sum_{\mu=1}^{m}\sum_{\nu=1}^{n} a_{\mu,\nu} \to s,$$

and the single series $\quad \sum_{m=1}^{\infty} a_{m,n}, \quad \sum_{n=1}^{\infty} a_{m,n}$

all converge, then the sum by rows and the sum by columns are also both equal to s. See Bromwich, *Infinite Series*, § 30.

(xvii) From the formula

$$1-\frac{1}{3}+\frac{1}{5}-\ldots=\frac{\pi}{4}$$

(*P.M.* § 214), deduce that

$$\frac{1}{1^2}+\frac{1}{3^2}+\frac{1}{5^2}+\ldots=\frac{\pi^2}{8}.$$

[The result follows from the formula

$$\sum_{m=-\infty}^{\infty}\sum_{n=-\infty}^{\infty}\frac{(-1)^n}{(m+\frac{1}{2})(m+n+\frac{1}{2})}=\sum_{n=-\infty}^{\infty}\sum_{m=-\infty}^{\infty}\frac{(-1)^n}{(m+\frac{1}{2})(m+n+\frac{1}{2})}.$$

For the left-hand side (putting $n = r-m$) is equal to

$$\sum_{m=-\infty}^{\infty}\frac{(-1)^{-m}}{m+\frac{1}{2}}\sum_{r=-\infty}^{\infty}\frac{(-1)^r}{r+\frac{1}{2}}=\left(\frac{2}{\frac{1}{2}}-\frac{2}{\frac{3}{2}}+\ldots\right)^2=\pi^2.$$

Also, if $n \neq 0$,

$$\sum_{m=-\infty}^{\infty}\frac{1}{(m+\frac{1}{2})(m+n+\frac{1}{2})}=\frac{1}{n}\sum_{m=-\infty}^{\infty}\left(\frac{1}{m+\frac{1}{2}}-\frac{1}{m+n+\frac{1}{2}}\right)=0,$$

so that the right-hand side is equal to

$$\sum_{m=-\infty}^{\infty}\frac{1}{(m+\frac{1}{2})^2}=8\left(\frac{1}{1^2}+\frac{1}{3^2}+\ldots\right).$$

We have therefore to justify the inversion of the repeated series. The associated double series is clearly not absolutely convergent, so that a special method is required.

We have, for every N,

$$\sum_{m=-\infty}^{\infty}\sum_{n=-N}^{N}=\sum_{n=-N}^{N}\sum_{m=-\infty}^{\infty},$$

and hence (see § 1.621) it is sufficient to prove that, as $N\to\infty$,

$$\sum_{m=-\infty}^{\infty}\sum_{n=N+1}^{\infty}\to 0, \qquad \sum_{m=-\infty}^{\infty}\sum_{n=-\infty}^{-N-1}\to 0.$$

Consider the former sum. If $m \geqslant -N-1$, we have

$$\left|\sum_{n=N+1}^{\infty} \frac{(-1)^n}{m+n+\frac{1}{2}}\right| < \frac{1}{m+N+\frac{3}{2}}.$$

This part of the sum is therefore less in absolute value than

$$\sum_{m=-N-1}^{\infty} \frac{1}{|m+\frac{1}{2}|(m+N+\frac{3}{2})} \leqslant \sum_{-N-1 \leqslant m < -\frac{1}{2}N-\frac{1}{2}} \frac{1}{\frac{1}{2}N(m+N+\frac{3}{2})} +$$

$$+ \sum_{-\frac{1}{2}N-\frac{1}{2} \leqslant m \leqslant N} \frac{1}{|m+\frac{1}{2}|\frac{1}{2}N} + \sum_{N+1}^{\infty} \frac{1}{(m+\frac{1}{2})^2},$$

which plainly tends to zero. In the rest of the sum, we write

$$\sum_{n=N+1}^{\infty} \frac{(-1)^n}{m+n+\frac{1}{2}} = (-1)^m\pi - \sum_{n=-\infty}^{N} \frac{(-1)^n}{m+n+\frac{1}{2}}.$$

The sum formed from the terms involving π plainly tends to zero, and the last term is similar to the one already considered, and so also gives a sum which tends to zero.

Finally, the other sum can be dealt with in the same way, and the result follows.]

1.7. Integration of series. Having completed our discussion of repeated summations, we now turn to a similar set of problems, in which one of the summations is replaced by an integration. Since a finite integral is itself a limit, whereas a finite sum is not, this makes everything one degree more complicated.

We first consider the term-by-term integration of a series over a finite range.

1.71. *A uniformly convergent series of continuous functions may be integrated term by term; that is, if $u_1(x)$, $u_2(x)$,... are continuous, and*

$$u_1(x)+u_2(x)+... = s(x)$$

converges uniformly over (a, b), then

$$\int_a^b u_1(x)\,dx + \int_a^b u_2(x)\,dx + ... = \int_a^b s(x)\,dx.$$

Since $s(x)$ is continuous (§ 1.14), it has a Riemann integral. Also the sum of the first n terms of the integrated series is (with our usual notation)

$$\int_a^b s_n(x)\,dx.$$

We have therefore to prove that

$$\int_a^b s_n(x)\,dx \to \int_a^b s(x)\,dx,$$

or that

$$\int_a^b \{s(x)-s_n(x)\}\,dx \to 0.$$

But, given ϵ, we can find n_0 such that

$$|s(x)-s_n(x)| < \epsilon$$

for $n > n_0$ and all values of x. Hence, by *P.M.* § 160, (7),

$$\left|\int_a^b \{s(x)-s_n(x)\}\,dx\right| < \epsilon(b-a),$$

and the result follows.

Examples. (i) If $0 < x < 1$,

$$\log\frac{1}{1-x} = \int_0^x \frac{dt}{1-t} = \int_0^x (1+t+t^2+\ldots)\,dt = x+\tfrac{1}{2}x^2+\tfrac{1}{3}x^2+\ldots.$$

(ii) Similarly, $$\arctan x = x-\frac{x^3}{3}+\frac{x^5}{5}-\ldots.$$

(iii) Prove that $$\int_0^1 \log\frac{1+x}{1-x}\,\frac{dx}{x} = \frac{\pi^2}{4}.$$

[Use Abel's continuity theorem.]

(iv) Show that, if $r < 1$, and n is a positive integer,

$$\int_0^\pi \frac{1-r^2}{1-2r\cos\theta+r^2}\cos n\theta\,d\theta = \pi r^n.$$

1.72. *A series may be differentiated term by term if the differentiated series is a uniformly convergent series of continuous functions; that is, if*

$$u_1(x)+u_2(x)+\ldots = s(x),$$

and the functions $u_1(x)$, $u_2(x)$,... *have continuous derivatives* $u_1'(x)$,... *such that the series*

$$u_1'(x)+u_2'(x)+\ldots$$

converges uniformly to $f(x)$ *in* (a, b), *then* $f(x) = s'(x)$ *for* $a < x < b$.

By the previous theorem, the second series may be integrated term by term over (a, x), so that

$$\{u_1(x)-u_1(a)\}+\{u_2(x)-u_2(a)\}+\ldots = \int_a^x f(t)\,dt.$$

But the left-hand side is also equal to $s(x)-s(a)$. Hence

$$s(x)-s(a) = \int_a^x f(t)\,dt,$$

and, differentiating, the result follows, since $f(x)$ is continuous.

Examples. (i) If $|x| < 1$,

$$\sum_{n=k}^{\infty} n(n-1)\ldots(n-k+1)x^{n-k} = \frac{k!}{(1-x)^{k+1}}.$$

(ii) If $s > 1$,

$$\zeta'(s) = -\sum_{n=2}^{\infty} n^{-s}\log n.$$

1.73. *A real power series may be integrated or differentiated any number of times within the interval of convergence.* That is to say, the result of any number of formal term-by-term integrations or differentiations is true, provided that we are inside the interval of convergence.

Let the power series be

$$f(x) = \sum_0^{\infty} a_n x^n \qquad (|x| < R).$$

We can integrate once, by uniform convergence (§ 1.21), and obtain

$$\int_0^x f(t)\,dt = \sum_{n=0}^{\infty} \frac{a_n}{n+1}\, x^{n+1} \qquad (|x| < R).$$

The interval of convergence, and so of uniform convergence, is plainly at least as wide for this series as for the previous one, and so the process may be repeated.

Term-by-term differentiation gives

$$f'(x) = \sum_1^{\infty} n a_n x^{n-1}.$$

This series also is convergent for $|x| < R$. For, if $0 < \rho < R$, $|a_n \rho^n| < K$. Hence

$$|na_n x^{n-1}| = n|a_n \rho^n| \cdot \left|\frac{x^{n-1}}{\rho^n}\right| < \frac{K}{\rho} n \left(\frac{|x|}{\rho}\right)^{n-1},$$

and hence the differentiated series is convergent for $|x| < \rho$, by comparison with the convergent series

$$\sum_1^{\infty} n\left(\frac{|x|}{\rho}\right)^{n-1} = \frac{1}{(1-|x|/\rho)^2}.$$

Hence the differentiated series is uniformly convergent over any interval included in $(-R, R)$, and so term-by-term differentiation is justified. The process can, of course, now be repeated.

It follows in particular that *a function represented by a power series has derivatives of all orders.*

It is also clear, since neither integration nor differentiation can decrease the interval of convergence, and the two processes are reciprocal, that neither can increase the interval.

Example. The Maclaurin expansion of $f(x) = \sum_{n=0}^{\infty} a_n x^n$ in powers of x is the original series.

1.74. *If x is real, and*

$$f(x) = \sum_{n=0}^{\infty} a_n x^n \qquad (|x| < R),$$

then $f(x+h)$ may be expanded by Taylor's theorem in powers of h, provided that $|x| < R$ and $|h| < R - |x|$.

The formal expansion is

$$f(x+h) = \sum_{m=0}^{\infty} \frac{h^m}{m!} f^{(m)}(x),$$

where, by the previous theorem,

$$f^{(m)}(x) = \sum_{n=m}^{\infty} n(n-1)...(n-m+1)a_n x^{n-m}.$$

To prove that this actually holds, we write

$$f(x+h) = \sum_{n=0}^{\infty} a_n(x+h)^n = \sum_{n=0}^{\infty} a_n \sum_{m=0}^{n} \frac{n(n-1)...(n-m+1)}{m!} x^{n-m}h^m$$

$$= \sum_{m=0}^{\infty} \sum_{n=m}^{\infty} a_n \frac{n(n-1)...(n-m+1)}{m!} x^{n-m}h^m = \sum_{m=0}^{\infty} \frac{h^m}{m!} f^{(m)}(x).$$

We have to justify the inversion; and it is justified by absolute convergence if

$$\sum_{n=0}^{\infty} |a_n| \sum_{m=0}^{n} \frac{n(n-1)...(n-m+1)}{m!} |x|^{n-m}|h|^m = \sum_{n=0}^{\infty} |a_n|(|x|+|h|)^n$$

is convergent. This is true if $|x|+|h| < R$, which proves the theorem.

Notice that the interval of convergence obtained for the new series extends just up to one end of the interval of convergence of the original series. The actual interval of convergence may be no larger, e.g. in the case of

$$f(x) = \frac{1}{1-x^2} = 1+x^2+x^4+\dots.$$

But it may in some cases extend further; e.g. if

$$f(x) = \frac{1}{1+x} = 1-x+x^2-\dots$$

we have
$$f^{(m)}(x) = \frac{(-1)^m m!}{(1+x)^{m+1}},$$

and
$$f(\tfrac{1}{2}+h) = \sum_{m=0}^{\infty} (-1)^m (\tfrac{2}{3})^{m+1} h^m,$$

which is convergent for $|h| < \frac{3}{2}$.

It is impossible to give a satisfactory account of this phenomenon so long as we consider *real* power series only, and we must postpone further discussion until we have considered functions of a complex variable.

1.75. Series which cannot be integrated term by term. A simple example of such a series is obtained by putting

$$s_n(x) = n^2 x(1-x)^n \qquad (0 \leqslant x \leqslant 1).$$

Then $s(x) = 0$ for all values of x, so that

$$\int_0^1 s(x)\,dx = 0.$$

But
$$\int_0^1 s_n(x)\,dx = \frac{n^2}{(n+1)(n+2)} \to 1,$$

so that term-by-term integration gives an incorrect result. The series is, of course, not uniformly convergent.

On the other hand, uniform convergence is not a necessary condition for term-by-term integration. Some of the non-uniformly convergent series of § 1.3, e.g. those for which

$$s_n(x) = \frac{1}{1+nx}, \qquad s_n(x) = nx(1-x)^n,$$

can be integrated term by term.

This leads us to consider more general classes of series which can be integrated term by term.

1.76. Boundedly convergent series. *A series*

$$u_1(x)+u_2(x)+\dots$$

is said to be boundedly convergent in an interval (a, b), *if it con-*

verges for every value of x in the interval, and if there is a constant M such that $|s_n(x)| \leqslant M$ for all values of n and x $(a \leqslant x \leqslant b)$.

It is clear that the sum of a boundedly convergent series is bounded. So far we have no method of integrating bounded functions in general, so boundedness by itself does not enable us (at this stage) to integrate term by term. We have to combine it with another condition.

We say that a series *is uniformly convergent over* (a, b), *except in the neighbourhood of the point* c, if it is uniformly convergent over the intervals $(a, c-\delta)$, $(c+\delta, b)$, however small δ may be. We can then justify term-by-term integration under the following conditions:

If the series is uniformly convergent over (a, b), except in the neighbourhood of a finite number of points, and also boundedly convergent over the whole interval, then it may be integrated term by term over (a, b).

To prove this, it is sufficient to suppose that there is one exceptional point, c. Suppose that $|s_n(x)| \leqslant M$. Then $|s(x)| \leqslant M$ also. The integral of $s(x)$ exists in the sense of § 1.5, and

$$\left|\int_a^b \{s(x)-s_n(x)\}\,dx\right| \leqslant \left|\int_a^{c-\delta} \{s(x)-s_n(x)\}\,dx\right| + \left|\int_{c+\delta}^b \{s(x)-s_n(x)\}\,dx\right| + \\ + \left|\int_{c-\delta}^{c+\delta} s(x)\,dx\right| + \left|\int_{c-\delta}^{c+\delta} s_n(x)\,dx\right|$$

$$\leqslant \left|\int_a^{c-\delta}\right| + \left|\int_{c+\delta}^b\right| + 4\delta M.$$

We can choose δ so small that the last term is less than a given ϵ, for all n. Then, having fixed δ, the other terms tend to zero, by uniform convergence. This proves the theorem.

Various extensions of the theorem are possible. It is not necessary that the series should be uniformly convergent over $(a, c-\delta)$ and $(c+\delta, b)$, if term-by-term integration over these intervals can be justified in some other way. A more important observation is that the theorem remains true if we insert in the integral a factor $\phi(x)$, which is integrable, but not necessarily bounded. Suppose, for example, that $\phi(x)$ is continuous over (a, b) except at $x = a$, in the neighbourhood of which it is unbounded; and that

$$\int_a^b |\phi(x)|dx$$

exists as an infinite integral. Then we may multiply the series by $\phi(x)$ and integrate term by term. For

$$\left|\int_a^{a+\delta} \{s(x)-s_n(x)\}\phi(x)\,dx\right| \leqslant 2M\int_a^{a+\delta} |\phi(x)|\,dx,$$

which can be made less than any given ϵ, by choice of δ, for all values of n; and the integral over $(a+\delta, b)$ may be dealt with as before.

We observe finally that later, when we have developed the theory of the Lebesgue integral, we can put all these theorems into a much more satisfactory form. All the restrictions involving continuity and uniform convergence are only necessary because of the limitations of the Riemann integral, and disappear in the final form of the theorem.

Examples. (i) The series for which

$$s_n(x) = \frac{1}{1+nx}, \qquad s_n(x) = nx(1-x)^n \quad (0 \leqslant x \leqslant 1),$$

are boundedly convergent.

(ii) Consider the series

$$\frac{\sin x}{1}+\frac{\sin 2x}{2}+\frac{\sin 3x}{3}+\dots.$$

It follows from one of the general tests (see § 1.12, ex. (i)) that this series is uniformly convergent except in the neighbourhood of the points $x = 0, \pm 2\pi, \pm 4\pi, \dots$. To show that it is boundedly convergent, and to sum it, we use a more special method.

Since each term has the period 2π, it is sufficient to consider the interval $0 \leqslant x < 2\pi$. Here we write

$$s_n(x) = \int_0^x (\cos t+\cos 2t+\dots+\cos nt)\,dt = \int_0^x \frac{\sin(n+\frac{1}{2})t-\sin\frac{1}{2}t}{2\sin\frac{1}{2}t}\,dt$$

$$= \int_0^x \frac{\sin(n+\frac{1}{2})t}{t}\,dt+\int_0^x \left(\frac{1}{2\sin\frac{1}{2}t}-\frac{1}{t}\right)\sin(n+\tfrac{1}{2})t\,dt-\tfrac{1}{2}x$$

$$= \int_0^{(n+\frac{1}{2})x} \frac{\sin u}{u}\,du+\int_0^x \left(\frac{1}{2\sin\frac{1}{2}t}-\frac{1}{t}\right)\sin(n+\tfrac{1}{2})t\,dt-\tfrac{1}{2}x.$$

Now
$$\int_0^h \frac{\sin u}{u}\,du$$
is always positive, and has an absolute maximum at $h = \pi$ (*P.M.* § 181, exs. LXXVI, 9–10). Hence
$$|s_n(x)| \leqslant \int_0^\pi \frac{\sin u}{u}\,du + \int_0^\pi \left(\frac{1}{2\sin\frac{1}{2}t} - \frac{1}{t}\right) dt + \tfrac{1}{2}\pi$$
for $0 \leqslant x \leqslant \pi$, i.e. the series is boundedly convergent in this interval. Since each term is odd, it is boundedly convergent in $(-\pi, 0)$, and so, by periodicity, in any interval.

To sum the series, let x be fixed, $0 < x < 2\pi$, and make $n \to \infty$. Then
$$\lim_{n\to\infty} \int_0^{(n+\frac{1}{2})x} \frac{\sin u}{u}\,du = \int_0^\infty \frac{\sin u}{u}\,du$$
exists (§ 1.5). Let us denote the value of this integral by I. Also
$$\int_0^x \left(\frac{1}{2\sin\frac{1}{2}t} - \frac{1}{t}\right) \sin(n+\tfrac{1}{2})t\,dt$$
$$= -\left(\frac{1}{2\sin\frac{1}{2}x} - \frac{1}{x}\right)\frac{\cos(n+\frac{1}{2})x}{n+\frac{1}{2}} + \frac{1}{n+\frac{1}{2}}\int_0^x \frac{d}{dt}\left(\frac{1}{2\sin\frac{1}{2}t} - \frac{1}{t}\right)\cos(n+\tfrac{1}{2})t\,dt,$$
which tends to zero on account of the factor $n+\frac{1}{2}$ in the denominator (the other factors being bounded). Hence, if $s(x)$ is the sum,
$$s(x) = I - \tfrac{1}{2}x \qquad (0 < x < 2\pi).$$
But plainly $s(\pi) = 0$, so that $I = \frac{1}{2}\pi$. This gives at the same time the sum of the series and the value of the infinite integral.

The reader should draw the graph of the sum of the series, noticing its discontinuities at the points $x = 0, \pm 2\pi, \pm 4\pi, \ldots$. [See also Ch. XIII, ex. 11.]

(iii) Prove, without using integrals, that the above series is boundedly convergent, using a method similar to that used in § 1.131.

(iv) Sum the series
$$\frac{1}{1^2} + \frac{1}{3^2} + \frac{1}{5^2} + \ldots$$
by integrating the above series over $(0, \pi)$.

(v) Prove that
$$\sum_{n=1}^\infty \frac{\cos nx}{n^2} = \frac{\pi^2}{6} - \frac{\pi x}{2} + \frac{x^2}{4} \qquad (0 < x < \pi).$$

1.77. Term-by-term integration when the integrals are infinite. We now pass to the general case of term-by-term integration over an infinite range, or of functions which become infinite in the range of integration. In each case the results are similar to those already obtained for repeated series. For convenience we state them as a single theorem.

Suppose that $u_n(x) \geqslant 0$ *for all values of* n *and* x, *and that*

$$\int_a^c \{\sum u_n(x)\}\,dx = \sum \int_a^c u_n(x)\,dx \tag{1}$$

for all values of c *less than* b (*or for all finite values of* c). *Then*

$$\int_a^b \{\sum u_n(x)\}\,dx = \sum \int_a^b u_n(x)\,dx \tag{2}$$

(*or*
$$\int_a^\infty \{\sum u_n(x)\}\,dx = \sum \int_a^\infty u_n(x)\,dx \tag{2a}$$

in the second case), *provided that either side of the resulting equation is convergent.*

The proof is the same in the two cases. Let us take the case of a finite interval (a, b).

Suppose that the series on the right of (2) is convergent, say

$$\sum \int_a^b u_n(x)\,dx = S.$$

Then, since $u_n(x) \geqslant 0$,

$$\int_a^c \{\sum u_n(x)\}\,dx = \sum \int_a^c u_n(x)\,dx \leqslant S$$

for all values of c less than b. Hence (see *P.M.*, 7th edition, § 185) the integral on the left of (2) exists, as an infinite integral at b; and, if its value is I, then $I \leqslant S$. On the other hand,

$$\sum_{n=1}^{N} \int_a^b u_n(x)\,dx = \int_a^b \left\{\sum_1^N u_n(x)\right\} dx \leqslant \int_a^b \left\{\sum_1^\infty u_n(x)\right\} dx = I,$$

and, making $N \to \infty$, we see that $S \leqslant I$. Hence in fact $S = I$.

A similar method may be used if the left-hand side of (2) or (2a) is assumed to exist. The reader should write it out in detail.

As in the case of series, we can omit our 'positive' condition if we assume instead that one of the sides of (2) or (2a) is 'absolutely convergent', i.e. remains convergent if $u_n(x)$ is replaced by its modulus.

The above results remain true for functions $u_n(x)$ which may have either sign, or be complex, provided that either of

$$\int_a^b \{\sum |u_n(x)|\}\,dx, \qquad \sum \int_a^b |u_n(x)|\,dx,$$

(*or*
$$\int_a^\infty \{\sum |u_n(x)|\}\,dx, \qquad \sum \int_a^\infty |u_n(x)|\,dx$$

in the second case), *is convergent.*

It follows from the theorem already proved that the two conditions are equivalent. We then obtain the result precisely as in the case of double series. If $u_n(x)$ is real, we consider the functions $|u_n(x)| \pm u_n(x)$, each of which is positive. If $u_n(x)$ is complex, say $u_n(x) = \alpha_n(x) + i\beta_n(x)$, we consider the four functions $|u_n(x)| \pm \alpha_n(x)$, $|u_n(x)| \pm \beta_n(x)$, each of which is positive.

1.78. Miscellaneous examples on term-by-term integration.

(i) Prove that
$$\int_0^1 \log \frac{1}{1-x}\,dx = 1$$

by expanding in powers of x and integrating term by term. [Notice that the series is not uniformly or even boundedly convergent in the neighbourhood of $x = 1$.]

(ii) We have

$$\int_0^\infty \frac{x^{s-1}}{e^x - 1}\,dx = \int_0^\infty \{\sum x^{s-1} e^{-nx}\}\,dx = \sum \int_0^\infty x^{s-1} e^{-nx}\,dx$$
$$= \sum n^{-s} \int_0^\infty y^{s-1} e^{-y}\,dy = \sum n^{-s}\Gamma(s) = \Gamma(s)\zeta(s).$$

Justify this process (a) for $s > 1$, (b) for s complex, $\mathbf{R}(s) > 1$.

(iii) Prove that
$$\int_0^\infty e^{-ax} \cos bx\,dx = \frac{a}{a^2+b^2}$$

by expanding $\cos bx$ in powers of x. [The process is justified by absolute convergence if $\mathbf{R}(a) > |b|$, though the result holds in a wider range than this.]

(iv) Prove that, if $p > 0$,

$$\int_0^1 \frac{x^{p-1}}{1-x} \log \frac{1}{x}\,dx = \frac{1}{p^2} + \frac{1}{(p+1)^2} + \dots.$$

(v) Prove that, if $p > 0$,

$$\int_0^1 \frac{x^{p-1}}{1+x}\,dx = \frac{1}{p} - \frac{1}{p+1} + \dots .$$

[Here the absolute convergence test fails. Integrate over $(0, \xi)$, where $0 < \xi < 1$, and use Abel's theorem.]

(vi) Prove that $$\int_0^\infty \frac{\sinh ax}{\sinh bx}\,dx = \frac{\pi}{2b}\tan\frac{a\pi}{2b} \qquad (0 < a < b).$$

[On expanding in powers of e^{-bx}, we obtain the series

$$\sum \frac{2a}{(2n-1)^2 b^2 - a^2}.$$

For the summation of this we must refer to Chapter III.]

(vii) Prove that $$\int_0^\infty \frac{\cosh ax}{\cosh bx}\,dx = \frac{\pi}{2b}\sec\frac{a\pi}{2b} \qquad (0 < a < b).$$

[The general test fails, but the integral can be evaluated by means of (v) above.]

(viii) If $u_n(x) = ae^{-nax} - be^{-nbx}$ $(0 < a < b)$, show that

$$\sum \int_0^\infty u_n(x)\,dx \neq \int_0^\infty \{\sum u_n(x)\}\,dx.$$

[We have $$\sum \int_0^\infty u_n(x)\,dx = \sum \left(\frac{1}{n} - \frac{1}{n}\right) = 0,$$

but $$\int_0^\infty \{\sum u_n(x)\}\,dx = \int_0^\infty \left(\frac{a}{e^{ax}-1} - \frac{b}{e^{bx}-1}\right)dx > 0,$$

since the integrand is positive for all values of x, $u/(e^u - 1)$ being a steadily decreasing function of u.

It is easy to prove directly that

$$\sum \int_0^\infty |u_n(x)|\,dx$$

is divergent.]

(ix) Consider the integral

$$\int_0^\infty e^{-x^{\frac{1}{4}}} \sin x^{\frac{1}{4}} \cos ax\,dx.$$

Here we shall anticipate for a moment some of the results of Chapter III.

If we expand $\cos ax$ in powers of x and integrate term by term, we obtain

$$\sum_{n=0}^{\infty} \frac{(-1)^n a^{2n}}{(2n)!} \int_0^{\infty} e^{-x^{\frac{1}{4}}} \sin x^{\frac{1}{4}} x^{2n}\, dx,$$

every term of which is zero (§ 3.125). But the given integral is not identically zero (see § 3.13).

The test of § 1.77 fails; for the integral obtained by replacing $u_n(x)$ by $|u_n(x)|$ is

$$\int_0^{\infty} e^{-x^{\frac{1}{4}}} |\sin x^{\frac{1}{4}}| \cosh ax\, dx,$$

which is divergent.

1.79. As a final example of term-by-term integration under special conditions, we prove the following theorem:

The power series $\sum_{n=0}^{\infty} a_n x^n$, *supposed convergent for* $x > 0$, *may be multiplied by* e^{-x} *and integrated term by term over* $(0, \infty)$, *provided only that the resulting series is convergent.**

The formula is

$$\int_0^{\infty} e^{-x} \left\{ \sum_{n=0}^{\infty} a_n x^n \right\} dx = \sum_{n=0}^{\infty} a_n \int_0^{\infty} e^{-x} x^n\, dx = \sum_{n=0}^{\infty} a_n n!,$$

and we have to justify the inversion on the assumption that $\sum a_n n!$ is convergent.

Put $a_n n! = b_n$, so that $\sum b_n$ is convergent. Then b_n is bounded, $|b_n| < B$, say, and

$$\left| \frac{b_n x^n}{n!} \right| < B \frac{X^n}{n!} \qquad (0 \leqslant x \leqslant X).$$

Hence $\sum b_n x^n/n!$ is uniformly convergent over $(0, X)$, and we may multiply by e^{-x} and integrate term by term over this range. Thus

$$\int_0^{X} e^{-x} \left\{ \sum_{n=0}^{\infty} \frac{b_n x^n}{n!} \right\} dx = \sum_{n=0}^{\infty} \frac{b_n}{n!} \int_0^{X} e^{-x} x^n\, dx. \qquad (1)$$

We are given that

$$\sum_{n=0}^{\infty} \frac{b_n}{n!} \int_0^{\infty} e^{-x} x^n\, dx = \sum_{n=0}^{\infty} b_n$$

* Hardy, (1), (6).

is convergent. Hence (1) may be written

$$\int_0^X e^{-x}\left\{\sum_{n=0}^{\infty}\frac{b_n x^n}{n!}\right\}dx = \sum_{n=0}^{\infty} b_n - \sum_{n=0}^{\infty}\frac{b_n}{n!}\int_X^{\infty} e^{-x}x^n\,dx, \tag{2}$$

and it remains to be proved that the last term tends to zero as $X \to \infty$. Now

$$\int_X^{\infty} e^{-x}x^n\,dx = e^{-X}\{X^n+nX^{n-1}+...+n!\},$$

so that the series in question is

$$e^{-X}\sum_{n=0}^{\infty} b_n \sum_{m=0}^{n}\frac{X^m}{m!}. \tag{3}$$

Let $r_n = \sum_{\nu=n}^{\infty} b_\nu$, so that $r_n \to 0$ as $n \to \infty$. Then $b_n = r_n - r_{n+1}$, and

$$\sum_{n=0}^{N} b_n \sum_{m=0}^{n}\frac{X^m}{m!} = \sum_{n=0}^{N}(r_n - r_{n+1})\sum_{m=0}^{n}\frac{X^m}{m!}$$

$$= \sum_{n=0}^{N} r_n \frac{X^n}{n!} - r_{N+1}\left(1+X+...+\frac{X^N}{N!}\right).$$

The last term tends to zero as $N \to \infty$, for any X, so that, making $N \to \infty$, (3) takes the form

$$e^{-X}\sum_{n=0}^{\infty} r_n \frac{X^n}{n!}.$$

The result now easily follows. Given any positive ϵ, we can find N so that $|r_n| < \epsilon$ for $n > N$. Then, since $|r_n| < A$ for all values of n,

$$\left|e^{-X}\sum_{n=0}^{\infty} r_n \frac{X^n}{n!}\right| < Ae^{-X}\sum_{0}^{N}\frac{X^n}{n!} + \epsilon e^{-X}\sum_{N+1}^{\infty}\frac{X^n}{n!}$$

$$< Ae^{-X}\sum_{0}^{N}\frac{X^n}{n!} + \epsilon,$$

and, having fixed N, we can choose X so large that the first term also is less than ϵ. Hence the result.

1.8. Repeated integrals. A repeated integral is one degree more complicated than a series of integrals. Even if the limits of each integration are finite, a repeated limit is involved, and its inversion requires justification. If the limits of each integral are infinite, four successive limiting operations are involved.

1.81. We consider first continuous functions and finite limits.

If $f(x,y)$ is a continuous function of x and y in the rectangle $a \leqslant x \leqslant b$, $\alpha \leqslant y \leqslant \beta$, then

$$\int_a^b dx \int_\alpha^\beta f(x,y)\,dy = \int_\alpha^\beta dy \int_a^b f(x,y)\,dx.$$

Since $f(x,y)$ is continuous, the integral

$$F(x) = \int_\alpha^\beta f(x,y)\,dy$$

is a continuous function of x (§ 1.52). We can therefore integrate it over (a,b). The result is the left-hand side of the equation. Similarly the right-hand side has a meaning.

To prove that the two sides are equal, divide up the ranges of integration by points x_μ and y_ν ($a = x_0$, $b = x_m$, $\alpha = y_0$, $\beta = y_n$), such that $x_{\mu+1} - x_\mu < \delta$, $y_{\nu+1} - y_\nu < \delta$, for all values of μ and ν. Let $m_{\mu,\nu}$, $M_{\mu,\nu}$ be the lower and upper bounds of $f(x,y)$ in the rectangle $(x_\mu, x_{\mu+1}; y_\nu, y_{\nu+1})$. Then for $y_\nu \leqslant y \leqslant y_{\nu+1}$

$$m_{\mu,\nu}(x_{\mu+1} - x_\mu) \leqslant \int_{x_\mu}^{x_{\mu+1}} f(x,y)\,dx \leqslant M_{\mu,\nu}(x_{\mu+1} - x_\mu),$$

and hence, integrating with respect to y,

$$m_{\mu,\nu}(x_{\mu+1} - x_\mu)(y_{\nu+1} - y_\nu) \leqslant \int_{y_\nu}^{y_{\nu+1}} dy \int_{x_\mu}^{x_{\mu+1}} f(x,y)\,dx$$
$$\leqslant M_{\mu,\nu}(x_{\mu+1} - x_\mu)(y_{\nu+1} - y_\nu).$$

Summing with respect to μ and ν, we have

$$\sum\sum m_{\mu,\nu}(x_{\mu+1} - x_\mu)(y_{\nu+1} - y_\nu) \leqslant \int_\alpha^\beta dy \int_a^b f(x,y)\,dx$$
$$\leqslant \sum\sum M_{\mu,\nu}(x_{\mu+1} - x_\mu)(y_{\nu+1} - y_\nu).$$

The same inequalities are also satisfied by the other repeated integral. Also, when $\delta \to 0$, the difference between the extreme terms of the inequality tends to zero; for we can choose δ so small that the maximum value of $M_{\mu,\nu}-m_{\mu,\nu}$ is less than any given ϵ; and then

$$\sum\sum (M_{\mu,\nu}-m_{\mu,\nu})(x_{\mu+1}-x_{\mu})(y_{\nu+1}-y_{\nu}) \leqslant \epsilon\sum\sum (x_{\mu+1}-x_{\mu})(y_{\nu+1}-y_{\nu}) = \epsilon(b-a)(\beta-\alpha).$$

Hence the two repeated integrals are equal.

1.82. Extension to discontinuous functions. Suppose next that the rectangle is crossed by a continuous monotonic curve $y=\phi(x)$, or $x=\psi(y)$, from $x=a$ to $x=c$; and let $f(x,y)$ be bounded, and continuous except on this curve. Then the repeated integrals still exist and are equal.

In the first place, the function $F(x)$ is still continuous. For if $a<x<c$,

$$F(x) = \int_{\alpha}^{\phi(x)} f(x,y)\,dy + \int_{\phi(x)}^{\beta} f(x,y)\,dy = F_1(x)+F_2(x),$$

say; and

$$F_1(x+h)-F_1(x) = \int_{\alpha}^{\phi(x)} \{f(x+h,y)-f(x,y)\}\,dy + \int_{\phi(x)}^{\phi(x+h)} f(x+h,y)\,dy,$$

which plainly tends to zero with h. Hence $F_1(x)$ is continuous, and similarly $F_2(x)$ is continuous. Hence the first repeated integral exists, and, similarly, so does the second one.

To prove that they are equal, consider the strip

$$\phi(x)-\eta<y<\phi(x)+\eta,$$

and suppose for simplicity that $\phi(x)$ is steadily increasing. Construct rectangles $(x_{\mu}, x_{\mu+1}; y_{\nu}, y_{\nu+1})$ with sides less than δ, as before. Then the area of those rectangles between x_{μ} and $x_{\mu+1}$ which contain any point of the strip is less than

$$(x_{\mu+1}-x_{\mu})[\{\phi(x_{\mu+1})+\eta\}-\{\phi(x_{\mu})-\eta\}+2\delta] < \delta\{\phi(x_{\mu+1})-\phi(x_{\mu})\}+(x_{\mu+1}-x_{\mu})(2\eta+2\delta);$$

the total area of such rectangles is therefore less than

$$\delta(\beta-\alpha)+(b-a)(2\eta+2\delta).$$

Hence, if $\sum_1$ denotes a summation over these rectangles, and $|f(x,y)|\leqslant M$,

$$\sum\nolimits_1 (M_{\mu,\nu}-m_{\mu,\nu})(x_{\mu+1}-x_{\mu})(y_{\nu+1}-y_{\nu}) < 2M\{\delta(\beta-\alpha)+(2\eta+2\delta)(b-a)\},$$

which can be made arbitrarily small by choice of η and δ.

Finally, since $f(x,y)$ is continuous in each of the remaining regions, we can choose δ so small that

$$\max(M_{\mu,\nu}-m_{\mu,\nu})<\epsilon$$

in the remaining rectangles. The result now follows as before.

We can, of course, now extend the result to functions which have any finite number of discontinuities of the above type. In particular, the result holds for an integral taken over a non-rectangular region bounded by curves of the above type; for this can be considered as an integral over a rectangle, the function being continuous in a limited part of it and zero elsewhere.

Notice finally the following inequality. Suppose that $f(x,y)$ is continuous, and $|f(x,y)|\leqslant M$, in a region of the above type, and that $f(x,y)=0$ elsewhere. Let $F(x,y)=M$ in the region, and $F(x,y)=0$ elsewhere. Then

$$\left|\int\limits_{\alpha}^{\beta} dy \int\limits_{a}^{b} f(x,y)\,dx\right| \leqslant \int\limits_{\alpha}^{\beta} dy \int\limits_{a}^{b} F(x,y)\,dx.$$

The reader should have no difficulty in deducing this from the above analysis.

1.83. Change of variables in a repeated integral. The formula by which the variables in a repeated integral are changed may be obtained as follows. Consider the integral

$$\int dy \int f(x,y)\,dx$$

over a given region, and let

$$x=\phi(u,v), \qquad y=\psi(u,v).$$

Suppose that these functions are such that, if y is constant, x is a monotonic differentiable function of u. If we transform the integral with respect to x into one with respect to u, we obtain

$$\int f(x,y)\,dx=\int f\frac{dx}{du}\,du.$$

But*

$$\frac{dx}{du}=\frac{\partial\phi}{\partial u}+\frac{\partial\phi}{\partial v}\frac{\partial v}{\partial u}, \qquad 0=\frac{\partial\psi}{\partial u}+\frac{\partial\psi}{\partial v}\frac{\partial v}{\partial u},$$

so that†

$$\frac{dx}{du}=\left(\frac{\partial\phi}{\partial u}\frac{\partial\psi}{\partial v}-\frac{\partial\phi}{\partial v}\frac{\partial\psi}{\partial u}\right)\Big/\frac{\partial\psi}{\partial v}=\frac{\partial(\phi,\psi)}{\partial(u,v)}\Big/\frac{\partial\psi}{\partial v}.$$

* See *P.M.* § 153.

† For the Jacobian notation see *P.M.* ch. VII, ex. 20.

Hence the repeated integral becomes

$$\int dy \int f \frac{\partial(\phi,\psi)}{\partial(u,v)} \Big/ \frac{\partial\psi}{\partial v}\, du = \int du \int f \frac{\partial(\phi,\psi)}{\partial(u,v)} \Big/ \frac{\partial\psi}{\partial v}\, dy$$

if the order of integration may be inverted. Finally, expressing y as a function of v (for a fixed u), and assuming that it is monotonic, we have

$$\frac{dy}{dv} = \frac{\partial\psi}{\partial v},$$

and the integral becomes

$$\int du \int F(u,v) \frac{\partial(\phi,\psi)}{\partial(u,v)}\, dv,$$

where $F(u,v) = f\{\phi(u,v), \psi(u,v)\}$.

The process is valid if, for example, the integrand at every stage is a continuous function, and the region is bounded by monotonic curves as in § 1.82. Some care is needed in verifying this in particular cases. Consider, for example, the integral

$$I = \int_0^a dy \int_0^{\sqrt{(a^2-y^2)}} f(x,y)\, dx,$$

where $f(x,y)$ is continuous, and transform to polar coordinates (r,θ) given by $x = r\cos\theta$, $y = r\sin\theta$. Transforming first to (r,y), we have $x = \sqrt{(r^2-y^2)}$, and

$$\frac{dx}{dr} = \frac{r}{\sqrt{(r^2-y^2)}},$$

which becomes infinite at $r = y$. To avoid this difficulty, consider instead the integral

$$I_\delta = \int_0^{\sqrt{(a^2-\delta^2)}} dy \int_\delta^{\sqrt{(a^2-y^2)}} f(x,y)\, dx,$$

where $0 < \delta < a$. Transforming first to (r,y), then to (r,θ), in the above manner, we obtain

$$I_\delta = \int_\delta^a r\, dr \int_0^{\arccos(\delta/r)} f(r\cos\theta, r\sin\theta)\, d\theta.$$

Now let $\delta \to 0$. Then $I_\delta \to I$, and the last integral tends to

$$\int_0^a r\, dr \int_0^{\frac{1}{2}\pi} f(r\cos\theta, r\sin\theta)\, d\theta.$$

Each of these statements is readily proved by means of the inequality noticed at the end of § 1.82.

For the general theory of these transformations see Goursat's *Cours d'Analyse*, t. 1, ch. 6.

1.84. Repeated integrals, one range being infinite. The most important theorem here is the analogue of the theorem on term-by-term integration of a uniformly convergent series.

Suppose that

$$\int_a^b dx \int_\alpha^\beta f(x,y)\,dy = \int_\alpha^\beta dy \int_a^b f(x,y)\,dx$$

for all values of b greater than a, and that

$$\int_a^\infty f(x,y)\,dx$$

is uniformly convergent in the range $\alpha \leqslant y \leqslant \beta$. *Then*

$$\int_a^\infty dx \int_\alpha^\beta f(x,y)\,dy = \int_\alpha^\beta dy \int_a^\infty f(x,y)\,dx.$$

For

$$\int_a^n f(x,y)\,dx = s_n(y) \to s(y)$$

uniformly in (α, β). Hence, using the result for sequences,

$$\int_a^n dx \int_\alpha^\beta f(x,y)\,dy = \int_\alpha^\beta dy \int_a^n f(x,y)\,dx$$

$$= \int_\alpha^\beta s_n(y)\,dy \to \int_\alpha^\beta s(y)\,dy = \int_\alpha^\beta dy \int_a^\infty f(x,y)\,dx,$$

the required result.

A similar theorem holds for infinite integrals of the second kind.

The same results also hold if the integral is not uniformly convergent in the neighbourhood of certain points, but is boundedly convergent. In fact the same proof holds under these conditions.

1.85. Repeated infinite integrals. The following theorem is the analogue for repeated integrals of the theorems of § 1.62 and § 1.77 for double series and series of integrals.

Suppose that $f(x,y)$ is positive, and that

$$\int_a^c dx \int_\alpha^\beta f(x,y)\,dy = \int_\alpha^\beta dy \int_a^c f(x,y)\,dx \tag{1}$$

for all values of c less than b, and that

$$\int_a^b dx \int_\alpha^\gamma f(x,y)\,dy = \int_\alpha^\gamma dy \int_a^b f(x,y)\,dx \tag{2}$$

for all values of γ less than β. Then

$$\int_a^b dx \int_\alpha^\beta f(x,y)\,dy = \int_\alpha^\beta dy \int_a^b f(x,y)\,dx, \tag{3}$$

provided that either side of this equation is convergent.

The theorem is still true if one or both of b and β is replaced by infinity.

Take, for example, the case of two finite intervals, and suppose that the left-hand side of (3) exists. Since $f(x,y) \geqslant 0$

$$\int_\alpha^\gamma f(x,y)\,dy \leqslant \int_\alpha^\beta f(x,y)\,dy \qquad (\alpha < \gamma < \beta),$$

and hence

$$\int_\alpha^\gamma dy \int_a^b f(x,y)\,dx = \int_a^b dx \int_\alpha^\gamma f(x,y)\,dy \leqslant \int_a^b dx \int_\alpha^\beta f(x,y)\,dy.$$

Making $\gamma \to \beta$, we see that the right-hand side of (3) exists, and that

$$\int_\alpha^\beta dy \int_a^b f(x,y)\,dx \leqslant \int_a^b dx \int_\alpha^\beta f(x,y)\,dy.$$

The same process may now be reversed, and it yields the reversed inequality. Hence the two sides are equal.

The same proof holds if b, or β, or both of them, are infinite.

Though the actual proof is simple, we have, of course, made rather far-reaching assumptions; and, in applying the theorem, we have to justify (1) and (2) on other grounds, for example, by uniform convergence. This is made necessary by the limitations of the Riemann integral; for from the mere fact that

$$\int_\alpha^\beta dy \int_a^c f(x,y)\,dx$$

is bounded as $c \to b$, we cannot deduce that

$$\int_a^b f(x,y)\,dx$$

is integrable in Riemann's sense. When we come to the Lebesgue integral, we shall see that difficulties of this kind disappear.

The theorem remains true for functions $f(x,y)$ which may have either sign, or be complex, provided that either of the integrals

$$\int_a^b dx \int_\alpha^\beta |f(x,y)|\,dy, \qquad \int_\alpha^\beta dy \int_a^b |f(x,y)|\,dx$$

is convergent.

The extension is made as in the case of series. If $f(x,y)$ is real, we consider the functions $|f(x,y)| \pm f(x,y)$, and if $f(x,y)$ is complex, we consider $|f(x,y)| \pm \mathbf{R}f(x,y)$, $|f(x,y)| \pm \mathbf{I}f(x,y)$.

1.86. The Gamma-function. The function

$$\Gamma(x) = \int_0^\infty t^{x-1}e^{-t}\,dt \tag{1}$$

is, as we have already observed (§§ 1.51, 1.52), continuous for $\mathbf{R}(x) > 0$. We are now in a position to investigate its properties more fully. We shall suppose throughout this section and the following one that x and y are real, leaving it to the reader to consider how far the results are true for complex values of the variables.

If $x > 1$, we may integrate by parts, and obtain

$$\Gamma(x) = [-t^{x-1}e^{-t}]_0^\infty + (x-1)\int_0^\infty t^{x-2}e^{-t}\,dt.$$

The integrated terms vanish, and we have

$$\Gamma(x) = (x-1)\Gamma(x-1) \qquad (x > 1). \tag{2}$$

Since $\Gamma(1) = \int_0^\infty e^{-t}\,dt = 1$, repeated application of (2) gives

$$\Gamma(n) = (n-1)! \tag{3}$$

if n is a positive integer. Thus $\Gamma(x)$ may be regarded as a generalization of a factorial.

Now consider the product

$$\Gamma(x)\Gamma(y)=\int_0^\infty t^{x-1}e^{-t}\,dt\int_0^\infty u^{y-1}e^{-u}\,du \quad (x>0,\ y>0).$$

We may regard this as a repeated integral. Putting $u=tv$, and inverting the order of integration, we have, formally,

$$\Gamma(x)\Gamma(y)=\int_0^\infty t^{x-1}e^{-t}\,dt\int_0^\infty t^y v^{y-1}e^{-tv}\,dv$$

$$=\int_0^\infty v^{y-1}\,dv\int_0^\infty t^{x+y-1}e^{-t(1+v)}\,dt$$

$$=\int_0^\infty v^{y-1}\,dv\int_0^\infty \frac{w^{x+y-1}e^{-w}\,dw}{(1+v)^{x+y}}$$

$$=\Gamma(x+y)\int_0^\infty \frac{v^{y-1}}{(1+v)^{x+y}}\,dv.$$

Hence $$\frac{\Gamma(x)\Gamma(y)}{\Gamma(x+y)}=\phi(x,y) \qquad (x>0,\ y>0), \tag{4}$$

where

$$\phi(x,y)=\int_0^\infty \frac{v^{y-1}}{(1+v)^{x+y}}\,dv=2\int_0^{\frac{1}{2}\pi}(\cos\theta)^{2x-1}(\sin\theta)^{2y-1}\,d\theta$$

$$=\int_0^1 \lambda^{x-1}(1-\lambda)^{y-1}\,d\lambda.$$

The difficulty of the proof lies in the inversion of the repeated integral

$$\int_0^\infty dt\int_0^\infty t^{x+y-1}v^{y-1}e^{-t(1+v)}\,dv.$$

Since each integral is infinite at each limit if the indices of t and v are negative, this requires several applications of the theorem of § 1.85. It is easily seen that the t- and v-integrals are both uniformly convergent over any finite range which excludes the origin. Hence the integrals over $(0, T; v_0, V)$ and $(t_0, T; 0, V)$ may be inverted if $t_0>0$, $v_0>0$. Hence, the integrand being positive, the integral over $(0, T; 0, V)$ may be inverted. Since also the integral over $(t_0, T; 0, \infty)$ may be inverted by uniform convergence, it follows that the integral over $(0, T; 0, \infty)$ may be inverted. Similarly the integral over $(0,\infty;0,V)$ may be inverted. Hence, finally the whole integral may be inverted.

Again, putting $x=y=\frac{1}{2}$ in (4), we have

$$\{\Gamma(\tfrac{1}{2})\}^2 = 2\Gamma(1)\int_0^{\frac{1}{2}\pi} d\theta = \pi,$$

or, since $\Gamma(\frac{1}{2})$ is plainly positive,

$$\Gamma(\tfrac{1}{2}) = \sqrt{\pi}. \tag{5}$$

Next, putting $y=x$ in (4),

$$\frac{\{\Gamma(x)\}^2}{\Gamma(2x)} = \int_0^1 \lambda^{x-1}(1-\lambda)^{x-1}\,d\lambda = 2\int_0^{\frac{1}{2}} \lambda^{x-1}(1-\lambda)^{x-1}\,d\lambda.$$

Putting $\lambda=\frac{1}{2}-\frac{1}{2}\sqrt{\mu}$, so that $\lambda(1-\lambda)=\frac{1}{4}-\frac{1}{4}\mu$, this gives

$$\tfrac{1}{2}\int_0^1 (\tfrac{1}{4}-\tfrac{1}{4}\mu)^{x-1}\mu^{-\frac{1}{2}}\,d\mu = 2^{1-2x}\int_0^1 (1-\mu)^{x-1}\mu^{-\frac{1}{2}}\,d\mu = 2^{1-2x}\frac{\Gamma(x)\Gamma(\frac{1}{2})}{\Gamma(x+\frac{1}{2})}.$$

Hence we obtain the 'duplication formula'

$$\Gamma(2x)\Gamma(\tfrac{1}{2}) = 2^{2x-1}\Gamma(x)\Gamma(x+\tfrac{1}{2}). \tag{6}$$

1.87. Asymptotic behaviour of $\Gamma(x)$ as $x\to\infty$. Consider first the case where x is an integer, n say, so that $\Gamma(x)=(n-1)!$. We use the well-known method of comparing a sum $\sum\phi(n)$ with the corresponding integral $\int\phi(t)\,dt$. We have

$$\log\{(n-1)!\} = \sum_{\nu=1}^{n-1}\log\nu.$$

Now
$$\int_{\nu-\frac{1}{2}}^{\nu+\frac{1}{2}} \log t\,dt = \int_0^{\frac{1}{2}} \{\log(\nu+t)+\log(\nu-t)\}\,dt$$

$$= \int_0^{\frac{1}{2}} \left\{\log\nu^2+\log\left(1-\frac{t^2}{\nu^2}\right)\right\}dt = \log\nu + C_\nu,$$

say, where clearly $C_\nu = O(1/\nu^2)$. Hence

$$\log\Gamma(n) = \log\{(n-1)!\} = \int_{\frac{1}{2}}^{n-\frac{1}{2}} \log t\,dt - \sum_{\nu=1}^{n-1} C_\nu$$

$$= (n-\tfrac{1}{2})\log(n-\tfrac{1}{2})-(n-\tfrac{1}{2})-\tfrac{1}{2}\log\tfrac{1}{2}+\tfrac{1}{2}-\sum_{\nu=1}^{\infty}C_\nu+o(1)$$

$$= (n-\tfrac{1}{2})\log n-n+C+o(1), \tag{1}$$

where C is a constant.

We can extend this result to non-integral values of x by means of the following lemma.

LEMMA. *If a is constant, as $x \to \infty$,*

$$\frac{\Gamma(x)}{\Gamma(x+a)} \sim x^{-a}. \tag{2}$$

Suppose that $a > 1$. Then

$$\frac{\Gamma(x)\Gamma(a)}{\Gamma(x+a)} = \int_0^1 (1-\lambda)^{a-1}\lambda^{x-1}\,d\lambda = \int_0^\infty (1-e^{-t})^{a-1}e^{-xt}\,dt$$

$$= \int_0^\infty t^{a-1}e^{-xt}\,dt - \int_0^\infty \{t^{a-1}-(1-e^{-t})^{a-1}\}e^{-xt}\,dt.$$

The first integral is $\Gamma(a)x^{-a}$, and the second is $O(x^{-a-1})$. For

$$1-e^{-t} < t \quad (t>0), \qquad 1-e^{-t} > t-\tfrac{1}{2}t^2 \quad (0<t<1).$$

Hence the second integral is positive, and less than

$$\int_0^1 \{1-(1-\tfrac{1}{2}t)^{a-1}\}t^{a-1}e^{-xt}\,dt + \int_1^\infty t^{a-1}e^{-xt}\,dt$$

$$< K\int_0^1 t^a e^{-xt}\,dt + \int_1^\infty t^a e^{-xt}\,dt < Kx^{-a-1},$$

where K depends on a only. This proves the lemma for $a > 1$. The result for other values of a then follows from 1.86 (2).

If now x is not an integer, let $x = n+a$, where n is an integer and $0 < a < 1$. Then

$$\begin{aligned}\log\Gamma(x) &= \log\Gamma(n+a) = \log\Gamma(n) + a\log n + o(1)\\ &= (n-\tfrac{1}{2})\log n - n + C + a\log n + o(1)\\ &= (x-a-\tfrac{1}{2})\log(x-a) - x + a + C + a\log(x-a) + o(1)\\ &= (x-\tfrac{1}{2})\log x - x + C + o(1),\end{aligned} \tag{3}$$

the required result.

To find the value of C, we use the duplication formula 1.86 (6). Taking logarithms and using (3), we have

$$(2x-\tfrac{1}{2})\log 2x - 2x + C + \log\sqrt{\pi} + o(1)$$

$$= (2x-1)\log 2 + (x-\tfrac{1}{2})\log x + x\log(x+\tfrac{1}{2}) - 2x - \tfrac{1}{2} + 2C + o(1),$$

and equating the constant terms, we obtain

$$C = \log\sqrt{(2\pi)}.$$

Hence finally

$$\Gamma(x) = x^{x-\frac{1}{2}}e^{-x}\sqrt{(2\pi)}\{1+o(1)\}. \tag{4}$$

This result is known as Stirling's theorem.

1.88. Differentiation of integrals. The two following theorems cover most of the cases that are ordinarily met with.

If $f(x,y)$ *and* $\dfrac{\partial f}{\partial y}$ *are continuous in the rectangle* $a \leqslant x \leqslant b$, $y_0 - \eta \leqslant y \leqslant y_0 + \eta$ $(\eta > 0)$, *then*

$$\frac{d}{dy}\int_a^b f(x,y)\,dx = \int_a^b \frac{\partial f}{\partial y}\,dx \tag{1}$$

for $y = y_0$.

Let $$\phi(y) = \int_a^b f(x,y)\,dx, \qquad g(x,y) = \frac{\partial f}{\partial y}.$$

Then $$\frac{\phi(y_0+k)-\phi(y_0)}{k} = \frac{1}{k}\int_a^b \{f(x,y_0+k)-f(x,y_0)\}\,dx$$
$$= \int_a^b g(x, y_0+\theta k)\,dx,$$

where $0 < \theta < 1$. Since $g(x,y)$ is uniformly continuous, we have, as in § 1.52,

$$\lim_{k\to 0}\int_a^b g(x,y_0+\theta k)\,dx = \int_a^b g(x,y_0)\,dx,$$

the required result.

If the equation (1) *is true for all values of b greater than a, and if the integral*

$$\int_a^\infty f\,dx$$

is convergent, and

$$\int_a^\infty \frac{\partial f}{\partial y}\,dx$$

is uniformly convergent in the interval $(y_0-\eta, y_0+\eta)$, *then*

$$\frac{d}{dy}\int_a^\infty f(x,y)\,dx = \int_a^\infty \frac{\partial f}{\partial y}\,dx.$$

We can deduce this from the corresponding theorem for series (§ 1.72). For let

$$\int_{a+n-1}^{a+n} f(x,y)\,dx = u_n(y).$$

Then
$$\int_a^\infty f(x,y)\,dx = \sum u_n(y),$$

and, by the previous theorem,

$$\int_a^\infty \frac{\partial f}{\partial y}\,dx = \sum \int_{a+n-1}^{a+n} \frac{\partial f}{\partial y}\,dx = \sum \frac{d}{dy} \int_{a+n-1}^{a+n} f(x,y)\,dx = \sum \frac{d}{dy} u_n(y).$$

The result now follows at once from the theorem for series.

MISCELLANEOUS EXAMPLES

1. Consider the uniform convergence of the series

$$\sum_{n=1}^\infty \frac{1}{x^2+n^2}, \quad \sum_{n=1}^\infty \frac{1}{x^2n^2}, \quad \sum_{n=1}^\infty \frac{1}{x^2-n^2},$$

and
$$\sum_{n=1}^\infty \frac{\sin nx}{n^\alpha} \quad (\alpha > 0).$$

2. Discuss with reference to uniformity the convergence of the series

$$\sum_{n=1}^\infty \frac{1}{n^2}\frac{z^n}{1+z^n}, \quad \sum_{n=1}^\infty \frac{(-1)^n}{n}\frac{z^n}{1+z^n}$$

(i) for z real and positive, (ii) for general complex values of z.

3. Consider the uniform convergence of the integrals

$$\int_0^\infty \frac{dx}{x^2+y^2}, \quad \int_0^\infty e^{-xy}\,dx, \quad \int_0^\infty \cosh xy\, e^{-x^2}\,dx.$$

4. Consider the uniform convergence of the integral

$$\int_0^\infty \frac{\sin^2 xy}{x^2}\,dx.$$

Evaluate the integral by differentiating with respect to y.

5. The Bessel function $J_\nu(z)$ is defined for $\nu > -1$ by the series

$$J_\nu(z) = \sum_{n=0}^\infty \frac{(-1)^n(\frac{1}{2}z)^{\nu+2n}}{n!\Gamma(\nu+n+1)}.$$

Prove that, if $\nu > -\frac{1}{2}$,

$$J_\nu(z) = \frac{z^\nu}{2^{\nu-1}\Gamma(\nu+\frac{1}{2})\Gamma(\frac{1}{2})} \int_0^{\frac{1}{2}\pi} \cos(z\cos\theta)\sin^{2\nu}\theta\,d\theta,$$

and that, if $\mu > -1$, $\nu > -1$,

$$J_{\mu+\nu+1}(z) = \frac{z^{\nu+1}}{2^\nu\Gamma(\nu+1)} \int\limits_0^{\frac{1}{2}\pi} J_\mu(z\sin\theta)\sin^{\mu+1}\theta\cos^{2\nu+1}\theta\,d\theta.$$

6. Prove that

$$\int\limits_0^\infty e^{-ax}J_0(bx)\,dx = \frac{1}{\sqrt{(a^2+b^2)}} \qquad (0 < b < a),$$

and that
$$\int\limits_0^\infty e^{-x}\,J_0(x)\,dx = \frac{1}{\sqrt{2}}.$$

7. Prove that
$$\int\limits_0^\infty J_\nu(at)e^{-t^2}t^{\nu+1}\,dt = \frac{a^\nu}{2^{\nu+1}}e^{-\frac{1}{4}a^2}$$

8. Prove that

$$\sum_{n=1}^\infty \int\limits_0^\infty \frac{\sin nx}{n^2x}\,dx = \int\limits_0^\infty \left\{\sum_{n=1}^\infty \frac{\sin nx}{n^2x}\right\}dx.$$

9. Show that the repeated integrals

$$\int\limits_1^\infty dx \int\limits_1^\infty \frac{x-y}{(x+y)^3}\,dy, \qquad \int\limits_0^1 dx \int\limits_0^1 \frac{x-y}{(x+y)^3}\,dy,$$

$$\int\limits_1^\infty dx \int\limits_1^\infty \frac{x^2-y^2}{(x^2+y^2)^2}\,dy, \qquad \int\limits_0^1 dx \int\limits_1^\infty (ae^{-axy}-be^{-bxy})\,dy,$$

are not equal to the integrals obtained by inverting the order of integration.

10. Prove that
$$\int\limits_0^\infty e^{-x^2}\cos 2xy\,dx = \tfrac{1}{2}\sqrt{\pi}e^{-y^2}$$

(i) by expanding $\cos 2xy$ in powers of x and integrating term by term, and (ii) by proving that the integral satisfies the differential equation

$$\frac{dI}{dy} = -2yI.$$

11. If
$$\phi(y) = \int\limits_0^\infty \frac{\sin xy}{x(a^2+x^2)}\,dx,$$

prove that
$$\phi''(y)-a^2\phi(y)$$

is a constant, and hence that

$$\phi(y) = \frac{\pi}{2a^2}(1-e^{-ay}) \qquad (y>0).$$

12. Deduce from the previous example the values of the integrals

$$\int_0^\infty \frac{\cos xy}{a^2+x^2}\,dx, \qquad \int_0^\infty \frac{x\sin xy}{a^2+x^2}\,dx.$$

13. If
$$\phi(p,q,a,b) = \Gamma(p)\int_0^\infty \frac{e^{-bx}x^{q-1}}{(a+cx)^p}\,dx,$$
where a, b, p, and q are positive, prove that
$$\phi(p,q,a,b) = \phi(q,p,b,a).$$

14. If
$$\psi(y) = \int_0^\infty e^{-x^2-y^2/x^2}\,dx,$$
prove that
$$\psi(y) = \tfrac{1}{2}\sqrt{\pi}e^{-2y} \qquad (y>0)$$
(i) by proving that $\psi'(y) = -2\psi(y)$, and (ii) by means of the substitution $w = x-y/x$.

15. Show that the repeated integral
$$\int_0^\infty dx \int_0^\infty \cos 2mx\,.\,ye^{-y^2(1+x^2)}\,dy$$
may be inverted; and hence deduce the value of the first integral of example 12 from those of examples 10 and 14.

16. Prove that if $\lambda>0$, $\mu>0$,
$$\int_0^\infty e^{-x^2}x^{2\lambda-1}\,dx \int_0^\infty e^{-y^2}y^{2\mu-1}\,dy = \int_0^\infty e^{-r^2}r^{2\lambda+2\mu-1}\,dr \int_0^{\frac{1}{2}\pi} \cos^{2\lambda-1}\theta\sin^{2\mu-1}\theta\,d\theta,$$
and hence obtain another proof of the formulae of § 1.86.

17. Show that, in the repeated integral
$$\int_0^\infty \sin ax\,dx \int_0^\infty f(y)e^{-xy}\,dy,$$
the order of integration may be inverted if the integrals
$$\int_0^1 |f(y)|\,dy, \qquad \int_1^\infty |f(y)|y^{-2}\,dy$$
are convergent. Hence show that
$$\int_0^\infty \frac{\sin ax}{\sqrt{(1+x^2)}}\,dx = a\int_0^\infty \frac{J_0(y)}{a^2+y^2}\,dy.$$

[The integral $\int_\xi^X \sin ax\,dx \int_0^\infty f(y)e^{-xy}\,dy$ may be inverted, since the

y-integral is uniformly convergent for $0 < \xi \leqslant x \leqslant X$. Hence it is sufficient to prove that the integrals

$$\int_0^\infty f(y)\,dy \int_0^\xi \sin ax\, e^{-xy}\,dx, \qquad \int_0^\infty f(y)\,dy \int_X^\infty \sin ax\, e^{-xy}\,dx$$

tend to zero as $\xi \to 0$ and $X \to \infty$. Since $|\sin ax| \leqslant ax$, the modulus of the former does not exceed

$$\int_0^\infty |f(y)|\,dy \int_0^\xi ax\, e^{-xy}\,dx < a\int_0^Y |f(y)|\,dy \int_0^\xi x\,dx + a\int_Y^\infty |f(y)|\,dy \int_0^\infty xe^{-xy}\,dx$$

$$= \tfrac{1}{2}a\xi^2 \int_0^Y |f(y)|\,dy + a\int_Y^\infty |f(y)|y^{-2}\,dy,$$

which can be made arbitrarily small by choosing first Y and then ξ. The modulus of the latter integral is

$$\left|\int_0^\infty f(y)\frac{y\sin aX + a\cos aX}{a^2+y^2}e^{-Xy}\,dy\right| < \frac{1}{a}\int_0^\infty |f(y)|e^{-Xy}\,dy,$$

which tends to zero.]

18. If $\alpha > 0$, $\beta > 0$,

$$\int_y^x (x-t)^{\alpha-1}(t-y)^{\beta-1}\,dt = \frac{\Gamma(\alpha)\Gamma(\beta)}{\Gamma(\alpha+\beta)}(x-y)^{\alpha+\beta-1}.$$

19. If $\alpha > 0$, $\beta > 0$, and $\lambda < y$ or $\lambda > x$,

$$\int_y^x \frac{(x-t)^{\alpha-1}(t-y)^{\beta-1}}{|t-\lambda|^{\alpha+\beta}}\,dt = \frac{\Gamma(\alpha)\Gamma(\beta)}{\Gamma(\alpha+\beta)}\,\frac{(x-y)^{\alpha+\beta-1}}{|x-\lambda|^\beta|y-\lambda|^\alpha}.$$

[Consider the general linear transformation of the interval (y, x) into itself, and use the previous example.]

20. Prove that if $c > b > 0$, $c-a-b > 0$,

$$\sum_{n=0}^\infty \frac{a(a+1)\ldots(a+n-1)}{n!}\int_0^1 t^{b+n-1}(1-t)^{c-b-1}\,dt = \int_0^1 t^{b-1}(1-t)^{c-a-b-1}\,dt.$$

Deduce that

$$\sum_{n=0}^\infty \frac{a\ldots(a+n-1)b\ldots(b+n-1)}{n!\,c\ldots(c+n-1)} = \frac{\Gamma(c)\Gamma(c-a-b)}{\Gamma(c-a)\Gamma(c-b)}.$$

CHAPTER II

ANALYTIC FUNCTIONS

2.1. Functions of a complex variable. It is just as easy to construct a function of a complex variable $z = x+iy$ as it is to construct a function of a real variable x. Any finite or infinite convergent expression involving z gives such a function. For example, z^2, $1/z$, e^z are functions of the complex variable z. The reader of Hardy's *Pure Mathematics* is already familiar with many such functions.

Throughout this chapter and the next all functions are supposed to be one-valued in the region in which they are defined.

Our first task is to give a general definition which will be appropriate to all such functions.

We might say that w is a function of z if to every value of z in a certain region corresponds one or more values of w. This is modelled on the usual definition of a function of a real variable. It is perfectly legitimate, but, as is explained in Hardy's *Pure Mathematics*, it is futile because it is too wide. It makes a function of the complex variable z exactly the same thing as a complex function

$$u(x,y)+iv(x,y)$$

of two real variables x and y. Of course this is not what we meant when we began to speak of functions of a complex variable.

Our method of procedure is to assign various properties to our function which appear to be desirable, and to see whether any such properties distinguish between what we feel to be 'proper' and 'improper' functions of z.

2.11. Continuity. Let $f(z)$ be a function of z defined in the above way. *It is said to be continuous at the point $z = z_0$ if, given any positive number ϵ, we can find a number δ such that*

$$|f(z)-f(z_0)| < \epsilon$$

provided that $|z-z_0| < \delta$.

This is quite satisfactory as far as it goes, but it does not go very far. A continuous function of z is merely a continuous complex function of the two variables x and y. For if

$$f(z) = u(x,y)+iv(x,y),$$

and $z_0 = x_0+iy_0$, then

$$|u(x,y)-u(x_0,y_0)| \leqslant |f(z)-f(z_0)| < \epsilon$$

if $|z-z_0| < \delta$, which is true if

$$|x-x_0| < \frac{\delta}{\sqrt{2}}, \qquad |y-y_0| < \frac{\delta}{\sqrt{2}}.$$

Hence $u(x,y)$ is continuous, and so is $v(x,y)$. Conversely, if u and v are continuous, so is $f(z)$.

2.12. Differentiability. From the class of continuous complex functions, we next select those which can be differentiated. The meaning of this term for complex functions must now be defined.

Following the suggestion of real differential calculus, we write

$$f'(z_0) = \lim_{z\to z_0}\frac{f(z)-f(z_0)}{z-z_0},$$

and we say that $f(z)$ is differentiable if the limit on the right exists. The limit is called the derivative or differential coefficient of $f(z)$. As in the definition of continuity, the approach of z to its limit z_0 can take place in all possible ways. More precisely, we interpret the above formula as meaning that, given any positive number ϵ, we can find a number δ such that

$$\left|f'(z_0)-\frac{f(z)-f(z_0)}{z-z_0}\right| < \epsilon$$

provided that $0 < |z-z_0| < \delta$. Thus we assert that, along whatever path z approaches z_0, the ratio

$$\frac{f(z)-f(z_0)}{z-z_0}$$

always tends to a limit, and that all the limiting values are the same. Our requirements are therefore somewhat exacting.

This property of differentiability is, however, one which belongs to many familiar functions. A constant is differentiable. A positive integral power of z is differentiable; for the familiar proof for x^n applies word for word to the case of z^n. Similarly, the sum or product of two (or any finite number) of differentiable functions is differentiable; and the quotient of two differentiable functions is differentiable provided that the denominator does not vanish. Finally, a differentiable function of a differentiable function is differentiable. All these theorems are proved in the same way for functions of z as for functions of x.

For example, any rational function of z is differentiable for all values of z, other than zeros of its denominator.

2.13. We now naturally ask whether this property of differentiability corresponds to a simple property of the functions $u(x,y)$ and $v(x,y)$ which are the real and imaginary parts of $f(z)$.

Suppose first that $z-z_0$ is purely real, so that

$$z_0 = x_0+iy_0, \qquad z = x+iy_0.$$

Then

$$\frac{f(z)-f(z_0)}{z-z_0} = \frac{\{u(x,y_0)+iv(x,y_0)\}-\{u(x_0,y_0)+iv(x_0,y_0)\}}{x-x_0}$$
$$= \frac{u(x,y_0)-u(x_0,y_0)}{x-x_0}+i\frac{v(x,y_0)-v(x_0,y_0)}{x-x_0}.$$

If this tends to a limit as $x \to x_0$, then its real and imaginary parts separately tend to limits. But this means simply that the partial differential coefficients

$$\frac{\partial u}{\partial x}, \quad \frac{\partial v}{\partial x}$$

exist at the point (x_0, y_0). Also

$$f'(z_0) = \left(\frac{\partial u}{\partial x}+i\frac{\partial v}{\partial x}\right)_{x=x_0,\, y=y_0} \tag{1}$$

Similarly, if we take $z-z_0$ to be purely imaginary, say

$$z_0 = x_0+iy_0, \qquad z = x_0+iy,$$

we obtain

$$\frac{f(z)-f(z_0)}{z-z_0} = \frac{\{u(x_0,y)+iv(x_0,y)\}-\{u(x_0,y_0)+iv(x_0,y_0)\}}{iy-iy_0}$$
$$= \frac{v(x_0,y)-v(x_0,y_0)}{y-y_0}-i\frac{u(x_0,y)-u(x_0,y_0)}{y-y_0}.$$

Hence the partial differential coefficients

$$\frac{\partial v}{\partial y}, \quad \frac{\partial u}{\partial y}$$

exist at the point (x_0, y_0); and

$$f'(z_0) = \left(\frac{\partial v}{\partial y}-i\frac{\partial u}{\partial y}\right)_{x=x_0,\, y=y_0} \tag{2}$$

Further, on comparing (1) and (2), and equating real and imaginary parts, we see that

$$\frac{\partial u}{\partial x} = \frac{\partial v}{\partial y}, \qquad \frac{\partial v}{\partial x} = -\frac{\partial u}{\partial y} \qquad (x = x_0,\ y = y_0). \tag{3}$$

We now see that the results of assuming *differentiability* are much more far-reaching than those of assuming *continuity*. Not only do the functions $u(x,y)$, $v(x,y)$ possess partial differential coefficients of the first order, but they are connected by the differential equations (3). These are called the Cauchy-Riemann equations.

Thus, even if u and v are functions of x and y with partial differential coefficients of the first order, $u+iv$ will not in general be a differentiable function of z.

Examples. (i) Let $f(z) = \mathbf{R}(z) = x$. Then

$$\frac{\partial u}{\partial x} = 1, \qquad \frac{\partial u}{\partial y} = 0, \qquad \frac{\partial v}{\partial x} = 0, \qquad \frac{\partial v}{\partial y} = 0.$$

The partial differential coefficients all exist, but the Cauchy-Riemann equations are not satisfied for any value of z.

(ii) Let $f(z) = |z|^2 = x^2+y^2$. Then

$$\frac{\partial u}{\partial x} = 2x, \qquad \frac{\partial u}{\partial y} = 2y, \qquad \frac{\partial v}{\partial x} = 0, \qquad \frac{\partial v}{\partial y} = 0.$$

The Cauchy-Riemann equations are satisfied at the point $z = 0$ only.

2.14. Analytic functions. Since the property which we have been discussing goes far beyond what we ordinarily think of as differentiability, we give it a special name. A function which is differentiable in this sense is said to be *analytic*.

The property of being analytic is in fact the distinguishing property for 'proper' functions of a complex variable for which we have been seeking.

We have seen that the truth of the Cauchy-Riemann equations is a *necessary* condition for the function to be analytic. But it is not a *sufficient* condition. This is perhaps to be expected, since we obtained the equations as particular cases only of the general property of differentiability.

Consider, for example, the function

$$f(z) = \sqrt{|xy|}.$$

This vanishes on both axes, so that at $z = 0$

$$\frac{\partial u}{\partial x} = \frac{\partial u}{\partial y} = \frac{\partial v}{\partial x} = \frac{\partial v}{\partial y} = 0,$$

and the Cauchy-Riemann equations are satisfied. But $f(z)$ is not differentiable at $z = 0$. For

$$\frac{f(z)}{z} = \frac{\sqrt{|xy|}}{x+iy},$$

and, if $x = \alpha r$, $y = \beta r$, where α and β are constants, this tends to

$$\frac{\sqrt{|\alpha\beta|}}{\alpha+i\beta}$$

as $r \to 0$. The limit is therefore not unique, and so the function is not analytic.

This example shows that $f(z)$ may not be analytic if we merely know that

$$\{f(z)-f(z_0)\}/(z-z_0)$$

tends to a limit along two straight lines at right angles. Actually the definition fails if we restrict ourselves to any special class of paths. Consider, for example, the function

$$f(z) = \frac{xy^2(x+iy)}{x^2+y^4} \qquad (z \neq 0), \qquad f(0) = 0.$$

Then it is easily seen that

$$\lim\{f(z)-f(0)\}/z = 0$$

as $z \to 0$ along any straight line. But, on the curve $x = y^2$,

$$\frac{f(z)-f(0)}{z} = \frac{y^4}{y^4+y^4} = \frac{1}{2}.$$

Hence $f(z)$ is not analytic at $z = 0$.

2.15. *Suppose, however, that the four partial derivatives of the first order exist throughout a region, and are continuous at all points of the region. Then the truth of the Cauchy-Riemann equations is a necessary and sufficient condition for $f(z)$ to be analytic throughout the region.*

We have seen already that the condition is necessary. To prove that it is sufficient, we use the mean-value theorem for functions of two variables (*P.M.* § 154). Consider a point (x, y) of the region, and a neighbouring point $(x+\delta x, y+\delta y)$. Then

$$\delta u = u(x+\delta x, y+\delta y)-u(x, y)$$
$$= \left(\frac{\partial u}{\partial x}+\epsilon\right)\delta x+\left(\frac{\partial u}{\partial y}+\eta\right)\delta y,$$

where ϵ and η tend to zero with δx and δy. Similarly

$$\delta v = \left(\frac{\partial v}{\partial x}+\epsilon'\right)\delta x+\left(\frac{\partial v}{\partial y}+\eta'\right)\delta y.$$

Hence, using the Cauchy-Riemann equations,

$$\delta u+i\delta v=\left(\frac{\partial u}{\partial x}+i\frac{\partial v}{\partial x}\right)(\delta x+i\delta y)+\rho,$$

where $$|\rho|\leqslant(|\epsilon|+|\epsilon'|)|\delta x|+(|\eta|+|\eta'|)|\delta y|.$$

Hence

$$\frac{f(z+\delta z)-f(z)}{\delta z}=\frac{\delta u+i\delta v}{\delta x+i\delta y}=\frac{\partial u}{\partial x}+i\frac{\partial v}{\partial x}+\frac{\rho}{\delta x+i\delta y},$$

and $$|\rho/(\delta x+i\delta y)|\leqslant|\epsilon|+|\epsilon'|+|\eta|+|\eta'|\rightarrow 0.$$

Hence $f(z)$ is analytic.

2.16. *A power series represents an analytic function inside its circle of convergence.*

We shall see later that this is merely a particular case of a general theorem on series which represent analytic functions. But the following direct proof may be inserted at this point. Let

$$f(z)=\sum_{n=0}^{\infty}a_n z^n$$

be convergent for $|z|<R$. Then, if $\rho<R$, $a_n\rho^n$ is bounded, say $|a_n\rho^n|\leqslant K$. Let

$$g(z)=\sum_{n=1}^{\infty}na_n z^{n-1}.$$

Then if $|z|<\rho$ and $|z|+|h|<\rho$,

$$\frac{f(z+h)-f(z)}{h}-g(z)=\sum_{n=0}^{\infty}a_n\left\{\frac{(z+h)^n-z^n}{h}-nz^{n-1}\right\}.$$

Now

$$\left|\frac{(z+h)^n-z^n}{h}-nz^{n-1}\right|=\left|\frac{n(n-1)}{1.2}z^{n-2}h+...+h^{n-1}\right|$$

$$\leqslant\frac{n(n-1)}{1.2}|z|^{n-2}|h|+...+|h|^{n-1}=\frac{(|z|+|h|)^n-|z|^n}{|h|}-n|z|^{n-1}.$$

Hence

$$\left|\frac{f(z+h)-f(z)}{h}-g(z)\right|\leqslant K\sum_{n=0}^{\infty}\frac{1}{\rho^n}\left\{\frac{(|z|+|h|)^n-|z|^n}{|h|}-n|z|^{n-1}\right\}$$

$$=K\left\{\frac{1}{|h|}\left(\frac{\rho}{\rho-|z|-|h|}-\frac{\rho}{\rho-|z|}\right)-\frac{\rho}{(\rho-|z|)^2}\right\}$$

$$=\frac{K\rho|h|}{(\rho-|z|-|h|)(\rho-|z|)^2},$$

which tends to zero with h. Hence $f(z)$ has the derivative $g(z)$.

2.17. Functions analytic in a region. A function is said to be analytic in a region if it is analytic at all points of the region. Henceforth we always consider functions which are analytic *in a region*. No particular interest attaches to the fact that a function (like $|z|^2$) happens to be analytic at certain points, or even on a certain curve. It is when it is analytic in a region that interesting consequences follow.

Examples. (i) The function

$$\log(1+z) = z - \frac{z^2}{2} + \frac{z^3}{3} - \ldots$$

is analytic for $|z| < 1$.

(ii) The functions

$$e^z = \sum_{n=0}^{\infty} \frac{z^n}{n!}, \qquad \cos z = \tfrac{1}{2}(e^{iz}+e^{-iz}), \qquad \sin z = \frac{1}{2i}(e^{iz}-e^{-iz}),$$

are analytic for all finite values of z.

(iii) The function

$$f(z) = e^{-z^{-4}} \quad (z \neq 0), \qquad f(0) = 0$$

is analytic for all finite values of z, except $z = 0$. At the point $z = 0$, the Cauchy-Riemann equations are satisfied; for at $z = 0$

$$\frac{\partial u}{\partial x} = \lim_{x\to 0} \frac{e^{-x^{-4}}}{x} = 0, \qquad \frac{\partial v}{\partial x} = \lim_{x\to 0} \frac{0}{x} = 0,$$

$$\frac{\partial u}{\partial y} = \lim_{y\to 0} \frac{e^{-y^{-4}}}{y} = 0, \qquad \frac{\partial v}{\partial y} = \lim_{y\to 0} \frac{0}{y} = 0,$$

so that

$$\frac{\partial u}{\partial x} = \frac{\partial v}{\partial y}, \qquad \frac{\partial u}{\partial y} = -\frac{\partial v}{\partial x}.$$

In spite of this, $f(z)$ is not analytic at $z = 0$. For suppose that $z = re^{\frac{1}{4}i\pi}$. Then

$$f(z) = \exp\{-(re^{\frac{1}{4}i\pi})^{-4}\} = e^{r^{-4}},$$

which tends to infinity as $r \to 0$.

2.2. The complex differential calculus. The reader might expect that we should now proceed after the manner of the real differential calculus. There, having distinguished the special class of differentiable functions, we next consider the still more special class of functions which have a second differential coefficient. Some of these functions have differential coefficients of the third order, and so on. Finally, from among functions which have differential coefficients of all orders, we pick out those which can be expanded in a power series by Taylor's theorem.

There is no such process of successive specialization for analytic functions of a complex variable. A function which is analytic in a region has differential coefficients of all orders at all points of the region, and the function can be expanded in a power series, after the manner of Taylor's theorem, about any point of the region.

All these facts follow from the definition of an analytic function by means of its first differential coefficient.

The reader would perhaps expect us to begin by proving that an analytic function has a second differential coefficient. We are unable to do this.

The results have of course been proved, or we should not have been able to announce what they were. But they have never been proved directly. They all depend on the complex integral calculus, and it is to this that we must now turn.

2.3. Complex integration. The reader of Hardy's *Pure Mathematics* should know what a complex integral is (*P.M.* § 222). We shall, however, introduce the subject in a slightly different way.

Let AB be an arc C of a curve defined by the equations

$$x = \phi(t), \qquad y = \psi(t),$$

where ϕ and ψ are functions of t with continuous differential coefficients $\phi'(t)$ and $\psi'(t)$, and suppose that, as t varies from t_A to t_B, the point (x, y) moves along the curve steadily from A to B.

Let $f(z)$ be any complex function of z, continuous along C. Let

$$f(z) = u(x,y)+iv(x,y).$$

Let $z_0, z_1,..., z_n$ be points on C, z_0 being A and z_n being B. Consider the sum

$$\sum_{m=1}^{n} f(\zeta_m)(z_m - z_{m-1}), \tag{1}$$

where ζ_m is a point of the curve between z_{m-1} and z_m. Writing $\zeta_m = \xi_m + i\eta_m$, $u_m = u(\xi_m, \eta_m)$, $v_m = v(\xi_m, \eta_m)$, this is

$$\sum_{m=1}^{n} (u_m + iv_m)(x_m + iy_m - x_{m-1} - iy_{m-1}).$$

Now
$$x_m - x_{m-1} = \phi(t_m) - \phi(t_{m-1}) = \phi'(\tau_m)(t_m - t_{m-1}),$$
$$y_m - y_{m-1} = \psi(t_m) - \psi(t_{m-1}) = \psi'(\tau_m')(t_m - t_{m-1}),$$

where $t_{m-1} \leqslant \tau_m \leqslant t_m$, $t_{m-1} \leqslant \tau'_m \leqslant t_m$. Hence the sum may be written

$$\sum_{m=1}^{n} (u_m+iv_m)\{\phi'(\tau_m)+i\psi'(\tau'_m)\}(t_m-t_{m-1}). \tag{2}$$

Since all the functions concerned are continuous (and therefore uniformly continuous), we can, given ϵ, find δ so that

$$|u_m \phi'(\tau_m)-u(x_m, y_m)\phi'(t_m)| < \epsilon$$

for every m, provided that each $|t_m-t_{m-1}| < \delta$. Also

$$\sum_{m=1}^{n} \epsilon(t_m-t_{m-1}) = \epsilon(t_B-t_A).$$

It follows that, as ϵ and δ tend to zero,

$$\sum_{m=1}^{n} u_m \phi'(\tau_m)(t_m-t_{m-1})$$

tends to the same limit as

$$\sum_{m=1}^{n} u(x_m, y_m)\phi'(t_m)(t_m-t_{m-1}),$$

viz. to the limit
$$\int_{t_A}^{t_B} u\{\phi(t), \psi(t)\}\phi'(t)\, dt.$$

Similarly the other terms of (2) tend to limits, and we find that the whole sum tends to the limit

$$\int_{t_A}^{t_B} (u+iv)\{\phi'(t)+i\psi'(t)\}\, dt, \tag{3}$$

this integral being interpreted in the obvious way as the sum of two real integrals, one of which is multiplied by i.

This limit is taken as the definition of the complex integral of $f(z)$ along C, and it is written

$$\int_C f(z)\, dz. \tag{4}$$

In particular, the above analysis holds for any function $f(z)$ which is analytic throughout a region including C.

Some of the most obvious properties of real integrals extend at once to complex integrals; for example,

$$\int_C \{f(z)+g(z)\}\, dz = \int_C f(z)\, dz + \int_C g(z)\, dz,$$

and, if k is a constant,

$$\int_C kf(z)\, dz = k \int_C f(z)\, dz.$$

Also, if C' denotes the contour C described in the opposite direction,

$$\int_{C'} f(z)\,dz = -\int_C f(z)\,dz.$$

Examples. (i) Let $f(z) = k$, a constant, and let C be any curve joining the points $z = a$ and $z = b$. Then

$$\sum_{m=1}^{n} f(\zeta_m)(z_m - z_{m-1}) = k\sum_{m=1}^{n}(z_m - z_{m-1}) = k(b-a).$$

Hence
$$\int_C k\,dz = k(b-a).$$

Since the result is independent of the particular curve C taken, we may write the result as

$$\int_a^b k\,dz = k(b-a).$$

(ii) Let $f(z) = z$, and let C be any curve joining the points $z = a$ and $z = b$.

First take $\zeta_m = z_m$. Then

$$\sum_{m=1}^{n} f(\zeta_m)(z_m - z_{m-1}) = \sum_{m=1}^{n} z_m(z_m - z_{m-1}).$$

Taking $\zeta_m = z_{m-1}$, the sum is

$$\sum_{m=1}^{n} z_{m-1}(z_m - z_{m-1}).$$

These sums tend to the same limit, and hence so does half their sum, viz.

$$\tfrac{1}{2}\sum (z_m^2 - z_{m-1}^2) = \tfrac{1}{2}(b^2 - a^2).$$

Hence
$$\int_C z\,dz = \tfrac{1}{2}(b^2 - a^2),$$

and the integral is again independent of the path.

(iii) Calculate the integral

$$\int_C \frac{dz}{z},$$

where C is the circle with centre the origin and radius ρ.

[Here we can put

$$x = \rho\cos\theta = \phi(\theta), \qquad y = \rho\sin\theta = \psi(\theta),$$

where θ varies from 0 to 2π. Now

$$\phi'(\theta) + i\psi'(\theta) = -\rho\sin\theta + \rho i\cos\theta = \rho i e^{i\theta}.$$

Hence the integral is

$$\int_0^{2\pi} \frac{1}{\rho e^{i\theta}}\rho i e^{i\theta}\,d\theta = i\int_0^{2\pi} d\theta = 2\pi i.]$$

(iv) Prove similarly that $\displaystyle\int_C z^n\,dz = 0,$

where n is any integer, positive or negative, other than -1.

2.31. An inequality for complex integrals. We may define the length of the curve $x = \phi(t)$, $y = \psi(t)$, where $\phi'(t)$ and $\psi'(t)$ are continuous, to be the integral

$$\int [\{\phi'(t)\}^2 + \{\psi'(t)\}^2]^{\frac{1}{2}}\, dt,$$

taken between appropriate limits. For the justification of this, see *P.M.* § 146.

If M is the upper bound of $|f(z)|$ on the curve C, and L is the length of C, then

$$\left|\int f(z)\, dz\right| \leqslant ML.$$

In the first place, if $F(t)$ is any continuous complex function of a real variable t,

$$\left|\int_a^b F(t)\, dt\right| \leqslant \int_a^b |F(t)|\, dt. \tag{1}$$

For $\quad |\sum F(t_m)(t_m - t_{m-1})| \leqslant \sum |F(t_m)|(t_m - t_{m-1}),$

and (1) follows on proceeding to the limit.

Hence, with our previous notation,

$$\begin{aligned} \left|\int_C f(z)\, dz\right| &= \left|\int_{t_A}^{t_B} (u+iv)\{\phi'(t)+i\psi'(t)\}\, dt\right| \\ &\leqslant \int_{t_A}^{t_B} M[\{\phi'(t)\}^2 + \{\psi'(t)\}^2]^{\frac{1}{2}}\, dt \\ &= ML. \end{aligned}$$

2.32. Contours. By a contour we mean a continuous curve consisting of a finite number of arcs of the type already considered, that is to say, arcs defined by equations $x = \phi(t)$, $y = \psi(t)$, where $\phi'(t)$ and $\psi'(t)$ are continuous. The contour is closed if the end-point of the last arc is the same as the starting-point of the first.

Let C be a closed contour. Suppose that there is an interval (a, b) such that, if $a < x' < b$, the line $x = x'$ meets C in just two points, say $y_1(x')$ and $y_2(x')$, where $y_1 < y_2$; while if $x' < a$ or $x' > b$, the line $x = x'$ does not meet C. Suppose similarly that there is an interval (α, β), such that if $\alpha < y' < \beta$, the line $y = y'$ meets C in just two points, say $x_1(y')$ and $x_2(y')$, where $x_1 < x_2$; while if $y' < \alpha$ or $y' > \beta$, the line $y = y'$ does not meet C.

Then the point (x, y) is said to be inside C if $a < x < b$ and

$y_1(x) < y < y_2(x)$. A point not inside C or on C is said to be outside C.

A contour which satisfies these conditions is called a simple closed contour. For example, a circle, square, or ellipse in any position is a simple closed contour. The definition of 'inside' and 'outside' which we have given may strike the reader as unnecessarily elaborate, and the class of curves considered unnecessarily restricted. But the general study of questions of this kind is not quite so easy as might be supposed, and we regard it as outside our scope. It forms the subject known as 'analysis situs', and is dealt with, for example, in Watson's *Complex Integration and Cauchy's Theorem*. On the other hand, the reader who prefers to ignore our explanations and trust to geometrical intuition will find that he gets on perfectly well.

We can extend the class of contours to which our theorems apply by 'addition' and 'subtraction' of simple closed contours. Suppose that C and C' are two simple closed contours having one or more arcs in common, but lying outside each other. We form a new closed contour C'' by deleting the common boundary. The interior of C'' consists of the interiors of C and C', together with points on the deleted boundary. Similarly, if all points inside C' are also inside C, we form a new closed contour C'', the interior of which consists of points inside C but outside C'.

A closed contour of this kind which is often useful is formed by the semi-circles $|z| = \rho$, $|z| = R$ $(0 < \rho < R)$, in the upper half-plane, joined by intervals of the real axis.

Still more complicated contours can be introduced by further additions; for example, add the contour just described to its reflection in the real axis, and delete the common boundary from $z = -R$ to $z = -\rho$. We obtain a closed contour with a definite inside and outside. The outside consists of the regions $|z| < \rho$ and $|z| > R$. In describing the contour, the interval from $z = \rho$ to $z = R$ is described twice in opposite directions.

2.33. Cauchy's theorem. The keystone of the theory of analytic functions is the following theorem of Cauchy:

If a function $f(z)$ is analytic and one-valued inside and on a simple closed contour C, then

$$\int_C f(z)\, dz = 0.$$

To prove this, we divide up the region inside C into a large number of small parts by a network of lines parallel to the real and imaginary axes. Suppose that this divides the inside of C into a number of squares, $C_1,\ldots, C_M$ say, and a number of irregular regions, $D_1,\ldots, D_N$ say, parts of whose boundaries are parts of C. Then

$$\int_C f(z)\,dz = \sum_{m=1}^{M} \int_{C_m} f(z)\,dz + \sum_{n=1}^{N} \int_{D_n} f(z)\,dz, \tag{1}$$

where each contour is described in the positive (anti-clockwise) direction. Consider, for example, two squares $ABCD$ and $DCEF$ with a common side CD. The side CD is described from C to D in the first square, and from D to C in the second. Hence the two integrals along CD cancel. So all the integrals cancel, except those which form part of C itself, since these are described once only. This proves (1).

We now use the fact that $f(z)$ is analytic at every point. This means that, if z_0 is any point inside or on C, then

$$\left|\frac{f(z)-f(z_0)}{z-z_0} - f'(z_0)\right| < \epsilon$$

provided that $0 < |z-z_0| < \delta = \delta(z_0)$; i.e. if $|z-z_0| < \delta$,

$$|f(z)-f(z_0)-(z-z_0)f'(z_0)| \leqslant \epsilon\,|z-z_0|. \tag{2}$$

If we consider any particular region C_m or D_n in the above construction, it is evident that we can choose its sides so small that (2) is satisfied if z_0 is a given point of the region, and z any other point. It is not, however, immediately obvious that we can choose the whole network so that the conditions are satisfied in all the partial regions at the same time. We shall prove that this is actually possible.

Having given ϵ, we can choose the network in such a way that, in every partial region C_m or D_n, there is a point z_0 such that (2) *holds for every z in this region.*

This means, substantially, that the function is *uniformly differentiable* throughout the interior of C.

Let us assume for the moment that this is true. Consider one of the squares C_m, of side l_m. Here, by (2),

$$f(z) = f(z_0) + (z-z_0)f'(z_0) + \phi(z),$$

where

$$|\phi(z)| \leqslant \epsilon\,|z-z_0|.$$

Hence

$$\int_{C_m} f(z)\,dz = \int_{C_m} \{f(z_0)+(z-z_0)f'(z_0)\}\,dz + \int_{C_m} \phi(z)\,dz.$$

The first integral on the right is zero, by § 2.3, examples (i) and (ii). Also, by § 2.31,

$$\left|\int_{C_m} \phi(z)\,dz\right| < \epsilon\sqrt{2}l_m.4l_m,$$

since $|z-z_0| \leqslant \sqrt{2}l_m$, and the length of C_m is $4l_m$.

In the case of one of the irregular regions D_n, the length is not greater than $4l_n+s_n$, where s_n is the length of the curved part of the boundary. Hence

$$\left|\int_{D_n} \phi(z)\,dz\right| < \epsilon\sqrt{2}l_n(4l_n+s_n).$$

Adding all the parts, we obtain

$$\left|\int_C f(z)\,dz\right| < 4\sqrt{2}\epsilon \sum (l_m^2+l_n^2)+\epsilon\sqrt{2}l \sum s_n. \tag{3}$$

Now $\sum (l_m^2+l_n^2)$ is the area of a region which just includes C, and is therefore bounded; in fact, if $(a, b; \alpha, \beta)$ is a rectangle including C,

$$\sum (l_m^2+l_n^2) \leqslant (b-a)(\beta-\alpha).$$

Also $\sum s_n$ is the length of the contour C. Hence the right-hand side of (3) is less than a constant multiple of ϵ. But the left-hand side is independent of ϵ. It must therefore be zero.

2.34. We have still to prove the assumption which we have made. This is done by the well-known process of subdivision. Suppose that we start with a network of parallel lines at constant distance l. Some of the squares formed by these lines may contain a point z_0 with the required property. We leave these unchanged. The rest we subdivide by lines midway between the previous ones. If there still remain any parts which have not the required property, we subdivide them again in the same way. There are now *a priori* two possibilities. The process may terminate after a finite number of steps, and then the result is obtained; or it may go on indefinitely. In the second case there is at least one region which we can subdivide indefinitely without obtaining the required result. Call this region (boundary included) R_1. After the first subdivision we obtain a part R_2, contained in R_1, with the same property. So we shall have an infinity of regions R_1, R_2,..., R_n,... each contained in the previous one, and for each of which (2) is impossible.

There is a point z_0 common to all the regions R_n; and, since the dimensions of R_n decrease indefinitely, $|z-z_0|<\delta$ if n is sufficiently large (say $n>n_0$) and z is in R_n. But $f(z)$ is analytic at z_0. Hence (2) holds with this z_0 in R_n, if $n>n_0$. We have thus arrived at a contradiction, and so proved the theorem.

2.35. Cauchy's theorem may obviously be extended at once to a closed contour of any of the types defined in § 2.32. It may also be expressed in slightly different forms. Suppose, for example, that z_0 and z_1 are points connected by two different curves C and C' such that C, and C' reversed, together make up a simple closed contour, or a closed contour of one of the other types described in § 2.32. Let $f(z)$ be a function analytic in the whole region between C and C', and on the curves themselves. Then Cauchy's theorem obviously gives

$$\int_C f(z)\,dz=\int_{C'} f(z)\,dz. \tag{1}$$

Suppose again that C is a simple closed contour, and C' another simple closed contour lying entirely inside C. Let $f(z)$ be analytic and one-valued at all points in the ring-shaped region between C and C'. Then

$$\int_C f(z)\,dz=\int_{C'} f(z)\,dz. \tag{2}$$

For we can join C to C' by a straight line l, parallel, say, to the real axis. Then the region between C and C', cut by l, is the inside of a closed contour Γ, formed by C described positively, C' negatively, and l described twice in opposite directions. Now

$$\int_\Gamma f(z)\,dz=\int_C f(z)\,dz-\int_{C'} f(z)\,dz+\int_l f(z)\,dz-\int_l f(z)\,dz.$$

Since the integral round Γ is zero, the result follows.

A similar result holds if there are any finite number of contours C', C'',... inside C, and $f(z)$ is analytic in the region between them. Then

$$\int_C f(z)\,dz=\int_{C'} f(z)\,dz+\int_{C''} f(z)\,dz+\dots. \tag{3}$$

Another important remark is that, for the truth of Cauchy's theorem, it is not necessary that $f(z)$ should be analytic on C, provided that it is analytic inside it and continuous up to and on C. For if $f(z)$ is continuous, it can be shown that

$$\int_C f(z)\,dz=\lim\int_{C'} f(z)\,dz, \tag{4}$$

where C' is a contour inside C and tending to it. It is perhaps not worth while describing how this is to be done in all cases—if C is a circle, C' is a concentric circle, and so on. Now the right-hand side of (4) is zero for all positions of C' inside C. Hence the left-hand side is zero.

2.36. A complex integral as a function of its upper limit. Let $f(z)$ be a function analytic in a region D. Let

$$F(z) = \int_{z_0}^{z} f(w)\, dw,$$

the path of integration being any contour lying entirely inside D. It follows from Cauchy's theorem that the value of $F(z)$ depends on z only, and not on the particular path of integration taken from z_0 to z. Our notation has, of course, anticipated this.

The function $F(z)$ *is analytic in* D. For

$$F(z+\delta z) - F(z) = \int_{z}^{z+\delta z} f(w)\, dw,$$

where (by Cauchy's theorem) we may suppose the integral to be taken along the straight line from z to $z+\delta z$. Hence

$$\frac{F(z+\delta z)-F(z)}{\delta z} - f(z) = \frac{1}{\delta z}\int_{z}^{z+\delta z} \{f(w)-f(z)\}\, dw.$$

Since $f(z)$ is continuous,

$$|f(w)-f(z)| < \epsilon \qquad (|w-z| < \delta).$$

Hence, if $0 < |\delta z| < \delta$,

$$\left|\frac{F(z+\delta z)-F(z)}{\delta z} - f(z)\right| < \epsilon.$$

This proves that $F(z)$ is analytic, and that its derivative is $f(z)$.

As in the theory of functions of a real variable, we call $F(z)$ the *indefinite integral* of $f(z)$.

Suppose, on the other hand, that we know an analytic function $G(z)$ such that

$$G'(z) = f(z)$$

throughout D. Then

$$\frac{d}{dz}\{F(z)-G(z)\} = 0.$$

Let

$$F(z)-G(z) = X+iY.$$

Then (as in the proof of the Cauchy-Riemann equations)

$$\frac{\partial X}{\partial x}=0,\qquad \frac{\partial X}{\partial y}=0,\qquad \frac{\partial Y}{\partial x}=0,\qquad \frac{\partial Y}{\partial y}=0.$$

Hence X and Y are constant, i.e. $F-G$ is constant. Hence

$$\int_a^b f(z)\,dz=G(b)-G(a).$$

2.37. Integration and differentiation of complex series. *A uniformly convergent series of analytic functions of a complex variable may be integrated term by term along any path in the region of uniform convergence.*

This may be proved precisely as in the case of real functions (using the inequality of § 2.31).

A series of analytic functions may be differentiated term by term at any point inside a region where the differentiated series is uniformly convergent.

This also may be proved in the same way as for real functions. It will, however, be superseded later (§ 2.8) by a much more useful theorem, which is a characteristic achievement of complex function theory, and which has no analogue in the theorems of Chapter I.

2.4. Cauchy's integral. Let $f(z)$ be a function analytic inside and on a simple closed contour C. Let z be any point inside C. Consider the function of w

$$\frac{f(w)}{w-z}.$$

This function is analytic except at $w=z$, where the denominator vanishes. Hence

$$\int_C \frac{f(w)}{w-z}\,dw=\int_\gamma \frac{f(w)}{w-z}\,dw,$$

where γ is any other closed contour inside C and including $w=z$. Let γ be the circle with centre z and radius ρ. Since $f(w)$ is continuous, we can take ρ so small that

$$|f(w)-f(z)|<\epsilon$$

on γ. Then

$$\int_\gamma \frac{f(w)}{w-z}\,dw=f(z)\int_\gamma \frac{dw}{w-z}+\int_\gamma \frac{f(w)-f(z)}{w-z}\,dw.$$

The first term is equal to $2\pi i f(z)$ (§ 2.3, ex. 3). Also, by § 2.31, the modulus of the second term does not exceed

$$\frac{\epsilon}{\rho}\,.\,2\pi\rho = 2\pi\epsilon.$$

Hence
$$\left|\int_C \frac{f(w)}{w-z}\,dw - 2\pi i f(z)\right| < 2\pi\epsilon.$$

But the left-hand side is independent of ϵ. It must therefore be zero. Thus

$$f(z) = \frac{1}{2\pi i}\int_C \frac{f(w)}{w-z}\,dw.$$

This is Cauchy's integral formula. It expresses the value of $f(z)$ at any point inside C in terms of its values on C.

2.41. The derivatives of an analytic function. Let z be any point inside C, and $z+h$ a neighbouring point, also inside C. Then

$$f(z+h) = \frac{1}{2\pi i}\int_C \frac{f(w)}{w-z-h}\,dw.$$

Subtracting the previous result from this, and dividing by h,

$$\frac{f(z+h)-f(z)}{h} = \frac{1}{2\pi i}\int_C \frac{f(w)}{(w-z)(w-z-h)}\,dw. \tag{1}$$

When $h \to 0$ the integrand tends to the limit $f(w)/(w-z)^2$. To prove that we can proceed to the limit under the integral sign, consider the difference:

$$\int_C \frac{f(w)}{(w-z)(w-z-h)}\,dw - \int_C \frac{f(w)}{(w-z)^2}\,dw$$
$$= h\int_C \frac{f(w)}{(w-z)^2(w-z-h)}\,dw. \tag{2}$$

Suppose that $|f(w)| \leqslant M$ on C, and that the distance from z to C (i.e. the minimum of $|w-z|$ as w describes C) is δ. Let the length of C be L. Then if $|h| < \delta$,

$$\left|\int_C \frac{f(w)}{(w-z)^2(w-z-h)}\,dw\right| \leqslant \frac{ML}{\delta^2(\delta-|h|)}$$

which is bounded as $|h| \to 0$. Hence the right-hand side of (2) tends to zero with $|h|$. Hence, by (1),

$$f'(z) = \frac{1}{2\pi i}\int_C \frac{f(w)}{(w-z)^2}\,dw. \tag{3}$$

This is Cauchy's formula for $f'(z)$.

The existence of $f'(z)$ was, of course, our original hypothesis. But now that we have obtained this formula for it, we can repeat the above process. We have

$$\frac{f'(z+h)-f'(z)}{h} = \frac{1}{2\pi i}\int_C \frac{2w-2z-h}{(w-z)^2(w-z-h)^2} f(w)\,dw$$

and we prove as before that when $h \to 0$ this tends to the limit

$$\frac{2}{2\pi i}\int_C \frac{f(w)}{(w-z)^3}\,dw.$$

Hence $f''(z)$, the derivative of $f'(z)$, exists, and is given by the formula

$$f''(z) = \frac{2}{2\pi i}\int_C \frac{f(w)}{(w-z)^3}\,dw.$$

The argument can obviously be repeated indefinitely; *hence $f(z)$ has derivatives of all orders, the nth being given by the formula*

$$f^{(n)}(z) = \frac{n!}{2\pi i}\int_C \frac{f(w)}{(w-z)^{n+1}}\,dw.$$

2.42. Morera's theorem. This is a sort of converse of Cauchy's theorem.

If $f(z)$ is a continuous function of z in a region D, and if the integral

$$\int f(z)\,dz$$

taken round any closed contour in D is zero, then $f(z)$ is analytic inside D.

In this theorem the precise sense of the word 'contour' does not matter. The result holds even if we restrict ourselves, say, to convex polygons.

Consider the function

$$F(z) = \int_{z_0}^{z} f(w)\,dw.$$

Its value is independent of the path of integration; and

$$\frac{F(z+h)-F(z)}{h} - f(z) = \frac{1}{h}\int_{z}^{z+h} \{f(w)-f(z)\}\,dw,$$

where the path of integration may be taken to be the straight line. This tends to zero with h, since $f(w)$ is continuous. Hence $F(z)$ is analytic, and has the derivative $f(z)$. But we have just proved that the derivative of an analytic function is analytic. Hence $f(z)$ is analytic.

2.43. Taylor's series. An analytic function can be expanded in powers of its argument by a formula similar to Taylor's series for a real function.

Suppose that $f(z)$ is analytic on and inside a simple closed contour C, and let a be a point inside C. Then

$$f(z)=f(a)+(z-a)f'(a)+...+\frac{(z-a)^n}{n!}f^{(n)}(a)+...,$$

the series being convergent if $|z-a|<\delta$, where δ is the distance from a to the nearest point of C.

We start from Cauchy's formula

$$f(z)=\frac{1}{2\pi i}\int_\Gamma \frac{f(w)}{w-z}\,dw,$$

where we take Γ to be a circle with centre a and radius $\rho<\delta$. The formula holds if z lies inside this circle, i.e. if $|z-a|<\rho$.

Now
$$\frac{1}{w-z}=\frac{1}{w-a}+\frac{z-a}{(w-a)^2}+...+\frac{(z-a)^n}{(w-a)^{n+1}}+...,$$

the series being uniformly convergent on Γ. Hence we may multiply by $f(w)/2\pi i$ and integrate term by term round Γ. We obtain

$$f(z)=\frac{1}{2\pi i}\int_\Gamma \frac{f(w)}{w-a}\,dw+\frac{z-a}{2\pi i}\int_\Gamma \frac{f(w)}{(w-a)^2}\,dw+$$
$$+...+\frac{(z-a)^n}{2\pi i}\int_\Gamma \frac{f(w)}{(w-a)^{n+1}}\,dw+...,$$

and, by the formulae of § 2.41, this is the desired result. It is sometimes known as the Cauchy-Taylor theorem.

There is one difference to be noted between this proof and the corresponding investigation for functions of a real variable. In the real variable theory we obtain the first n terms of the expansion, and a remainder term, and a special investigation is required to see whether this term tends to zero. In the complex theory the fact that it tends to zero follows from our original hypotheses. This state of affairs is quite natural; for combining the above theorem with § 2.16, we see that *the necessary and sufficient condition that a function should be expansible in a power series is that it should be analytic in a region.* We cannot define an analytic function of a real variable except as one which can be expanded in a power series. If therefore we start from other hypotheses, we may expect to meet with difficulties. For example, the radius of convergence of the series depends on the extent of the region where the function is analytic; and it may therefore be controlled by the existence of singularities off the real axis, of which, if we confine ourselves to real variables, we can have no knowledge.

Thus the expansion

$$\frac{1}{1+x^2} = 1-x^2+x^4-\ldots$$

holds for $|x| < 1$ only. There seems to be nothing in the nature of the function, considered as a function of the *real* variable x, to account for this restriction. But if we make x complex, it is accounted for by the fact that the function is not analytic at the points $x = \pm i$.

2.5. Cauchy's inequality. *If*

$$f(z) = \sum_{n=0}^{\infty} a_n z^n \qquad (|z| < R),$$

and $M(r)$ *is the upper bound of* $|f(z)|$ *on the circle* $|z| = r$, $(r < R)$, *then*

$$|a_n| r^n \leqslant M(r)$$

for all values of n.

For

$$a_n = \frac{1}{2\pi i} \int_{|z|=r} \frac{f(z)}{z^{n+1}}\, dz,$$

and the theorem of § 2.31 gives at once

$$|a_n| \leqslant \frac{M(r)}{r^n}.$$

Cauchy's inequality may also be proved as follows. Let $\bar{a}_n$ be the conjugate of a_n. Then if $r < R$,

$$|f(z)|^2 = f(z)\overline{f(z)} = \sum_{m=0}^{\infty} a_m r^m e^{im\theta} \sum_{n=0}^{\infty} \bar{a}_n r^n e^{-in\theta}.$$

Both series being absolutely convergent, we may multiply by the usual rule (§ 1.65). The resulting series is uniformly convergent for $0 \leqslant \theta \leqslant 2\pi$, and we may therefore integrate term by term over this interval. On integration, all the terms for which $m \neq n$ vanish, and we obtain

$$\int_0^{2\pi} |f(z)|^2\, d\theta = \sum_{n=0}^{\infty} a_n \bar{a}_n r^{2n} \int_0^{2\pi} d\theta = 2\pi \sum_{n=0}^{\infty} |a_n|^2 r^{2n},$$

or $$\sum_{n=0}^{\infty} |a_n|^2 r^{2n} = \frac{1}{2\pi} \int_0^{2\pi} |f(z)|^2 \, d\theta \leqslant \{M(r)\}^2.$$

The result clearly follows from this.

Example. Show that Cauchy's inequality reduces to an equality if, and only if, $f(z)$ is a constant multiple of a power of z.

2.51. Liouville's theorem. *A function which is analytic for all finite values of z, and is bounded, is a constant.*

We give two proofs of this important theorem.

FIRST PROOF. If $f(z)$ is analytic for all finite values of z, the Taylor's series
$$f(z) = \sum_{n=0}^{\infty} a_n z^n$$
is convergent for all z. Also, if $|f(z)| \leqslant M$, then by Cauchy's inequality
$$|a_n| \leqslant Mr^{-n}$$
for all values of n and r. Making $r \to \infty$, the right-hand side tends to zero if $n > 0$. Hence $a_n = 0$ for $n > 0$, and $f(z) = a_0$.

SECOND PROOF. If z_1, z_2 are any two numbers,
$$f(z_1) - f(z_2) = \frac{1}{2\pi i} \int_C \frac{f(z)}{z - z_1} \, dz - \frac{1}{2\pi i} \int_C \frac{f(z)}{z - z_2} \, dz$$
$$= \frac{1}{2\pi i} \int_C \frac{z_1 - z_2}{(z - z_1)(z - z_2)} f(z) \, dz,$$
where C is a contour including both z_1 and z_2. Taking C to be a circle with centre the origin, and radius R greater than $|z_1|$ or $|z_2|$, we have
$$|f(z_1) - f(z_2)| \leqslant \frac{|z_1 - z_2| MR}{(R - |z_1|)(R - |z_2|)}.$$
The right-hand side tends to zero as $R \to \infty$. Hence $f(z_1) = f(z_2)$. Since this holds for all values of z_1 and z_2, $f(z)$ is a constant.

The same result holds if the function is bounded on any sequence of contours tending to infinity. This is clear from either proof.

2.52. The following is a more general result of the same kind. *If $f(z)$ is analytic for all finite values of z, and as $|z| \to \infty$*
$$f(z) = O(|z|^k),$$
then $f(z)$ is a polynomial of degree $\leqslant k$.

For by Cauchy's inequality

$$|a_n| \leqslant M(r)r^{-n} = O(r^{k-n}),$$

and the right-hand side tends to zero as $r \to \infty$ if $n > k$. Hence $a_n = 0$ for $n > k$, and the result follows.

2.53. The function $A(r)$. Let $A(r)$ denote the upper bound of the real part of $f(z)$ on $|z| = r$. We next prove an inequality similar to Cauchy's inequality, but involving $A(r)$ instead of $M(r)$.

We have

$$|a_n|r^n \leqslant \max\{4A(r), 0\} - 2\mathbf{R}\{f(0)\}$$

for all values of $n > 0$ and r.

Let $z = re^{i\theta}$, and

$$f(z) = \sum_{n=0}^{\infty} a_n z^n = u(r,\theta) + iv(r,\theta),$$

$$a_n = \alpha_n + i\beta_n.$$

Then

$$u(r,\theta) = \sum_{n=0}^{\infty} (\alpha_n \cos n\theta - \beta_n \sin n\theta)r^n.$$

The series converges uniformly with respect to θ. Hence we may multiply by $\cos n\theta$ or $\sin n\theta$ and integrate term by term; and we obtain

$$\frac{1}{\pi}\int_0^{2\pi} u(r,\theta)\cos n\theta\, d\theta = \alpha_n r^n, \qquad \frac{1}{\pi}\int_0^{2\pi} u(r,\theta)\sin n\theta\, d\theta = -\beta_n r^n$$

for $n > 0$, while

$$\frac{1}{2\pi}\int_0^{2\pi} u(r,\theta)\, d\theta = \alpha_0.$$

Hence

$$a_n r^n = (\alpha_n + i\beta_n)r^n = \frac{1}{\pi}\int_0^{2\pi} u(r,\theta)e^{-in\theta}\, d\theta \qquad (n > 0),$$

and

$$|a_n|r^n \leqslant \frac{1}{\pi}\int_0^{2\pi} |u(r,\theta)|\, d\theta.$$

Hence

$$|a_n|r^n + 2\alpha_0 \leqslant \frac{1}{\pi}\int_0^{2\pi} \{|u(r,\theta)| + u(r,\theta)\}\, d\theta.$$

Now $|u|+u$ is zero if $u < 0$. Hence if $A(r) < 0$ the right-hand side is zero. If $A(r) \geqslant 0$, the right-hand side does not exceed

$$\frac{1}{\pi}\int_0^{2\pi} 2A(r)\,d\theta = 4A(r).$$

This proves the theorem.

There are, of course, similar results involving the lower bound of $\mathbf{R}f(z)$, and the upper and lower bounds of $\mathbf{I}f(z)$.

2.54. The analogue of Liouville's theorem for $A(r)$. *If $f(z)$ is analytic for all finite values of z, and $A(r)$ is bounded as $r \to \infty$, then $f(z)$ is a constant. If $A(r) < Ar^k$, where A and k are constants, then $f(z)$ is a polynomial of degree $\leqslant k$.*

In the first case, it follows from the above theorem that $|a_n|r^n$ is bounded as $r \to \infty$ for every $n > 0$. Hence $a_n = 0$ for $n > 0$, and $f(z) = a_0$. Similarly, in the second case, $a_n = 0$ for $n > k$, and $f(z)$ is of degree $\leqslant k$. It is sufficient that the conditions should hold for some arbitrarily large values of r.

The first part of the theorem may also be proved directly as follows. Consider the function $\phi(z) = \exp\{f(z)\}$. Then

$$|\phi(z)| = e^{u(r,\theta)}.$$

Hence, if $u(r,\theta) \leqslant A$, then

$$|\phi(z)| \leqslant e^A.$$

Hence, by Liouville's theorem, $\phi(z)$ is a constant. Hence $f(z)$ is a constant.

2.6. The zeros of an analytic function. A zero of an analytic function $f(z)$ is a value of z such that $f(z) = 0$. If $f(z)$ is analytic in the neighbourhood of $z = a$, then

$$f(z) = \sum_{n=0}^{\infty} a_n(z-a)^n$$

for $|z-a|$ small enough; and if $z = a$ is a zero, the first one or more of the coefficients $a_0, a_1,\ldots$ vanish. If $a_n = 0$ for $n < m$, but $a_m \neq 0$, then $f(z)$ is said to have a zero of the mth order. Thus every zero is of some definite integral order—a function cannot have a zero of fractional order inside a region where it is analytic.

At a zero of order m, we have

$$f(a) = f'(a) = \ldots = f^{(m-1)}(a) = 0,$$

while $f^{(m)}(a) \neq 0$. This is clear from the form of Taylor's series.

The zeros of an analytic function are isolated points; that is to say, *if a function $f(z)$ is not identically zero, and is analytic in a region including $z = a$, then there is a circle $|z-a| = \rho$ $(\rho > 0)$ inside which $f(z)$ has no zeros except possibly $z = a$ itself.*

The theorem may also be stated as follows:

Let $f(z)$ be a function analytic in a region D, and let $P_1, P_2, \ldots, P_n, \ldots$ be a set of points having a limit-point P inside D. Then if $f(z) = 0$ at every point P_n, it follows that $f(z) = 0$ at all points of D.

It may be supposed without loss of generality that P is $z = 0$. Then $f(z)$ is analytic in a region including $z = 0$, and hence

$$f(z) = \sum_{n=0}^{\infty} a_n z^n$$

for $|z| < R$, say. It will be proved that all the coefficients in this series are zero. If this is not so, there is a first coefficient which is not zero, say a_k. Then

$$f(z) = z^k(a_k + a_{k+1}z + \ldots) \qquad (|z| < R).$$

If $0 < \rho < R$, the series is convergent for $z = \rho$, and so $a_n \rho^n$ is bounded, say $|a_n|\rho^n \leqslant K$. Hence

$$|f(z)| \geqslant |z|^k\left(|a_k| - \frac{K|z|}{\rho^{k+1}} - \frac{K|z|^2}{\rho^{k+2}} - \ldots\right)$$

$$= |z|^k\left\{|a_k| - \frac{K|z|}{\rho^k(\rho - |z|)}\right\},$$

and the right-hand side is positive if $|z|$ is sufficiently small, except at the point $z = 0$ itself. This contradicts the hypothesis that $f(z)$ has zeros arbitrarily near to, but not coincident with, $z = 0$. Hence in fact all the coefficients vanish. Hence $f(z) = 0$ inside the circle of convergence of the above series.

We can now, however, repeat the process, starting from any point inside this circle; for the data now hold for any such point, by what has just been proved.* In this way the result may be extended to any point interior to D.

2.61. The theorem has the following obvious corollaries:

(i) If a function is analytic in a region, and vanishes at all points of any smaller region included in the given region, or along any arc of a continuous curve in the region, then it must vanish identically.

* For a more detailed discussion of such chains of circles see Ch. IV.

(ii) If two functions are analytic in a region, and have the same value at an infinity of points which have a limit-point in the region, then they must be equal throughout the region.

Examples. (i) The function $\sin z$ has zeros of the first order at the points $z = 0, \pm\pi, \pm 2\pi,...,$ and no other zeros. [The formula

$$|\sin(x+iy)| = \sqrt{(\sin^2 x+\sinh^2 y)},$$

P.M. § 233, ex. 2, shows that $\sin z$ has these zeros and no others.]

(ii) The function $\cos z$ has zeros of the first order at $z = \pm\frac{1}{2}\pi, \pm\frac{3}{2}\pi,...,$ and no other zeros.

(iii) If $f(z)$ and $g(z)$ are both analytic functions in a region D, and $f(z)g(z) = 0$ in D, then either $f(z) = 0$ throughout D, or $g(z) = 0$ throughout D.

2.7. Laurent series. *Let $f(z)$ be a function analytic in the ring-shaped region between two concentric circles C and C', of radii R and R' ($R' < R$), and centre a, and on the circles themselves.*

Then $f(z)$ can be expanded in a series of positive and negative powers of $z-a$, convergent at all points of the ring-shaped region.

We must remind the reader that all functions considered here are one-valued. This assumption excludes certain functions, any one value of which is analytic at all points of the ring. Consider, for example, the function

$$f(z) = z^p$$

where p is real. This is analytic except possibly at $z = 0$. Now, if $z = re^{i\theta}$,

$$f(z) = r^p e^{ip\theta}.$$

As we pass round a circle with centre the origin, starting at $\theta = 0$, say, $f(z)$ changes from r^p to $r^p e^{2ip\pi}$, and so does not return to its original value unless p is an integer.

To prove the theorem, let z be a point of the ring, and consider the integral

$$\frac{1}{2\pi i}\int \frac{f(w)}{w-z}\,dw$$

taken round the outer circle C in the positive direction, then along a radius vector (which we may suppose does not pass through z) to the inner circle C', then round C' in the negative direction, then back along the radius vector to the starting-point. This is a closed contour to which we may apply our previous results—the fact that part of it is described twice does not affect any of the arguments. The value of the integral is therefore $f(z)$.

Since $f(z)$ is one-valued, the two integrals along the radius vector joining the circles cancel, and we obtain

$$f(z) = \frac{1}{2\pi i}\int_C \frac{f(w)}{w-z}\,dw - \frac{1}{2\pi i}\int_{C'} \frac{f(w)}{w-z}\,dw,$$

where now each integral is taken in the positive direction. As in the proof of the Cauchy-Taylor theorem

$$\frac{1}{2\pi i}\int_C \frac{f(w)}{w-z}\,dw = \sum_{n=0}^{\infty} a_n(z-a)^n,$$

where
$$a_n = \frac{1}{2\pi i}\int_C \frac{f(w)}{(w-a)^{n+1}}\,dw.$$

In this case, however, a_n is not in general equal to $f^{(n)}(a)/n!$, since $f(z)$ is not necessarily analytic throughout the interior of C.

Again

$$\frac{1}{z-w} = \frac{1}{z-a} + \frac{w-a}{(z-a)^2} + \ldots + \frac{(w-a)^{n-1}}{(z-a)^n} + \ldots,$$

this series being uniformly convergent on C'. Hence

$$-\frac{1}{2\pi i}\int_{C'} \frac{f(w)}{w-z}\,dw = \frac{1}{z-a}\cdot\frac{1}{2\pi i}\int_{C'} f(w)\,dw + \ldots +$$
$$+\frac{1}{(z-a)^n}\cdot\frac{1}{2\pi i}\int_{C'} (w-a)^{n-1}f(w)\,dw + \ldots = \sum_{n=1}^{\infty} \frac{b_n}{(z-a)^n},$$

where
$$b_n = \frac{1}{2\pi i}\int_{C'} (w-a)^{n-1}f(w)\,dw.$$

These two series together form Laurent's expansion. They may be written together in the form

$$f(z) = \sum_{n=-\infty}^{\infty} a_n(z-a)^n,$$

where
$$a_n = \frac{1}{2\pi i}\int \frac{f(w)}{(w-a)^{n+1}}\,dw$$

for all values of n, the integral being taken round any simple closed contour which passes round the ring.

In the particular case where $f(z)$ is analytic inside C', all the coefficients b_n are zero (by Cauchy's theorem), and the series reduces to Taylor's series.

Notice that the series of positive powers of $z-a$ converges, not merely in the ring, but everywhere inside the circle C. Similarly the series of negative powers converges everywhere outside C'.

Examples. (i) Show that

$$e^{\frac{1}{2}c(z-\frac{1}{z})} = \sum_{n=-\infty}^{\infty} a_n z^n,$$

where
$$a_n = \frac{1}{2\pi}\int_0^{2\pi} \cos(n\theta - c\sin\theta)\,d\theta.$$

(ii) Show that, if $c > 0$,

$$e^{z+\frac{c^2}{2z^2}} = \sum_{n=-\infty}^{\infty} a_n z^n$$

where
$$a_n = \frac{e^{-\frac{1}{2}c}}{2\pi c^n}\int_0^{2\pi} e^{c(\cos\theta+\cos^2\theta)}\cos\{c\sin\theta(1-\cos\theta)-n\theta\}\,d\theta.$$

2.71. Isolated singularities of an analytic function. Suppose that a function $f(z)$ is analytic throughout the neighbourhood of a point a, say for $|z-a| < R$, except at the point a itself. Then the point a is called an isolated singularity of the function.

Suppose that $f(z)$ is one-valued. We may then expand $f(z)$ in a Laurent series of powers of $z-a$, and the inner circle C' of § 2.7 may be taken as small as we please. Thus

$$f(z) = \sum_{n=0}^{\infty} a_n(z-a)^n + \sum_{n=1}^{\infty} b_n(z-a)^{-n} \qquad (0 < |z-a| < R).$$

There are now three possible cases. All the coefficients b_n may be zero. The function $f(z)$ is then equal to a function analytic for $|z-a| < R$, except at the point a; for example, we might define $f(z)$ to be 1 except at $z = a$, and $f(a) = 0$. This is a rather artificial sort of singularity, and of no further interest in the theory.

Secondly, the series of negative powers of $z-a$ may contain a finite number of terms only. Then $f(z)$ is said to have a *pole* at the point $z = a$. If b_m is the last coefficient which does not vanish, then

$$f(z) = \sum_{n=0}^{\infty} a_n(z-a)^n + \sum_{n=1}^{m} b_n(z-a)^{-n},$$

and the pole is said to be of order m, or to be a simple, double,... pole in the cases $m = 1, 2, \ldots$.

If $f(z)$ has a pole of order m, then plainly $(z-a)^m f(z)$ is analytic and not zero at $z=a$. Hence also

$$\phi(z) = \frac{1}{(z-a)^m f(z)}$$

is analytic and not zero at $z=a$. Hence

$$\frac{1}{f(z)} = (z-a)^m \phi(z),$$

so that $1/f(z)$ has a zero of order m at $z=a$.

Conversely, a similar argument shows that, if $f(z)$ has a zero of order m, then $1/f(z)$ has a pole of order m.

The finite series

$$\sum_{n=1}^{m} b_n(z-a)^{-n}$$

is called the *principal part* of $f(z)$ at $z=a$.

If $f(z)$ has a pole at $z=a$, then $|f(z)| \to \infty$ as $z \to a$. For

$$\left|\sum_{n=1}^{m} b_n(z-a)^{-n}\right| = |z-a|^{-m}\left|\sum_{n=1}^{m} b_n(z-a)^{m-n}\right|$$

$$\geqslant |z-a|^{-m}\left\{|b_m| - \sum_{n=1}^{m-1} |b_n||z-a|^{m-n}\right\},$$

and the expression in brackets tends to $|b_m|$, so that the whole tends to infinity—in fact the function is dominated by the last term in the principal part.

If $f(z) = O(|z-a|^{-k})$ as $|z-a| \to 0$, the singularity is at most a pole of order k; in particular, if $f(z) = O(1)$, there is no singularity (except of the trivial type first mentioned).

An argument similar to that of § 2.52, but with b_n, and r tending to 0, shows that $b_n = 0$ for $n > k$. Again it is clearly sufficient that the data should hold on a sequence of contours tending to 0.

Examples. (i) The functions $\cot z$ and $\operatorname{cosec} z$ have simple poles at the points $z = 0, \pm\pi, \pm 2\pi, \ldots$.

(ii) The functions $\tan z$ and $\sec z$ have simple poles at the points $z = \pm\frac{1}{2}\pi, \pm\frac{3}{2}\pi, \ldots$.

(iii) Find the poles of

$$\frac{1}{\sin z \pm \sin a}, \qquad \frac{1}{\cos z \pm \cos a}.$$

(iv) The function $\operatorname{cosec} z^2$ has one double pole and an infinity of simple poles.

(v) Find the poles of the functions

$$\frac{1}{z^2+1}, \quad \frac{1}{z^4+1}, \quad \frac{1}{z^4+2z^2+1}.$$

2.72. Essential singularities. The third possibility is that in the expansion of $f(z)$ in powers of $z-a$, the series of negative powers may not terminate. The point $z=a$ is then called an *essential singularity* of $f(z)$. In this case

$$f(z) = \sum_{n=0}^{\infty} a_n(z-a)^n + \sum_{n=1}^{\infty} \frac{b_n}{(z-a)^n},$$

where the last series does not terminate, but is convergent for all values of z except $z=a$.

The complicated behaviour of a function in the neighbourhood of an essential singularity is shown by the following theorem of Weierstrass.

Given any positive numbers ρ, ϵ, *and any number* c, *there is a point* z *in the circle* $|z-a| < \rho$ *at which* $|f(z)-c| < \epsilon$.

That is to say, $f(z)$ tends to any given limit as z tends to a through a suitable sequence of values.

We begin by proving that, if ρ and M are any positive numbers, then there are values of z in the circle $|z-a|<\rho$ at which $|f(z)|>M$. If this is not true, then $|f(z)| \leqslant M$ for $|z-a|<\rho$. Hence, if the radius of C' is R',

$$|b_n| = \left|\frac{1}{2\pi i}\int_{C'} (w-a)^{n-1} f(w)\, dw\right| \leqslant MR'^n,$$

by § 2.31. This holds for all positive values of n and R', and, making $R' \to 0$, we see that $b_n = 0$ for $n \geqslant 1$. Hence there is no essential singularity, contrary to hypothesis.

Now consider any finite value of c. If $f(z)-c$ has zeros inside every circle $|z-a|=\rho$, the result follows at once. If not, we can choose ρ so small that $f(z)-c$ has no zero for $|z-a|<\rho$.

Then
$$\phi(z) = \frac{1}{f(z)-c}$$

is regular for $|z-a|<\rho$, except at $z=a$. The point $z=a$ is an essential singularity of $\phi(z)$; for

$$f(z) = \frac{1}{\phi(z)} + c,$$

and if $\phi(z)$ had a pole, $f(z)$ would be analytic; while, if $\phi(z)$ were analytic, $f(z)$ would be analytic or have a pole.

It now follows from the first part that there is a point z in the circle $|z-a| < \rho$ such that

$$|\phi(z)| > \frac{1}{\epsilon},$$

i.e.
$$|f(z)-c| < \epsilon,$$

and this is the result stated.

This theorem distinguishes clearly between poles and essential singularities. While at a pole $f(z)$ tends to infinity, at an essential singularity $f(z)$ has no unique limiting value, and in fact comes arbitrarily near to any assigned value an infinity of times.

Examples. (i) The functions

$$e^{1/z}, \qquad \sin\frac{1}{z}, \qquad \cos\frac{1}{z}$$

have isolated essential singularities at $z = 0$.

(ii) The function $\operatorname{cosec}(1/z)$ has a singularity at $z = 0$, but it is not an isolated singularity, being the limit-point of the poles at the points $z = 1/(n\pi)$. We call such a point an essential singularity also.

(iii) The function $e^{1/z}$ actually takes every value except 0 an infinity of times in the neighbourhood of $z = 0$; and it tends to the limit 0 as $z \to 0$ along the negative real axis.

2.73. The 'point at infinity'. We may consider 'infinity' as a point by making the substitution $z = 1/w$. Then the behaviour of $f(z)$ 'at infinity' depends on the behaviour of $f(1/w)$ at $w = 0$. We say that $f(z)$ is analytic, has a simple pole, etc., at infinity, if $f(1/w)$ has the same property at $w = 0$. Thus $f(z) = z^2$ has a double pole at infinity.

A function which is analytic everywhere, including infinity, is a constant. For by Laurent's theorem, since $f(z)$ is regular for all finite values of z,

$$f(z) = \sum_{n=0}^{\infty} a_n z^n,$$

$$f\left(\frac{1}{w}\right) = \sum_{n=0}^{\infty} \frac{a_n}{w^n}.$$

Since $f(1/w)$ is regular at $w = 0$, $a_n = 0$ for $n = 1, 2,...$, so that $f(z) = a_0$.

A function which has no singularities other than poles is a rational function.

In the first place, there can only be a finite number of such poles; otherwise the poles would have a limit-point either at a finite point or at infinity, and at such a limit-point the function would not be analytic or have a pole, contrary to hypothesis.

Suppose, then, that the poles of $f(z)$ at finite points are at $a, b, \ldots, k$, with multiplicities $\alpha, \beta, \ldots, \kappa$. Then

$$g(z) = f(z)(z-a)^{\alpha} \ldots (z-k)^{\kappa}$$

is analytic except at infinity, where it has at most a pole. Hence

$$g(z) = \sum_{n=0}^{\infty} a_n z^n,$$

$$g(1/w) = \sum_{n=0}^{\infty} a_n w^{-n}.$$

Since the singularity of $g(1/w)$ at the origin, if there is one, is a pole, this series must terminate, i.e. $g(z)$ is a polynomial. Hence $f(z)$ is the quotient of two polynomials, i.e. a rational function.

Conversely, *a rational function has no singularities other than poles.*

2.8. Uniformly convergent series of analytic functions. *Suppose that*

(i) *each member of a sequence of functions*

$$u_1(z),\ u_2(z), \ldots, u_n(z), \ldots$$

is analytic inside a region D,

(ii) *the series*

$$\sum_{n=1}^{\infty} u_n(z)$$

is uniformly convergent throughout every region D' interior to D. Then the function

$$f(z) = \sum_{n=1}^{\infty} u_n(z)$$

is analytic inside D, and all its derivatives may be calculated by term-by-term differentiation.

Let C be a simple closed contour lying entirely inside D, and let z be a point inside C.

If we knew already that $f(z)$ was an analytic function, we should have

$$f(z) = \frac{1}{2\pi i}\int_C \frac{f(w)}{w-z}\,dw. \tag{1}$$

Actually we obtain this result from our data, and then use it to prove that $f(z)$ is analytic.

We have

$$u_n(z) = \frac{1}{2\pi i}\int_C \frac{u_n(w)}{w-z}\,dw$$

for each function $u_n(z)$. Hence

$$f(z) = \sum_{n=1}^{\infty} u_n(z) = \sum_{n=1}^{\infty} \frac{1}{2\pi i}\int_C \frac{u_n(w)}{w-z}\,dw.$$

But, since $\sum u_n(w)$ is uniformly convergent on C, we may multiply by $1/(w-z)$ and integrate term by term. Thus

$$\int_C \left\{ \sum \frac{u_n(w)}{w-z} \right\} dw = \sum \int_C \frac{u_n(w)}{w-z}\,dw,$$

and we obtain

$$f(z) = \frac{1}{2\pi i}\int_C \left\{ \sum \frac{u_n(w)}{w-z} \right\} dw = \frac{1}{2\pi i}\int_C \frac{f(w)}{w-z}\,dw,$$

i.e. we have proved (1).

We can now deduce from (1), as in § 2.41, that $f(z)$ has a derivative $f'(z)$, given by the formula

$$f'(z) = \frac{1}{2\pi i}\int_C \frac{f(w)}{(w-z)^2}\,dw.$$

In this case the boundedness of $f(z)$ follows from the uniform convergence of the series. Hence $f(z)$ is analytic.

Also

$$\frac{1}{2\pi i}\int_C \frac{f(w)}{(w-z)^2}\,dw = \frac{1}{2\pi i}\int_C \sum_{n=1}^{\infty} u_n(w)\,\frac{dw}{(w-z)^2}$$

$$= \sum_{n=1}^{\infty} \frac{1}{2\pi i}\int_C \frac{u_n(w)}{(w-z)^2}\,dw = \sum_{n=1}^{\infty} u_n'(z),$$

using the uniformity of convergence again. Hence the series may be differentiated term by term. Also the differentiated series is uniformly convergent in any region interior to D. For if D' is such a region, we can suppose that the curve C includes

D' in its interior, and that the least distance of points of D' from C is δ. Then for any point z of D'

$$\left|\sum_{n=N}^{N'} u_n'(z)\right| = \left|\sum_{N}^{N'} \frac{1}{2\pi i}\int_C \frac{u_n(w)}{(w-z)^2}\,dw\right|$$

$$= \left|\frac{1}{2\pi i}\int_C \sum_{N}^{N'} u_n(w)\,\frac{dw}{(w-z)^2}\right| \leqslant \frac{\epsilon l}{2\pi\delta^2},$$

where l is the length of C, and ϵ the maximum modulus of

$$\sum_{N}^{N'} u_n(w)$$

on C. Since the right-hand side is independent of z, and tends to zero when N and N' tend independently to infinity, the result follows.

The whole process may now be repeated, starting from the differentiated series, and the general result thus follows.

The theorem, in a slightly different form, is known as 'Weierstrass's double-series theorem'.*

2.81. Remarks on the above theorem.

(i) We have already pointed out (§ 2.37) the contrast between the conditions for term-by-term differentiation of real series, and of series of analytic functions. In the case of real series, we have to assume that the differentiated series is uniformly convergent. In the above theorem no such assumption is necessary; actually the differentiated series is uniformly convergent, but this is one of the conclusions of the theorem.

(ii) If we merely assumed that the given series was uniformly convergent on a certain closed curve C, we could prove as before that $f(z)$ was analytic at all points inside C.

(iii) Even if we assume that each $u_n(z)$ is analytic on the boundary of D, and that the series is uniformly convergent on the boundary, we cannot prove that $f(z)$ is analytic on the boundary, or that the differentiated series converges on the boundary. Consider, for example, the series

$$\sum_{n=1}^{\infty} \frac{z^n}{n^2}.$$

This is uniformly convergent for $|z| \leqslant 1$; but the differentiated

* See Knopp's *Infinite Series*, § 56.

series is not uniformly convergent in the neighbourhood of $z = 1$, nor is the function represented by the series analytic at $z = 1$. In fact

$$f'(z) = -\frac{\log(1-z)}{z}.$$

(iv) The theorem may be stated as a theorem on sequences of functions: if $f_n(z)$ is analytic in D for each value of n, and tends to $f(z)$ uniformly in any region interior to D, then $f(z)$ is analytic inside D, and $f_n'(z)$ tends to $f'(z)$ uniformly in any region interior to D.

Examples. (i) The function $\zeta(s) = \sum_{n=1}^{\infty} n^{-s}$ is analytic for $\mathbf{R}(s) > 1$. [For the series is uniformly convergent in any finite region to the right of $\mathbf{R}(s) = 1$, see § 1.21, example.]

(ii) We have, for $\mathbf{R}(s) > 1$,

$$\zeta'(s) = -\sum_{n=2}^{\infty} n^{-s} \log n,$$

and generally

$$\zeta^{(k)}(s) = (-1)^k \sum_{n=2}^{\infty} n^{-s} \log^k n.$$

(iii) In what region does the series

$$\sum_{n=1}^{\infty} \frac{\sin nz}{2^n}$$

represent an analytic function?

(iv) The series

$$\sum_{n=1}^{\infty} \frac{\sin nz}{n^2}$$

is uniformly convergent on the real axis, but not in any *region* of the z-plane; so we can deduce nothing about the analytic character of the function which it represents.

2.82. Another proof of the theorem. We can also deduce Weierstrass's theorem from Morera's theorem (§ 2.42). For, since $\sum u_n(z)$ is uniformly convergent, we may integrate it term by term round any contour C. Thus

$$\int_C f(z)\, dz = \sum_{n=1}^{\infty} \int_C u_n(z)\, dz.$$

But, since each $u_n(z)$ is analytic, every term on the right is zero. Hence

$$\int_C f(z)\, dz = 0.$$

Hence, by Morera's theorem, $f(z)$ is analytic.

2.83. Definition of analytic functions by means of integrals. *Let $f(z,w)$ be a continuous function of the complex variables z and w, where z ranges over a region D, and w lies on a contour C. Let $f(z,w)$ be an analytic function of z in D, for every value of w on C. Then*

$$F(z) = \int_C f(z,w)\,dw$$

is an analytic function of z in D; and

$$F'(z) = \int_C \frac{\partial f}{\partial z}\,dw,$$

and similarly for higher derivatives.

We may suppose that the contour C consists of a single regular curve, on which $w = u+iv$, $u = u(t)$, $v = v(t)$, $t_0 \leqslant t \leqslant t_1$, and $u'(t)$ and $v'(t)$ are continuous.

Let Γ be a contour lying in D, on which $z = x+iy$, $x = x(s)$, $y = y(s)$, $s_0 \leqslant s \leqslant s_1$, and $x'(s)$ and $y'(s)$ are continuous. Let ζ be a point inside Γ. Then

$$f(\zeta,w) = \frac{1}{2\pi i}\int_\Gamma \frac{f(z,w)}{z-\zeta}\,dz,$$

$$F(\zeta) = \frac{1}{2\pi i}\int_C dw \int_\Gamma \frac{f(z,w)}{z-\zeta}\,dz.$$

We may invert the order of these two integrations. For we can express each of these complex integrals as a sum of real integrals, as in § 2.3; and we clearly obtain an expression of the form

$$\int_{t_0}^{t_1} dt \int_{s_0}^{s_1} \{\phi(s,t)+i\psi(s,t)\}\,ds,$$

where ϕ and ψ are real continuous functions of s and t. Now we know that a repeated integral of this type may be inverted (§ 1.81). Hence

$$F(\zeta) = \frac{1}{2\pi i}\int_\Gamma \frac{dz}{z-\zeta}\int_C f(z,w)\,dw$$

$$= \frac{1}{2\pi i}\int_\Gamma \frac{F(z)}{z-\zeta}\,dz.$$

Thus $F(z)$ satisfies Cauchy's integral formula, and from this point the proof that $F(z)$ is analytic, and that we can differentiate under the integral sign, proceeds as in the theorem on uniformly convergent series.

Examples. (i) If $f(t)$ is continuous in (a, b), then

$$F(z) = \int_a^b \cos zt\, f(t)\, dt, \qquad G(z) = \int_a^b \sin zt\, f(t)\, dt$$

are analytic functions for all finite values of z.

(ii) Under the same conditions

$$\int_a^b \frac{f(t)}{z-t}\, dt$$

is analytic, except possibly when z is real and lies in the interval (a, b).

2.84. Infinite integrals. *Let C be a contour going to infinity, any bounded part of which is regular. Suppose that the conditions of the previous theorem are satisfied on any bounded part of C, and that*

$$\int_C f(z, w)\, dw$$

is uniformly convergent. Then the results of the previous theorem still hold.

Let C_n be the part of C inside the circle $|z| = n$, and let

$$F_n(z) = \int_{C_n} f(z, w)\, dw.$$

Then $F_n(z)$ is analytic for every n, by the theorem on finite integrals. Also

$$F_n(z) \to F(z)$$

uniformly as $n \to \infty$. Hence, by the theorem on uniformly convergent sequences, $F(z)$ is analytic. Finally

$$F'(z) = \lim_{n\to\infty} F_n'(z) = \lim_{n\to\infty} \int_{C_n} \frac{\partial f}{\partial z}\, dw = \int_C \frac{\partial f}{\partial z}\, dw.$$

2.85. Infinite integrals of the second kind. There is a similar theorem for the case of a finite contour C, at one end of which $f(z, w) \to \infty$. Such an integral represents an analytic function, provided that the convergence of the integral is uniform. The formal statement and the proof are practically the same as those of the previous theorem.

Examples. (i) The function

$$\Gamma(z) = \int_0^\infty e^{-w} w^{z-1}\, dw$$

is an analytic function for $\mathbf{R}(z) > 0$. [The uniform convergence of this integral has been discussed in § 1.51, ex. (i). It converges uniformly in any finite region in which $\mathbf{R}(z) \geqslant a > 0$; and any point at which $\mathbf{R}(z) > 0$ is an internal point of such a region.]

(ii) In what regions do the integrals

$$\int_0^\infty e^{-zw^2}\,dw, \qquad \int_0^\infty \frac{\sin w}{w^z}\,dw, \qquad \int_0^\infty \frac{\cos w}{w^z}\,dw,$$

represent analytic functions?

(iii) The integral

$$\int_0^\infty \frac{\sin wz}{w}\,dw$$

converges uniformly in certain intervals of real values of z, but not in any *region*; so we cannot deduce anything about the analytic character of the function which it represents.

2.9. Remark on Laurent series. Suppose that we have obtained in any manner, or as the definition of $f(z)$, the formula

$$f(z) = \sum_{n=-\infty}^{\infty} A_n(z-a)^n \qquad (R' < |z-a| < R).$$

Is the series necessarily identical with the Laurent series of $f(z)$? Yes; for if C is the circle $|z-a| = \rho$, $R' < \rho < R$, the Laurent coefficient a_n is

$$a_n = \frac{1}{2\pi i}\int_C \frac{f(z)}{(z-a)^{n+1}}\,dz = \sum_{m=-\infty}^{\infty} \frac{A_m}{2\pi i}\int_C \frac{(z-a)^m}{(z-a)^{n+1}}\,dz,$$

by uniform convergence; and the right-hand side is A_n, by § 2.3, exs. (iii) and (iv).

CHAPTER III

RESIDUES, CONTOUR INTEGRATION, ZEROS

3.1. The residue at a singularity. We know (§ 2.71) that, in the neighbourhood of an isolated singularity $z=a$, a one-valued analytic function $f(z)$ may be expanded in the form

$$f(z)=\sum_{n=0}^{\infty}a_n(z-a)^n+\sum_{n=1}^{\infty}b_n(z-a)^{-n}.$$

The coefficient b_1 is of particular importance, and is called the *residue* of $f(z)$ at the point $z=a$. By the formulae of Laurent's expansion,

$$b_1=\frac{1}{2\pi i}\int_{\gamma}f(z)\,dz,$$

where γ is any circle with centre $z=a$, which excludes all other singularities of the function.

It is easily seen that, if $z=a$ is a simple pole,

$$b_1=\lim_{z\to a}(z-a)f(z).$$

3.11. The theorem of residues. *Let $f(z)$ be one-valued and analytic inside and on a simple closed contour C, except at a finite number of singularities $z_1, z_2,..., z_n$. Let the residues of $f(z)$ at these points be $R_1, R_2,..., R_n$. Then*

$$\int_C f(z)\,dz=2\pi i(R_1+R_2+...+R_n).$$

Let $\gamma_1, \gamma_2,..., \gamma_n$ be circles with centres $z_1, z_2,..., z_n$, and radii so small that they lie entirely inside C and do not overlap. Then $f(z)$ is analytic in the region between C and these circles, so that, by Cauchy's theorem (see § 2.35),

$$\int_C f(z)\,dz=\int_{\gamma_1}f(z)\,dz+...+\int_{\gamma_n}f(z)\,dz.$$

But

$$\int_{\gamma_1}f(z)\,dz=2\pi iR_1,$$

etc., and the result follows.

3.12. Contour integration. The theorem of residues may be used to evaluate a large number of real definite integrals. To do this we take a contour, part of which consists of the real axis, and the remaining part of which is usually made to tend to infinity. The process is called contour integration. It is best made clear by means of examples.

3.121. It is well known that

$$\int_0^\infty \frac{dx}{1+x^2} = \frac{\pi}{2}.$$

To prove this by contour integration, consider the integral

$$\int \frac{dz}{1+z^2}$$

taken round the contour consisting of the real axis from $-R$ to R, with a semi-circle, on this line as diameter, above it. Since

$$\frac{1}{1+z^2} = \frac{1}{2i}\left(\frac{1}{z-i} - \frac{1}{z+i}\right),$$

the integrand has a pole at $z = i$, which is inside the contour if $R > 1$, with residue $1/2i$. Hence, by the theorem of residues, the integral is equal to π.

Now, on the semicircle, $|1+z^2| \geqslant R^2-1$, so that the integral round the semicircle does not exceed

$$\frac{\pi R}{R^2-1},$$

and so it tends to zero as $R \to \infty$. Hence

$$\lim_{R\to\infty} \int_{-R}^{R} \frac{dx}{1+x^2} = \pi.$$

Since the integrand is an even function, the result now follows.

The integral of any even rational function which behaves suitably at infinity can be evaluated in a similar way.

Of course we know the indefinite integral of $1/(1+x^2)$, viz. $\arctan x$, and can evaluate the integral from this. The method shows to better advantage in cases where we do not know the indefinite integral.

3.122. It has been shown in § 1.76 that

$$\int_0^\infty \frac{\sin x}{x}\,dx = \frac{\pi}{2}.$$

To prove this by contour integration, consider the integral

$$\int \frac{e^{iz}}{z}\,dz$$

taken round the contour consisting of the real axis from $z=\rho$ to $z=R$, where $0<\rho<R$; a semicircle Γ of radius R above the real axis; the real axis again from $-R$ to $-\rho$; and finally a semicircle γ of radius ρ above the real axis. We take ρ small and R large. The small semicircle is necessary to avoid the singularity of the integrand at $z=0$, and the large semicircle is necessary to close up the contour.

The function e^{iz}/z has no singularity inside the contour, and the value of the integral is therefore zero. Thus

$$\int\limits_{\rho}^{R}\frac{e^{ix}}{x}\,dx+\int_{\Gamma}\frac{e^{iz}}{z}\,dz+\int\limits_{\rho}^{R}\frac{e^{-ix}}{-x}\,dx+\int_{\gamma}\frac{e^{iz}}{z}\,dz=0.$$

The two integrals along the real axis are together equal to

$$\int\limits_{\rho}^{R}\frac{e^{ix}-e^{-ix}}{x}\,dx=2i\int\limits_{\rho}^{R}\frac{\sin x}{x}\,dx.$$

The integral along Γ tends to 0 when $R\to\infty$. For

$$\left|\int_{\Gamma}\frac{e^{iz}}{z}\,dz\right|=\left|\int\limits_{0}^{\pi}e^{iRe^{i\theta}}i\,d\theta\right|\leqslant\int\limits_{0}^{\pi}e^{-R\sin\theta}\,d\theta$$

$$\leqslant\int\limits_{0}^{\delta}d\theta+\int\limits_{\delta}^{\pi-\delta}e^{-R\sin\delta}\,d\theta+\int\limits_{\pi-\delta}^{\pi}d\theta<2\delta+\pi e^{-R\sin\delta}.$$

We first take δ arbitrarily small, and then, having fixed δ, the second term may be made as small as we please by choosing R sufficiently large. Hence the integral along Γ tends to 0.

Finally,

$$\int_{\gamma}\frac{e^{iz}}{z}\,dz=\int_{\gamma}\frac{dz}{z}+\int_{\gamma}\frac{e^{iz}-1}{z}\,dz.$$

The integrand in the last integral is bounded as $\rho\to 0$, and so, by § 2.31, the integral tends to zero. Also

$$\int_{\gamma}\frac{dz}{z}=\int\limits_{\pi}^{0}i\,d\theta=-i\pi.$$

Hence, making $\rho\to 0$ and $R\to\infty$, we obtain

$$2i\int\limits_{0}^{\infty}\frac{\sin x}{x}\,dx-i\pi=0$$

and the result follows.

Notice that the integral in the negative direction round the semicircle γ tends to $-i\pi$ into the residue at $z = 0$. It is easily verified that this is true of any *simple* pole, but not of a pole of higher order.

Notice also that we do not consider the integral

$$\int \frac{\sin z}{z}\, dz$$

because the integrand does not behave suitably at infinity.

3.123. *If* $0 < a < 1$,

$$\int_0^\infty \frac{x^{a-1}}{1+x}\, dx = \frac{\pi}{\sin a\pi}.$$

Consider the integral

$$\int \frac{z^{a-1}}{1+z}\, dz$$

taken along the real axis from $z=\rho$ to $z=R$; then in the positive direction along the circle Γ with centre the origin and radius R; then back along the real axis to $z=\rho$; and finally round the circle γ with centre the origin and radius ρ in the negative direction. This is a closed contour which excludes the origin. It is necessary to do so, because the function is not one-valued in a region which includes the origin, so that the theorem of residues would not apply to such a contour. The many-valued function z^{a-1} is taken to be real on the first part of the contour. It is then given at all other points by the formula $r^{a-1}e^{(a-1)i\theta}$, where $0 \leqslant \theta \leqslant 2\pi$.

There is one pole inside the contour, at $z = -1$, the residue there being $e^{(a-1)i\pi}$. Hence

$$\int_\rho^R \frac{x^{a-1}}{1+x}\, dx + \int_\Gamma \frac{z^{a-1}}{1+z}\, dz + \int_R^\rho \frac{(xe^{2i\pi})^{a-1}}{1+x}\, dx + \int_\gamma \frac{z^{a-1}}{1+z}\, dz$$
$$= 2\pi i\, e^{(a-1)i\pi}.$$

The two integrals along the real axis together give

$$(1-e^{2i\pi(a-1)}) \int_\rho^R \frac{x^{a-1}}{1+x}\, dx = -2ie^{ia\pi}\sin a\pi \int_\rho^R \frac{x^{a-1}}{1+x}\, dx.$$

The other two integrals tend to 0. For on Γ

$$\left|\frac{z^{a-1}}{1+z}\right| \leqslant \frac{R^{a-1}}{R-1},$$

so that
$$\left|\int_{\Gamma} \frac{z^{a-1}}{1+z}\,dz\right| \leqslant \frac{R^{a-1}}{R-1}\cdot 2\pi R = \frac{2\pi R^{a}}{R-1},$$

which tends to zero since $a < 1$. Similarly

$$\left|\int_{\gamma} \frac{z^{a-1}}{1+z}\,dz\right| \leqslant \frac{2\pi\rho^{a}}{1-\rho},$$

which tends to zero since $a > 0$. The result therefore follows on making $R \to \infty$ and $\rho \to 0$.

3.124. The above result has an application to the theory of the Γ-function. Putting $y = 1-x$ in § 1.86 (4), we have

$$\Gamma(x)\Gamma(1-x) = \int_0^\infty \frac{u^{-x}}{1+u}\,du$$

$$= \frac{\pi}{\sin(1-x)\pi},$$

or
$$\Gamma(x)\Gamma(1-x) = \frac{\pi}{\sin x\pi}$$

where $0 < x < 1$.

3.125. *For* $n = 0, 1, 2,\ldots$

$$\int_0^\infty x^n e^{-x^{\frac{1}{4}}} \sin x^{\frac{1}{4}}\,dx = 0.$$

Putting $x = t^4$, the integral becomes

$$4\int_0^\infty t^{4n+3}e^{-t}\sin t\,dt.$$

Consider the integral
$$\int z^{4n+3}e^{(i-1)z}\,dz$$

taken along the real axis from 0 to R, then along a quadrant of a circle of radius R to the positive imaginary axis, and then back to the origin along the imaginary axis. On the arc of the circle
$$|e^{(i-1)z}| = e^{-R\cos\theta - R\sin\theta} \leqslant e^{-R},$$

so that
$$\left|\int z^{4n+3}e^{(i-1)z}\,dz\right| \leqslant \tfrac{1}{2}\pi R^{4n+4}e^{-R} \to 0.$$

Hence $$\int_0^\infty x^{4n+3}e^{(i-1)x}\,dx-\int_0^\infty (iy)^{4n+3}e^{(i-1)iy}i\,dy=0,$$

or, replacing y by x in the last integral,

$$\int_0^\infty x^{4n+3}e^{-x}(e^{ix}-e^{-ix})\,dx=0,$$

and the result follows.

3.126. *If* $c>0$, *then*

$$\frac{1}{2\pi i}\int_{c-i\infty}^{c+i\infty}\frac{a^z}{z}\,dz=\begin{matrix}1 & (a>1),\\ 0 & (0<a<1).\end{matrix}$$

If $a>1$, i.e. $\log a>0$, we consider the integral round the contour consisting of the line from $c-iR$ to $c+iR$, completed by a semicircle on the left. If R is sufficiently large, this contour includes the pole at $z=0$, with residue 1; and it may be proved as in § 3.121 that the integral round the large semicircle tends to zero as $R\to\infty$.

If $a<1$, we complete the contour by a semicircle on the right. There is now no pole in the contour, and the second result follows.

3.127. The Γ-function integral. We have

$$\int_0^\infty x^{p-1}e^{-ax}\,dx=\frac{\Gamma(p)}{a^p}\qquad(a>0,\,p>0).$$

If we could make the substitution $x=it$ in the integral, we should obtain

$$\int_0^\infty (it)^{p-1}e^{-ait}i\,dt=\frac{\Gamma(p)}{a^p},$$

and, multiplying by $e^{-\frac{1}{2}ip\pi}$ and separating real and imaginary parts, we find

$$\int_0^\infty t^{p-1}\begin{matrix}\cos\\ \sin\end{matrix}\,at\,dt=\frac{\Gamma(p)}{a^p}\begin{matrix}\cos\\ \sin\end{matrix}\,\tfrac{1}{2}p\pi. \tag{1}$$

The ordinary rules of integration by substitution, of course, do not cover a 'complex substitution' of this kind. The process is really an application of Cauchy's theorem. Consider the integral

$$\int z^{p-1}e^{-az}\,dz$$

taken round the contour consisting of the real axis from $z=\rho$ to $z=R$, the arc of $|z|=R$ to the imaginary axis, the imaginary axis from $z=iR$ to $z=i\rho$, and the arc of $|z|=\rho$ back to the starting-point. By Cauchy's theorem, the integral round this contour is zero. It may be proved as in previous cases that the integral along $|z|=\rho$ tends to 0 as $\rho\to 0$ if $p>0$, and that along $|z|=R$ tends to 0 as $R\to\infty$ if $p<1$. Hence the integral along the imaginary axis is minus that along the real axis, and on evaluating it we obtain (1) again, for $0<p<1$.

3.128. Occasionally we use the converse process, and deduce the residue from the value of the integral.

If p is an even positive integer, the residue of $\tan^{p-1}\pi z$ *at* $z=\frac{1}{2}$ *is* $(-1)^{\frac{1}{2}p}/\pi$.

The residue is equal to

$$\frac{1}{2\pi i}\left\{\int\limits_{-iR}^{1-iR}+\int\limits_{1-iR}^{1+iR}+\int\limits_{1+iR}^{iR}+\int\limits_{iR}^{-iR}\right\}\tan^{p-1}\pi z\,dz,$$

and

$$\int\limits_{1-iR}^{1+iR}=\int\limits_{-iR}^{iR}=-\int\limits_{iR}^{-iR},$$

since $\tan\pi z$ is periodic with period 1. Hence the residue is equal to

$$\frac{1}{2\pi i}\left\{\int\limits_{-iR}^{1-iR}+\int\limits_{1+iR}^{iR}\right\}\tan^{p-1}\pi z\,dz.$$

Now

$$\tan\pi z=\frac{1}{i}\frac{e^{2i\pi x-2\pi y}-1}{e^{2i\pi x-2\pi y}+1}\to\begin{matrix}-1/i\\+1/i\end{matrix}\qquad\left(y\to\begin{matrix}+\infty\\-\infty\end{matrix}\right).$$

Hence as $R\to\infty$

$$\int\limits_{-iR}^{1-iR}\tan^{p-1}\pi z\,dz\to\left(\frac{1}{i}\right)^{p-1},$$

$$\int\limits_{1+iR}^{iR}\tan^{p-1}\pi z\,dz\to-\left(-\frac{1}{i}\right)^{p-1}=\left(\frac{1}{i}\right)^{p-1},$$

and the residue is

$$\frac{1}{\pi i}\left(\frac{1}{i}\right)^{p-1}=\frac{1}{\pi i^{p}}=\frac{(-1)^{\frac{1}{2}p}}{\pi}.$$

3.13. *Consider the behaviour of the integral*

$$f(t) = \int_0^\infty e^{ixt - x^{\frac{1}{4}}} \sin x^{\frac{1}{4}} \, dx$$

as $t \to \infty$.

It is convenient, for reasons which will appear later, to begin by integrating by parts. Integrating the factor e^{ixt}, the integrated term vanishes at both limits, and we obtain

$$f(t) = \frac{i}{4t} \int_0^\infty e^{ixt - x^{\frac{1}{4}}} (\cos x^{\frac{1}{4}} - \sin x^{\frac{1}{4}}) x^{-\frac{3}{4}} \, dx.$$

As in previous examples, we replace the circular functions by exponentials, and consider, instead of $f(t)$, the function

$$\phi(t) = \tfrac{1}{4} \int_0^\infty e^{ixt - x^{\frac{1}{4}} + ix^{\frac{1}{4}}} x^{-\frac{3}{4}} \, dx.$$

Putting $x = u^4/t$, we obtain

$$\phi(t) = t^{-\frac{1}{4}} \int_0^\infty e^{iu^4 - (1-i)ut^{-\frac{1}{4}}} \, du.$$

Next, turn the line of integration through an angle λ, i.e. use Cauchy's theorem as in § 3.125. We obtain

$$\phi(t) = t^{-\frac{1}{4}} \int_0^\infty e^{iv^4 e^{4i\lambda} - (1-i)ve^{i\lambda} t^{-\frac{1}{4}}} e^{i\lambda} \, dv.$$

This process is valid if the real part of the coefficient of v^4 is negative for all values of λ through which the line of integration turns; i.e. if $\sin 4\lambda > 0$, or $\lambda < \frac{1}{4}\pi$.

Actually we take $\lambda = \frac{1}{8}\pi$. This has the effect of making the term in e^{-v^4} tend to zero as rapidly as possible. It gives

$$\phi(t) = t^{-\frac{1}{4}} e^{i\pi/8} \int_0^\infty e^{-v^4 - (1-i)ve^{i\pi/8} t^{-\frac{1}{4}}} \, dv.$$

When $t \to \infty$, this last integral, being uniformly convergent, tends to the limit

$$\int_0^\infty e^{-v^4} \, dv = \tfrac{1}{4} \int_0^\infty e^{-w} w^{-\frac{3}{4}} \, dw = \tfrac{1}{4}\Gamma(\tfrac{1}{4}).$$

Hence
$$\phi(t) \sim \tfrac{1}{4}\Gamma(\tfrac{1}{4}) e^{i\pi/8} t^{-\frac{1}{4}}.$$

Similarly, if $$\psi(t) = \tfrac{1}{4}\int\limits_0^\infty e^{ixt-x^{\frac{1}{4}}-ix^{\frac{1}{4}}}x^{-\frac{3}{4}}\,dx,$$

we obtain the same asymptotic formula for $\psi(t)$ as for $\phi(t)$. Hence finally

$$f(t) = \frac{i}{t}\left\{\frac{\phi(t)+\psi(t)}{2}-\frac{\phi(t)-\psi(t)}{2i}\right\}$$

$$\sim \tfrac{1}{4}i\Gamma(\tfrac{1}{4})e^{i\pi/8}t^{-\frac{5}{4}}.$$

A similar process might, of course, have been applied to the integral before integrating by parts. The reader may verify that it only leads to the result $f(t) = o(t^{-1})$.

3.2. Expansion of a meromorphic function. A function is said to be *meromorphic* in a region if it is analytic in the region except at a finite number of poles. The expression is used in contrast to *holomorphic*, which is sometimes used instead of analytic.

The simplest meromorphic functions are rational functions. We know that a rational function can be expressed in a simple way by means of partial fractions; we shall now obtain a similar expression for a more general class of meromorphic functions.

Let $f(z)$ be a function whose only singularities, except at infinity, are poles. We shall suppose for simplicity that all these poles are simple. Let them be $a_1, a_2, \ldots$, where

$$0 < |a_1| \leqslant |a_2| \leqslant |a_3| \leqslant \ldots,$$

and let the residues at the poles be $b_1, b_2, \ldots$ respectively. Suppose that there is a sequence of closed contours C_n, such that C_n includes $a_1, a_2, \ldots, a_n$, but no other poles; such that the minimum distance R_n of C_n from the origin tends to infinity with n, while L_n, the length of C_n, is $O(R_n)$; and such that, on C_n, $f(z) = o(R_n)$. This last condition will be satisfied if, for example, $f(z)$ is bounded on the system of contours C_n taken as a whole.

Under these conditions

$$f(z) = f(0) + \sum_{n=1}^{\infty} b_n\left(\frac{1}{z-a_n}+\frac{1}{a_n}\right)$$

for all values of z except the poles.

To prove this, consider the integral

$$I = \frac{1}{2\pi i} \int_{C_n} \frac{f(w)}{w(w-z)}\, dw,$$

where z is a point inside C_n. The integrand has poles at the points a_m, with residues $b_m/\{a_m(a_m-z)\}$; at $w=z$, with residue $f(z)/z$; and at $w=0$, with residue $-f(0)/z$. In particular cases these last two residues may of course vanish. Hence

$$I = \sum_{m=1}^{n} \frac{b_m}{a_m(a_m-z)} - \frac{f(0)}{z} + \frac{f(z)}{z}.$$

On the other hand

$$|I| \leqslant \frac{L_n}{2\pi R_n(R_n-|z|)} \max_{C_n} |f(w)|,$$

which tends to 0 as $n \to \infty$, under the conditions stated. Hence

$$\frac{f(z)}{z} = \frac{f(0)}{z} - \lim_{n\to\infty} \sum_{m=1}^{n} \frac{b_m}{a_m(a_m-z)},$$

and the result stated follows.

It is also obvious from the proof that *the series converges uniformly inside any closed contour such that all the poles are outside it.*

3.21. We leave to the reader the modifications which are necessary if $f(z)$ has poles of higher order than the first. A more important extension can be made to functions which do not satisfy the condition $f(z) = o(R_n)$ on C_n. Suppose now that this is not satisfied, but that there is a positive integer p such that $f(z) = O(R_n^p)$, or, more generally, $f(z) = o(R_n^{p+1})$, on C_n. Consider the integral

$$I = \frac{1}{2\pi i} \int_{C_n} \frac{f(w)}{w^{p+1}(w-z)}\, dw.$$

The calculations proceed as before, except that the residue at $w=0$ is now

$$-\frac{1}{z}\left\{\frac{f(0)}{z^p} + \frac{f'(0)}{z^{p-1}} + \ldots + \frac{f^{(p)}(0)}{p!}\right\}.$$

The integral again tends to 0 as $n \to \infty$, and we obtain

$$f(z) = f(0) + zf'(0) + \ldots + \frac{z^p f^{(p)}(0)}{p!} + \sum_{n=0}^{\infty} \frac{b_n z^{p+1}}{a_n^{p+1}(z-a_n)},$$

or

$$f(z) = \sum_{q=0}^{p} \frac{f^{(q)}(0)z^q}{q!} + \sum_{n=0}^{\infty} b_n \left(\frac{1}{z-a_n} + \frac{1}{a_n} + \frac{z}{a_n^2} + \dots + \frac{z^p}{a_n^{p+1}} \right).$$

3.22. Application to trigonometrical functions. Consider the function

$$f(z) = \operatorname{cosec} z - 1/z \ \ (z \neq 0), \qquad f(0) = 0.$$

At the point $z = n\pi$, where n is any positive or negative integer, $\sin z$ has a simple zero, so that $f(z)$ has a simple pole. The residue is

$$\lim_{z \to n\pi} (z-n\pi)\left(\operatorname{cosec} z - \frac{1}{z}\right) = \lim_{\zeta \to 0} \frac{\zeta}{\sin(\zeta+n\pi)} = \lim_{\zeta \to 0} \frac{(-1)^n \zeta}{\sin \zeta} = (-1)^n.$$

But there is no singularity at $z = 0$, since

$$\frac{z - \sin z}{z \sin z} = \frac{O(|z|^3)}{z^2 + O(|z|^4)} = O(1).$$

Let C_n be the square with corners at the points

$$(n+\tfrac{1}{2})(\pm 1 \pm i)\pi.$$

The function $1/z$ is obviously bounded on these squares. To prove that $\operatorname{cosec} z$ is bounded, consider separately the regions (i) $y > \frac{1}{2}\pi$, (ii) $-\frac{1}{2}\pi \leqslant y \leqslant \frac{1}{2}\pi$, (iii) $y < -\frac{1}{2}\pi$. In the first region

$$|\operatorname{cosec} z| = \left| \frac{2i}{e^{iz} - e^{-iz}} \right| \leqslant \frac{2}{e^{\frac{1}{2}\pi} - e^{-\frac{1}{2}\pi}},$$

and a similar result holds for the third region. Also $|\operatorname{cosec} z|$ is evidently bounded on the straight line joining $\frac{1}{2}(1-i)\pi$ to $\frac{1}{2}(1+i)\pi$, and so, since it has the period π, on all the lines $(n+\frac{1}{2}-\frac{1}{2}i)\pi$, $(n+\frac{1}{2}+\frac{1}{2}i)\pi$. Hence $\operatorname{cosec} z$ is bounded on the parts of C_n which lie in (ii), and so on the whole square.

The theorem of § 3.2 therefore gives

$$\operatorname{cosec} z - \frac{1}{z} = \sum_{n=-\infty}^{\infty}{}' (-1)^n \left(\frac{1}{z-n\pi} + \frac{1}{n\pi} \right),$$

the accent indicating that the term $n = 0$ is omitted from the sum. Since, when we pass from C_{n-1} to C_n, we include the two poles $\pm n\pi$ together, we should, in the first place, bracket the corresponding residues together in the sum. However, the series with $n > 0$ and $n < 0$ converge separately, so that the brackets may be omitted.

If we add together the terms corresponding to $\pm n$, the expansion takes the form

$$\operatorname{cosec} z = \frac{1}{z} + 2z \sum_{n=1}^{\infty} \frac{(-1)^{n-1}}{n^2\pi^2 - z^2}.$$

Examples. (i) Obtain the expansions

$$\sec z = 2\pi \sum_{n=0}^{\infty} \frac{(-1)^n(n+\frac{1}{2})}{(n+\frac{1}{2})^2\pi^2 - z^2},$$

$$\tan z = 2z \sum_{n=0}^{\infty} \frac{1}{(n+\frac{1}{2})^2\pi^2 - z^2},$$

and

$$\cot z = \frac{1}{z} + 2z \sum_{n=1}^{\infty} \frac{1}{z^2 - n^2\pi^2}.$$

(ii) Obtain the corresponding expressions for the hyperbolic functions.

(iii) Prove that

$$\frac{1}{e^z - 1} = \frac{1}{z} - \frac{1}{2} + 2z \sum_{n=1}^{\infty} \frac{1}{z^2 + 4n^2\pi^2}.$$

(iv) Prove that $$\operatorname{cosec}^2 z = \sum_{n=-\infty}^{\infty} \frac{1}{(z-n\pi)^2}.$$

3.23. Expansion of an integral function as an infinite product. An integral function is a function which is analytic for all finite values of z. For example, e^z, $\cos z$, $\sin z$, are integral functions. An integral function may be regarded as a generalization of a polynomial; and, just as we can extend the partial fraction formula to certain meromorphic functions, so we can extend the expression of a polynomial as a product of factors to certain integral functions.

Let $f(z)$ be an integral function of z. Suppose that it has simple zeros at the points $a_1, a_2, \ldots$. In the neighbourhood of a_n,

$$f(z) = (z - a_n)g(z)$$

where $g(z)$ is analytic and not zero. Hence

$$\frac{f'(z)}{f(z)} = \frac{1}{z - a_n} + \frac{g'(z)}{g(z)},$$

and the last term is analytic at a_n. Hence $f'(z)/f(z)$ has a simple pole at $z = a_n$, with residue 1.

Suppose now that $f'(z)/f(z)$ is a function of the type considered in § 3.2. Then

$$\frac{f'(z)}{f(z)}=\frac{f'(0)}{f(0)}+\sum_{n=1}^{\infty}\left(\frac{1}{z-a_n}+\frac{1}{a_n}\right).$$

Integrating from 0 to z along a path not passing through any of the poles, we obtain

$$\log f(z)-\log f(0)=z\frac{f'(0)}{f(0)}+\sum_{n=1}^{\infty}\left\{\log(z-a_n)-\log(-a_n)+\frac{z}{a_n}\right\}.$$

The values of the logarithms will depend on the path chosen; but when we take exponentials all ambiguity disappears, and we obtain

$$f(z)=f(0)e^{z\frac{f'(0)}{f(0)}}\prod_{n=1}^{\infty}\left(1-\frac{z}{a_n}\right)e^{\frac{z}{a_n}}.$$

For example, the function $f(z)=\sin z/z$ satisfies our condition, and we obtain the well-known formula

$$\frac{\sin z}{z}=\prod_{n=-\infty}^{\infty}{}'\left(1-\frac{z}{n\pi}\right)e^{\frac{z}{n\pi}},$$

or

$$\sin z=z\prod_{n=1}^{\infty}\left(1-\frac{z^2}{n^2\pi^2}\right).$$

Similarly,

$$\cos z=\prod_{n=1}^{\infty}\left\{1-\frac{z^2}{(n-\frac{1}{2})^2\pi^2}\right\}.$$

If $f'(z)/f(z)$ satisfies the conditions of § 3.21, we obtain for $f(z)$ a product formula of the form

$$f(z)=f(0)e^{c_1z+\ldots+c_{p+1}z^{p+1}}\prod_{n=1}^{\infty}\left(1-\frac{z}{a_n}\right)e^{\frac{z}{a_n}+\frac{1}{2}\frac{z^2}{a_n^2}+\ldots+\frac{1}{(p+1)}\frac{z^{p+1}}{a_n^{p+1}}}$$

3.3. Summation of certain series. The method of contour integration is often effective in summing series of the form

$$\sum f(n)$$

where $f(z)$ is an analytic function of z of a fairly simple kind.

Let C be a closed contour including the points $m, m+1,\ldots,n$, and suppose that $f(z)$ is analytic in this contour, except for poles at a finite number of points $a_1,\ldots,a_k$, say simple poles with residues $b_1,\ldots,b_k$. Consider the integral

$$\int_C \pi\cot\pi z f(z)\,dz.$$

The function $\pi \cot \pi z$ has in C simple poles at $z = m, m+1,..., n$, with residue 1 at each pole. Hence $\pi \cot \pi z\, f(z)$ has the residues $f(m), f(m+1),..., f(n)$. Including the residues due to poles of $f(z)$, we find that

$$\int_C \pi \cot \pi z f(z)\, dz = 2\pi i\{f(m)+f(m+1)+...+f(n)+ \\ +b_1\pi \cot \pi a_1+...+b_k\pi \cot \pi a_k\}.$$

Suppose, for example, that $f(z)$ is a rational function, none of whose poles are integers, and which is $O(|z|^{-2})$ at infinity. Take the contour C to be the square with corners $(n+\frac{1}{2})(\pm 1 \pm i)$. Then, as in § 3.22, the integral round C tends to zero as $n \to \infty$, and we have

$$\lim_{n\to\infty} \sum_{m=-n}^{n} f(m) = -\pi\{b_1 \cot \pi a_1+...+b_k \cot \pi a_k\}.$$

Similarly, by using $\pi \operatorname{cosec} \pi z$ instead of $\pi \cot \pi z$, we can obtain expressions for sums of the form

$$\sum (-1)^m f(m).$$

Consider, for example, the series

$$\sum_{n=-\infty}^{\infty} \frac{1}{(a+n)^2}.$$

Here $f(z) = 1/(a+z)^2$ has a double pole at $z = -a$. By Taylor's theorem

$$\cot \pi z = \cot(-\pi a)+(\pi z+\pi a)\{-\operatorname{cosec}^2(-\pi a)\}+...,$$

so that the residue of $\cot \pi z/(z+a)^2$ at $z = -a$ is $-\pi \operatorname{cosec}^2 \pi a$. Hence

$$\sum_{n=-\infty}^{\infty} \frac{1}{(a+n)^2} = \pi^2 \operatorname{cosec}^2 \pi a.$$

3.4. Poles and zeros of a meromorphic function. *If $f(z)$ is analytic inside and on a closed contour C, apart from a finite number of poles, and is not zero on the contour, then*

$$\frac{1}{2\pi i} \int_C \frac{f'(z)}{f(z)}\, dz = N-P$$

where N is the number of zeros inside the contour (a zero of order m being counted m times), and P is the number of poles (a pole of order m being counted m times).

Suppose that $z=a$ is a zero of order m. Then in the neighbourhood of this point

$$f(z)=(z-a)^m g(z),$$

where $g(z)$ is analytic and not zero. Hence

$$\frac{f'(z)}{f(z)}=\frac{m}{z-a}+\frac{g'(z)}{g(z)}.$$

The last term is analytic at $z=a$, so that $f'(z)/f(z)$ has a simple pole at $z=a$ with residue m. Hence the sum of the residues at the zeros of $f(z)$ is N.

Similarly the sum of the residues at the poles of $f(z)$ is $-P$ (we need merely change the sign of m).

It may be proved similarly that *if $\phi(z)$ is analytic in and on C and $f(z)$ has zeros at $a_1,\ldots,a_m$ and poles at $b_1,\ldots,b_n$, then*

$$\frac{1}{2\pi i}\int_C \frac{f'(z)}{f(z)}\phi(z)\,dz=\sum_{\mu=1}^{m}\phi(a_\mu)-\sum_{\nu-1}^{n}\phi(b_\nu).$$

3.41. *If $f(z)$ is analytic in C, then the above formula reduces to*

$$\frac{1}{2\pi i}\int_C \frac{f'(z)}{f(z)}\,dz=N.$$

This result can also be expressed in another way. Since

$$\frac{d}{dz}\log\{f(z)\}=\frac{f'(z)}{f(z)},$$

we have
$$\int_C \frac{f'(z)}{f(z)}\,dz=\Delta_C\log\{f(z)\},$$

where Δ_C denotes the variation of $\log\{f(z)\}$ round the contour C. The value of the logarithm with which we start is clearly indifferent. Also

$$\log\{f(z)\}=\log|f(z)|+i\arg\{f(z)\},$$

and $\log|f|$ is one-valued. Hence the formula may be written

$$N=\frac{1}{2\pi}\Delta_C\arg\{f(z)\}.$$

3.42. Rouché's theorem. *If $f(z)$ and $g(z)$ are analytic inside and on a closed contour C, and $|g(z)|<|f(z)|$ on C, then $f(z)$ and $f(z)+g(z)$ have the same number of zeros inside C.*

In the first place it is clear that neither $f(z)$ nor $f(z)+g(z)$ has a zero on C. Hence, if N is the number of zeros of $f(z)$, and

N' the number of zeros of $f(z)+g(z)$,

$$2\pi N = \Delta_C \arg f,$$

$$2\pi N' = \Delta_C \arg(f+g) = \Delta_C \arg f + \Delta_C \arg\left(1+\frac{g}{f}\right).$$

To prove that $N = N'$, we have therefore to prove that

$$\Delta_C \arg\left(1+\frac{g}{f}\right) = 0.$$

Since $|g| < |f|$, the point $w = 1+g/f$ is always an interior point of the circle in the w-plane with centre 1 and radius 1; thus, if $w = \rho e^{i\phi}$, ϕ always lies between $-\frac{1}{2}\pi$ and $\frac{1}{2}\pi$; and therefore $\arg(1+g/f) = \phi$ must return to its original value when z describes C—it cannot increase or decrease by a multiple of 2π. This proves the theorem.

Another proof is as follows. Let $\phi(z) = g(z)/f(z)$. Then

$$N' = \frac{1}{2\pi i}\int_C \frac{f'(z)+g'(z)}{f(z)+g(z)}\,dz = \frac{1}{2\pi i}\int_C \frac{f'+f'\phi+f\phi'}{f(1+\phi)}\,dz$$

$$= \frac{1}{2\pi i}\int_C \left(\frac{f'}{f}+\frac{\phi'}{1+\phi}\right)dz = N + \frac{1}{2\pi i}\int_C \frac{\phi'}{1+\phi}\,dz,$$

and the last integral is zero, as we see by expanding in powers of ϕ and integrating term by term.

3.43. The following is an example of the type of problem which can be solved by means of the above theorems.

In which quadrants do the roots of the equation

$$z^4+z^3+4z^2+2z+3 = 0$$

lie?

The equation has no real roots; for obviously it has no positive root: putting $z = -x$ it is

$$x^4-x^3+4x^2-2x+3 = 0.$$

For $0 < x < 1$ the first three terms together are positive, and so are the last two. For $x > 1$ the first two terms together are positive, and so are the last three.

Putting $z = iy$ the equation becomes

$$y^4-iy^3-4y^2+2iy+3 = 0,$$

and the real and imaginary parts of this do not vanish together. Hence there are no purely imaginary roots.

Now consider $\Delta \arg(z^4+...+3)$ taken round the part of the first quadrant bounded by $|z| = R$, where R is large. The variation along the real axis is zero. On the arc of the circle, $z = Re^{i\theta}$, and we have

$$\begin{aligned}\Delta \arg(z^4+\ldots) &= \Delta \arg(R^4 e^{4i\theta})+\Delta \arg\{1+O(R^{-1})\}\\ &= 2\pi+O(R^{-1}).\end{aligned}$$

On the imaginary axis we have

$$\arg(z^4+\ldots) = \arc\tan\left(\frac{-y^3+2y}{y^4-4y^2+3}\right).$$

The numerator of the expression in brackets vanishes at $y = \sqrt{2}$, and the denominator at $y = \sqrt{3}$ and $y = 1$. Hence the rational fraction varies as follows as y varies from ∞ to 0:

$$\begin{array}{ccccccccc} y=\infty & & \sqrt{3} & & \sqrt{2} & & 1 & & 0 \\ 0, & -, & \infty, & +, & 0, & -, & \infty, & +, & 0. \end{array}$$

Hence $\arc\tan(z^4+\ldots)$ decreases by 2π, and therefore the total variation of $\arg(z^4+\ldots)$ round the quadrant is zero, if R is large enough.

Hence there are no zeros in the first quadrant.

Since zeros occur in conjugate pairs, it follows that *there are no zeros in the fourth quadrant, and two in each of the second and third quadrants.*

Any algebraic equation may be treated in the same way.

3.44. The fundamental theorem of algebra. *Every polynomial of degree n has n zeros.*

In the first place, z^n has n zeros, all at the origin. Now consider any polynomial

$$a_0+a_1 z+\ldots+a_n z^n,$$

where $a_n \neq 0$. Let

$$f(z) = a_n z^n, \qquad g(z) = a_0+a_1 z+\ldots+a_{n-1} z^{n-1},$$

and take the contour C of Rouché's theorem to be a circle with centre the origin and radius $R > 1$. On C

$$|f(z)| = |a_n| R^n,$$

$$|g(z)| \leqslant |a_0|+|a_1|R+\ldots+|a_{n-1}|R^{n-1} \leqslant (|a_0|+\ldots+|a_{n-1}|)R^{n-1}.$$

Hence $|g| < |f|$ on C provided that

$$R > (|a_0|+\ldots+|a_{n-1}|)/|a_n|.$$

Hence, by Rouché's theorem, $f(z)+g(z)$ has n zeros in a circle with centre the origin whose radius R satisfies this condition.

The theorem can also be proved as follows. Suppose that the above polynomial has no zeros; then the function

$$\frac{1}{a_0+\ldots+a_n z^n}$$

is analytic for all values of z, since its only possible singularities are the zeros of the denominator; and it is bounded as $|z| \to \infty$. Hence, by Liouville's theorem, it is a constant. Hence the polynomial reduces to the single term a_0.

This proves only that the polynomial has one zero, and the fact that there are n has to be deduced by the familiar process of algebra.

3.45. A theorem of Hurwitz.* *Let $f_n(z)$ be a sequence of functions, each analytic in a region D bounded by a simple closed contour, and let $f_n(z) \to f(z)$ uniformly in D. Suppose that $f(z)$ is not identically zero. Let z_0 be an interior point of D. Then z_0 is a zero of $f(z)$ if, and only if, it is a limit-point of the set of zeros of the functions $f_n(z)$, points which are zeros for an infinity of values of n being counted as limit-points.*

This easily follows from Rouché's theorem. We can choose ρ so small that the circle $|z-z_0| = \rho$ lies entirely in D, and contains or has on it no zero of $f(z)$ except possibly the point z_0 itself. Then $|f(z)|$ has a positive lower bound on the circle, say $|f(z)| \geqslant m > 0$. Having fixed ρ and m, we can choose n_0 so large that

$$|f_n(z)-f(z)| < m \qquad (n > n_0)$$

on the circle. Since $f_n(z) = f(z)+\{f_n(z)-f(z)\}$, it follows from Rouché's theorem that, for $n > n_0$, $f_n(z)$ has the same number of zeros in the circle as $f(z)$; that is, if z_0 is a zero of $f(z)$, it has at least one, and otherwise it has none. This proves the theorem.

The example $f_n(z) = e^z/n$ shows that it is necessary to assume that $f(z)$ is not identically zero. The example in which $f_n(z) = 1-z^n/n$, and D is the unit circle, shows that the theorem does not apply to points on the boundary of D. For $f_n(z) \to 1$ uniformly in D and on the boundary, but every point of the boundary is a limit-point of zeros of the functions $f_n(z)$.

3.5. The functions $|f(z)|$, $\mathbf{R}\{f(z)\}$, $\mathbf{I}\{f(z)\}$. Let $f(z)$ be a function analytic in a given region, and let $u(x,y)$, $v(x,y)$ be its

* Hurwitz (1).

real and imaginary parts. We write

$$u_x = \frac{\partial}{\partial x} u(x,y), \quad u_y = \frac{\partial}{\partial y} u(x,y),$$

and similarly for derivatives of higher order.

We have already shown that the Cauchy-Riemann equations

$$u_x = v_y, \qquad u_y = -v_x,$$

hold at all points of the region.

Since $f''(z)$ exists, so do all the partial derivatives of u and v of the second order. Hence

$$u_{xx} = \frac{\partial}{\partial x}(v_y) = \frac{\partial}{\partial y}(v_x) = -u_{yy},$$

i.e. u satisfies the partial differential equation (Laplace's equation)

$$u_{xx} + u_{yy} = 0.$$

Similarly v satisfies the same equation.

A function which satisfies this differential equation is called a *harmonic function* or a *potential function*. The modulus $|f(z)|$ is not in general a harmonic function; but $\log|f(z)|$ is, since it is the real part of the function $\log\{f(z)\}$.

3.51. The loci $|f| = \text{const.}$, $\mathbf{R}\{f\} = \text{const.}$, $\mathbf{I}\{f\} = \text{const.}$, are curves in the z-plane.

If $|f(z)| = $ constant throughout a whole region where $f(z)$ is analytic, then $f(z) = $ constant.

For if $|f(z)| = c$, then

$$u^2 + v^2 = c^2.$$

Hence

$$uu_x + vv_x = 0,$$

$$uu_y + vv_y = 0,$$

or, by the Cauchy-Riemann equations,

$$uu_x - vu_y = 0,$$

$$uu_y + vu_x = 0.$$

Eliminating u_y we obtain

$$(u^2 + v^2)u_x = 0.$$

Hence $u_x = 0$, and similarly u_y, v_x, and v_y are zero. Hence u and v are constants, i.e. $f(z)$ is constant.

If $u = c$ or $v = c$ the proof (which we leave to the reader) that $f(z)$ is a constant is even simpler.

3.52. *The zeros of $f(z)$ are the intersections of the curves $u = 0$, $v = 0$.* This is obvious.

At a simple zero, the curves $u = 0$, $v = 0$ intersect at right angles. This follows at once from the Cauchy-Riemann equations; or it may be seen by taking the zero to be at $z = 0$, and writing

$$f(z) = ae^{i\alpha}z + O(|z|^2),$$

so that

$$u = ar\cos(\alpha+\theta) + O(r^2),$$
$$v = ar\sin(\alpha+\theta) + O(r^2).$$

Then the directions of the tangents to $u = 0$, $v = 0$, are given by $\theta = \frac{1}{2}\pi - \alpha$, $\theta = -\alpha$.

At a point where $f(z)$ is real, and $f'(z) = 0$, the curve $v = 0$ has a double point.

For at such a point $v = 0$, $v_x = 0$, $v_y = 0$, which are the conditions for a double point.

The curves $|f(z)| =$ constant are called *level curves.*

Example. Prove that, at a double zero, each of the curves $u = 0$, $v = 0$, has a double point, and the two curves intersect at an angle $\frac{1}{4}\pi$.

3.53. *A level curve has a double point, if, and only if, it passes through a zero of $f'(z)$.*

The equation of a level curve is

$$u^2 + v^2 = c^2,$$

and this has a double point if, and only if,

$$uu_x + vv_x = 0,$$
$$uu_y + vv_y = 0.$$

Both these conditions are satisfied if $f'(z) = 0$. Conversely, the second equation may be written

$$-uv_x + vu_x = 0,$$

and squaring and adding we have

$$(u_x^2 + v_x^2)(u^2 + v^2) = 0.$$

Hence $u_x = 0$ and $v_x = 0$, i.e. $f'(z) = 0$.

3.54. The level curves and the zeros of $f(z)$. *If C is a simple closed level curve, and $f(z)$ is analytic inside and on it, then $f(z)$ has at least one zero inside C or is a constant.*

Let

$$f(z) = u + iv = ce^{i\phi}$$

on C, so that c is a constant. Then

$$c = \sqrt{(u^2+v^2)}, \qquad \phi = \arctan(v/u).$$

Let s be the length of C measured from some fixed point on it. Then

$$0 = \frac{dc}{ds} = \left(u\frac{du}{ds}+v\frac{dv}{ds}\right)\frac{1}{c}, \tag{1}$$

$$\frac{d\phi}{ds} = \left(u\frac{dv}{ds}-v\frac{du}{ds}\right)\frac{1}{c^2}. \tag{2}$$

Now $d\phi/ds$ cannot vanish on C. For if it did we should have, on squaring and adding the above equations,

$$(u^2+v^2)\left\{\left(\frac{du}{ds}\right)^2+\left(\frac{dv}{ds}\right)^2\right\} = 0,$$

i.e.

$$\frac{du}{ds} = 0, \qquad \frac{dv}{ds} = 0.$$

Now

$$\frac{du}{ds} = u_x\frac{dx}{ds}+u_y\frac{dy}{ds},$$

$$\frac{dv}{ds} = v_x\frac{dx}{ds}+v_y\frac{dy}{ds} = -u_y\frac{dx}{ds}+u_x\frac{dy}{ds},$$

so that, squaring and adding,

$$(u_x^2+u_y^2)\left\{\left(\frac{dx}{ds}\right)^2+\left(\frac{dy}{ds}\right)^2\right\} = 0.$$

The last factor is 1, so that $u_x = 0$, $u_y = 0$, i.e. $f'(z) = 0$. This is impossible on a level curve without double points.

It follows that $d\phi/ds$ has the same sign at all points of C, i.e. that ϕ increases or decreases steadily round the contour. Hence its variation round the contour is not zero.

But the variation of ϕ round the contour is equal to 2π multiplied by the number of zeros inside C. Hence there is at least one such zero.

3.55. *If $f(z)$ has n zeros inside C, then $f'(z)$ has $n-1$ zeros inside C.*

Let

$$f(z) = ce^{i\phi}$$

on C. Then

$$f'(z) = cie^{i\phi}\frac{d\phi}{dz}.$$

Hence

$$\arg\{f'(z)\} = \text{const.}+\phi+\arg\frac{d\phi}{dz}.$$

Hence, if Δ_C denotes variation round C,

$$\Delta_C \arg\{f'(z)\} = \Delta_C \arg\{f(z)\} + \Delta_C \arg\frac{d\phi}{dz}.$$

Let n' be the number of zeros of $f'(z)$. Then

$$2\pi n' = 2\pi n + \Delta_C \arg\frac{d\phi}{dz}. \tag{1}$$

Now
$$\frac{d\phi}{dz} = \frac{d\phi}{ds}\frac{ds}{dz},$$

and, as we have already seen, $d\phi/ds$ is real and of constant sign on C. Hence

$$\Delta_C \arg\frac{d\phi}{dz} = \Delta_C \arg\frac{ds}{dz},$$

Also
$$\frac{dz}{ds} = \frac{dx}{ds} + i\frac{dy}{ds} = \cos\psi + i\sin\psi = e^{i\psi},$$

where ψ is the angle the tangent to C makes with the x-axis. Hence

$$\Delta_C \arg\frac{ds}{dz} = -\Delta_C\psi = -2\pi,$$

so that on dividing (1) by 2π we obtain $n' = n-1$, the required result.

3.56. The following theorem sometimes gives useful information about the zeros of a function.*

Let C be a simple closed contour, inside and on which $f(z)$ is analytic. Then if $\mathbf{R}\{f(z)\}$ vanishes at $2k$ distinct points on C, $f(z)$ has at most k zeros inside C.

If $f(z) = u+iv$, the number n of zeros of $f(z)$ inside C is given by

$$n = \frac{1}{2\pi}\Delta_C\left(\arctan\frac{v}{u}\right).$$

Starting at a point where $u \neq 0$, we may take the initial value of $\arctan(v/u)$ to lie between $-\frac{1}{2}\pi$ and $\frac{1}{2}\pi$. We can only pass out of this range, say to $(\frac{1}{2}\pi, \frac{3}{2}\pi)$, if u vanishes, and only pass on to $(\frac{3}{2}\pi, \frac{5}{2}\pi)$, if u vanishes again. Thus, if u vanishes twice on C, $\Delta_C(\arctan v/u)$ is at most equal to 2π, and n is at most equal to 1. The general result obviously follows from the same argument.

* See, for example, Backlund (1).

3.6. Poisson's integral formula. *Let $f(z)$ be analytic in a region including the circle $|z| \leqslant R$, and let $u(r,\theta)$ be its real part. Then for $0 \leqslant r < R$*

$$u(r,\theta) = \frac{1}{2\pi}\int_0^{2\pi} \frac{R^2-r^2}{R^2-2Rr\cos(\theta-\phi)+r^2} u(R,\phi)\, d\phi.$$

There is a similar formula for the imaginary part $v(r,\theta)$ of $f(z)$.

These formulae are analogous to Cauchy's formula giving the value of $f(z)$ at any point inside a contour in terms of its values on the contour. They cannot, however, be obtained merely by separating Cauchy's formula into real and imaginary parts.

We shall give two proofs.

FIRST PROOF. We can suppose without loss of generality that $f(z) = \sum a_n z^n$, where all the coefficients a_n are real. For, in the general case, $a_n = \alpha_n + i\beta_n$, and

$$f(z) = \sum \alpha_n z^n + i \sum \beta_n z^n = f_1(z) + i f_2(z),$$

so that $$\mathbf{R}(f) = \mathbf{R}(f_1) - \mathbf{I}(f_2).$$

Also, since $|\alpha_n| \leqslant |a_n|$, $|\beta_n| \leqslant |a_n|$, f_1 and f_2 are analytic for $|z| \leqslant R$. Hence the general result follows from the special case.

In the special case, if $f(re^{i\theta}) = u+iv$, then $f(re^{-i\theta}) = u-iv$.

Let z_1 be a point on the circle $|z| = R$, and let $f(z_1) = u_1+iv_1$. Then, by Cauchy's formula,

$$u+iv = \frac{1}{2\pi i}\int \frac{u_1+iv_1}{z_1-z}\, dz_1 = \frac{1}{2\pi}\int_0^{2\pi} \frac{(u_1+iv_1)Re^{i\phi}\, d\phi}{Re^{i\phi}-re^{i\theta}}.$$

Since the point R^2/z is outside the circle, we have

$$0 = \frac{1}{2\pi i}\int \frac{u_1+iv_1}{z_1-R^2/z}\, dz_1 = \frac{1}{2\pi}\int_0^{2\pi} \frac{(u_1+iv_1)Re^{i\phi}\, d\phi}{Re^{i\phi}-R^2r^{-1}e^{-i\theta}}.$$

Replacing ϕ by $-\phi$, and so iv_1 by $-iv_1$, we obtain

$$\frac{1}{2\pi}\int_0^{2\pi} \frac{(u_1-iv_1)Re^{-i\phi}\, d\phi}{Re^{-i\phi}-R^2r^{-1}e^{-i\theta}} = 0,$$

or $$\frac{1}{2\pi}\int_0^{2\pi} \frac{(u_1-iv_1)re^{i\theta}\, d\phi}{re^{i\theta}-Re^{i\phi}} = 0.$$

Subtracting this formula from the previous one, we obtain

$$u+iv=\frac{1}{2\pi}\int_0^{2\pi}\left\{u_1\frac{Re^{i\phi}+re^{i\theta}}{Re^{i\phi}-re^{i\theta}}+iv_1\right\}d\phi,$$

and now, taking real parts, the result follows.

SECOND PROOF. Let

$$f(z)=\sum_{n=0}^{\infty}(\alpha_n+i\beta_n)r^n e^{in\theta} \qquad (r\leqslant R).$$

Then, as in § 2.53,

$$\alpha_n R^n=\frac{1}{\pi}\int_0^{2\pi}u(R,\phi)\cos n\phi\,d\phi, \qquad \beta_n R^n=-\frac{1}{\pi}\int_0^{2\pi}u(R,\phi)\sin n\phi\,d\phi,$$

for $n>0$, while $\quad \alpha_0=\dfrac{1}{2\pi}\displaystyle\int_0^{2\pi}u(R,\phi)\,d\phi.$

Hence

$$\begin{aligned}u(r,\theta)&=\sum_{n=0}^{\infty}(\alpha_n\cos n\theta-\beta_n\sin n\theta)r^n\\ &=\frac{1}{2\pi}\int_0^{2\pi}u(R,\phi)\,d\phi+\\ &\qquad+\frac{1}{\pi}\sum_{n=1}^{\infty}\frac{r^n}{R^n}\int_0^{2\pi}u(R,\phi)(\cos n\theta\cos n\phi+\sin n\theta\sin n\phi)\,d\phi\\ &=\frac{1}{\pi}\int_0^{2\pi}u(R,\phi)\left\{\frac{1}{2}+\sum_{n=1}^{\infty}\cos n(\theta-\phi)\left(\frac{r}{R}\right)^n\right\}d\phi,\end{aligned}$$

the inversion being justified by uniform convergence. The result now follows on summing the series in brackets.

3.61. Jensen's theorem. *Let $f(z)$ be analytic for $|z|<R$. Suppose that $f(0)$ is not zero, and let $r_1, r_2,\ldots, r_n,\ldots$ be the moduli of the zeros of $f(z)$ in the circle $|z|<R$, arranged as a non-decreasing sequence. Then, if $r_n\leqslant r\leqslant r_{n+1}$,*

$$\log\frac{r^n|f(0)|}{r_1 r_2\ldots r_n}=\frac{1}{2\pi}\int_0^{2\pi}\log|f(re^{i\theta})|\,d\theta. \tag{1}$$

Here a zero of order p is counted p times. The interest of this formula is that it connects the modulus of the function with the moduli of the zeros.

It can be put in another form, in some ways more useful. Let $n(x)$ denote the number of zeros of $f(z)$ for $|z| \leqslant x$. Then, if $r_n \leqslant r \leqslant r_{n+1}$,

$$\log \frac{r^n}{r_1 \ldots r_n} = n \log r - \sum_{m=1}^{n} \log r_m$$

$$= \sum_{m=1}^{n-1} m(\log r_{m+1} - \log r_m) + n(\log r - \log r_n)$$

$$= \sum_{m=1}^{n-1} m \int_{r_m}^{r_{m+1}} \frac{dx}{x} + n \int_{r_n}^{r} \frac{dx}{x}.$$

Now $m = n(x)$ for $r_m \leqslant x < r_{m+1}$, $n = n(x)$ for $r_n \leqslant x < r$. Hence the right-hand side is equal to

$$\int_0^r \frac{n(x)}{x}\, dx,$$

and Jensen's formula takes the form

$$\int_0^r \frac{n(x)}{x}\, dx = \frac{1}{2\pi} \int_0^{2\pi} \log|f(re^{i\theta})|\, d\theta - \log|f(0)|. \tag{2}$$

We shall give two proofs of the theorem.

First Proof. If $f(z)$ has no zero on $|z| = r$, then

$$n(r) = \frac{1}{2\pi i} \int \frac{f'(z)}{f(z)}\, dz = \frac{1}{2\pi} \int_0^{2\pi} \frac{f'(re^{i\theta})}{f(re^{i\theta})} re^{i\theta}\, d\theta. \tag{3}$$

Jensen's formula is obtained formally by dividing by r, integrating with respect to r, and taking real parts. This process is not obviously valid, owing to the infinities of the integrand. We therefore adopt a slightly different method.

In an interval between the moduli r_n, r_{n+1}, of two zeros, each side of Jensen's formula has a continuous derivative; the derivative of the left-hand side is n/r, and that of the right-hand side is

$$\frac{1}{2\pi} \int_0^{2\pi} \frac{d}{dr}\{\log|f(re^{i\theta})|\}\, d\theta = \frac{1}{4\pi} \int_0^{2\pi} \frac{d}{dr}\{\log f(re^{i\theta}) + \log \bar{f}(re^{-i\theta})\}\, d\theta$$

$$= \frac{1}{4\pi} \int_0^{2\pi} \left\{\frac{f'(re^{i\theta})}{f(re^{i\theta})} e^{i\theta} + \frac{\bar{f}'(re^{-i\theta})}{\bar{f}(re^{i\theta})} e^{-i\theta}\right\} d\theta = \mathbf{R}\left\{\frac{1}{2\pi} \int_0^{2\pi} \frac{f'(re^{i\theta})}{f(re^{i\theta})} e^{i\theta}\, d\theta\right\},$$

which is also equal to n/r, by (3). Hence the two derivatives are equal in any such interval. Hence the two sides of Jensen's formula differ by a constant in any such interval.

Secondly, the two sides are obviously equal when $r = 0$.

Hence it is sufficient to prove that each side is continuous when r passes through a value r_n.

This is obvious in the case of the left-hand side. For the right-hand side, it will be sufficient to suppose that there is one zero of modulus r_n, and that its amplitude is zero. Then

$$\log|f(re^{i\theta})| = \log\left|1 - \frac{r}{r_n}e^{i\theta}\right| + \psi(r, \theta),$$

where ψ is continuous in the neighbourhood of $r = r_n$. Hence it is sufficient to show that the integral

$$\int_0^{2\pi} \log\left|1 - \frac{r}{r_n}e^{i\theta}\right| d\theta$$

is continuous at $r = r_n$. Now for $r/r_n < 2$

$$9 \geqslant \left|1 - \frac{r}{r_n}e^{i\theta}\right|^2 = 1 - 2\frac{r}{r_n}\cos\theta + \frac{r^2}{r_n^2}$$

$$= \sin^2\theta + \left(\cos\theta - \frac{r}{r_n}\right)^2 \geqslant \sin^2\theta.$$

Hence, if $\delta < \pi$,

$$\left|\int_{-\delta}^{\delta} \log\left|1 - \frac{r}{r_n}e^{i\theta}\right| d\theta\right| < \int_{-\delta}^{\delta} \left\{\log 3 + \left|\log|\sin\theta|\right|\right\} d\theta$$

$$< \int_{-\delta}^{\delta} \left\{A + \left|\log|\theta|\right|\right\} d\theta < A\delta\log\frac{1}{\delta}.$$

We can choose δ so that this is arbitrarily small, for all values of r in the neighbourhood of r_n. Having fixed δ, the remainder of the integral is evidently continuous. Hence the whole integral is continuous.

SECOND PROOF. We obtain the result in a number of stages.

(i) If $f(z)$ has no zeros for $|z| \leqslant r$, then $\log f(z)$ is analytic for $|z| \leqslant r$, and

$$\log f(0) = \frac{1}{2\pi i}\int_{|z|=r} \frac{\log f(z)}{z}\,dz = \frac{1}{2\pi}\int_0^{2\pi} \log\{f(re^{i\theta})\}\,d\theta,$$

and, taking real parts, we have the result.

(ii) If $a_1 = r_1 e^{i\theta_1}$, $0 < r_1 < r$, we have

$$\int\limits_{|w|=1/r} \log(1-w\bar{a}_1)\frac{dw}{w} = 0$$

by Cauchy's theorem, the logarithm having its principal value. Hence, with suitable determinations of the logarithms,

$$\frac{1}{2\pi i}\int\limits_{|w|=1/r} \log\left(1-\frac{1}{w\bar{a}_1}\right)\frac{dw}{w} = \frac{1}{2\pi i}\int\limits_{|w|=1/r} \log\left(-\frac{1}{w\bar{a}_1}\right)\frac{dw}{w}$$

$$= \log\left(-\frac{1}{\bar{a}_1}\right)-\frac{1}{4\pi i}[\log^2 w]_{\arg w=0}^{\arg w=2\pi}$$

$$= \log\left(-\frac{1}{\bar{a}_1}\right)-\frac{1}{4\pi i}\left(\log\frac{1}{r}+2\pi i\right)^2+\frac{1}{4\pi i}\log^2\frac{1}{r}.$$

Taking real parts,

$$\frac{1}{2\pi}\int\limits_0^{2\pi} \log\left|1-\frac{r}{r_1}e^{i(\theta_1-\theta)}\right| d\theta = \log\frac{r}{r_1}.$$

This is Jensen's formula for

$$f(z) = 1-\frac{z}{a_1}.$$

(iii) The above result may be extended to the case $r = r_1$ by applying Cauchy's theorem to the circle $|w| = 1/r$ with a small circular indentation so that the point $w = 1/\bar{a}_1$ is excluded. The integral round the indentation tends to 0 with the radius, and the proof concludes as before.

(iv) In the general case

$$f(z) = \left(1-\frac{z}{a_1}\right)\left(1-\frac{z}{a_2}\right)\dots\left(1-\frac{z}{a_n}\right)\phi(z),$$

where $\phi(z)$ is not zero for $|z| < r_{n+1}$, and $\phi(0) = f(0)$. The general result then follows by addition of the previous ones.

The theorem may be extended at once to a function which has poles as well as zeros. Let $f(z)$ satisfy the same conditions as before, but now let it have zeros $a_1,\ldots, a_m$ and poles $b_1,\ldots, b_n$ with moduli not exceeding r. Then

$$\log\left\{\left|\frac{b_1\ldots b_n}{a_1\ldots a_m}f(0)\right|r^{m-n}\right\} = \frac{1}{2\pi}\int_0^{2\pi}\log|f(re^{i\theta})|\,d\theta. \tag{4}$$

For if $$f(z) = g(z)\Big/\left(1-\frac{z}{b_1}\right)\ldots\left(1-\frac{z}{b_n}\right) = g(z)/h(z),$$

we have $$\log\frac{r^m g(0)}{|a_1\ldots a_m|} = \frac{1}{2\pi}\int_0^{2\pi}\log|g(re^{i\theta})|\,d\theta$$

and $$\log\frac{r^n}{|b_1\ldots b_n|} = \frac{1}{2\pi}\int_0^{2\pi}\log|h(re^{i\theta})|\,d\theta.$$

The result therefore follows on subtracting.

3.62. The Poisson-Jensen formula. *Let $f(z)$ have zeros at the points $a_1, a_2,\ldots, a_m$, and poles at $b_1, b_2,\ldots, b_n$, inside the circle $|z| \leqslant R$, and be analytic elsewhere inside and on the circle. Then*

$$\log|f(re^{i\theta})| = \frac{1}{2\pi}\int_0^{2\pi}\frac{R^2-r^2}{R^2-2Rr\cos(\theta-\phi)+r^2}\log|f(Re^{i\phi})|\,d\phi -$$
$$-\sum_{\mu=1}^{m}\log\left|\frac{R^2-\bar{a}_\mu re^{i\theta}}{R(re^{i\theta}-a_\mu)}\right| + \sum_{\nu=1}^{n}\log\left|\frac{R^2-\bar{b}_\nu re^{i\theta}}{R(re^{i\theta}-b_\nu)}\right|.$$

This contains both Poisson's formula and Jensen's formula as particular cases. If there are no zeros or poles, it reduces to Poisson's formula for the real part of the function $\log f(z)$. On the other hand, if $r = 0$, we obtain the general Jensen formula

$$\log|f(0)| = \frac{1}{2\pi}\int_0^{2\pi}\log|f(Re^{i\phi})|\,d\phi - \log\left\{\left|\frac{b_1 b_2\ldots b_n}{a_1 a_2\ldots a_m}\right|R^{m-n}\right\}.$$

(i) Let $f(z) = z-a$, where $|a| < R$. Then we have to prove that

$$\log|re^{i\theta}-a| = \frac{1}{2\pi}\int_0^{2\pi}\frac{R^2-r^2}{R^2-2Rr\cos(\theta-\phi)+r^2}\log|Re^{i\phi}-a|\,d\phi -$$
$$-\log\left|\frac{R^2-\bar{a}re^{i\theta}}{R(re^{i\theta}-a)}\right|,$$

or

$$\log\left|R-\frac{\bar{a}re^{i\theta}}{R}\right|=\frac{1}{2\pi}\int_0^{2\pi}\frac{R^2-r^2}{R^2-2Rr\cos(\theta-\phi)+r^2}\log|Re^{i\phi}-a|\,d\phi.$$

But this is equivalent to Poisson's formula for the real part of the function

$$\log\left(R-\frac{\bar{a}z}{R}\right),$$

which is analytic for $|z|\leqslant R$.

(ii) Similarly, if $f(z)=1/(z-b)$, the formula is equivalent to Poisson's formula for the real part of

$$\log\left(R-\frac{\bar{b}z}{R}\right).$$

(iii) If $f(z)$ is analytic and has no zeros or poles in $|z|\leqslant R$, the formula is Poisson's formula for the real part of $\log f(z)$.

The general case can now be obtained by addition of these particular cases.

3.7. In all the above theorems the region considered is a circle. We shall conclude the chapter by proving two theorems of the same general type as Jensen's theorem, but applying to a half-plane and a rectangle respectively.

Carleman's theorem.* *Let $f(z)$ be analytic for $|z|\geqslant\rho$, $-\frac{1}{2}\pi\leqslant\arg z\leqslant\frac{1}{2}\pi$, and suppose that it has the zeros $r_1e^{i\theta_1}$, $r_2e^{i\theta_2}$,..., $r_ne^{i\theta_n}$ inside the contour consisting of the semicircles $|z|=\rho$, $|z|=R$, $-\frac{1}{2}\pi\leqslant\arg z\leqslant\frac{1}{2}\pi$, and the parts of the imaginary axis joining them, and that it has no zeros on the contour. Then*

$$\sum_{\nu=1}^{n}\left(\frac{1}{r_\nu}-\frac{r_\nu}{R^2}\right)\cos\theta_\nu=\frac{1}{\pi R}\int_{-\frac{1}{2}\pi}^{\frac{1}{2}\pi}\log|f(Re^{i\theta})|\cos\theta\,d\theta+$$

$$+\frac{1}{2\pi}\int_\rho^R\left(\frac{1}{y^2}-\frac{1}{R^2}\right)\log|f(iy)f(-iy)|\,dy+O(1),$$

where $O(1)$ denotes a function of ρ and R which, for fixed ρ, is bounded as $R\to\infty$.

Consider the integral

$$I=\frac{1}{2\pi i}\int\log f(z)\left(\frac{1}{z^2}+\frac{1}{R^2}\right)dz$$

* Carleman (1).

taken round the contour in the positive direction, starting from the point $z = i\rho$ with a fixed determination of the logarithm. The integral along the small semicircle is bounded. On the negative imaginary axis $z = -iy$, and we obtain

$$\frac{1}{2\pi}\int_{\rho}^{R} \log\{f(-iy)\}\left(\frac{1}{y^2}-\frac{1}{R^2}\right) dy.$$

On the large semicircle, $z = Re^{i\theta}$, and we obtain

$$\frac{1}{2\pi i}\int_{-\frac{1}{2}\pi}^{\frac{1}{2}\pi} \log\{f(Re^{i\theta})\}\left(\frac{e^{-2i\theta}}{R^2}+\frac{1}{R^2}\right)iRe^{i\theta}\, d\theta$$

$$= \frac{1}{\pi R}\int_{-\frac{1}{2}\pi}^{\frac{1}{2}\pi} \log\{f(Re^{i\theta})\}\cos\theta\, d\theta.$$

The integral along the positive imaginary axis gives

$$\frac{1}{2\pi}\int_{\rho}^{R} \log\{f(iy)\}\left(\frac{1}{y^2}-\frac{1}{R^2}\right) dy,$$

and, taking the real part of I, we obtain the right-hand side of Carleman's formula.

Again, integrating by parts, we have

$$I = \frac{1}{2\pi i}\left[\log f(z)\left(\frac{z}{R^2}-\frac{1}{z}\right)\right] + \frac{1}{2\pi i}\int \frac{f'(z)}{f(z)}\left(\frac{1}{z}-\frac{z}{R^2}\right) dz.$$

As we describe the contour, $\log f(z)$ increases by $2\pi i n$. The integrated term is therefore purely imaginary. By the theorem of residues, the last integral is equal to

$$\sum_{\nu=1}^{n}\left(\frac{1}{r_\nu e^{i\theta_\nu}}-\frac{r_\nu e^{i\theta_\nu}}{R^2}\right),$$

and, taking real parts, the theorem follows.

The result is easily extended to the case where $f(z)$ has zeros on the imaginary axis; we make small indentations round these zeros, and proceed to the limit.

3.71. *Let $f(z)$ be analytic and bounded for $x \geqslant 0$, and let its zeros in the right half-plane be $r_1 e^{i\theta_1}$, $r_2 e^{i\theta_2}$,.... Then the series*

$$\sum_{n=1}^{\infty} \frac{\cos\theta_n}{r_n}$$

is convergent.

Under the conditions stated, the right-hand side of Carleman's formula is bounded above, say $< M$. Hence

$$\sum_{\nu=1}^{n}\left(\frac{1}{r_\nu}-\frac{r_\nu}{R^2}\right)\cos\theta_\nu < M$$

for all values of R. Every term on the left is positive, and, if $r_\nu < \frac{1}{2}R$,

$$\frac{1}{r_\nu}-\frac{r_\nu}{R^2} > \frac{3}{4r_\nu}.$$

Hence

$$\sum_{r_\nu<\frac{1}{2}R}\frac{\cos\theta_\nu}{r_\nu} < \frac{4M}{3},$$

and the result follows.

It is easily seen that the theorem remains true if, instead of $f(z) = O(1)$, we suppose that $f(z) = O(e^{|z|^\alpha})$, where $\alpha < 1$. But if $\alpha = 1$ the theorem fails, as the example $f(z) = \cos z$ shows.

3.72. The above theorem may be used to prove the following result.

Let $f(z)$ be analytic for $x \geqslant 0$, and of the form $O(e^{-a|z|})$ as $z \to \infty$, where $a > 0$, uniformly for $|\arg z| \leqslant \frac{1}{2}\pi$. Then $f(z) = 0$ identically.

For consider the function $F(z) = f(z)\sin bz$, where $0 < b < a$. Here $F(z)$ is analytic and bounded for $x \geqslant 0$; it has zeros at the points $z = n\pi/b$, and $\sum b/n\pi$ is divergent. This is inconsistent with the result of the previous section, unless $F(z)$ is identically zero. Hence $F(z) = 0$, and so $f(z) = 0$.

A more complete form of this result will be obtained in § 5.8.

3.8. A theorem of Littlewood.* Let C denote the rectangle bounded by the lines $x = x_1$, $x = x_2$, $y = y_1$, $y = y_2$, where $x_1 < x_2$, $y_1 < y_2$. Let $f(z)$ be analytic and not zero on C, and meromorphic inside it. Let $F(z) = \log f(z)$, the logarithm being defined as follows: we start with a particular determination on $x = x_2$, and obtain the value at other points by continuous variation along $y =$ constant from $\log(x_2+iy)$. If, however, this path would cross a zero or pole of $f(z)$, we take $F(z)$ to be $F(z\pm i0)$ according as we approach the path from above or below.

* Littlewood (4).

Let $\nu(x')$ denote the excess of the number of zeros of $f(z)$ over the number of poles in the part of the rectangle where $x > x'$. Then

$$\int_C F(z)\,dz = -2\pi i \int_{x_1}^{x_2} \nu(x)\,dx.$$

Consider first the function $f(z) = z-a$, where $a = \alpha+i\beta$ is a point of the rectangle. Let C' be the contour obtained by describing C in the positive direction from (x_2, y_1) as far as (x_1, β), then the straight line $y = \beta$ as far as $\alpha-\epsilon+i\beta$, then a circle of radius ϵ about $z = a$ described in the negative direction, and then returning along $y = \beta$ and the rest of C to the starting-point. Then $F(z)$ is analytic in C', so that

$$\int_{C'} F(z)\,dz = 0.$$

The integral round the small circle tends to zero with the radius, and it follows that

$$\int_C F(z)\,dz = -\int_{x_1}^{\alpha} \{F_1(z)-F_2(z)\}\,dx,$$

where F_1 and F_2 are the values of F on the two paths joining $x_1+i\beta$ to $\alpha+i\beta$. Since we obtain F_2 from F_1 by passing in the negative direction round a simple zero of $f(z)$ at $z = a$, we have

$$F_2(z) = F_1(z)-2\pi i.$$

Hence

$$\int_C F(z)\,dz = -2\pi i \int_{x_1}^{\alpha} dx = -2\pi i \int_{x_1}^{x_2} \nu(x)\,dx,$$

where $\nu(x) = 1$ $(x_1 < x < \alpha)$, 0 $(\alpha < x < x_2)$, i.e. $\nu(x)$ is the ν-function for the case considered.

The general theorem now easily follows by addition of terms corresponding to the various poles and zeros of $f(z)$.

MISCELLANEOUS EXAMPLES

1. Evaluate the integrals

$$\int_0^\infty \frac{dx}{x^4+1}, \qquad \int_0^\infty \frac{x^2\,dx}{x^4+1}, \qquad \int_0^\infty \frac{dx}{x^6+1}$$

by contour integration.

2. Evaluate the integrals

$$\int_0^\infty \frac{\cos x}{a^2+x^2}\,dx, \qquad \int_0^\infty \frac{\cos x}{(a^2+x^2)^2}\,dx, \qquad \int_0^\infty \frac{\sin^2 x}{x^2}\,dx, \qquad \int_0^\infty \frac{\sin^3 x}{x^3}\,dx$$

by contour integration.

3. Prove that, if $c > 0$,

$$\frac{1}{2\pi i}\int_{c-i\infty}^{c+i\infty} \frac{a^z}{z^2}\,dz = \begin{matrix} \log a & (a > 1), \\ 0 & (0 < a < 1). \end{matrix}$$

4. Prove that the integral

$$\int \frac{dz}{\sqrt{(4z^2+4z+3)}}$$

taken round the unit circle, starting with the positive value of the square root at $z = 1$, is equal to $i\pi$.

5. By integrating $\log^2 z/(1+z^2)$ round the usual semi-circular contour, prove that

$$\int_0^\infty \frac{\log^2 z}{1+z^2}\,dz = \frac{\pi^3}{8}.$$

6. By evaluating the integral

$$\frac{1}{2\pi i}\int \frac{dz}{(z-a)(z-1/a)}$$

round the unit circle, prove that, if $0 < a < 1$,

$$\int_0^{2\pi} \frac{d\theta}{1+a^2-2a\cos\theta} = \frac{2\pi}{1-a^2}.$$

What is the value of the integral if $a > 1$?

7. Prove that, if $b > a > -1$,

$$\int_0^{\frac{1}{2}\pi} \cos^a\theta \cos b\theta\, d\theta = \frac{\pi\Gamma(a+1)}{2^{a+1}\Gamma(\frac{1}{2}a+\frac{1}{2}b+1)\Gamma(\frac{1}{2}a-\frac{1}{2}b+1)}.$$

[Take the integral $\int (z+1/z)^a z^{b-1}\,dz$ round the right-hand half of the unit circle.]

8. By integrating

$$\int \frac{z\,dz}{a-e^{-iz}}$$

round the rectangle with corners at $-\pi$, π, $\pi+in$, $-\pi+in$, and making $n \to \infty$, show that

$$\int_0^\pi \frac{x\sin x\,dx}{1+a^2-2a\cos x} = \frac{\pi}{a}\log(1+a) \quad (0 < a < 1), \qquad \frac{\pi}{a}\log\frac{1+a}{a} \quad (a > 1).$$

[Lindelöf, *Calcul des Résidus*, pp. 48–9.]

9. Show that the function $f(x) = \operatorname{sech}\{x\sqrt{(\frac{1}{2}\pi)}\}$ satisfies the equation

$$f(t) = \sqrt{\left(\frac{2}{\pi}\right)} \int_0^\infty f(x)\cos xt\, dx.$$

[Take the integral $\int \cos tz \operatorname{sech} az\, dz$ round the rectangle with corners at $\pm n$, $\pm n + i\pi/a$, and make $n \to \infty$.]

10. Show that the function

$$f(x) = \frac{1}{e^{x\sqrt{(2\pi)}}-1} - \frac{1}{x\sqrt{(2\pi)}}$$

satisfies the equation

$$f(t) = \sqrt{\left(\frac{2}{\pi}\right)} \int_0^\infty f(x)\sin xt\, dx.$$

[Take the integral $$\int \frac{\sin zt}{e^{az}-1}\, dz,$$

where $a > 0$, round the rectangle with corners at 0, n, $n+2i\pi/a$, $2i\pi/a$, and make $n \to \infty$.]

11. Prove that, if $0 < a < 1$ and $0 < c < 1$,

$$\frac{1}{2\pi i} \int_{c-i\infty}^{c+i\infty} \frac{dz}{a^z \sin \pi z} = \frac{1}{\pi(1+a)}.$$

12. Prove that if $a > 0$, $-\frac{1}{2}\pi < a\lambda < \frac{1}{2}\pi$,

$$\int_0^\infty e^{-r^a \cos a\lambda} \frac{\cos}{\sin}(r^a \sin a\lambda)\, dr = \frac{\cos}{\sin}(\lambda) \frac{1}{a} \Gamma\left(\frac{1}{a}\right).$$

13. Sum the series

$$\sum_{n=1}^\infty \frac{1}{n^4+a^4}, \qquad \sum_{n=1}^\infty \frac{n^2}{n^4+a^4}.$$

14. Prove that if $-\pi < a < \pi$, and x is not an integer,

$$\sum_{n=1}^\infty (-1)^n \frac{n \sin na}{x^2-n^2} = \tfrac{1}{2}\pi \frac{\sin ax}{\sin \pi x}.$$

15. Prove that*

$$\frac{\coth \pi}{1^7} + \frac{\coth 2\pi}{2^7} + \frac{\coth 3\pi}{3^7} + \ldots = \frac{19\pi^7}{56700}.$$

16. Prove that
$$\sum_{n=1}^\infty \frac{1}{n^3 \sin n\pi\sqrt{2}} = -\frac{13\pi^3}{360\sqrt{2}}.$$

[Hardy. Consider the integral

* Ramanujan; see Watson (1).

$$\frac{1}{2\pi i}\int \frac{1}{\sin \pi z \sin \theta\pi z}\frac{dz}{z^3}$$

where $\theta = \surd 2 - 1$. The series is convergent; for if m is the nearest integer to $n\surd 2$,

$$|n\surd 2 - m| = \frac{|2n^2 - m^2|}{n\surd 2 + m} \geqslant \frac{1}{n\surd 2 + m} > \frac{A}{n},$$

and hence $\operatorname{cosec} n\pi\surd 2 = O(n)$.]

17. Show that if $f(z) = \sum_{n=0}^{\infty} a_n/z^{n+1}$ $(|z| > 0)$, C is a closed contour including the origin, and $\phi(z)$ is regular in a sufficiently wide region, then

$$\frac{1}{2\pi i}\int_C f(w)\phi(z-w)\,dw = a_0\phi(z) - a_1\phi'(z) + \frac{a_2}{2!}\phi''(z) - \dots .$$

18. Prove that

$$e^{az} - e^{bz} = (a-b)ze^{\frac{1}{2}(a+b)z}\prod_{n=1}^{\infty}\left\{1 + \frac{(a-b)^2z^2}{4n^2\pi^2}\right\}.$$

19. Show that, however small ρ is, all the zeros of the function

$$1 + \frac{1}{z} + \frac{1}{2!\,z^2} + \dots + \frac{1}{n!\,z^n}$$

lie in the circle $|z| \leqslant \rho$, if n is sufficiently large.

20. If $a > e$, the equation $e^z = az^n$ has n roots inside the unit circle. [Take $f(z) = az^n$, $g(z) = e^z$, in Rouché's theorem.]

21. Show that, if α and β are real, the equation

$$z^{2n} + \alpha^2 z^{2n-1} + \beta^2 = 0$$

has $n-1$ roots with positive real parts if n is odd, and n roots with positive real parts if n is even.

22. Prove that, if α is not an even integer,

$$\int_0^{\infty} e^{-x^{\alpha}}\cos xt\,dx \sim \frac{\Gamma(\alpha+1)\sin\frac{1}{2}\pi\alpha}{t^{\alpha+1}}$$

as $t \to \infty$ through real values.*

23. If $f(z) = u + iv$ is an analytic function of $z = x + iy$, and ψ is any function of x and y with differential coefficients of the first two orders, then

$$\left(\frac{\partial\psi}{\partial x}\right)^2 + \left(\frac{\partial\psi}{\partial y}\right)^2 = \left\{\left(\frac{\partial\psi}{\partial u}\right)^2 + \left(\frac{\partial\psi}{\partial v}\right)^2\right\}|f'(z)|^2,$$

and

$$\frac{\partial^2\psi}{\partial x^2} + \frac{\partial^2\psi}{\partial y^2} = \left(\frac{\partial^2\psi}{\partial u^2} + \frac{\partial^2\psi}{\partial v^2}\right)|f'(z)|^2.$$

[See Hardy (8), p. 270.]

* Pólya (1).

24. If $f(z) = u+iv$ is an analytic function of $z = x+iy$, show that

$$\left(\frac{\partial^2}{\partial x^2}+\frac{\partial^2}{\partial y^2}\right)|f(z)|^p = p^2|f(z)|^{p-2}|f'(z)|^2,$$

and that
$$\left(\frac{\partial^2}{\partial x^2}+\frac{\partial^2}{\partial y^2}\right)|u|^p = p(p-1)|u|^{p-2}|f'(z)|^2.$$

25. Let $\phi(t)$ be a real integrable function in the interval (a, b), and let

$$f(z) = \int_a^b e^{zt}\phi(t)\,dt$$

have zeros at the points $r_1e^{i\theta_1}$, $r_2e^{i\theta_2}$,.... Then the series

$$\sum \frac{\cos\theta_n}{r_n}$$

is absolutely convergent.

[The function $e^{-bz}f(z)$ is bounded for $x \geqslant 0$, and $e^{az}f(z)$ is bounded for $x \leqslant 0$.]

CHAPTER IV

ANALYTIC CONTINUATION

4.1. General theory. It is natural to think of the aggregate of all values of z^2, say, or $\log z$, for all values of z, as a single entity, and each such aggregate we describe as *an analytic function*. We have, however, not yet encountered the general idea of an analytic function as a whole. What we have always been concerned with is the idea of a function associated with a region, and defined in that region by a formula. Thus

$$1+z+z^2+\ldots \qquad (|z|<1) \tag{1}$$

and

$$\int_0^\infty e^{-t(1-z)}\,dt \qquad (\mathbf{R}(z)<1) \tag{2}$$

appear as different functions, whose values happen to be the same for certain values of z. But it is obviously more natural to regard (1) as a part of (2), and (2) as part of the function defined for all values of z other than 1 as $1/(1-z)$.

This particular function is one-valued, i.e. has just one value for each value of z (except $z = 1$). But it is also natural to regard the two values of $\sqrt{z}$ as parts of the same function, and our definition must include cases of this kind also.

To connect these new ideas with our previous theory, we require a process by which we can extend the definition of a function beyond a limited region in which it is originally defined. This process is called analytic continuation. It is characteristic of analytic functions of a complex variable, and has no counterpart in the theory of functions of a real variable.

4.11. Analytic continuation. Suppose that $f_1(z)$ and $f_2(z)$ are functions analytic in regions D_1 and D_2 respectively, and that D_1 and D_2 have a common part, throughout which $f_1(z)=f_2(z)$. Then we consider the aggregate of values of $f_1(z)$ and $f_2(z)$ at points interior to D_1 or D_2 as a single analytic function $f(z)$. Thus $f(z)$ is analytic in $D=D_1+D_2$, and $f(z)=f_1(z)$ in D_1, $f(z)=f_2(z)$ in D_2.

The function $f_2(z)$ may be considered as extending the domain in which $f_1(z)$ is defined, and it is called an analytic continuation of $f_1(z)$. Of course in the same way $f_1(z)$ is an analytic continuation of $f_2(z)$. This process of extending the definition of a given function is known as analytic continuation.

For the process to have any value it is necessary that it should, under suitable conditions, give a unique result, and we shall show that this is so. Before giving the proof it may be interesting to note the difficulties which we encounter if we try to define a similar process for functions of a real variable.

It would be natural to suggest that if, say, $f(x) = \frac{1}{2}(\pi - x)$ for $0 < x < \pi$, then we should extend the definition of $f(x)$ to other values of x by using the same formula. The difficulty is that two formulae may represent the same function in one interval, but different functions in another interval, and there may be no obvious way of deciding which is the 'proper' formula. For example, the above function is also represented by the series

$$\frac{\sin x}{1} + \frac{\sin 2x}{2} + \dots$$

for $0 < x < \pi$; but if we define the function as the sum of this series, we find that its value in the interval $(-\pi, 0)$ is $-\frac{1}{2}(\pi + x)$.

This series is not uniformly convergent, but even if we restrict ourselves to uniformly convergent series, the same sort of thing happens. For example, the series

$$\frac{x \sin x}{1} + \frac{x \sin 2x}{2} + \dots$$

is uniformly convergent in an interval including $x = 0$; yet if we use it to continue its sum from positive to negative values of x, we obtain the undesirable conclusion that the continuation of $\frac{1}{2}x(\pi - x)$ is $-\frac{1}{2}x(\pi + x)$.

4.12. Uniqueness of analytic continuation. Suppose that we have a region D, overlapped by regions D_1 and D_2, which have a common part D_3, itself overlapping D. Let $f(z)$ be analytic in D, and let $f_1(z)$ be a continuation of $f(z)$ to D_1, and $f_2(z)$ a continuation of $f(z)$ to D_2. Then either of these functions provides a continuation of $f(z)$ to D_3. To show that the results of the two processes of continuation are the same, we have to show that $f_1(z) = f_2(z)$ throughout D_3. This follows from the theorem of § 2.6, which itself depends on the fact that an analytic function can be expressed as a power series. The function $f_1(z) - f_2(z)$ is analytic throughout D_3; it is zero in the part of D_3 which overlaps D, since there $f_1(z)=f_2(z)=f(z)$. Hence it is zero throughout D_3.

The proof depends on the existence of a region common to D and D_3, and if there is no such region, the result no longer necessarily holds. We may now have $f_1(z)=f(z)$ in DD_1, $f_2(z)=f(z)$ in DD_2, but $f_1(z)\neq f_2(z)$ in D_3. This does not contradict the principle of uniqueness, since it only applies to regions throughout which the function is analytic; and now D, D_1, and D_2 may surround, without including, a point where the function is not analytic.*

4.13. In the second case considered above, where $f_1(z)\neq f_2(z)$ in D_3, we still consider the aggregate of values of $f_1(z)$ and $f_2(z)$ as a single analytic function of z, but now the function is not one-valued, and in fact is at least two-valued in D_3. In the same way, different methods of continuation may lead to many different results, and the function is then many-valued.

The reader of Hardy's *Pure Mathematics* is already familiar with the different values taken by the function $\log z$ (though, of course, there even the idea of a function analytic at a point does not appear). The properties of some other many-valued functions, such as $z^a = e^{a\log z}$, may be derived from those of $\log z$.

4.14. Definition of an analytic function as a whole. An analytic function is usually defined originally in some restricted region of the plane. The principle of continuation enables us to define *an analytic function*, without reference to any particular region in which it is defined. It consists of the original function, and all continuations thereof, and all continuations of these continuations, and so on. In this way we may succeed in defining the function $f(z)$ for all values of z, or everywhere except at certain special points; or only in some restricted region of the plane beyond which we are unable to pass. In the last case the region is referred to as the *region of existence* of the function, and its boundary as a *natural boundary* of the function. In the case of many-valued functions we shall obtain many values of the function for some or all values of z.

* This case may be illustrated by a figure in which the regions D, D_1, and D_2 are circles with centres at the vertices of an equilateral triangle, and each radius just exceeds half the side of the triangle. The function may not be analytic at the centre of the triangle.

The definition depends *prima facie* on the particular definition of the function from which we start. Since, however, the relation between two functions which are continuations of each other is reciprocal, all these processes may be reversed; and it will appear from the general theory that the definition is really independent of any particular starting-point.

4.15. The standard method of continuation. The standard method of continuation is the method of power series. Suppose that we start with the series

$$f(z) = \sum_{n=0}^{\infty} a_n(z-a)^n$$

convergent in a circle $|z-a| < R$. Taking any point b in this circle other than a, we calculate the value of the function $f(b)$ and the derivatives $f'(b)$, $f''(b)$,..., and so obtain the expansion of the function in powers of $z-b$. This series will certainly converge in any circle, centre b, which lies in the original circle, and it may converge in a larger circle, and so provide an analytic continuation of the function. So the whole function may be constructed by means of power series. Each of the power series, or, what comes to the same thing, each set of values $f(a)$, $f'(a)$, $f''(a)$,..., is called an *element* of the function.

The adoption of this particular method as a standard is justified by the following theorem: *All values of the function obtained by any method of continuation can also be obtained by means of power series.*

Let C be a contour joining two points $z = a$ and $z = b$, along which we have continued the function $f(z)$ by any means; that is, we have a sequence of formulae which define $f(z)$ in a sequence of regions D_n, such that (i) every point of C is an interior point of one or more D_n's, and (ii) consecutive D_n's overlap, and the different definitions of $f(z)$ agree in the common parts.

We now attempt to carry out the same process by means of power series; i.e. we try to find a sequence of points z_1, z_2,... on C such that the circle of convergence about each of them includes the next, such that the values found from the power series are the same as those found in the other way, and such that we reach b in this way in a finite number of steps.

With each point z on C is associated a positive radius of convergence ρ, and ρ *is a continuous function of* z. For take two neighbouring points z, $z+h$, and let ρ and ρ' be the corresponding radii of convergence. Let $|h| < \rho$. Since $f(z)$ is regular in the circle with centre $z+h$ and radius $\rho-|h|$, it follows from the Cauchy-Taylor theorem that

$$\rho' \geqslant \rho - |h|. \tag{1}$$

If $|h| < \rho'$ we can use the same argument with z and $z+h$ interchanged, so that

$$\rho \geqslant \rho' - |h|,$$

i.e.

$$\rho' \leqslant \rho + |h|. \tag{2}$$

Since the alternative to $|h| < \rho'$ is $\rho' \leqslant |h|$, (2) holds in any case. But (1) and (2) together show that $\rho' \to \rho$ as $h \to 0$, which is what is required.

Since ρ is continuous it attains its lower bound, and so, since it is always positive, its lower bound is positive. Let the lower bound be δ.

We now start at $z = a$ with a power series. Let z_1 be the point at distance $\frac{1}{2}\delta$ along the contour. It lies inside the circle of convergence about a, so that we can expand in powers of $z-z_1$. The new radius of convergence is at least δ, so that we can go on to the point z_2 distant δ from a along the curve. Proceeding in this way we plainly reach $z = b$ in a finite number of steps. The fact that we obtain the same value at b in this way as in the other way follows from the general uniqueness theorem.

4.16. Branches of a many-valued function. We have defined an analytic function as the aggregate of all values which can be obtained by continuation from any element of the function. In general the function will be many-valued, i.e. starting from z_0, say, we can, by taking suitable paths, arrive at z_1 with many different values of $f(z_1)$. We may, however, make this impossible by restricting ourselves to the interior of some particular region. We then say that there is a *branch* of the function in this region. Consider, for example, the function $\sqrt{z}$. The system of values defined by $\sqrt{r}e^{\frac{1}{2}i\theta}$ $(-\pi < \theta < \pi)$ is a branch in the plane cut along the negative real axis from the origin to infinity; and the system $-\sqrt{r}e^{\frac{1}{2}i\theta}$ is another branch in the same region. Similarly the function $\log z$ has in this region an infinity of branches defined by

$$\log r+i(\theta+2n\pi) \qquad (-\pi<\theta<\pi),$$

every integer n giving a branch.

It should be understood that there is no unique way of dividing up a function into branches; for example, we might, in the above cases, cut the plane along any other line from the origin to infinity. But, however we do it, we obtain a definite number of branches, e.g. $\sqrt{z}$ has two. The question of the number of branches will be considered again later.

4.2. Singularities of an analytic function. The only singularities which we have so far defined are isolated singularities of functions analytic and one-valued in a given region, or limit-points of such singularities. These were classified as poles and essential singularities. This classification now proves to be inadequate.

We shall now say that a one-valued analytic function is *regular* at any point which is interior to one of the circles used in continuation from the original element; and that it is singular at any limit-point of regular points which is not a regular point. A point where the function is singular is called a singular point or singularity. This definition includes the poles and essential singularities which we considered before; but there may also be singularities which are not isolated. In § 4.7 we shall construct functions for which every point of the unit circle is singular. A point of this kind is usually called an essential singularity also.

The expression 'regular', as we have used it here, means more than 'analytic'. A function may be analytic at a point, in accordance with the definition of § 2.14, without being regular there; for example, let $f(z)=e^{-1/z}$ for $-\frac{1}{4}\pi\leqslant\arg z\leqslant\frac{1}{4}\pi, |z|>0$, and let $f(z)=0$ elsewhere. It is easily seen that this function is analytic at $z=0$, and $f'(0)=0$. Consider, however, the contour consisting of the triangle with vertices at 0 and $1\pm\frac{1}{2}i$. The function is analytic everywhere inside and on the contour, but it is evidently not regular at $z=0$. The distinction is, however, not very important, since it has to be made only for somewhat artificial functions like the one considered.

In the theory of many-valued functions we have another kind of singularity, known as a *branch-point*. Suppose that, on continuing the function $f(z)$ round any sufficiently small circle with centre z_0, we return to the starting-point with a value of the function different from the one with which we started. Then z_0 is said to be a branch-point of $f(z)$. For example, if we continue $\sqrt{z} = \sqrt{r}e^{\frac{1}{2}i\theta}$ round the circle of centre 0 and radius r from $\theta = 0$ to $\theta = 2\pi$, the value of the function changes from $\sqrt{r}$ to $-\sqrt{r}$. Hence $z = 0$ is a branch-point of $\sqrt{z}$. Similarly it is a branch-point of $1/\sqrt{z}$ and $\log z$.

Notice that a branch-point is not necessarily an 'infinity' of the function.

A branch of a many-valued function may, of course, have poles and essential singularities; and a point may be a singularity for one branch of the function but not for another. For example, the point $z = 1$ is a pole of the branch of $1/\log z$ corresponding to the value of $\log z$ which is zero at $z = 1$, but not for any other value of the logarithm. A general definition of regular and singular points is not quite so simple for many-valued functions as for one-valued functions, and it is usually sufficient to consider particular branches separately. We define a regular point of a branch in the same way as for a one-valued function; but a singularity such as a branch-point cannot be assigned to one particular branch.

Examples. (i) The function z^a, defined as $e^{a\log z}$, has an infinity of values unless a is real and rational, when it has a finite number of values.

(ii) The function $z^{\frac{1}{2}}(1-z)^{\frac{1}{3}}$ has six values.

(iii) One branch of the function

$$\frac{1}{z}\log\frac{1}{1-z}$$

is given for $|z| < 1$ by the series

$$\frac{1}{z}(z+\tfrac{1}{2}z^2+\tfrac{1}{3}z^3+\dots),$$

and $z = 0$ is a regular point for this branch; but it is a pole for every other branch.

(iv) The function $\{\log 1/(1-z)\}^{\frac{1}{2}}$ has singularities at $z = 0$ and $z = 1$; $z = 0$ is a branch-point for one determination of the logarithm.

(v) Consider the singularities of the function $\log\log z$.

4.21. *If the radius of convergence of the series*

$$f(z) = \sum_{n=0}^{\infty} a_n z^n$$

is finite, $f(z)$ has at least one singularity on the circle of convergence.

Let C be the circle of convergence, of radius R, and C' a concentric circle of radius $R' < R$. Let ρ be the radius of convergence of the power series about a point z on C'. As in § 4.15, ρ is a continuous function of z. Also $\rho \geqslant R-R'$. Let δ be the lower bound of ρ for z on C'. Then $\delta \geqslant R-R'$.

If $\delta > R-R'$, the circles of convergence about points on C' together cover the region $|z| < R'+\delta$, and so $f(z)$ is regular in this larger circle. Hence, by the Cauchy-Taylor theorem, the radius of convergence of $\sum a_n z^n$ is greater than R, contrary to hypothesis.

It follows that $\delta = R-R'$. Since a continuous function attains its lower bound, there is a point $R'e^{i\alpha}$, say, on C', at which $\rho = R-R'$. Then $Re^{i\alpha}$ is a singularity of $f(z)$. For if it were a regular point, $f(z)$ would be regular in a circle with centre $z = Re^{i\alpha}$, and then the radius of convergence about $R'e^{i\alpha}$ would be greater than $R-R'$.

Since we have established the existence of a singularity on the circle of convergence, we may speak of it as the singularity nearest to the origin, or one of the nearest. We may then say that the circle of convergence passes through the nearest singularity of the function to the origin.

4.22. *If we continue an analytic function $f(z)$ along two different routes from z_0 to z_1, and obtain two different values of $f(z_1)$, then $f(z)$ must have a singularity somewhere between the two routes.*

We construct two chains of regions, say $D_1,\ldots, D_m$ and $D_1',\ldots, D_n'$, such that two consecutive regions of either chain overlap. D_1 and D_1' include z_0, D_m and D_n' include z_1; $f_k(z)$ is analytic in D_k, and $g_k(z)$ in D_k'; $f_k(z) = f_{k-1}(z)$ in the common part of D_k and D_{k-1}, and $f_1(z) = g_1(z)$ in the common part of D_1 and D_1'.

We have then to prove that, if we can continue the function to every point between the two routes, then $f_m(z_1) = g_n(z_1)$.

If δ is small enough, we can construct a polygonal line, starting at a point a in $D_1 D_1'$, and ending at b in $D_m D_n'$, with vertices at points $(p\delta, q\delta)$, such that the circles of radius 2δ with these points as centres lie entirely in the first chain, and each contains the centre of the following one. This chain of circles can be substituted for the first chain of regions. A similar chain of circles, with the same δ, can be substituted for the second chain of regions.

We can now replace the first route by a succession of new routes, consisting of circles of radius 2δ with centres at points $(p\delta, q\delta)$, such that each circle of each route overlaps the previous route, and the circles of the same route on each side of it, without leaving any space uncovered. No circle has a radius smaller than 2δ, or the previous theorem would show the existence of a singularity. It follows from the general principle of uniqueness of continuation that, with each such route, we arrive at z_1 with the same value of $f(z_1)$. Also, in a finite number of steps we pass from one of our original routes to the other; since the function is regular at every point between the two routes, the process of continuation is never stopped.

4.3. Riemann surfaces. The function $\sqrt{z}$ is a two-valued function of z; but, if we put $z = re^{i\theta}$ and distinguish between equal values of z arising from different values of θ, it is possible to represent it as a one-valued function. Suppose we consider the values of z corresponding to $\pi < \theta < 3\pi$ as distinct from those corresponding to $-\pi < \theta < \pi$; but those corresponding to $3\pi < \theta < 5\pi$ as the same again, and so on. This is equivalent to replacing the ordinary z-plane by two planes. We may think of them as superposed, each of them being cut along the negative real axis, and the planes being joined cross-wise along the cut. The configuration thus obtained is called a Riemann surface.

If now we pass along a path encircling the origin, starting on the upper plane from the negative real axis, we pass round the upper plane once, then cross to the lower plane, pass round it once, and then return to the upper plane.

This corresponds to the way in which we obtain the two different values of $\sqrt{z}$. On the upper plane, say with $-\pi < \theta < \pi$, we have $\sqrt{z} = \sqrt{r}e^{\frac{1}{2}i\theta}$ $(-\frac{1}{2}\pi < \frac{1}{2}\theta < \frac{1}{2}\pi)$, and on the lower plane $\sqrt{z} = \sqrt{r}e^{\frac{1}{2}i\theta}$ $(\frac{1}{2}\pi < \frac{1}{2}\theta < \frac{3}{2}\pi)$; and if θ is increased further we return to the upper plane again, and the values are repeated. Thus $\sqrt{z}$ is a one-valued function on the Riemann surface.

We represent the function $\log z$ in a similar way upon an infinity of superposed planes, each cut along the negative real axis and joined to the opposite edge of the one below. In this case there is no return to the starting-point.

For a function such as $\sqrt{\{(z-a)(z-b)\}}$ we may make a cut on each plane along the straight line joining the points $z = a$ and $z = b$, and join cross-wise along the cut.

The number of branches of a many-valued function may be defined as the least number of planes which are required to form a Riemann surface on which the function is one-valued.

Considerable ingenuity is required in constructing Riemann surfaces for more complicated functions. They are of great importance in the general theory of many-valued functions, but it is beyond the scope of this chapter to pursue the subject further.

4.4. Integrals containing a complex parameter. We know that if z is real and positive, then

$$\int_0^\infty e^{-zt}\,dt = \frac{1}{z}. \tag{1}$$

Now the integral is uniformly convergent in any finite region to the right of the imaginary axis, and therefore represents an analytic function of z, regular for $\mathbf{R}(z) > 0$. Hence the function

$$F(z) = \int_0^\infty e^{-zt}\,dt - \frac{1}{z}$$

is regular for $\mathbf{R}(z) > 0$, and $F(z) = 0$ on the real axis. Hence $F(z) = 0$ wherever it is regular, i.e. (1) holds for complex values of z whose real part is positive. Thus we may put $z = x+iy$ $(x > 0)$ and separate real and imaginary parts, and obtain the well-known results

$$\int_0^\infty e^{-xt}\cos yt\,dt = \frac{x}{x^2+y^2}, \qquad \int_0^\infty e^{-xt}\sin yt\,dt = \frac{y}{x^2+y^2}. \tag{2}$$

Examples. (i) Prove that

$$\int_0^\infty e^{-zt^2}\,dt = \frac{\sqrt{\pi}}{2\sqrt{z}} \qquad (\mathbf{R}(z) > 0).$$

[Assuming the result for real values of z, the general result may be obtained either by analytic continuation, or by using Cauchy's theorem to turn the line of integration through an angle $-\frac{1}{2}\arg z$.]

(ii) Prove that

$$\int_0^\pi \frac{dt}{1-z\cos t} = \frac{\pi}{\sqrt{(1-z^2)}}$$

except when z is real and $z \geqslant 1$ or $z \leqslant -1$.

4.41. The Gamma-function. The formula

$$\Gamma(z) = \int_0^\infty e^{-w} w^{z-1}\, dw \tag{1}$$

defines $\Gamma(z)$ as an analytic function, regular for $\mathbf{R}(z) > 0$ (§ 2.85). As it stands it tells us nothing about $\Gamma(z)$ on the imaginary axis or to the left of it.

Consider, however, the function

$$f(z) = \int_C e^{-w}(-w)^{z-1}\, dw, \tag{2}$$

where C consists of the real axis from ∞ to δ, the circle $|w| = \delta$ described in the positive direction, and the real axis from δ to ∞ again. The many-valued function $(-w)^{z-1} = e^{(z-1)\log(-w)}$ is made definite by taking $\log(-w)$ to be real at $w = -\delta$. The contour integral is uniformly convergent in any finite region of the z-plane, for the question of convergence now arises at infinity only, a case already discussed in § 1.51. Hence $f(z)$ is regular for all finite values of z.

If $w = \rho e^{i\phi}$, then $\log w = \log\rho + i(\phi - \pi)$ on the contour. The integrals along the real axis therefore give

$$\int_\delta^\infty \{-e^{-\rho+(z-1)(\log\rho - i\pi)} + e^{-\rho+(z-1)(\log\rho+i\pi)}\}\, d\rho$$

$$= -2i\sin z\pi \int_\delta^\infty e^{-\rho}\rho^{z-1}\, d\rho.$$

On the circle of radius δ

$$|(-w)^{z-1}| = |e^{(z-1)\{\log\delta + i(\phi-\pi)\}}| = e^{(x-1)\log\delta - y(\phi-\pi)} = O(\delta^{x-1}).$$

The integral round the circle is therefore $O(\delta^x) = o(1)$ as $\delta \to 0$ if $x > 0$. Hence, making $\delta \to 0$, we obtain

$$f(z) = -2i\sin z\pi \int_0^\infty e^{-\rho}\rho^{z-1}\, d\rho = -2i\sin z\pi\, \Gamma(z) \quad (\mathbf{R}(z) > 0).$$

Now the function $\frac{1}{2}if(z)\operatorname{cosec} z\pi$ is regular for all values of z except possibly at the poles of $\operatorname{cosec} z\pi$, viz. $z = 0, \pm 1, \pm 2,...$; and it equals $\Gamma(z)$ for $\mathbf{R}(z) > 0$. We can therefore take this function as a continuation of $\Gamma(z)$ over the whole z-plane. But we know already that $\Gamma(z)$ is regular at $z = 1, 2,...$. Hence the only possible poles are at $z = 0, -1, -2,...$.

These points are actually poles of $\Gamma(z)$; for if z is a negative integer, $(-w)^{z-1}$ is one-valued, and the integral (2) can be evaluated by the calculus of residues. We obtain

$$f(-n) = -2\pi i/n!,$$

and the residue of $\Gamma(z)$ is

$$\lim_{z\to -n} \frac{2\pi i}{n!}\,\frac{z+n}{2i\sin z\pi} = \frac{(-1)^n}{n!}.$$

All the gamma-function formulae can now be extended to general complex values of z. For example, the functional equation

$$\Gamma(z)\Gamma(1-z) = \pi \operatorname{cosec} z\pi,$$

proved on the assumption that z is real and $0 < z < 1$, holds for all non-integral values of z.

A consequence of this formula is that $1/\Gamma(z)$ is an integral function. For in the above formula the poles of $\Gamma(1-z)$ are all cancelled by zeros of $\sin z\pi$.

We can now prove for $\Gamma(z)$ formulae similar to those of §§ 3.22–3. By § 1.86 (4)

$$\frac{\Gamma(z-h)\Gamma(h)}{\Gamma(z)} = \frac{1}{h}+\int_0^1 \{(1-t)^{z-h-1}-1\}t^{h-1}\,dt \qquad (0 < h < x)$$

$$= \frac{1}{h}+\int_0^1 \{(1-t)^{z-1}-1\}t^{-1}\,dt + o(1)$$

as $h \to 0$. The left-hand side is

$$\frac{1}{\Gamma(z)}\{\Gamma(z)-h\Gamma'(z)+...\}\left\{\frac{1}{h}+A+...\right\},$$

where A is a constant. Equating the constant terms, we obtain

$$\frac{\Gamma'(z)}{\Gamma(z)} = \int_0^1 \{1-(1-t)^{z-1}\}\frac{dt}{t} - A \qquad (x > 0).$$

Putting $1/t = \sum (1-t)^n$ and integrating term by term,

$$\frac{\Gamma'(z)}{\Gamma(z)} = \sum_{n=0}^{\infty}\left(\frac{1}{n+1}-\frac{1}{n+z}\right) - A.$$

The process is justified by § 1.77 if $z > 1$; the result holds by analytic continuation for any z except a negative integer.

It is easily seen that the formula can be rearranged as

$$\frac{\Gamma'(z)}{\Gamma(z)}+\frac{1}{z}=\sum_{n=1}^{\infty}\left(\frac{1}{n}-\frac{1}{n+z}\right)-C,$$

where C is another constant. Integrating and taking exponentials,

$$\frac{1}{\Gamma(z)}=e^{Cz}z\prod_{n=1}^{\infty}\left(1+\frac{z}{n}\right)e^{-z/n}.$$

Putting $z=1$, $$1=e^{C}\prod_{1}^{\infty}\left(1+\frac{1}{n}\right)e^{-1/n}.$$

Hence

$$C=-\log\prod_{1}^{\infty}\left(1+\frac{1}{n}\right)e^{-1/n}$$

$$=\lim_{N\to\infty}\left(1+\frac{1}{2}+\ldots+\frac{1}{N}-\log N\right)=\gamma,$$

γ being Euler's constant.

4.42. Stirling's formula for complex values of z. The formula of the previous section gives

$$\log\Gamma(z)=\sum_{n=1}^{\infty}\left\{\frac{z}{n}-\log\left(1+\frac{z}{n}\right)\right\}-\gamma z-\log z, \tag{1}$$

each logarithm having its principal value. Now it is easily verified that

$$\int_{0}^{N}\frac{[u]-u+\frac{1}{2}}{u+z}\,du=\sum_{n=0}^{N-1}\int_{n}^{n+1}\left(\frac{n+\frac{1}{2}+z}{u+z}-1\right)du$$

$$=\sum_{n=1}^{N-1}\left\{\frac{z}{n}-\log\left(1+\frac{z}{n}\right)\right\}-\log\{(N-1)!\}-z\left(1+\frac{1}{2}+\ldots+\frac{1}{N-1}\right)-$$
$$-(z+\tfrac{1}{2})\log z+(N-\tfrac{1}{2}+z)\log(N+z)-N.$$

Using 1.87 (1), and the relations

$$1+\frac{1}{2}+\ldots+\frac{1}{N-1}=\log N+\gamma+o(1),$$

$$\log(N+z) = \log N + \frac{z}{N} + O\left(\frac{1}{N^2}\right),$$

and making $N \to \infty$, it follows that

$$\log\Gamma(z) = (z-\tfrac{1}{2})\log z - z + \tfrac{1}{2}\log 2\pi + \int_0^\infty \frac{[u]-u+\frac{1}{2}}{u+z}\,du. \quad (2)$$

Writing $\phi(u) = \int_0^u ([v]-v+\frac{1}{2})\,dv$, $\phi(u)$ is bounded, since clearly $\phi(n+1) = \phi(n)$ if n is an integer. Hence the last term in (2) is

$$\int_0^\infty \frac{\phi'(u)}{u+z}\,du = \int_0^\infty \frac{\phi(u)}{(u+z)^2}\,du = O\left\{\int_0^\infty \frac{du}{u^2+r^2-2ur\cos\delta}\right\} = O\left(\frac{1}{r}\right)$$

uniformly for $-\pi+\delta \leqslant \arg z \leqslant \pi-\delta$. This is the extension of Stirling's formula to complex values of z.

Examples. (i) For any constant a

$$\log\Gamma(z+a) = (z+a-\tfrac{1}{2})\log z - z + \tfrac{1}{2}\log 2\pi + O(1/|z|)$$

as $|z| \to \infty$, uniformly for $-\pi+\delta \leqslant \arg z \leqslant \pi-\delta$.

(ii) For any fixed value of x, as $y \to \pm\infty$

$$|\Gamma(x+iy)| \sim e^{-\frac{1}{2}\pi|y|}|y|^{x-\frac{1}{2}}\sqrt{(2\pi)}.$$

(iii) Show that the series $\phi(u) = \sum_{\nu=1}^{\infty}(1-\cos 2\nu\pi u)/(2\pi^2\nu^2)$ can be inserted in the above formula and integrated term by term. Hence prove that the integral in (2) is

$$\frac{1}{12z} + O\left(\frac{1}{|z|^2}\right).$$

This process can be carried to any number of terms by repeated partial integrations.

4.43. The Zeta-function. The function $\zeta(z)$, defined originally by the series

$$\zeta(z) = \frac{1}{1^z} + \frac{1}{2^z} + \ldots \qquad (\mathbf{R}(z) > 1), \quad (1)$$

has been shown (§ 1.78 (ii)) to be also given by the formula

$$\zeta(z) = \frac{1}{\Gamma(z)} \int_0^\infty \frac{w^{z-1}}{e^w - 1}\,dw \qquad (\mathbf{R}(z) > 1). \quad (2)$$

We can use this formula to continue $\zeta(z)$ across the line $x = 1$, in the same way that we continued $\Gamma(z)$ across $x = 0$. In fact we can prove in precisely the same way as before that, if $\mathbf{R}(z) > 1$,

$$\zeta(z) = -\frac{1}{2i\sin z\pi\,\Gamma(z)}\int_C \frac{(-w)^{z-1}}{e^w-1}\,dw, \tag{3}$$

where, as before, the contour C comes from positive infinity and encircles the origin once in the positive direction. The only difference is that C must now exclude all the poles of $1/(e^w-1)$ other than $w = 0$, viz. the points $w = \pm 2i\pi,\ \pm 4i\pi,\ldots$.

Using the functional equation for the Γ-function, we may write the result in the form

$$\zeta(z) = \frac{i\Gamma(1-z)}{2\pi}\int_C \frac{(-w)^{z-1}}{e^w-1}\,dw.$$

As in the case of the Γ-function, the contour integral is an integral function of z. This formula therefore provides the continuation of $\zeta(z)$ over the whole plane. The only possible singularities are at the poles of $\Gamma(1-z)$, viz. at $z = 1, 2,\ldots$. But we know already that $\zeta(z)$ is regular at $z = 2, 3,\ldots$. Hence the only possible pole is at $z = 1$. This is actually a simple pole, with residue 1. For at $z = 1$ the contour integral is equal to

$$\int_C \frac{dw}{e^w-1} = 2\pi i$$

by the theorem of residues, and $\Gamma(1-z)$ has a simple pole with residue -1, whence the result.

Again, it is well known that

$$\frac{1}{e^w-1} = \frac{1}{w}-\frac{1}{2}+\sum_{n=1}^{\infty}\frac{(-1)^{n-1}B_n w^{2n-1}}{(2n)!}$$

where the coefficients B_n (Bernoulli's numbers) are rational numbers. Hence we can evaluate $\zeta(-n)$, where n is any positive integer, by the theorem of residues. We find that

$$\zeta(0) = -\tfrac{1}{2},$$

$$\zeta(-2m) = 0 \qquad (m = 1, 2,\ldots),$$

$$\zeta(-2m-1) = \frac{(-1)^{m+1}B_{m+1}}{2m+2} \qquad (m = 0, 1,\ldots).$$

4.44. The functional equation for the ζ-function. The ζ-function satisfies the functional equation

$$\zeta(1-z) = 2^{1-z}\pi^{-z}\cos\tfrac{1}{2}\pi z\,\Gamma(z)\zeta(z).$$

To prove this, we take the formula (3) of the previous section, where now z may have any value, and deform the contour into the contour C_n consisting of the square with centre the origin and sides parallel to the axes, length of side $(4n+2)\pi$, together with the positive real axis from $(2n+1)\pi$ to infinity. In so doing we pass over poles of the integrand at the points $w = 2i\pi, 4i\pi, \ldots, 2ni\pi$, and $-2i\pi, \ldots, -2ni\pi$. The residue at $2\nu i\pi$ $(\nu > 0)$ is

$$e^{(z-1)(\log 2\nu\pi - \frac{1}{2}i\pi)} = (2\nu\pi)^{z-1}ie^{-\frac{1}{2}i\pi z},$$

and at $-2\nu i\pi$ it is

$$e^{(z-1)(\log 2\nu\pi + \frac{1}{2}i\pi)} = -(2\nu\pi)^{z-1}ie^{\frac{1}{2}i\pi z}.$$

The sum of these two residues is

$$(2\nu\pi)^{z-1}2\sin\tfrac{1}{2}\pi z.$$

Hence § 4.43 (3) gives

$$\sin\pi z\,\Gamma(z)\zeta(z) = -\frac{1}{2i}\int_{C_n}\frac{(-w)^{z-1}}{e^w-1}\,dw + 2\pi\sin\tfrac{1}{2}\pi z\sum_{\nu=1}^{n}(2\nu\pi)^{z-1}.$$

Suppose now that $\mathbf{R}(z) < 0$. On the square

$$|(-w)^{z-1}| = e^{(x-1)\log|w| - y\arg(-w)} = O(n^{x-1}),$$

and $|e^w - 1| > A$, while the length of the square is $O(n)$. Hence this part of the integral is $O(n^x)$, and so it tends to zero. The remaining part of the integral plainly tends to zero also. Hence, making $n \to \infty$, we obtain

$$\begin{aligned}\sin\pi z\,\Gamma(z)\zeta(z) &= 2\pi\sin\tfrac{1}{2}\pi z(2\pi)^{z-1}\sum_{\nu=1}^{\infty}\nu^{z-1}\\ &= 2\pi\sin\tfrac{1}{2}\pi z(2\pi)^{z-1}\zeta(1-z),\end{aligned}$$

which is equivalent to the result stated. This proves the functional equation for $\mathbf{R}(z) < 0$, and so, by § 4.43, for all values of z.

4.45. An alternative proof. The following proof* proceeds on quite different lines. Let

$$f(x) = \sum_{n=0}^{\infty}\frac{\sin(2n+1)x}{2n+1}. \tag{1}$$

* Hardy (15).

This series is boundedly convergent, and $f(x) = (-1)^m \frac{1}{4}\pi$ for $m\pi < x < (m+1)\pi$, $m = 0, 1, 2,...$; for

$$f(x) = \sum_{n=1}^{\infty} \frac{\sin nx}{n} - \sum_{n=1}^{\infty} \frac{\sin 2nx}{2n},$$

and the results easily follow from those of § 1.76, ex. (ii). We may therefore multiply (1) by x^{p-1} $(0 < p < 1)$ and integrate term by term over any finite interval $(0, X)$. Thus

$$\int_0^X x^{p-1} f(x)\, dx = \sum_{n=0}^{\infty} \frac{1}{2n+1} \int_0^X x^{p-1} \sin(2n+1)x\, dx. \tag{2}$$

We may then replace X by ∞ provided that

$$\lim_{X\to\infty} \sum_{n=0}^{\infty} \frac{1}{2n+1} \int_X^{\infty} x^{p-1} \sin(2n+1)x\, dx = 0. \tag{3}$$

Integrating by parts, the integral is

$$X^{p-1} \frac{\cos(2n+1)X}{2n+1} + \frac{p-1}{2n+1} \int_X^{\infty} x^{p-2} \cos(2n+1)x\, dx$$

$$= O\left(\frac{X^{p-1}}{2n+1}\right) + O\left(\frac{1}{2n+1} \int_X^{\infty} x^{p-2}\, dx\right) = O\left(\frac{X^{p-1}}{2n+1}\right),$$

and (3) clearly follows.

Inserting the value of $f(x)$, and evaluating the integrals on the right-hand side by § 3.127, we therefore obtain

$$\tfrac{1}{4}\pi \sum_{m=0}^{\infty} (-1)^m \int_{m\pi}^{(m+1)\pi} x^{p-1}\, dx = \Gamma(p) \sin \tfrac{1}{2}p\pi \sum_{n=0}^{\infty} \frac{1}{(2n+1)^{p+1}}.$$

The series on the right-hand side converges to

$$(1 - 2^{-p-1})\zeta(p+1).$$

That on the left is

$$\frac{\pi^p}{p}\left[1 + \sum_{m=1}^{\infty} (-1)^m \{(m+1)^p - m^p\}\right].$$

This series is convergent for $p < 1$, and, as a little consideration of the above argument shows, uniformly convergent for $\mathbf{R}(p) \leqslant 1 - \delta < 1$. Its sum is therefore an analytic function of p, regular for $\mathbf{R}(p) < 1$. But for $p < 0$ it is

$$2(1^p - 2^p + 3^p - ...) = 2(1 - 2^{p+1})\zeta(-p).$$

By the theory of analytic continuation its sum is the same analytic function of p for $\mathbf{R}(p) < 1$.

Hence, for $0 < p < 1$,

$$\frac{\pi^{p+1}}{2p}(1-2^{p+1})\zeta(-p) = \Gamma(p)\sin\tfrac{1}{2}p\pi(1-2^{-p-1})\zeta(p+1),$$

and putting $p = z-1$ we have the same functional equation as before. The proof holds for $1 < z < 2$ only, but the result, proved for these values of z, holds for all values by analytic continuation.

4.5. The principle of reflection. *Let $f(z)$ be an analytic function, regular in a region D intersected by the real axis, and real on the real axis. Then $f(z)$ takes conjugate values for conjugate values of z.*

Let z_0 be an interior point of D on the real axis. Then

$$f(z) = \sum_{n=0}^{\infty} a_n(z-z_0)^n$$

for sufficiently small values of $|z-z_0|$.

All the coefficients a_n are real; for

$$a_0 = f(z_0), \qquad a_1 = f'(z_0), \ldots .$$

Clearly a_0 is real. a_1 may be calculated as the limit of

$$\frac{f(z)-f(z_0)}{z-z_0}$$

as $z \to z_0$ by real values. Hence a_1 is real. So they are all real.

The result now follows inside the circle of convergence of the above series. The general result then follows by continuation, since the power series about conjugate points will always have conjugate coefficients.

4.51. A method of obtaining the analytic continuation of certain functions is given by the Riemann-Schwarz 'principle of reflection'. This is contained in the following theorem, which is a sort of converse of the previous one.

Suppose that a region D of the z-plane has as part of its boundary a segment l of a straight line; and that $w = f(z)$ is an analytic function, regular in D and continuous on l, and such that, as z describes l, w describes a straight line λ in the w-plane. Let z be a point of D, z_1 its reflection in l, and let w_1 be the reflection of w in λ. Then $w_1 = w_1(z_1)$ is an analytic continuation of w.

In the first place, w_1 is an analytic function of z_1; for it is easily seen from a figure that, if w' corresponds to z', and w'_1, z'_1 are their reflections,

$$|z'_1-z_1| = |z'-z|, \qquad |w'_1-w_1| = |w'-w|;$$

and

$$\arg(z'_1-z_1) = 2\alpha-\arg(z'-z), \qquad \arg(w'_1-w_1) = 2\beta-\arg(w'-w),$$

where α and β are the angles between l and λ respectively and the real axis. Now when $z' \to z$, the limit

$$\lim \frac{w'-w}{z'-z}$$

exists, i.e. the limits

$$\lim\left|\frac{w'-w}{z'-z}\right|, \qquad \lim\{\arg(w'-w)-\arg(z'-z)\}$$

exist. Hence the limits

$$\lim\left|\frac{w'_1-w_1}{z'_1-z_1}\right|, \qquad \lim\{\arg(w'_1-w_1)-\arg(z'_1-z_1)\}$$

exist, and so

$$\lim \frac{w'_1-w_1}{z'_1-z_1}$$

exists, i.e. w_1 is an analytic function of z_1.

Secondly, it is clear that, on l, $w_1 = w$.

To prove that the two functions are analytic continuations of each other, take any point of the line l, and describe round it a circle C so small that it lies entirely inside D and its reflection D_1. Let c be the boundary of the part of C in D, c_1 of the part in D_1. Let $\phi(z) = w$ in the part of D inside C, and $\phi(z) = w_1$ in the remainder of C. Then $\phi(z)$ is continuous, and it is sufficient to prove that $\phi(z)$ is an analytic function of z.

Let z_0 be a point inside C and D. Then*

$$\phi(z_0) = \frac{1}{2\pi i}\int_c \frac{\phi(z)}{z-z_0}\,dz,$$

and, since $\phi(z)/(z-z_0)$ is regular in D_1,

$$0 = \frac{1}{2\pi i}\int_{c_1} \frac{\phi(z)}{z-z_0}\,dz.$$

* See the end of § 2.35.

Adding, we obtain

$$\phi(z_0) = \frac{1}{2\pi i} \int_C \frac{\phi(z)}{z-z_0}\, dz, \tag{1}$$

the integrals along l cancelling.

We clearly obtain the same formula if z_0 is any point inside C and D_1; and also, since each side of (1) is continuous, if z_0 is inside C and on l. But, as in § 2.8, the right-hand side of (1) is an analytic function of z_0, regular inside C. This proves the theorem.

The method of proof also gives the following general theorem: *if two functions $f(z)$, $f_1(z)$, are analytic and regular in regions D and D_1 separated by a contour C, and continuous on C, and $f(z) = f_1(z)$ along C, then the two functions are analytic continuations of each other.*

4.6. Hadamard's multiplication theorem.* The following problem, which was considered by Hadamard, is a good example of the principles of analytic continuation. Suppose that

$$f(z) = \sum_{n=0}^{\infty} a_n z^n$$

is convergent for $|z| < R$, and

$$g(z) = \sum_{n=0}^{\infty} b_n z^n$$

is convergent for $|z| < R'$, and that the singularities of $f(z)$ and $g(z)$ are known. What can be said about the singularities of the function

$$F(z) = \sum_{n=0} a_n b_n z^n, \tag{1}$$

whose coefficients are the products of those in the given series?

The general result is that, *if $f(z)$ has singularities at $\alpha_1, \alpha_2,\ldots$, and $g(z)$ at $\beta_1, \beta_2,\ldots$, then the singularities of $F(z)$ are to be found among the points $\alpha_m \beta_n$.*

Let us suppose, to take the simplest possible case, that $f(z)$ has just one singularity, $z = \alpha$, and $g(z)$ just one singularity, $z = \beta$.

In the first place, $F(z)$ is regular for sufficiently small values of z, and in fact for $|z| < RR'$. For if $\epsilon > 0$

$$|a_n(R-\epsilon)^n| < K, \qquad |b_n(R'-\epsilon)^n| < K,$$

so that

$$|a_n b_n| < \frac{K}{\{(R-\epsilon)(R'-\epsilon)\}^n},$$

* Hadamard (3).

and by taking ϵ small enough we see that the radius of convergence of the series for $F(z)$ is at least equal to RR'.

Hadamard's theorem depends on the following representation of $F(z)$ as an integral:

$$F(z) = \frac{1}{2\pi i} \int_C f(w) g\left(\frac{z}{w}\right) \frac{dw}{w}, \tag{2}$$

where C is a contour, including the origin, on which $|w| < R$, $|z/w| < R'$. To prove this, write

$$g\left(\frac{z}{w}\right) = \sum_{n=0}^{\infty} b_n \left(\frac{z}{w}\right)^n$$

in the integral, and integrate term by term, as we may by uniform convergence. We obtain

$$\frac{1}{2\pi i} \int_C f(w) g\left(\frac{z}{w}\right) \frac{dw}{w} = \sum_{n=0}^{\infty} \frac{b_n z^n}{2\pi i} \int_C \frac{f(w)}{w^{n+1}} dw$$
$$= \sum_{n=0}^{\infty} a_n b_n z^n,$$

the required result. In order that the inequalities $|w| < R$, $|z/w| < R'$, should be consistent, it is, of course, necessary that $|z| < RR'$. If this condition is satisfied, C may, for example, be any circle between $|w| = R$ and $|w| = |z|/R'$.

In the case where each function has just one singularity, $R = |\alpha|$ and $R' = |\beta|$.

We next continue the function $F(z)$ beyond the circle $|z| = RR'$ by deforming the contour C. As long as C remains fixed, z may, in (2), take any value such that z/β remains inside the contour C. For, by § 2.83, the right-hand side of (2) is an analytic function of z for all such values of z, and the continuation of $F(z)$ to all such values follows at once.

Suppose on the other hand that we deform C into another contour C_1, which includes $z = 0$ and excludes $z = \alpha$. Let

$$F_1(z) = \frac{1}{2\pi i} \int_{C_1} f(w) g\left(\frac{z}{w}\right) \frac{dw}{w}. \tag{3}$$

Then by Cauchy's theorem $F_1(z) = F(z)$, provided that the point $w = z/\beta$ lies within both C and C_1; for then the integrand is an analytic function of w, regular between C and C_1.

The formula (3) therefore provides the continuation of $F(z)$ to all values of z such that z/β lies within C_1.

The only restriction on z, therefore, is that z/β must lie within a contour which excludes $z = \alpha$. But we can choose such a contour for every z except $z = \alpha\beta$.

Thus $F(z)$ is regular except at $z = \alpha\beta$. The proof, however, applies only to what we may call the principal branch of $F(z)$, viz. that obtained by continuation from the original element without encircling any of the points $\alpha\beta$. In the above example the contour C_1 cannot have a loop going round α and enclosing the origin again on the other side, since the integrand has a singularity at $w = 0$. Hence if $z/\beta \to 0$ along a path encircling the point α, it is impossible to choose C_1 appropriately. The point $z = 0$ may therefore be a singularity of other branches of $F(z)$.

In the general case the details of the proof are, of course, more complicated, but the general method is the same.

Examples. (i) If

$$f(z) = \frac{1}{a-z}, \qquad g(z) = \frac{1}{b-z},$$

then

$$F(z) = \frac{1}{ab-z}.$$

(ii) If

$$f(z) = \frac{1}{1-z^2}, \qquad g(z) = \frac{z}{1-z^2},$$

then $F(z) = 0$, so that the points $\alpha\beta$ are not *necessarily* singularities of $F(z)$.

4.7. Functions with natural boundaries. Let

$$f(z) = \sum_{n=0}^{\infty} z^{n!}.$$

Then $f(z)$ is an analytic function, regular for $|z| < 1$. Let

$$z = re^{2p\pi i/q},$$

and consider the behaviour of $f(z)$ as $r \to 1$ through real values.

Now

$$f(z) = \sum_{n=0}^{q-1} z^{n!} + \sum_{n=q}^{\infty} z^{n!} = f_1(z) + f_2(z),$$

say. Then $f_1(z)$ is a polynomial, and tends to a finite limit as $r \to 1$. When $n \geqslant q$, q is a divisor of $n!$, and so

$$z^{n!} = r^{n!} \qquad (n \geqslant q).$$

Hence

$$f_2(z) = \sum_{n=q}^{\infty} r^{n!},$$

which tends to infinity as $r \to 1$. Hence $f(z) \to \infty$, and so $z = e^{2p\pi i/q}$ is a singularity of $f(z)$. But points of this kind are dense everywhere round the unit circle, so that there is no arc, however small, on which $f(z)$ is regular. It is therefore impossible to continue $f(z)$ across the unit circle, and so the unit circle is a natural boundary of the function.

A similar result holds for the function

$$f(z) = \sum_{n=0}^{\infty} z^{2^n}$$

—we put $z = re^{2p\pi i/2^s}$ and proceed as before.

4.71. Lambert's series.* Let

$$f(z) = \sum_{n=1}^{\infty} d(n)z^n \qquad (|z| < 1),$$

where $d(n)$ denotes the number of divisors of n. We shall prove that the unit circle is a natural boundary of this function.

Consider the double series

$$\sum_{\mu=1}^{\infty} \sum_{\nu=1}^{\infty} z^{\mu\nu}.$$

If we arrange it as a single power series we obtain $f(z)$, and if we sum it by rows we obtain

$$f(z) = \sum_{\mu=1}^{\infty} \frac{z^\mu}{1-z^\mu} \qquad (|z| < 1).$$

The double series is absolutely convergent for $|z| < 1$, so that the transformation is justified.

Let $$z = re^{2p\pi i/q}$$

where p and q are positive integers, $p > 0$, $q > 1$, and p is prime to q. Then we shall prove that, as $r \to 1$,

$$(1-r)f(z) \to \infty.$$

For let $$f(z) = \sum\nolimits_1 \frac{z^\mu}{1-z^\mu} + \sum\nolimits_2 \frac{z^\mu}{1-z^\mu},$$

where, in $\sum_1$, μ takes all values $\equiv 0 \pmod q$, in $\sum_2$ all other values. In $\sum_1$, putting $\mu = mq$,

$$z^\mu = (re^{2\pi ip/q})^{mq} = r^{mq},$$

so that

$$(1-r)\sum\nolimits_1 = (1-r)\sum_{m=1}^{\infty} \frac{r^{mq}}{1-r^{mq}}$$

$$= \frac{1-r}{1-r^q} \sum_{m=1}^{\infty} \frac{1-r^q}{1-r^{mq}} r^{mq}$$

* See Knopp (1).

$$= \frac{1}{1+r+\ldots+r^{q-1}} \sum_{m=1}^{\infty} \frac{r^{mq}}{1+r^q+\ldots+r^{(m-1)q}}$$

$$\geqslant \frac{1}{q} \sum_{m=1}^{\infty} \frac{r^{mq}}{m} = \frac{1}{q} \log \frac{1}{1-r^q} \to \infty.$$

On the other hand, if $\mu \not\equiv 0 \pmod q$,

$$|1-z^\mu|^2 = |1-r^\mu e^{2\pi i p\mu/q}|^2 = (1-r^\mu)^2 + 4r^\mu \sin^2 \frac{p\mu\pi}{q}$$

$$\geqslant 4r^\mu \sin^2 \frac{\pi}{q}.$$

Hence

$$|(1-r) \textstyle\sum_2| \leqslant \frac{(1-r)}{2\sin \pi/q} \displaystyle\sum_{\mu=0}^{\infty} r^{\frac{1}{2}\mu} = \frac{1+\sqrt{r}}{2\sin \pi/q} \leqslant \frac{1}{\sin \pi/q}.$$

Hence, as in the previous cases, the unit circle is a natural boundary of $f(z)$.

MISCELLANEOUS EXAMPLES

1. The power series $\qquad z - \frac{1}{2}z^2 + \frac{1}{3}z^3 - \ldots$

may be continued to a wider region by means of the series

$$\log 2 - \frac{1-z}{2} - \frac{(1-z)^2}{2.2^2} - \frac{(1-z)^3}{3.2^3} - \ldots.$$

2. The power series $\qquad z + \frac{1}{2}z^2 + \frac{1}{3}z^3 + \ldots$

and $\qquad i\pi - (z-2) + \frac{1}{2}(z-2)^2 - \frac{1}{3}(z-2)^3 + \ldots$

have no common region of convergence, but are analytic continuations of the same function.

3. The functions defined by the series

$$1 + az + a^2z^2 + \ldots$$

and $$\frac{1}{1-z} - \frac{(1-a)z}{(1-z)^2} + \frac{(1-a)^2z^2}{(1-z)^3} - \ldots$$

are analytic continuations of each other.

4. If $f(z)$ and $g(z)$ are integral functions, the integral

$$\frac{1}{2\pi i} \int \left\{ \frac{f(w)}{w-z} + \frac{zg(1/w)}{zw-w^2} \right\} dw,$$

taken round the unit circle, represents $f(z)$ inside the circle and $g(1/z)$ outside it.

5. If $F(\alpha, \beta; \gamma; z)$ denotes the series

$$1 + \frac{\alpha\beta}{1.\gamma} z + \frac{\alpha(\alpha+1)\beta(\beta+1)}{1.2\gamma(\gamma+1)} z^2 + \ldots,$$

show that the series $\qquad f(z) = F(a, 1; c; z)$

and
$$g(z) = \frac{1}{1-z} F\left(c-a, 1; c; \frac{z}{z-1}\right)$$
have a common region of convergence; that in this region they both satisfy the differential equation
$$z(1-z)\frac{d^2u}{dz^2} + \{c-(a+2)z\}\frac{du}{dz} - au = 0;$$
that $f(0) = g(0)$ and $f'(0) \doteq g'(0)$; and hence that the two functions are analytic continuations of each other.

6. The function
$$\frac{1}{\sqrt{(2-z)}+1}$$
has two power series about $z = 0$, with radii of convergence 1 and 2 respectively.

7. Consider the singularities of the functions
$$\exp\left\{\frac{1}{\sqrt{(2-z)}+1}\right\}, \qquad \log\left\{\frac{1}{\sqrt{(2-z)}+1}\right\}.$$

8. Show that the formulae (2) of § 4.4, which were proved there for real values of x and y, hold also for all complex values such that $|\mathbf{I}(y)| < \mathbf{R}(x)$.

9. Prove that
$$\Gamma(z) = \int_1^\infty e^{-w}w^{z-1}\,dw + \sum_{n=0}^\infty \frac{(-1)^n}{n!(z+n)},$$
and hence give another proof of the analytic properties of $\Gamma(z)$.

10. Prove that
$$\int_0^\infty t^{z-1}\cos t\,dt = \Gamma(z)\cos\tfrac{1}{2}\pi z$$
if $0 < \mathbf{R}(z) < 1$; and that
$$\int_0^\infty t^{z-1}\sin t\,dt = \Gamma(z)\sin\tfrac{1}{2}\pi z$$
if $-1 < \mathbf{R}(z) < 1$.

11. Prove that if $0 < \mathbf{R}(z) < 1$
$$\zeta(z) = \frac{1}{\Gamma(z)}\int_0^\infty w^{z-1}\left(\frac{1}{e^w-1} - \frac{1}{w}\right)dw,$$
and that if $-1 < \mathbf{R}(z) < 0$
$$\zeta(z) = \frac{1}{\Gamma(z)}\int_0^\infty w^{z-1}\left(\frac{1}{e^w-1} - \frac{1}{w} + \frac{1}{2}\right)dw.$$

[Consider the corresponding contour integrals as in § 4.43.]

12. Deduce the functional equation for the zeta-function from the first formula of ex. 11, and the formula of Ch. III, Misc. Ex. 10.

13. Deduce the functional equation for the zeta-function from the second formula of ex. 11, and ex. (iii) of § 3.22.

14. The function $L(z)$ is defined for $\mathbf{R}(z) > 1$ by the series

$$L(z) = \frac{1}{1^z} - \frac{1}{3^z} + \frac{1}{5^z} - \dots.$$

Show that $L(z)$ is an integral function of z, and that it satisfies the functional equation

$$L(1-z) = 2^z \pi^{-z} \sin \tfrac{1}{2}\pi z \, \Gamma(z) L(z).$$

15. A function $f(z)$ is defined for $|z| < 1$ by the series

$$f(z) = \sum_{n=0}^{\infty} \frac{z^n}{(n+1)^s} \qquad (s > 0).$$

Show that

$$f(z) = \frac{1}{\Gamma(s)} \int_0^{\infty} \frac{t^{s-1}}{e^t - z} \, dt,$$

and hence that $f(z)$ is regular except possibly on the positive real axis.

By deforming the line of integration into a suitable curve, show that the principal branch of $f(z)$ is regular except at the point $z = 1$.

16. Show that the singularities of the principal branch of

$$\sum_{n=0}^{\infty} \frac{a_n z^n}{(n+1)^s}$$

are the same for all real values of s.

17. Two functions $f(x)$, $g(x)$, are connected by the formulae

$$f(x) = \int_{-\infty}^{\infty} e^{ixt} g(t) \, dt, \qquad g(x) = \frac{1}{2\pi} \int_{-\infty}^{\infty} e^{-ixt} f(t) \, dt,$$

for real values of x. Show that there cannot be a finite interval outside which both $f(x) = 0$ and $g(x) = 0$, unless both functions are everywhere 0.

[If $f(x) = 0$ for $x < a$ and $x > b$, $g(x)$ is analytic.]

18. If

$$f(z) = \sum_{n=0}^{\infty} z^{2^n},$$

show that

$$f(z) = f(z^2) + z,$$

and deduce from this that $|z| = 1$ is a natural boundary of the function.

19. If α is a real irrational number, the series

$$\sum_{n=1}^{\infty} \frac{1}{2^n (z - e^{2in\alpha\pi})}$$

represents two different analytic functions, one inside the unit circle and one outside it, the unit circle being a natural boundary of each. If α is rational, the series represents a single rational function. [In the first case the function is unbounded as $z \to e^{2im\alpha\pi}$ along the radius vector.]

20. The function $$\prod_{n=1}^{\infty}\left(1+\frac{1}{n^z+n}\right)$$

has the line $\mathbf{R}(z) = 1$ as a natural boundary.

[Every point of this line is a limit-point of zeros.]

21. Let $$f(z) = \sum_{n=0}^{\infty} a_n z^n \qquad (|z| < 1)$$

and $$\phi(z) = \sum_{n=0}^{\infty} \frac{a_n z^n}{n!}.$$

Then the integral $$F(z) = \int_0^{\infty} e^{-t}\phi(zt)\,dt$$

provides the continuation of $f(z)$ across any arc of the circle of convergence where $f(z)$ is regular.

[This is Borel's method of continuation—see his *Leçons sur les séries divergentes*, p. 94. We have $F(z) = f(z)$ wherever the series for $f(z)$ is convergent; for by § 1.79 we may then insert the series for $\phi(zt)$ and integrate term by term. But if $f(z)$ can be continued at all, $F(z)$ exists in a wider region. Let z be a regular point of $f(z)$. As in § 4.6 we have

$$\phi(zt) = \frac{1}{2\pi i}\int_C f(w)e^{zt/w}\frac{dw}{w},$$

where C is a contour including the origin, and excluding the singularities of $f(w)$. Hence

$$|\phi(zt)| < Ke^{Mt}$$

where K is independent of z and t, and M is the maximum of $\mathbf{R}(z/w)$ for values of w on C. To prove the integral for $F(z)$ convergent, we must have $M < 1$. Now $\mathbf{R}(z/w) < 1$ if w lies outside the circle on Oz as diameter. We therefore take C to be a concentric circle of slightly larger radius—say $\frac{1}{2}|z|+\delta$; on it $M = |z|/(|z|+\delta)$. Also C must exclude all singularities of $f(w)$, and it does so if z lies in a region D formed as follows: Through each singular point of $f(w)$, draw a line perpendicular to the line joining the point to the origin. The unit circle is included in a polygon D formed by these lines. It is easily seen from a figure that the conditions are fulfilled if z is inside D and δ is small enough. It now follows without difficulty that Borel's integral gives the continuation of $f(z)$ to all points inside D.]

22. Verify Borel's theorem for the functions

$$\frac{1}{1-z}, \qquad \frac{1}{1-z^2}, \qquad \frac{1}{1-z^4}.$$

CHAPTER V

THE MAXIMUM-MODULUS THEOREM

5.1. The maximum-modulus theorem. Let $f(z)$ be an analytic function, regular in a region D and on its boundary C, which we take to be a simple closed contour. Then $|f(z)|$ is continuous in D, since

$$\big||f(z+h)|-|f(z)|\big| \leqslant |f(z+h)-f(z)|,$$

which tends to zero with h. Hence $|f(z)|$ has a maximum value, which is attained at one or more points. The fundamental theorem of the chapter is that $|f(z)|$ reaches its maximum on the boundary C, and not at any interior point of D. We may express it by saying that *if* $|f(z)| \leqslant M$ *on* C, *then the same inequality holds at all points of* D.

A more precise form of the theorem is as follows:

If $|f(z)| \leqslant M$ *on* C, *then* $|f(z)| < M$ *at all interior points of* D, *unless* $f(z)$ *is a constant (when of course* $|f(z)| = M$ *everywhere).*

We shall give a number of different proofs of this theorem.

5.11. FIRST PROOF. This depends on the lemma that *if* $\phi(x)$ *is continuous,* $\phi(x) \leqslant k$, *and*

$$\frac{1}{b-a}\int_a^b \phi(x)\,dx \geqslant k, \tag{1}$$

then $\phi(x) = k$. For if $\phi(\xi) < k$, there is an interval $(\xi-\delta, \xi+\delta)$ in which $\phi(x) \leqslant k-\epsilon$, say; and

$$\int_a^b \phi(x)\,dx \leqslant 2\delta(k-\epsilon)+(b-a-2\delta)k = (b-a)k-2\delta\epsilon,$$

contradicting (1).

To prove the theorem, suppose that, at an interior point z_0 of D, $|f(z)|$ has a value at least equal to its value anywhere else. Let Γ be a circle with centre z_0 lying entirely in D. Then

$$f(z_0) = \frac{1}{2\pi i}\int_\Gamma \frac{f(z)}{z-z_0}\,dz. \tag{2}$$

Putting $z-z_0 = re^{i\theta}$, $f(z)/f(z_0) = \rho e^{i\phi}$, so that ρ and ϕ are functions of θ, we may write (2) as

$$1 = \frac{1}{2\pi}\int_0^{2\pi} \rho e^{i\phi}\,d\theta. \tag{3}$$

Hence $$1 \leqslant \frac{1}{2\pi}\int_0^{2\pi} \rho\, d\theta.$$

But by hypothesis $\rho \leqslant 1$. Hence, by the lemma, $\rho = 1$ for all values of θ.

Taking the real part of (3) we now obtain

$$1 = \frac{1}{2\pi}\int_0^{2\pi} \cos\phi\, d\theta,$$

so that, by the lemma, $\cos\phi = 1$. Hence $f(z) = f(z_0)$ on Γ, and so everywhere; that is, $f(z)$ is a constant.

5.12. Second proof. This is similar in principle to the first proof, but, instead of Cauchy's integral, we use the fact (§ 2.5) that, if

$$f(z) = \sum_{n=0}^{\infty} a_n(z-z_0)^n,$$

then $$\frac{1}{2\pi}\int_0^{2\pi} |f(z_0+re^{i\theta})|^2\, d\theta = \sum_{n=0}^{\infty} |a_n|^2 r^{2n}.$$

Under the same hypotheses as before, the left-hand side does not exceed $|f(z_0)|^2$, i.e. $|a_0|^2$. Hence

$$|a_0|^2 + |a_1|^2 r^2 + |a_2|^2 r^4 + \ldots \leqslant |a_0|^2$$

for a positive value of r. Hence $a_1 = a_2 = \ldots = 0$, and $f(z)$ is a constant.

5.13. Third proof. If z_0 is an interior point of D, we may expand $f(z)$ in a series of powers of $z-z_0$,

$$f(z) = \sum_{n=0}^{\infty} a_n(z-z_0)^n,$$

with a positive radius of convergence. Putting

$$z-z_0 = re^{i\theta}, \qquad a_n = A_n e^{i\alpha_n},$$

this is $$f(z) = \sum_{n=0}^{\infty} A_n r^n e^{i(\alpha_n+n\theta)}.$$

Hence $$|f(z)|^2 = \sum_{m=0}^{\infty}\sum_{n=0}^{\infty} A_m A_n r^{m+n} e^{i(\alpha_m+m\theta-\alpha_n-n\theta)}. \tag{1}$$

Suppose first that $a_0 \neq 0$. Since the series is absolutely convergent, we may rearrange it as a power series in r with a positive radius of convergence. Let k be the smallest positive value of

n for which $a_n \neq 0$. Then

$$|f(z)|^2 = A_0^2 + 2A_0 A_k r^k \cos(\alpha_0 - \alpha_k - k\theta) + \sum_{n=k+1}^{\infty} c_n r^n, \tag{2}$$

where $|c_n| < c^n$ for some value of c. Hence

$$\left| \sum_{n=k+1}^{\infty} c_n r^n \right| < \sum_{n=k+1}^{\infty} c^n r^n = c^{k+1} r^{k+1}/(1-cr),$$

which is less than $A_0 A_k r^k$ if r is small enough. For such a value of r, $|f(z)|^2 - A_0^2$ takes both positive and negative values as θ varies between 0 and 2π, the middle term on the right of (2) varying between $-2A_0 A_k r^k$ and $2A_0 A_k r^k$. Hence A_0 is neither a maximum nor a minimum of $|f(z)|$.

The proof breaks down if there is no a_n $(n > 0)$ which is not zero. But then $f(z) = a_0$ for all values of z.

Finally, if $a_0 = 0$, $|f(z_0)| = 0$, which cannot be a maximum, but is a minimum.

This proves the theorem. We have also shown incidentally that $|f(z)|$ *cannot have, in* D, *a minimum other than* 0. This may also be proved by applying the general theorem to the function $1/f(z)$.

5.14. Harmonic functions. The corresponding theorem for harmonic functions is that *a function which is harmonic in a region cannot have a maximum at an interior point of the region.* For let u be the real part of $f(z)$. If u is a maximum at an interior point, so is e^u; i.e. so is $|F(z)| = |e^{f(z)}|$. This has been shown to be impossible.

The theorem can also be proved by an argument similar to that of § 5.13. Without going into complete detail, this may be seen in a general way as follows. Let $u(x, y)$ be the real part of an analytic function $f(z) = \sum a_n z^n$, regular at $z = 0$. Then

$$u(x, y) = \mathbf{R} \sum a_n (x+iy)^n,$$

and we obtain for $u(x, y)$ a double series of powers of x and y. The coefficients being those of Taylor's theorem, we have

$$u(x, y) - u(0, 0) = u_x x + u_y y + \tfrac{1}{2}(u_{xx} x^2 + 2u_{xy} xy + u_{yy} y^2) + \dots.$$

A necessary condition for a maximum is that $u_x = u_y = 0$. But since, for a harmonic function, $u_{xx} + u_{yy} = 0$, u_{xx} and u_{yy} have opposite signs, and we can make $u(x, y) - u(0, 0)$ either positive or negative by taking $x = 0$ and y small, or $y = 0$ and x small.

Examples. (i) If $|f(z)| > m$ on $|z| = a$, $f(z)$ is regular for $|z| \leqslant a$, and $|f(0)| < m$, then $f(z)$ has at least one zero in $|z| < a$.

[For $|f(z)|$ has a minimum inside the circle, and the minimum value must be zero.]

(ii) Use the previous example to show that every algebraic equation has a root.

5.15. The maximum-modulus theorem is also true for a function $f(z)$ which is regular but not one-valued in a region, provided that $|f(z)|$ is one-valued (e.g. $\sqrt{z}$ in a ring-shaped region surrounding the origin); for the above proofs hold for any branch of the function.

5.16. Let $f(z)$ be an analytic function regular for $|z| < R$, and let $M(r)$ denote the maximum of $|f(z)|$ on $|z| = r$. *Then $M(r)$ is a steadily increasing function of r for $r < R$.* For it follows at once from the above theorem that $M(r_1) \leqslant M(r_2)$ if $r_1 < r_2$, and $M(r_1)$ can only be equal to $M(r_2)$ if $f(z)$ is a constant.

Similarly the function $A(r)$, defined in § 2.53 as the maximum of $\mathbf{R}\{f(z)\}$, is an increasing function of r. For

$$e^{A(r)} = \max_{|z|=r} |e^{f(z)}|.$$

5.2. Schwarz's lemma. *If $f(z)$ is an analytic function, regular for $|z| \leqslant R$, and $|f(z)| \leqslant M$ for $|z| = R$, and $f(0) = 0$, then*

$$|f(re^{i\theta})| \leqslant \frac{Mr}{R} \qquad (0 \leqslant r \leqslant R).$$

Let $\phi(z) = f(z)/z$. Then $\phi(z)$ is regular for $|z| \leqslant R$, and

$$|\phi(z)| \leqslant M/R$$

on the circle $|z| = R$. The same inequality therefore holds inside the circle also, and since $|\phi(z)| = |f(z)|/r$ the result follows.

5.21. Vitali's convergence theorem.* *Let $f_n(z)$ be a sequence of functions, each regular in a region D; let*

$$|f_n(z)| \leqslant M$$

for every n and z in D; and let $f_n(z)$ tend to a limit, as $n \to \infty$, at a set of points having a limit-point inside D. Then $f_n(z)$ tends uniformly to a limit in any region bounded by a contour interior to D, the limit being, therefore, an analytic function of z.

It is sufficient to consider the case where D is a circle, and the limit-point is its centre. For then, returning to the general

* This proof is given by Jentzsch (1).

case, we can prove uniform convergence in a circle with centre the limit-point interior to D. Then we can repeat the process with any point of this circle; and so, by the method used in analytic continuation, extend the domain of uniform convergence to any region bounded by a contour interior to D.

We may take the limit-point as origin. Let the radius of the circle D be R. Let

$$f_n(z) = a_{0,n} + a_{1,n}z + \ldots \qquad (|z| \leqslant R). \tag{1}$$

Then $\qquad |f_n(z) - f_n(0)| \leqslant |f_n(z)| + |f_n(0)| \leqslant 2M.$

But $f_n(z) - f_n(0)$ is zero at $z = 0$. Hence, by Schwarz's lemma,

$$|f_n(z) - f_n(0)| \leqslant \frac{2M|z|}{R} \qquad (|z| \leqslant R).$$

Let z' $(\neq 0)$ be a point where the sequence converges. Then

$$\begin{aligned} |f_n(0) - f_{n+m}(0)| &\leqslant |f_n(0) - f_n(z')| + |f_n(z') - f_{n+m}(z')| + \\ &\qquad + |f_{n+m}(z') - f_{n+m}(0)| \\ &\leqslant 4M|z'|/R + |f_n(z') - f_{n+m}(z')|. \end{aligned}$$

We can choose z' so that the first term is arbitrarily small, and then, since $f_n(z')$ tends to a limit, we can choose n so large that the second term is arbitrarily small for all positive values of m. Hence $f_n(0)$, i.e. $a_{0,n}$, tends to a limit, say a_0.

Next consider the function

$$g_n(z) = \{f_n(z) - a_{0,n}\}/z = a_{1,n} + a_{2,n}z + \ldots.$$

This also tends to a limit at z', since, as we have just proved, $a_{0,n}$ tends to a limit. Also

$$|g_n(z)| \leqslant 2M/R$$

for $|z| = R$, and so also for $|z| < R$. Thus $g_n(z)$ satisfies the same conditions as $f_n(z)$ (except for the value of its upper bound), and hence $a_{1,n}$ tends to a limit, say a_1. Similarly $a_{\nu,n}$ tends to a limit for all values of ν.

Finally, the convergence of (1) is uniform with respect to n and z for $|z| \leqslant R - \epsilon$, since, by Cauchy's inequality, $|a_{\nu,n}| \leqslant M/R^\nu$. So, since every term tends to a limit, the sum tends to a limit uniformly for $|z| \leqslant R - \epsilon$. This proves the theorem.

5.22. *From any sequence of functions regular and bounded in D, in the sense of the above theorem, we can select a sub-sequence which converges uniformly in any region interior to D.*

Let $f_n(z)$ be the sequence of functions, and let $|f_n(z)| \leqslant M$ in D. Let $z_1, z_2, \ldots$ be a sequence of points having a limit-point inside D. Then the points $w_n = f_n(z_1)$ all lie inside the circle $|w| \leqslant M$ in the w-plane. Hence they have at least one limit-point; i.e. there is a sequence of values of n, say $n_1, n_2, \ldots$, such that the sequence of functions

$$f_{n_1}(z), f_{n_2}(z), \ldots \tag{1}$$

converges at the point z_1.

Similarly from this sequence of functions we can select a sub-sequence

$$f_{p_1}(z), f_{p_2}(z), \ldots \tag{2}$$

which converges at z_2; and then from this a sub-sequence

$$f_{q_1}(z), f_{q_2}(z), \ldots \tag{3}$$

which converges at z_3; and so on.

Now consider the sequence

$$f_{n_1}(z), f_{p_2}(z), f_{q_3}(z), \ldots$$

formed by taking the diagonal terms of the above double array. Each of these functions belongs to the sequence (1), and so the sequence converges at z_1; each function after the first belongs to (2), and so the sequence converges at z_2; and so on. Therefore the sequence converges at each of the points $z_1, z_2, \ldots$, and so, by Vitali's theorem, uniformly in any region interior to D.

5.23. Montel's theorem.* *Let $f(z)$ be an analytic function of z, regular in the half-strip S defined by $a < x < b$, $y > 0$. If $f(z)$ is bounded in S, and tends to a limit l, as $y \to \infty$, for a certain fixed value ξ of x between a and b, then $f(z)$ tends to this limit l on every line $x = x_0$ in S, and indeed $f(z) \to l$ uniformly for $a+\delta \leqslant x_0 \leqslant b-\delta$.*

Consider the sequence of functions $f_n(z) = f(z+in)$, where $n = 0, 1, 2, \ldots$, in the rectangle R defined by $a < x < b$, $0 < y < 2$. Then $f_n(z) \to l$ at every point of the line $x = \xi$. Hence, by Vitali's theorem, $f_n(z) \to l$ uniformly in any region interior to R, and in particular in the rectangle $a+\delta \leqslant x \leqslant b-\delta$, $\frac{1}{2} \leqslant y \leqslant \frac{3}{2}$. This proves the theorem.

The result may be generalized by means of conformal transformations. For example, let $z = i \log w$. Then the strip in the z-plane becomes an angle in the w-plane, and the theorem states

* Montel (1), Hardy (18), Bohr (4).

that if $\phi(w)$ is bounded in the angle $\alpha < \arg w < \beta$, and $\phi(w) \to l$ as $w \to \infty$ along any line $\arg w =$ constant between α and β, then $\phi(w) \to l$ uniformly in any angle $\alpha+\delta \leqslant \arg w \leqslant \beta-\delta$.

5.24. The following theorem illustrates another way in which the maximum-modulus theorem can be applied.

Let $f(z)$ be regular, and $|f(z)| \leqslant M$, in the circle $|z-a| \leqslant R$, and suppose that $f(a) \neq 0$. Then the number of zeros of $f(z)$ in the circle $|z-a| \leqslant \frac{1}{3}R$ does not exceed $A \log\{M/|f(a)|\}$.

We may suppose that $a = 0$. Let $z_1, z_2,..., z_n$ be the zeros of $f(z)$ in $|z| \leqslant \frac{1}{3}R$, and let

$$g(z) = f(z) \Big/ \prod_{m=1}^{n} \left(1 - \frac{z}{z_m}\right).$$

Then $g(z)$ is regular for $|z| \leqslant R$, and on $|z| = R$ we have $|z/z_m| \geqslant 3$ for $m = 1, 2,..., n$. Hence

$$|g(z)| \leqslant M \Big/ \prod_{m=1}^{n} (3-1) = 2^{-n}M$$

for $|z| = R$, and so also for $|z| < R$. In particular this is true for $z = 0$. Since $g(0) = f(0)$, it follows that

$$|f(0)| \leqslant 2^{-n}M,$$

and hence
$$n \leqslant \frac{1}{\log 2} \log \frac{M}{|f(0)|},$$

the desired result.

The factor $\frac{1}{3}$ can clearly be replaced by any number less than $\frac{1}{2}$. Actually a more complete result can be obtained from Jensen's theorem (§ 3.61). If the zeros in $|z| \leqslant R$ are $r_1, r_2,..., r_N$, then

$$\log \frac{R^N}{r_1 r_2 ... r_N} = \frac{1}{2\pi} \int_0^{2\pi} \log|f(Re^{i\theta})| \, d\theta - \log|f(0)|$$

$$\leqslant \log M - \log|f(0)|.$$

Let the zeros in the circle $|z| \leqslant \delta R$, where $0 < \delta < 1$, be $r_1, r_2,..., r_n$. Then the left-hand side is not less than

$$\log \frac{R^n}{r_1 r_2 ... r_n} \geqslant \log\left(\frac{1}{\delta}\right)^n = n \log \frac{1}{\delta}.$$

Hence
$$n \leqslant \frac{1}{\log 1/\delta} \log \frac{M}{|f(0)|}.$$

5.3. Hadamard's three-circles theorem. *Let $f(z)$ be an analytic function, regular for $r_1 \leqslant |z| \leqslant r_3$. Let $r_1 < r_2 < r_3$, and let M_1, M_2, M_3 be the maxima of $|f(z)|$ on the three circles $|z| = r_1, r_2, r_3$ respectively. Then*

$$M_2^{\log(r_3/r_1)} \leqslant M_1^{\log(r_3/r_2)} M_3^{\log(r_2/r_1)}. \tag{1}$$

Let $\phi(z) = z^\lambda f(z)$, where λ is a constant to be determined later. Then $\phi(z)$ is regular in the ring-shaped region between $|z| = r_1$ and $|z| = r_3$, and $|\phi(z)|$ is one-valued. Hence the maximum of $|\phi(z)|$ occurs on one of the bounding circles, i.e.

$$|\phi(z)| \leqslant \max(r_1^\lambda M_1, r_3^\lambda M_3).$$

Hence on $|z| = r_2$

$$|f(z)| \leqslant \max(r_1^\lambda r_2^{-\lambda} M_1, r_3^\lambda r_2^{-\lambda} M_3). \tag{2}$$

We have now to choose λ to the best advantage, and this is done by making the two expressions in the bracket equal. We therefore define λ by the equation

$$r_1^\lambda M_1 = r_3^\lambda M_3.$$

Thus $$\lambda = -\{\log(M_3/M_1)\}/\{\log(r_3/r_1)\}.$$

With this value of λ, (2) gives

$$M_2 \leqslant (r_2/r_1)^{-\lambda} M_1,$$

and hence $$M_2^{\log(r_3/r_1)} \leqslant (r_2/r_1)^{\log(M_3/M_1)} M_1^{\log(r_3/r_1)}$$
$$= M_1^{\log(r_3/r_2)} M_3^{\log(r_2/r_1)},$$

the required result.

Notice that the case of equality can occur only if $\phi(z)$ is a constant, i.e. if $f(z)$ is a constant multiple of a power of z.

5.31. Convex functions. A function $\phi(x)$ of a real variable x is said to be *convex downwards*, or simply *convex*, if the curve $y = \phi(x)$ between x_1 and x_2 always lies below the chord joining the points $\{x_1, \phi(x_1)\}$ and $\{x_2, \phi(x_2)\}$. Analytically the condition is

$$\phi(x) < \frac{x_2 - x}{x_2 - x_1}\phi(x_1) + \frac{x - x_1}{x_2 - x_1}\phi(x_2) \qquad (x_1 < x < x_2). \tag{1}$$

The function is said to be convex in the wide sense if the sign of equality can also occur.

A convex function is continuous. For if we make $x \to x_1$ in (1), we obtain $\phi(x_1 + 0) \leqslant \phi(x_1)$; and if we make $x_2 \to x$ we obtain $\phi(x) \leqslant \phi(x + 0)$. Hence $\phi(x) = \phi(x + 0)$ for all values of x. Similarly $\phi(x-0) = \phi(x)$ for all values of x. Hence the function is continuous.

If we put $x = \frac{1}{2}(x_1+x_2)$ in (1), we obtain

$$\phi(\tfrac{1}{2}x_1+\tfrac{1}{2}x_2) < \tfrac{1}{2}\{\phi(x_1)+\phi(x_2)\}. \tag{2}$$

This is sometimes taken as the definition of convexity* instead of (1). It is less restrictive than the definition adopted here, and does not involve continuity.

A sufficient condition for $\phi(x)$ to be convex is that $\phi''(x) > 0$; for then $\phi'(x)$ is increasing, and

$$\frac{1}{x-x_1}\int_{x_1}^{x} \phi'(t)\,dt < \phi'(x) < \frac{1}{x_2-x}\int_{x}^{x_2} \phi'(t)\,dt \qquad (x_1 < x < x_2),$$

which gives (1).

5.32. The three-circles theorem as a convexity theorem. Hadamard's three-circles theorem may be expressed by saying that $\log M(r)$ *is a convex function of* $\log r$. For we may write it in the form

$$\log M(r_2) \leqslant \frac{\log r_3 - \log r_2}{\log r_3 - \log r_1}\log M(r_1) + \frac{\log r_2 - \log r_1}{\log r_3 - \log r_1}\log M(r_3),$$

and the sign of equality occurs only if the function is a constant multiple of a power of z.

5.4. Mean values of $|f(z)|$. The mean values

$$I_1(r) = \frac{1}{2\pi}\int_0^{2\pi} |f(re^{i\theta})|\,d\theta, \qquad I_2(r) = \frac{1}{2\pi}\int_0^{2\pi} |f(re^{i\theta})|^2\,d\theta,$$

have properties similar to those of $M(r)$.

5.41. *$I_2(r)$ increases steadily with r, and $\log I_2(r)$ is a convex function of $\log r$.*

Let
$$f(z) = \sum_{n=0}^{\infty} a_n z^n.$$

The fact that $I_2(r)$ is steadily increasing is then obvious from the formula

$$I_2(r) = \sum_{n=0}^{\infty} |a_n|^2 r^{2n}$$

of § 2.5.

* e.g. in Pólya and Szegö, *Aufgaben*, 1, p. 52.

To prove convexity, let $u = \log r$, and let I_2', I_2'' denote derivatives with respect to u. Then

$$\frac{d^2}{du^2}(\log I_2) = \frac{I_2 I_2'' - I_2'^2}{I_2^2},$$

and by Schwarz's inequality

$$I_2'^2 = (\sum |a_n|^2 2ne^{2nu})^2 \leqslant (\sum |a_n|^2 e^{2nu})(\sum |a_n|^2 4n^2 e^{2nu}) = I_2 I_2''.$$

Hence the result.

5.42. *$I_1(r)$ increases steadily with r, and $\log I_1(r)$ is a convex function of $\log r$.*

It is possible to prove this in the same sort of way as the previous theorem,* but the proof is not so easy, since there is no simple expression for I_1 in terms of the coefficients a_n. So we adopt an entirely different method.†

Let $0 < r_1 < r_2 < r_3$, and let $k(\theta)$ and $F(z)$ be defined by

$$k(\theta)f(r_2 e^{i\theta}) = |f(r_2 e^{i\theta})| \qquad (0 \leqslant \theta \leqslant 2\pi),$$

$$F(z) = \frac{1}{2\pi}\int_0^{2\pi} f(ze^{i\theta})k(\theta)\,d\theta.$$

Then $F(z)$ is regular for $|z| \leqslant r_3$, and attains its maximum in this circle on the boundary, say at $z = r_3 e^{i\lambda}$. Hence

$$I_1(r_2) = F(r_2) \leqslant |F(r_3 e^{i\lambda})| \leqslant I_1(r_3),$$

which proves the first part.

Now choose α so that

$$r_1^\alpha I_1(r_1) = r_3^\alpha I_1(r_3).$$

Then

$$r_2^\alpha I_1(r_2) = r_2^\alpha F(r_2) \leqslant \max_{r_1 \leqslant |z| \leqslant r_3} |z^\alpha F(z)| \leqslant r_1^\alpha I_1(r_1) = r_3^\alpha I_1(r_3),$$

and the result follows as in Hadamard's three-circles theorem.

5.5 Theorem of Borel and Carathéodory.‡ This result enables us to deduce an upper bound for the modulus of a function on a circle $|z| = r$, from bounds for its real or imaginary parts on a larger concentric circle $|z| = R$.

Let $f(z)$ be an analytic function regular for $|z| \leqslant R$, and let $M(r)$

* See Hardy (8), and Landau, *Ergebnisse der Funktionentheorie*, § 23.

† Pólya and Szegö, *Aufgaben*, Dritter Abschnitt, No. 308.

‡ See Borel, *Acta M.* 20, and Landau, *Ergebnisse*, § 24.

and $A(r)$ *denote, as usual, the maxima of* $|f(z)|$ *and* $\mathbf{R}\{f(z)\}$ *on* $|z| = r$. *Then for* $0 < r < R$

$$M(r) \leqslant \frac{2r}{R-r}A(R)+\frac{R+r}{R-r}|f(0)|.$$

The result is obvious if $f(z)$ is a constant. If $f(z)$ is not constant, suppose first that $f(0) = 0$. Then $A(R) > A(0) = 0$.

Let
$$\phi(z) = \frac{f(z)}{2A(R)-f(z)}.$$

Then $\phi(z)$ is regular for $|z| \leqslant R$, since the real part of the denominator does not vanish; $\phi(0) = 0$; and, if $f(z) = u+iv$,

$$|\phi(z)|^2 = \frac{u^2+v^2}{\{2A(R)-u\}^2+v^2} \leqslant 1$$

since $-2A(R)+u \leqslant u \leqslant 2A(R)-u$. Hence Schwarz's lemma gives

$$|\phi(z)| \leqslant \frac{r}{R}.$$

Hence
$$|f(z)| = \left|\frac{2A(R)\phi(z)}{1+\phi(z)}\right| \leqslant \frac{2A(R)r}{R-r},$$

and the result stated follows.

If $f(0)$ is not zero, we apply the result already obtained to $f(z)-f(0)$. Thus

$$|f(z)-f(0)| \leqslant \frac{2r}{R-r}\max_{|z|=R}\mathbf{R}\{f(z)-f(0)\} \leqslant \frac{2r}{R-r}\{A(R)+|f(0)|\},$$

and the result again follows. If $A(R) \geqslant 0$, we deduce

$$M(r) \leqslant \frac{R+r}{R-r}\{A(R)+|f(0)|\}.$$

By arguing with $-f(z)$, or with $\pm if(z)$ we obtain similar results in which $A(r)$ is replaced by $\min \mathbf{R}\{f(z)\}$, $\max \mathbf{I}\{f(z)\}$, or $\min \mathbf{I}\{f(z)\}$.

The inequality is thus proved. The form of the right-hand side may be varied to a certain extent. It must, however, contain, besides $A(R)$, a term involving $f(0)$, or we could falsify the result by replacing $f(z)$ by $f(z)+ik$, where k is a sufficiently large real number. Also it must contain a factor, such as $1/(R-r)$, which tends to infinity as $r \to R$. To show this, consider the function $f(z) = -i\log(1-z)$, and let $0 < r < R < 1$. Then $A(R) < \frac{1}{2}\pi$, however near R is to 1; and $f(0) = 0$. But $M(r) \to \infty$ as $r \to 1$.

5.51. The same principle can be extended to the derivatives of $f(z)$. *Under the conditions of the above theorem, with $A(R) \geqslant 0$,*

$$\max_{|z|=r} |f^{(n)}(z)| \leqslant \frac{2^{n+2} n! R}{(R-r)^{n+1}} \{A(R)+|f(0)|\}.$$

For
$$f^{(n)}(z) = \frac{n!}{2\pi i} \int_C \frac{f(w)}{(w-z)^{n+1}}\, dw, \tag{1}$$

where C is the circle with centre $w = z$ and radius $\delta = \frac{1}{2}(R-r)$. On this circle
$$|w| \leqslant r+\tfrac{1}{2}(R-r) = \tfrac{1}{2}(R+r),$$
so that Carathéodory's theorem gives

$$\max|f(w)| \leqslant \frac{R+\frac{1}{2}(R+r)}{R-\frac{1}{2}(R+r)}\{A(R)+|f(0)|\} < \frac{4R}{R-r}\{A(R)+|f(0)|\}.$$

Hence, by (1),

$$|f^{(n)}(z)| \leqslant \frac{n!}{\delta^n} \frac{4R}{(R-r)}\{A(R)+|f(0)|\} = \frac{2^{n+2} n! R}{(R-r)^{n+1}}\{A(R)+|f(0)|\}.$$

5.6. The theorems of Phragmén and Lindelöf.* The following important extension of the maximum-modulus theorem was given by Phragmén and Lindelöf:

Let C be a simple closed contour, and let $f(z)$ be regular inside and on C, except at one point P of C. Let $|f(z)| \leqslant M$ on C, except at P.

Suppose further that there is a function $\omega(z)$, regular and not zero in C, such that $|\omega(z)| \leqslant 1$ inside C, and such that, if ϵ is any given positive number, we can find a system of curves, arbitrarily near to P and connecting the two sides of C round P, on which

$$|\{\omega(z)\}^\epsilon f(z)| \leqslant M.$$

Then $|f(z)| \leqslant M$ at all points inside C.

To prove this, consider the function

$$F(z) = \{\omega(z)\}^\epsilon f(z),$$

which is regular in C. If z_0 is any point inside C, we can, by the hypothesis about $\omega(z)$, find a curve surrounding z_0 on which

$$|F(z)| \leqslant M.$$

Hence
$$|F(z_0)| \leqslant M,$$
and so
$$|f(z_0)| \leqslant M|\omega(z_0)|^{-\epsilon}$$
Making $\epsilon \to 0$,
$$|f(z_0)| \leqslant M.$$

This proves the theorem.

* Phragmén and Lindelöf-(1).

It is not difficult to see that the exceptional point P may be replaced by any finite number, or even by an infinity, of points, provided that functions $\omega(z)$ corresponding to them with suitable properties can be found.

In the following sections we give a number of theorems of this type. Instead of actually using the above theorem, it is usually simpler to start again with a special auxiliary function adapted to the region considered. In practice the exceptional point P is always at infinity.

5.61. The above theorem gives many important results about the behaviour of a function in the neighbourhood of an essential singularity. By making a preliminary transformation, we can always suppose that the exceptional point is at infinity. The fundamental theorem then takes the following form:

Let $f(z)$ be an analytic function of $z = re^{i\theta}$, regular in the region D between two straight lines making an angle π/α at the origin, and on the lines themselves. Suppose that

$$|f(z)| \leqslant M \tag{1}$$

on the lines, and that, as $r \to \infty$,

$$f(z) = O(e^{r^\beta}), \tag{2}$$

where $\beta < \alpha$, uniformly in the angle. Then actually the inequality (1) *holds throughout the region D.*

We may suppose without loss of generality that the two lines are $\theta = \pm\frac{1}{2}\pi/\alpha$. Let

$$F(z) = e^{-\epsilon z^\gamma} f(z),$$

where $\beta < \gamma < \alpha$ and $\epsilon > 0$. Then

$$|F(z)| = e^{-\epsilon r^\gamma \cos\gamma\theta} |f(z)|. \tag{3}$$

On the lines $\theta = \pm\frac{1}{2}\pi/\alpha$, $\cos\gamma\theta > 0$, since $\gamma < \alpha$. Hence on these lines

$$|F(z)| \leqslant |f(z)| \leqslant M.$$

Also on the arc $|\theta| \leqslant \frac{1}{2}\pi/\alpha$ of the circle $|z| = R$,

$$|F(z)| \leqslant e^{-\epsilon R^\gamma \cos\frac{1}{2}\gamma\pi/\alpha} |f(z)| < Ae^{R^\beta - \epsilon R^\gamma \cos\frac{1}{2}\gamma\pi/\alpha},$$

and the right-hand side tends to 0 as $R \to \infty$. Hence, if R is sufficiently large, $|F(z)| \leqslant M$ on this arc also. Hence, by the maximum-modulus theorem, $|F(z)| \leqslant M$ throughout the interior of the region $|\theta| \leqslant \frac{1}{2}\pi/\alpha$, $r \leqslant R$; i.e., since R is arbitrarily large, throughout the region D. Hence, by (3),

$$|f(z)| \leqslant Me^{\epsilon r^{\gamma}}$$

in D; and making $\epsilon \to 0$ the result stated follows.

It is evidently unnecessary to suppose that the function $f(z)$ is regular in the region $|z| \leqslant r_0$, if there is an arc $|z| = r_1 > r_0$ on which (1) is satisfied. With this extension the theorem is significant for $\alpha < \frac{1}{2}$, the angle including part of the plane more than once, and the function not being necessarily one-valued. We can also replace the straight lines of the theorem by curves extending to infinity; the reader should have no difficulty in supplying the details of such extensions.

5.62. It is important to notice the relation between the 'angle' of the theorem, and the order of $f(z)$ at infinity. The wider the angle is, the smaller the order of $f(z)$ must be for the theorem to be true.

In the following theorem, the order is just not small enough for the previous proof to apply, and a more subtle argument is required.

The conclusion of the previous theorem still holds, if we are only given that

$$f(z) = O(e^{\delta r^{\alpha}})$$

for every positive δ, *uniformly in the angle.*

As before we take the angle to be $-\frac{1}{2}\pi/\alpha \leqslant \theta \leqslant \frac{1}{2}\pi/\alpha$. Let

$$F(z) = e^{-\epsilon z^{\alpha}} f(z).$$

Then $F(z)$ tends to zero on the real axis, and so has an upper bound M' on the real axis. Let

$$M'' = \max(M, M').$$

We may now apply the previous theorem to each of the two angles $(-\frac{1}{2}\pi/\alpha, 0)$ and $(0, \frac{1}{2}\pi/\alpha)$, and we thus find that

$$|F(z)| \leqslant M''$$

throughout the whole given angle.

But in fact $M' \leqslant M$; for $|F(z)|$ attains the value M' at a point of the real axis; hence, if $M' = M''$, $F(z)$ must reduce to a constant, and $M'' = M$. Otherwise $M' < M''$, so that $M'' = M$ in any case. It therefore follows that

$$|F(z)| \leqslant M.$$

Hence $$|f(z)| \leqslant M|e^{\epsilon z^{\alpha}}|,$$

and the result follows on making $\epsilon \to 0$.

5.63. *If $f(z) \to a$ as $z \to \infty$ along two straight lines, and $f(z)$ is regular and bounded in the angle between them, then $f(z) \to a$ uniformly in the whole angle.*

We may suppose without loss of generality that the limit a is 0. We may also suppose that the angle between the two lines is less than π, since the general case can be reduced to this by a substitution of the form $z = w^k$. We may thus suppose that the lines are $\theta = \pm\theta'$, where $\theta' < \frac{1}{2}\pi$.

Let $$F(z) = \frac{z}{z+\lambda} f(z),$$

where $\lambda > 0$. Then

$$|F(z)| = \frac{r}{\sqrt{(r^2+2r\lambda\cos\theta+\lambda^2)}}|f(z)| < \frac{r}{\sqrt{(r^2+\lambda^2)}}|f(z)|.$$

Now $|f(z)| \leqslant M$, say, everywhere, and $|f(z)| < \epsilon$ for $r > r_1 = r_1(\epsilon)$ and $\theta = \pm\theta'$. Let $\lambda = r_1 M/\epsilon$. Then for $r \leqslant r_1$

$$|F(z)| < \frac{r}{\lambda} M < \epsilon$$

and $|F(z)| < |f(z)| < \epsilon$ for $r > r_1$ and $\theta = \pm\theta'$. Hence, by the main Phragmén-Lindelöf theorem, $|F(z)| \leqslant \epsilon$ in the whole region. Hence

$$|f(z)| \leqslant \left(1+\frac{\lambda}{r}\right)|F(z)| < 2\epsilon$$

if $r > \lambda$. This proves the theorem.

5.64. *If $f(z) \to a$ as $z \to \infty$ along a straight line, and $f(z) \to b$ as $z \to \infty$ along another straight line, and $f(z)$ is regular and bounded in the angle between, then $a = b$, and $f(z) \to a$ uniformly in the angle.*

Let $f(z) \to a$ along $\theta = \alpha$, and $f(z) \to b$ along $\theta = \beta$, where $\alpha < \beta$. The function

$$\{f(z)-\tfrac{1}{2}(a+b)\}^2$$

is regular and bounded in the angle, and tends to $\frac{1}{4}(a-b)^2$ on each of the straight lines. Hence it tends to this limit uniformly in the angle; that is,

$$\{f(z)-\tfrac{1}{2}(a+b)\}^2-\tfrac{1}{4}(a-b)^2 = \{f(z)-a\}\{f(z)-b\}$$

tends uniformly to zero. Thus to any ϵ corresponds an arc on which

$$|\{f(z)-a\}\{f(z)-b\}| \leqslant \epsilon.$$

At every point of this arc either $|f(z)-a| \leqslant \sqrt{\epsilon}$ or $|f(z)-b| \leqslant \sqrt{\epsilon}$ (or both), and we may suppose that the former inequality holds at $\theta = \alpha$, the latter at $\theta = \beta$; let θ_0 be the upper bound of values of θ for which the former holds; then θ_0 is a limit of points where the former holds, and is either a point where the latter holds, or a limit of such points; hence, since $f(z)$ is continuous, both inequalities hold at θ_0. Taking z to be this point, we have

$$|a-b| \leqslant |f(z)-a|+|f(z)-b| \leqslant 2\sqrt{\epsilon},$$

and, making $\epsilon \to 0$, it follows that $a = b$. Finally $f(z) \to a$ uniformly, by the previous theorem.

These theorems have obvious affinities with Montel's theorem (§ 5.23). But in Montel's theorem the line along which the function tends to a limit must be interior to the region of boundedness, so that these theorems become corollaries of Montel's only if we assume a slightly wider region of boundedness.

5.65. The Phragmén-Lindelöf theorem for other regions. The angle of the above theorem may be transformed into other regions, for example into a strip.

Take, for example, the theorem of § 5.61, applied to the region $r \geqslant 1$, $|\theta| \leqslant \frac{1}{2}\pi/\alpha$, and put $s = i\log z$, $f(z) = \phi(s)$. If $s = \sigma+it$, the lines $\arg z = \pm\frac{1}{2}\pi/\alpha$ become parallel lines $\sigma = \pm\frac{1}{2}\pi/\alpha$, and $t = \log|z|$. Hence, *if* $|\phi(s)| \leqslant M$ *on the upper half of the two parallel lines and on the segment of the real axis joining them, while*

$$\phi(\sigma+it) = O(e^{e^{\rho t}}) \qquad (\rho < \alpha) \tag{1}$$

in the strip between them, then actually $|\phi(s)| \leqslant M$ *throughout the strip.*

Another theorem of this type, which we shall require in the theory of Dirichlet series, is as follows:

If $\phi(s)$ *is regular and* $O(e^{\epsilon|t|})$, *for every positive* ϵ, *in the strip* $\sigma_1 \leqslant \sigma \leqslant \sigma_2$, *and*

$$\phi(\sigma_1+it) = O(|t|^{k_1}), \qquad \phi(\sigma_2+it) = O(|t|^{k_2}),$$

then

$$\phi(\sigma+it) = O(|t|^{k(\sigma)})$$

uniformly for $\sigma_1 \leqslant \sigma \leqslant \sigma_2$, $k(\sigma)$ *being the linear function of* σ *which takes the values* k_1, k_2 *for* $\sigma = \sigma_1, \sigma_2$.

The result is true more generally if $\phi(s)$ satisfies a condition of the form (1). With the given condition it may be proved directly as follows.

Suppose first that $k_1 = 0$, $k_2 = 0$, so that $\phi(s)$ is bounded for $\sigma = \sigma_1$, $\sigma = \sigma_2$. Let M be the upper bound of $\phi(s)$ on these two lines and on the segment of the real axis between σ_1 and σ_2. Let

$$g(s) = e^{\epsilon s i}\phi(s).$$

Then
$$|g(s)| = e^{-\epsilon t}|\phi(s)| \leqslant |\phi(s)| \leqslant M$$

for $\sigma = \sigma_1$, $\sigma = \sigma_2$. Also $|g(s)| \to 0$ as $t \to \infty$ for $\sigma_1 \leqslant \sigma \leqslant \sigma_2$; and so, if T is large enough, $|g(s)| \leqslant M$ on $t = T$, $\sigma_1 \leqslant \sigma \leqslant \sigma_2$. Hence $|g(s)| \leqslant M$ at all points of the rectangle (σ_1, σ_2), $(0, T)$. Hence $|g(s)| \leqslant M$ at all points in the half-strip, i.e.

$$|\phi(s)| \leqslant e^{\epsilon t}M.$$

Making $\epsilon \to 0$, it follows that $|\phi(s)| \leqslant M$ for $t > 0$, and similarly for $t < 0$. This proves the theorem in the particular case considered.

In the general case, let

$$\psi(s) = (-is)^{k(s)} = e^{k(s)\log(-is)},$$

where the logarithm has its principal value. This function is regular for $\sigma_1 \leqslant \sigma \leqslant \sigma_2$, $t \geqslant 1$; also, if $k(s) = as+b$,

$$\mathbf{R}\{k(s)\log(-is)\} = \mathbf{R}[\{k(\sigma)+iat\}\log(t-i\sigma)]$$
$$= k(\sigma)\log t + O(1).$$

Hence
$$|\psi(s)| = t^{k(\sigma)}e^{O(1)}.$$

The function $\Phi(s) = \phi(s)/\psi(s)$ therefore satisfies the same conditions as $\phi(s)$ did in the first part. Hence $\Phi(s)$ is bounded in the strip, and

$$\phi(s) = O\{|\psi(s)|\} = O(t^{k(\sigma)}).$$

5.7. The Phragmén-Lindelöf function $h(\theta)$. In several of the preceding theorems we have been considering the way in which a function behaves as z tends to infinity in different directions. We shall now make a more systematic study of this question.

Consider first the function

$$f(z) = e^{(a+ib)z^\rho}.$$

Then
$$|f(z)| = e^{r^\rho(a\cos\rho\theta - b\sin\rho\theta)}.$$

The behaviour of $\log|f(z)|$ depends in the first place on the factor r^ρ, which is independent of θ. The different behaviour in different directions is determined by the factor

$$h(\theta) = a\cos\rho\theta - b\sin\rho\theta = r^{-\rho}\log|f(z)|.$$

This is of course a very special case; but the general case is not so different from it as might be expected.

We shall suppose throughout the following sections that $f(z)$ is regular for $\alpha < \theta < \beta$, $|z| \geqslant r_0$, and that $f(z)$ is 'of order ρ' in this angle, i.e. that

$$\overline{\lim}\frac{\log|f(re^{i\theta})|}{r^{\rho+\epsilon}} = 0$$

uniformly in θ, for every positive value of ϵ, but not for any negative value. (For example, the above function is of order ρ.)

We define $h(\theta)$ in general as

$$h(\theta) = \overline{\lim_{r\to\infty}}\frac{\log|f(re^{i\theta})|}{V(r)},$$

where $V(r)$ depends on the function considered. We should naturally choose $V(r)$ so that $h(\theta)$ is finite and not identically zero. Here we shall consider the simplest case $V(r) = r^\rho$; but our argument would apply almost unchanged to any function such as

$$r^\rho(\log r)^p(\log\log r)^q\ldots.$$

5.701. It is convenient to introduce at this point an expression containing the word 'infinity', or the symbol ∞, which is not used in elementary analysis. We shall use $\lim\phi_n = \infty$ to mean the same thing as $\phi_n \to \infty$; and we shall say that $\phi(x)$ has an infinite value, or $\phi(x) = \infty$, if, and only if, $\phi(x)$ is defined as the limit of a sequence $\phi_n(x)$, and the sequence diverges to infinity for the particular value of x in question. We use $-\infty$ in the same way. For example, we might write

$$\int_0^1 \frac{dt}{t} = \infty$$

if the left-hand side is defined as $\lim_{\epsilon\to 0}\int_\epsilon^1$; and $h(\theta) = \infty$ means that $r^{-\rho}\log|f(re^{i\theta})|$ takes arbitrarily large values as $r \to \infty$.

The novelty consists in writing '$=\infty$', as if we had defined a number '∞'; but it should be remembered that we have not done so, and that 'infinity' remains an incomplete symbol.*

5.71. *Let $\alpha < \theta_1 < \theta_2 < \beta$, and $\theta_2 - \theta_1 < \pi/\rho$, and let*

$$h(\theta_1) \leqslant h_1, \qquad h(\theta_2) \leqslant h_2.$$

Let $H(\theta)$ be the function of the form $a\cos\rho\theta + b\sin\rho\theta$ which takes the values h_1, h_2 at θ_1, θ_2. Then

$$h(\theta) \leqslant H(\theta) \qquad (\theta_1 \leqslant \theta \leqslant \theta_2).$$

It is easily seen that

$$H(\theta) = \frac{h_1 \sin\rho(\theta_2 - \theta) + h_2 \sin\rho(\theta - \theta_1)}{\sin\rho(\theta_2 - \theta_1)},$$

but we do not require this expression in the proof.

Let $$H_\delta(\theta) = a_\delta \cos\rho\theta + b_\delta \sin\rho\theta$$

be the H-function which is equal to $h_1 + \delta$, $h_2 + \delta$ $(\delta > 0)$ for $\theta = \theta_1$, $\theta = \theta_2$ respectively. Let

$$F(z) = f(z)e^{-(a_\delta - ib_\delta)z^\rho}.$$

Then $$|F(z)| = |f(z)|e^{-H_\delta(\theta)r^\rho}, \tag{1}$$

and so, if r is large enough,

$$|F(re^{i\theta_1})| = O(e^{(h_1+\delta)r^\rho - H_\delta(\theta_1)r^\rho}) = O(1).$$

A similar result holds for $F(re^{i\theta_2})$. Hence, by the theorem of § 5.61, $F(z)$ is bounded in the angle (θ_1, θ_2). Hence, by (1),

$$f(z) = O(e^{H_\delta(\theta)r^\rho}) \tag{2}$$

uniformly in the angle. Hence $h(\theta) \leqslant H_\delta(\theta)$ for $\theta_1 \leqslant \theta \leqslant \theta_2$. Since $H_\delta(\theta) \to H(\theta)$ as $\delta \to 0$, the result follows.

5.711. As a particular case of the above theorem, one or both of $h(\theta_1)$, $h(\theta_2)$, may be $-\infty$. The conclusion is then that $h(\theta) = -\infty$ for $\theta_1 < \theta < \theta_2$. The same proof still applies, one or both of the numbers h_1, h_2 now being arbitrarily large and negative.

5.712. *If $\alpha < \theta_1 < \theta_2 < \theta_3 < \beta$, $\theta_2 - \theta_1 < \pi/\rho$, $\theta_3 - \theta_2 < \pi/\rho$; and $h(\theta_1)$, $h(\theta_2)$ are finite, and $H(\theta)$ is an H-function such that*

$$h(\theta_1) \leqslant H(\theta_1), \qquad h(\theta_2) = H(\theta_2),$$

then $$h(\theta_3) \geqslant H(\theta_3). \tag{1}$$

* *P.M.* § 55.

Choose θ_1' so that $\theta_3-\pi/\rho < \theta_1' < \theta_2$. Then $h(\theta_1') \leqslant H(\theta_1')$ by §5.71. Hence, by §5.711, $h(\theta_3)$ is not $-\infty$. If (1) is false, we can choose δ so that $h(\theta_3) \leqslant H(\theta_3)-\delta$. Let

$$H_\delta(\theta) = H(\theta)-\delta\frac{\sin\rho(\theta-\theta_1')}{\sin\rho(\theta_3-\theta_1')}.$$

Then

$$h(\theta_1') \leqslant H(\theta_1') = H_\delta(\theta_1'), \qquad h(\theta_3) \leqslant H(\theta_3)-\delta = H_\delta(\theta_3).$$

Hence $$h(\theta_2) \leqslant H_\delta(\theta_2) < H(\theta_2),$$

contrary to hypothesis.

5.713. *If* $\theta_1 < \theta_2 < \theta_3$, $\theta_2-\theta_1 < \pi/\rho$, $\theta_3-\theta_2 < \pi/\rho$, *then*

$$h(\theta_1)\sin\rho(\theta_3-\theta_2)+h(\theta_2)\sin\rho(\theta_1-\theta_3)+h(\theta_3)\sin\rho(\theta_2-\theta_1) \geqslant 0.$$

For any $H(\theta)$

$$H(\theta_1)\sin\rho(\theta_3-\theta_2)+H(\theta_2)\sin\rho(\theta_1-\theta_3)+H(\theta_3)\sin\rho(\theta_2-\theta_1) = 0,$$

and choosing $H(\theta)$ so that $H(\theta_1) = h(\theta_1)$, $H(\theta_2) = h(\theta_2)$, and observing that, by the above theorem, $h(\theta_3) \geqslant H(\theta_3)$, we have the result stated.

The function $h(\theta)$ *is continuous in any interval where it is finite.*

Let $h(\theta)$ be finite in the interval $\theta_1 \leqslant \theta \leqslant \theta_3$, and let $\theta_1 < \theta_2 < \theta_3$. Let $H_{1,2}(\theta)$ be the h-function which takes the values $h(\theta_1)$, $h(\theta_2)$ at θ_1, θ_2; and define $H_{2,3}(\theta)$ similarly. Then by the above theorems

$$H_{2,3}(\theta) \leqslant h(\theta) \leqslant H_{1,2}(\theta) \qquad (\theta_1 \leqslant \theta \leqslant \theta_2)$$

$$H_{1,2}(\theta) \leqslant h(\theta) \leqslant H_{2,3}(\theta) \qquad (\theta_2 \leqslant \theta \leqslant \theta_3).$$

Hence, in whichever of these intervals θ lies,

$$\frac{H_{1,2}(\theta)-H_{1,2}(\theta_2)}{\theta-\theta_2} \leqslant \frac{h(\theta)-h(\vartheta_2)}{\theta-\theta_2} \leqslant \frac{H_{2,3}(\theta)-H_{2,3}(\theta_2)}{\theta-\theta_2}.$$

The extreme terms tend to limits as $\theta \to \theta_2$; hence the middle term is bounded, and so $h(\theta) \to h(\theta_2)$.

It also follows that $|f(re^{i\theta})| < \exp[r^\rho\{h(\theta)+\epsilon\}]$ uniformly for $r > r_0(\epsilon)$; (divide the θ-range into $n = n(\epsilon)$ parts).

5.72. Geometrical interpretation of the property of $h(\theta)$. In the case $\rho = 1$, the property of the function $h(\theta)$ has a simple geometrical interpretation.

For every value of θ in an interval where $h(\theta)$ is finite and positive, consider the radius vector of length $h(\theta)$ making an angle θ with an initial line, and the perpendicular to this radius vector at its end. (Consider, for example, the cases $f(z) = \cosh z$, $f(z) = \cos z + \cosh z$.)

Let h_1, h_2, h_3 be the values of $h(\theta)$ at θ_1, θ_2, θ_3, where $\theta_1 < \theta_2 < \theta_3$. Then the three perpendiculars are

$$x\cos\theta_1 + y\sin\theta_1 = h_1,$$
$$x\cos\theta_2 + y\sin\theta_2 = h_2,$$
$$x\cos\theta_3 + y\sin\theta_3 = h_3.$$

The first and third meet at a point (X, Y) given by

$$X = \frac{h_1\sin\theta_3 - h_3\sin\theta_1}{\sin(\theta_3-\theta_1)}, \qquad Y = \frac{h_3\cos\theta_1 - h_1\cos\theta_3}{\sin(\theta_3-\theta_1)}.$$

Now the condition that (X, Y) should lie on the opposite side from the origin of the second perpendicular (or on it) is

$$X\cos\theta_2 + Y\sin\theta_2 - h_2 \geqslant 0,$$

or

$$(h_1\sin\theta_3 - h_3\sin\theta_1)\cos\theta_2 + {}$$
$$+ (h_3\cos\theta_1 - h_1\cos\theta_3)\sin\theta_2 - h_2\sin(\theta_3-\theta_1) \geqslant 0,$$

or $$h_1\sin(\theta_3-\theta_2) + h_2\sin(\theta_1-\theta_3) + h_3\sin(\theta_2-\theta_1) \geqslant 0.$$

This is precisely the condition which the function $h(\theta)$ satisfies.

If the perpendiculars envelope a curve, then two tangents to it meet on the opposite side to the origin of any tangent at a point between them. It is easily seen geometrically that this means that *the curve is always concave to the origin.*

5.8. The following interesting applications of the Phragmén-Lindelöf principle are due to Carlson.*

Let $f(z)$ be regular and of the form $O(e^{k|z|})$ for $\mathbf{R}(z) \geqslant 0$; and let $f(z) = O(e^{-a|z|})$, where $a > 0$, on the imaginary axis. Then $f(z) = 0$ identically.

We apply the argument of § 5.71 to $f(z)$, with $\rho = 1$, $\theta_1 = 0$, $\theta_2 = \frac{1}{2}\pi$, $h_1 = k$, $h_2 = -a$; and here we can take $\delta = 0$ throughout the argument. Then § 5.71 (2) gives

$$f(z) = O\{e^{(k\cos\theta - a|\sin\theta|)r}\} \tag{1}$$

for $0 \leqslant \theta \leqslant \frac{1}{2}\pi$; and a similar argument shows that (1) also holds for $-\frac{1}{2}\pi \leqslant \theta \leqslant 0$.

Let $$F(z) = e^{\omega z}f(z)$$

where ω is a (large) positive number. Then by (1) there is a

* In an Upsala thesis (1914). See M. Riesz (1), Hardy (14).

constant M, independent of ω, such that

$$|F(z)| \leqslant Me^{\{(k+\omega)\cos\theta - a|\sin\theta|\}r} \qquad (-\tfrac{1}{2}\pi \leqslant \theta \leqslant \tfrac{1}{2}\pi). \tag{2}$$

In particular we have

$$|F(z)| \leqslant M \tag{3}$$

for $\theta = \pm\frac{1}{2}\pi$ and $\theta = \pm\alpha$, where $\alpha = \arctan\{k+\omega)/a\}$.

We can now apply the theorem of § 5.61 to each of the three angles $(-\frac{1}{2}\pi, -\alpha)$, $(-\alpha, \alpha)$, and $(\alpha, \frac{1}{2}\pi)$. It follows that (3) actually holds for $-\frac{1}{2}\pi \leqslant \theta \leqslant \frac{1}{2}\pi$.

Hence $|f(z)| \leqslant Me^{-\omega r\cos\theta}$, and making $\omega \to \infty$ it follows that $|f(z)| = 0$. This proves the theorem.

5.81. *If $f(z)$ is regular and of the form $O(e^{k|z|})$, where $k < \pi$, for $\mathbf{R}(z) \geqslant 0$, and $f(z) = 0$ for $z = 0, 1, 2, 3,...,$ then $f(z) = 0$ identically.*

Consider the function $F(z) = f(z)\operatorname{cosec}\pi z$. On the circles $|z| = n+\frac{1}{2}$, $\operatorname{cosec}\pi z$ is bounded. Hence $F(z) = O(e^{k|z|})$ on these circles, and also on the imaginary axis. Since $F(z)$ is regular it follows that, if $n-\frac{1}{2} < |z| < n+\frac{1}{2}$,

$$F(z) = O(e^{k(n+\frac{1}{2})}) = O(e^{k|z|}),$$

and so $F(z)$ is of this form throughout $\mathbf{R}(z) \geqslant 0$. Also

$$F(z) = O(e^{(k-\pi)|z|})$$

on the imaginary axis. The result therefore follows from the previous theorem.

MISCELLANEOUS EXAMPLES

1. A function $f(z)$ is regular inside and on a simple closed contour C, and $|f(z)| \leqslant M$ on C. Deduce from Cauchy's integral for $\{f(z)\}^n$ that, if z is inside C,

$$|f(z)|^n \leqslant KM^n,$$

where K is independent of n. Hence show that $|f(z)| \leqslant M$ inside C. [Landau.]

2. Use Poisson's integral to show that a function which is harmonic in a region cannot have a maximum at an interior point of the region.

3. If $f(z)$ is regular and $O(e^{r^{1-\epsilon}})$ for $\mathbf{R}(z) \geqslant 0$, $|f(z)| \leqslant M$ on the imaginary axis, and $f(1) = 0$, then for $x > 0$

$$|f(x+iy)| \leqslant \left\{\frac{(1-x)^2+y^2}{(1+x)^2+y^2}\right\}^{\frac{1}{2}} M.$$

[Consider $(1+z)/(1-z).f(z)$].

4. A function $f(z)$ is regular and satisfies the inequalities

$$e^{-r^{\rho+\epsilon}} < |f(z)| < e^{r^{\rho+\epsilon}}$$

in an angle $\theta_1 \leqslant \theta \leqslant \theta_2$, where $\theta_2 - \theta_1 < \pi/\rho$. As $r \to \infty$, $r^{-\rho}\log|f(z)|$ tends to the limits h_1 and h_2 for $\theta = \theta_1, \theta_2$. Let $H(\theta)$ be the function of the form $a \cos \rho\theta + b \sin \rho\theta$ which takes the values h_1, h_2 for $\theta = \theta_1, \theta_2$. Then $h(\theta) = H(\theta)$ throughout the interval $\theta_1 \leqslant \theta \leqslant \theta_2$.

5. Show that, if $f(z) = O(e^{r^{1+\epsilon}})$ in a given angle, the function

$$h(\theta) = \overline{\lim} \frac{\log|f(re^{i\theta})|}{r \log r}$$

has properties similar to those of the h-functions considered in the text.

Show that if $f(z) = 1/\Gamma(\frac{1}{2}+z)$, then $h(\theta) = -\cos\theta$ for all values of θ.

6. An analytic function $f(z)$ is regular and not zero in the half-strip defined by $a < x < b$, $y > 0$; $f(z) = O(y^A)$ as $y \to \infty$ uniformly in the strip, and $|\log f(z)|$ is bounded on the middle line $x = \frac{1}{2}(a+b)$. Prove that $\log f(z) = O(\log y)$ uniformly for $a+\delta < x < b-\delta$.

[Apply Carathéodory's theorem to $\log f(z)$ in a circle with centre at $\frac{1}{2}(a+b)+iy$.]

7. A function $f(z)$ is regular, and $|f(z)| \leqslant M$, for $\mathbf{R}(z) \geqslant 0$, and $f(z)$ has zeros at $z_1, z_2, \ldots$ in this half-plane. Prove that

$$|f(z)| \leqslant \left|\frac{z_1-z}{\bar{z}_1+z}\frac{z_2-z}{\bar{z}_2+z}\cdots\frac{z_n-z}{\bar{z}_n+z}\right| M$$

for $\mathbf{R}(z) > 0$; and deduce that, if $f(z)$ is not identically zero, the series

$$\sum \mathbf{R}\left(\frac{1}{z_n}\right)$$

is convergent. [See Pólya and Szegö, Absch. III, Nos. 295, 298.]

CHAPTER VI

CONFORMAL REPRESENTATION

6.1. Conformal representation. If w is an analytic function of z, then to values of z, which we represent as points in the z-plane, correspond values of w, which we represent as points in the w-plane. We also speak of the point in the w-plane representing its corresponding point in the z-plane; and of regions of the z-plane being represented, or mapped, on corresponding regions of the w-plane. The object of this chapter is to discuss in more detail the nature of this representation or mapping.

Let $w = f(z)$ be an analytic function of z, regular and one-valued in a region D of the z-plane. Let z_0 be an interior point of D; and let C_1 and C_2 be two continuous curves passing through z_0, and having definite tangents at this point, making angles α_1, α_2, say, with the real axis.

We have to discover what is the representation of this figure in the w-plane. Before we go any further, we shall make a restriction, the reason for which will appear in a moment. *We shall suppose that $f'(z_0)$ is not zero.*

Let z_1 and z_2 be points of the curves C_1 and C_2 near to z_0. We shall suppose that they are at the same distance r from z_0, so that we can write

$$z_1 - z_0 = re^{i\theta_1}, \qquad z_2 - z_0 = re^{i\theta_2}.$$

Then as $r \to 0$, $\theta_1 \to \alpha_1$, and $\theta_2 \to \alpha_2$.

The point z_0 corresponds to a point w_0 in the w-plane, and z_1 and z_2 correspond to points w_1 and w_2 which describe curves C_1' and C_2'. Let

$$w_1 - w_0 = \rho_1 e^{i\phi_1}, \qquad w_2 - w_0 = \rho_2 e^{i\phi_2}.$$

Then, by the definition of an analytic function,

$$\lim \frac{w_1 - w_0}{z_1 - z_0} = f'(z_0).$$

Since $f'(z_0)$ is not zero, we may write it in the form $Re^{i\delta}$. Then

$$\lim \frac{\rho_1 e^{i\phi_1}}{re^{i\theta_1}} = Re^{i\delta}.$$

Hence $\lim(\phi_1 - \theta_1) = \delta$, i.e.

$$\lim \phi_1 = \alpha_1 + \delta.$$

Hence the curve C_1' has a definite tangent at w_0, making an angle $\alpha_1+\delta$ with the real axis.

Similarly C_2' has a definite tangent at w_0, making an angle $\alpha_2+\delta$ with the real axis.

Hence the curves C_1', C_2' intersect at the same angle as the curves C_1, C_2. Also the angle between the curves has the same sense in the two figures.

Because of this property of the conservation of angles, an analytic representation is called 'conformal'. Any small figure in one plane corresponds to an approximately similar figure in the other plane, since all angles are approximately the same. To obtain one figure from the other we must rotate it through a certain angle—the angle $\delta = \arg\{f'(z_0)\}$ of the above notation —and subject it to a certain magnification, viz.

$$\lim \frac{\rho_1}{r} = R = |f'(z_0)|.$$

It is clear from the above analysis that the magnification is the same in all directions.

6.11. The case $f'(z)=0$. Suppose now that $f'(z)$ has a zero of order n at the point z_0. Then in the neighbourhood of this point

$$f(z) = f(z_0)+a(z-z_0)^{n+1}+\dots$$

where $a \neq 0$. Hence

$$w_1-w_0 = a(z_1-z_0)^{n+1}+\dots,$$

i.e.

$$\rho_1 e^{i\phi_1} = |a| r^{n+1} e^{i\{\delta+(n+1)\theta_1\}}+\dots,$$

where $\delta = \arg a$. Hence

$$\lim \phi_1 = \lim\{\delta+(n+1)\theta_1\}$$
$$= \delta+(n+1)\alpha_1.$$

Similarly

$$\lim \phi_2 = \delta+(n+1)\alpha_2.$$

Thus the curves C_1', C_2' still have definite tangents at w_0, but the angle between the tangents is

$$\lim(\phi_2-\phi_1) = (n+1)(\alpha_2-\alpha_1).$$

Also the linear magnification, $\lim \rho_1/r$, is zero. The conformal property therefore does not hold at such a point.

6.12. In the above conformal representations we have, not merely conservation of angles, but conservation of the sign of angles; if we get from C_1 to C_2 by a rotation through an angle

α in the positive sense, we also get from C_1' to C_2' by a rotation through α in the positive sense.

There are also conformal representations in which the magnitude of angles is conserved, but their sign is changed. Consider for example, the transformation

$$w = \bar{z},$$

where $\bar{z}$ is the complex number conjugate to z. This replaces every point by its reflection in the real axis. Hence angles are conserved, but their signs are changed. And this is true generally for every transformation of the form

$$w = f(\bar{z}),$$

where $f(z)$ is an analytic function of z; for this is the product of two transformations:

$$\text{(i)}\ Z = \bar{z}, \qquad \text{(ii)}\ w = f(Z).$$

In (i) angles are conserved, their signs changed. In (ii) angles and signs are conserved. Hence in the resulting transformation angles are conserved and their signs changed.

6.2. Linear* transformation. The function

$$w = \frac{az+b}{cz+d}$$

is called a linear function of z. We shall suppose that

$$ad-bc \neq 0,$$

for otherwise the numerator and denominator are proportional, and w is merely a constant.

To every value of z corresponds just one value of w. This is apparent except, in the case $c \neq 0$, for the value $z = -d/c$, which makes the denominator vanish. But as $z \to -d/c$, $|w| \to \infty$; and we may regard the point at infinity in the w-plane as corresponding to the point $z = -d/c$ in the z-plane.

If $c = 0$, then

$$w = \frac{a}{d}z + \frac{b}{d}$$

and (since $a \neq 0$) the points at infinity in the two planes correspond.

Conversely

$$z = \frac{dw-b}{-cw+a},$$

so that z is a linear function of w.

* Or bilinear.

Example. Prove that in general there are two values of z ('invariant points') for which $w = z$, but that there is one only if

$$(a-d)^2+4bc = 0.$$

Show that, if there are distinct invariant points p and q, the transformation may be put in the form

$$\frac{w-p}{w-q} = k\frac{z-p}{z-q};$$

and that, if there is only one invariant point p, the transformation may be put in the form

$$\frac{1}{w-p} = \frac{1}{z-p}+k.$$

6.21. Circles.

The equation

$$|z-z_0| = \rho$$

represents a circle with centre z_0 and radius ρ.

Two points p, q are said to be inverse with respect to the circle if they are collinear with the centre and on the same side of it, and if the product of their distances from the centre is equal to ρ^2. Thus, if

$$p = z_0+le^{i\lambda},$$

then

$$q = z_0+\frac{\rho^2}{l}e^{i\lambda}.$$

If

$$z = z_0+\rho e^{i\theta}$$

is any point of the circle, then

$$\left|\frac{z-p}{z-q}\right| = \left|\frac{\rho e^{i\theta}-le^{i\lambda}}{\rho e^{i\theta}-\rho^2 l^{-1}e^{i\lambda}}\right| = \frac{l}{\rho}\left|\frac{\rho e^{i\theta}-le^{i\lambda}}{le^{i\theta}-\rho e^{i\lambda}}\right| = \frac{l}{\rho}.$$

This is therefore a new form of the equation of the circle.

Conversely, any equation

$$\left|\frac{z-p}{z-q}\right| = k \qquad (k \neq 1)$$

represents a circle* with respect to which p and q are inverse points. For the equation gives

$$|z|^2-2\mathbf{R}(\bar{p}z)+|p|^2 = k^2\{|z|^2-2\mathbf{R}(\bar{q}z)+|q|^2\},$$

or

$$|z|^2-2\frac{\mathbf{R}\{(\bar{p}-k^2\bar{q})z\}}{1-k^2}+\frac{|p|^2-k^2|q|^2}{1-k^2} = 0,$$

or

$$\left|z-\frac{p-k^2q}{1-k^2}\right|^2 = \frac{|p-k^2q|^2}{(1-k^2)^2}-\frac{|p|^2-k^2|q|^2}{1-k^2}.$$

Since

$$|p-k^2q|^2-(1-k^2)(|p|^2-k^2|q|^2) = k^2|p-q|^2,$$

* The 'circle of Apollonius'.

as is easily verified, we obtain

$$\left|z-\frac{p-k^2q}{1-k^2}\right|=\frac{k|p-q|}{|1-k^2|}.$$

The equation therefore represents a circle, with centre

$$z_0=\frac{p-k^2q}{1-k^2},$$

and radius

$$\rho=\frac{k|p-q|}{|1-k^2|}.$$

Also

$$p-z_0=\frac{k^2(q-p)}{1-k^2}, \qquad q-z_0=\frac{q-p}{1-k^2},$$

so that $(p-z_0)/(q-z_0)$ is real and positive, and

$$|p-z_0||q-z_0|=\rho^2.$$

Hence p and q are inverse points.

In the particular case $k=1$, z is equidistant from the points p and q, and therefore lies on the perpendicular bisector of the line joining them.

6.22. Linear transformation of a circle. *In a linear transformation, a circle transforms into a circle, and inverse points transform into inverse points. In the particular case in which the circle becomes a straight line, inverse points become points symmetrical about the line.*

For let

$$\left|\frac{z-p}{z-q}\right|=k$$

be a circle (or straight line), with p and q as inverse points. Let

$$w=\frac{az+b}{cz+d}, \qquad z=\frac{dw-b}{-cw+a}.$$

Then the circle transforms into

$$\left|\frac{dw-b-p(-cw+a)}{dw-b-q(-cw+a)}\right|=k$$

or

$$\left|\frac{w-\dfrac{ap+b}{cp+d}}{w-\dfrac{aq+b}{cq+d}}\right|=k\left|\frac{cq+d}{cp+d}\right|.$$

The result is obvious from this equation.

Example. Prove that the linear transformation in which only one point p is invariant may be considered as the result of (i) an inversion in a circle (centre z_0, say), through the point p, (ii) an inversion in the circle with centre w_0 corresponding to z_0 in the transformation, and touching the previous circle at p.

6.23. *To find all linear transformations of the half-plane* $\mathbf{I}(z) \geqslant 0$ *into the unit circle* $|w| \leqslant 1$.

To points z, $\bar{z}$, symmetrical about the real z-axis correspond points w, $\dfrac{1}{\bar{w}}$, inverse with respect to the unit w-circle. In particular, the origin and the point at infinity in the w-plane correspond to conjugate values of z.

Let
$$w = \frac{az+b}{cz+d}$$
be the required transformation. Plainly $c \neq 0$, or the points at infinity would correspond. Now $w = 0$, $w = \infty$ correspond to $z = -b/a$, $-d/c$. Hence we may write
$$-\frac{b}{a} = \alpha, \qquad -\frac{d}{c} = \bar{\alpha},$$
and
$$w = \frac{a}{c}\,\frac{z-\alpha}{z-\bar{\alpha}}.$$
The point $z = 0$ must correspond to a point of the circle $|w| = 1$, so that
$$\left|\frac{a}{c}\cdot\frac{-\alpha}{-\bar{\alpha}}\right| = \left|\frac{a}{c}\right| = 1.$$

Hence we put
$$a = ce^{i\lambda},$$
where λ is real, and obtain
$$w = e^{i\lambda}\,\frac{z-\alpha}{z-\bar{\alpha}}. \tag{1}$$
Since $z = \alpha$ gives $w = 0$, α must be a point of the upper half-plane, i.e. $\mathbf{I}(\alpha) > 0$. With this condition the function (1) gives the required representation. For if z is real, obviously $|w| = 1$; and if $\mathbf{I}(z) > 0$, then z is nearer to α than to $\bar{\alpha}$, and $|w| < 1$.

There are three arbitrary constants in the transformation, λ, $\mathbf{R}(\alpha)$, and $\mathbf{I}(\alpha)$. We can therefore make three given points of the real axis correspond to three given points of the unit circle.

Example. The general linear transformation of the half-plane $\mathbf{R}(z) \geqslant 0$ on the circle $|w| \leqslant 1$ is
$$w = e^{i\lambda}\,\frac{z-\alpha}{z+\bar{\alpha}} \qquad (\mathbf{R}(\alpha) > 0).$$

6.24. *To find all linear transformations of the unit circle* $|z| \leqslant 1$ *into the unit circle* $|w| \leqslant 1$.

Let

$$w = \frac{az+b}{cz+d}.$$

Here $w = 0$, $w = \infty$, must correspond to inverse points $z = \alpha$, $z = 1/\bar{\alpha}$, where $|\alpha| < 1$. Hence

$$-\frac{b}{a} = \alpha, \qquad -\frac{d}{c} = \frac{1}{\bar{\alpha}},$$

$$w = \frac{a}{c} \frac{z-\alpha}{z-1/\bar{\alpha}} = \frac{a\bar{\alpha}}{c} \cdot \frac{z-\alpha}{\bar{\alpha}z-1}.$$

The point $z = 1$ corresponds to a point on $|w| = 1$. Hence

$$\left| \frac{a\bar{\alpha}}{c} \cdot \frac{1-\alpha}{\bar{\alpha}-1} \right| = \left| \frac{a\bar{\alpha}}{c} \right| = 1.$$

Hence
$$w = e^{i\lambda} \frac{z-\alpha}{\bar{\alpha}z-1},$$

where λ is real.

This is the required result; for if $z = e^{i\theta}$, $\alpha = be^{i\beta}$, then

$$|w| = \left| \frac{e^{i\theta} - be^{i\beta}}{be^{i(\theta-\beta)} - 1} \right| = 1.$$

If $z = re^{i\theta}$, where $r < 1$, then

$$\begin{aligned} &|z-\alpha|^2 - |\bar{\alpha}z-1|^2 \\ &\quad = r^2 - 2rb\cos(\theta-\beta) + b^2 - \{b^2r^2 - 2br\cos(\theta-\beta) + 1\} \\ &\quad = (r^2-1)(1-b^2) < 0. \end{aligned}$$

Hence $|w| < 1$.

If we are also given that $z = 0$ corresponds to $w = 0$, then $\alpha = 0$, and the transformation becomes

$$w = e^{i\lambda} z.$$

If also $\dfrac{dw}{dz} = 1$ at $z = 0$, then

$$w = z.$$

Example. *The general linear transformation of the circle* $|z| \leqslant \rho$ *into the circle* $|w| \leqslant \rho'$ *is*

$$w = \rho\rho' e^{i\lambda} \frac{z-\alpha}{\bar{\alpha}z - \rho^2} \qquad (|\alpha| < \rho).$$

6.25. *If* $f(z)$ *is regular for* $|z| < 1$, $\mathbf{R}\{f(z)\} > 0$, *and* $f(0) = a > 0$, *then* $|f'(0)| \leqslant 2a$.

A result of this type follows from Carathéodory's theorem

and its corollary (§§ 5.5–5.51). The following argument is essentially the same, but can now be put in a form which throws some light on the general method.

Suppose that we can find a linear transformation $g = \phi(f)$ such that $\mathbf{R}(f) = 0$ corresponds to $|g| = 1$, while $f = a$ corresponds to $g = 0$. Then we shall have $|g(z)| < 1$ for $\mathbf{R}\{f(z)\} > 0$, i.e. for $|z| < 1$, and $g(0) = 0$. In this form the data are easy to deal with. We have

$$|g'(0)| = \left| \frac{1}{2\pi i} \int\limits_{|z|=1-\epsilon} \frac{g(z)}{z^2}\, dz \right| \leqslant \frac{1}{1-\epsilon},$$

and hence, making $\epsilon \to 0$, $|g'(0)| \leqslant 1$.

We find, as in § 6.23, that the required linear transformation is

$$g(z) = \frac{f(z)-a}{f(z)+a},$$

or
$$f(z) = a\frac{1+g(z)}{1-g(z)}.$$

Hence
$$f'(z) = \frac{2ag'(z)}{\{1-g(z)\}^2},$$

and
$$|f'(0)| = 2a|g'(0)| \leqslant 2a.$$

6.3. Various transformations. We shall now consider some examples of functions which are not linear.

6.31. The function $w = z^2$. If $z = re^{i\theta}$ and $w = \rho e^{i\phi}$, then $\rho e^{i\phi} = r^2 e^{2i\theta}$, so that

$$\rho = r^2, \qquad \phi = 2\theta.$$

The distance from the origin is therefore squared, and the polar angle is doubled. An angular region $\alpha < \arg z < \beta$ is represented on an angular region $2\alpha < \arg w < 2\beta$; if $\beta - \alpha > \pi$, the angular region in the w-plane covers part of this plane twice. The ambiguity arising from this is removed if we replace the w-plane by the Riemann surface described in § 4.3.

If $z = x+iy$, $w = u+iv$, then

$$u+iv = (x+iy)^2 = x^2-y^2+2ixy,$$

or
$$u = x^2-y^2, \qquad v = 2xy.$$

Hence the straight lines $u = a$, $v = b$ correspond to the rectangular hyperbolas

$$x^2-y^2 = a, \qquad 2xy = b.$$

These cut at right angles except in the case $a = 0$, $b = 0$, when they intersect at the origin at an angle $\frac{1}{4}\pi$. Since dw/dz has a simple zero at the origin, this is in accordance with the general theorems on the transformation of angles.

Examples. (i) Prove that the straight lines x = const., y = const., correspond to systems of confocal parabolas.

(ii) Consider in the same way the function $w = z^n$ for $n = 3, 4, \dots$.

6.32. The function $w=\frac{1}{2}\left(z+\frac{1}{z}\right)$. Here w becomes infinite at $z=0$, while

$$\frac{dw}{dz}=\frac{1}{2}\left(1-\frac{1}{z^2}\right),$$

which vanishes at $z=\pm 1$. These points may therefore be expected to play a special part in the transformation.

Putting $z=re^{i\theta}$, $w=u+iv$, we have

$$u=\frac{1}{2}\left(r+\frac{1}{r}\right)\cos\theta, \qquad v=\frac{1}{2}\left(r-\frac{1}{r}\right)\sin\theta,$$

and, eliminating θ,

$$\frac{u^2}{\frac{1}{4}\left(r+\frac{1}{r}\right)^2}+\frac{v^2}{\frac{1}{4}\left(r-\frac{1}{r}\right)^2}=1.$$

This is an ellipse in the w-plane, and it corresponds to each of the two circles $|z|=r$, $|z|=\frac{1}{r}$. As $r\to 1$, the major semi-axis of the ellipse tends to 1, while the minor semi-axis tends to zero. As $r\to 0$ or as $r\to\infty$, both semi-axes tend to infinity. From this it is clear that the inside and the outside of the unit circle in the z-plane both correspond to the whole w-plane, cut along the real axis from -1 to 1. The unit circle $|z|=1$ itself corresponds to the straight line from -1 to 1 described twice.

On solving the equation for z, we see that the inverse function is a two-valued function of w. We can remove all ambiguity from the representation by replacing the w-plane by a Riemann surface of two sheets, each slit from -1 to 1, and joined crossways along the slit. If we pass round one or other of the points $w=\pm 1$, a different value of z is reached, but, if we pass round both, z returns to its original value.

Examples. To what curve in the w-plane does the line $x = 1$ correspond? Consider the result as an example of § 6.11.

6.33. The logarithmic function. *If $w = \log z$, the angular region $\alpha < \arg z < \beta$ corresponds to the infinite strip $\alpha < v < \beta$ in the w-plane.*

For if $z = re^{i\theta}$, then a value of w is

$$\log r + i\theta.$$

Hence $$u = \log r, \qquad v = \theta.$$

As r goes from 0 to ∞, u goes from $-\infty$ to ∞, and the result follows.

If we consider the general value of $\log z$,

$$w = \log r + i(\theta + 2k\pi),$$

where k is any integer, we obtain, not one strip, but an infinity of strips in the w-plane, corresponding to the infinity of values of the logarithm.

On the other hand, a strip $\alpha < v < \beta$ corresponds to an angle in the z-plane; but, if $\beta - \alpha > 2\pi$, part of the plane will be covered more than once. We can, however, avoid any ambiguity by replacing the single z-plane by a Riemann surface consisting of an infinity of sheets, each cut along the real axis from 0 to $-\infty$, and the upper half-plane of each joined to the lower half-plane of the next along the slit. Then a strip of the w-plane of breadth 2π corresponds to one complete sheet of the Riemann surface, and every point of the Riemann surface corresponds to just one point of the w-plane.

Examples. (i) Investigate the properties of the transformation $w = \tan z$ by considering it as the result of the two transformations

$$w = \frac{1}{i}\frac{\zeta - 1}{\zeta + 1}, \qquad \zeta = e^{2iz},$$

and hence obtain a Riemann surface for the inverse function

$$z = \text{arc} \tan w.$$

[Hurwitz-Courant, *Funktionentheorie*, p. 293.]

(ii) Consider the properties of the transformation

$$w = z^{\alpha}$$

for general values of α.

[The function z^{α} is defined as $e^{\alpha \log z}$. Consider separately rational, irrational, and complex values of α.]

6.34. *If $w = \tan^2 \frac{1}{2}z$, the strip in the z-plane between the lines $x = 0$, $x = \frac{1}{2}\pi$ is represented on the interior of the unit circle in the w-plane, cut along the real axis from $w = -1$ to $w = 0$.*

We have $$w = \frac{1-\cos z}{1+\cos z}.$$

If $z = \frac{1}{2}\pi + iy$, then $\cos z = -i\sinh y$, and $|w| = 1$. It is easily seen that, as y goes from $-\infty$ to ∞, $\operatorname{am} w$ goes from $-\pi$ to π, so that w describes the whole unit circle once.

If $z = iy$, $\cos z = \cosh y$, and w is real. As y goes from $+\infty$ to 0, w goes from -1 to 0, and as y goes from 0 to $-\infty$, w retraces its path from 0 to -1.

The boundary of the strip therefore corresponds to the boundary of the cut circle, and there should be no difficulty in verifying that the interiors correspond.

Example. Prove that the line $x = \frac{1}{4}\pi$ corresponds to a loop of a closed curve, cutting the real axis at $w = -1$ and $w = 1/(3+2\sqrt{2})$.

6.4. Simple ('schlicht') functions.* We shall say that a function $f(z)$ is *simple* in a region D if it is analytic, one-valued, and does not take any value more than once in D.

The function $w = f(z)$ then represents the region D of the z-plane on a region D' of the w-plane, in such a way that there is a one-one correspondence between the points of the two regions.

If $f(z)$ is simple in D, $f'(z) \neq 0$ in D. For suppose, on the contrary, that $f'(z_0) = 0$. Then $f(z)-f(z_0)$ has a zero of order n ($n \geqslant 2$) at z_0. Since $f(z)$ is not constant, we can find a circle $|z-z_0| = \delta$ on which $f(z)-f(z_0)$ does not vanish, and inside which $f'(z)$ has no zeros except z_0. Let m be the lower bound of $|f(z)-f(z_0)|$ on this circle. Then by Rouché's theorem, if $0 < |a| < m$, $f(z)-f(z_0)-a$ has n zeros in the circle (it cannot have a double zero, since $f'(z)$ has no other zeros in the circle). This is contrary to the hypothesis that $f(z)$ does not take any value more than once.

A simple function of a simple function is simple. If $f(z)$ is simple in D, and $F(w)$ in D', then $F\{f(z)\}$ is simple in D; for $F\{f(z_1)\} = F\{f(z_2)\}$ implies $f(z_1) = f(z_2)$, since F is simple; and this implies $z_1 = z_2$, since f is simple.

6.41. Inverse functions. In the above relationship, to every point of D' corresponds just one point of D. We can therefore consider z as a function of w, say $z = \phi(w)$. This is called the inverse function of $w = f(z)$.

* The German word is *schlicht*; Dienes, *The Taylor Series*, uses *biuniform*.

The inverse function is simple in D'. For it is one-valued; and it does not take any value more than once, since $f(z)$ is one-valued. Finally, it is analytic; for if $w_0 = f(z_0)$, then it is easily seen by considering $\int f'(z)/\{f(z)-w\}\,dz$ that, in any given neighbourhood of z_0, $f(z)$ takes every value w sufficiently near to w_0. Hence $z = \phi(w)$ is continuous, and

$$(z-z_0)/(w-w_0) \to 1/f'(z_0)$$

as $w \to w_0$, since $f'(z_0) \neq 0$.

6.42. Uniqueness of conformal transformation. *A simple function $w = f(z)$ which represents the unit circle on itself, so that the centre and a given direction through it remain unaltered, is the identical transformation $w = z$.*

We have $|f(z)| = 1$ for $|z| = 1$, and $f(0) = 0$. Hence, by Schwarz's lemma (§ 5.2),

$$|w| = |f(z)| \leqslant |z|.$$

But, applying Schwarz's lemma to the inverse function, we have $|z| \leqslant |w|$. Hence $|w| = |z|$, i.e.

$$|f(z)/z| = 1 \qquad (|z| \leqslant 1).$$

Since a function of constant modulus is itself constant, it follows that

$$f(z) = az,$$

where $|a| = 1$. The remaining conditions then show that $a = 1$.

Other functions, such as $w = z^2$, satisfy the conditions except that they are not simple.

A simple function which represents the unit circle on itself is a linear function.

If $w = f(z)$ represents the unit circle on itself, and $f(0) = w_0$, we can, by § 6.24, find a linear function $l(w)$ which represents the unit circle on itself, and is such that $l(w_0) = 0$. Then $l\{f(z)\}$ represents the unit circle on itself, and $l\{f(0)\} = 0$. Hence, by the above theorem, $l\{f(z)\} = az$. Since the inverse function of a linear function is linear, $f(z)$ is a linear function of z.

6.43. *Let $f(z)$ be regular at $z = 0$, and $f'(0) \neq 0$. Then $f(z)$ is simple in the immediate neighbourhood of $z = 0$; i.e. in the circle $|z| \leqslant \rho$, if ρ is small enough.*

We may suppose that $f(0) = 0$. Since $f'(0) \neq 0$, the origin is a zero of $f(z)$ of the first order. We can therefore find a circle

C, with centre $z=0$, on which $f(z)\neq 0$, and inside which $f(z)$ has no zero other than $z=0$. Let m be the lower bound of $|f(z)|$ on C.

Since $f(z)$ is continuous and vanishes at $z=0$, we can find a circle $|z|=\rho$ inside which $|f(z)|<m$. Then $w=f(z)$ is simple in this circle. For let w' be any number such that $|w'|<m$. Then by Rouché's theorem (§ 3.42) the number of zeros of $f(z)-w'$ in C is the same as the number of zeros of $f(z)$, that is, one. Hence there is just one point z' in C corresponding to each such value of w'. The region consisting of these values of z' is therefore represented simply on the circle $|w|<m$; and this region includes the circle $|z|=\rho$.

An alternative proof may be obtained by considering the power series

$$f(z)=\sum_{n=0}^{\infty}a_n z^n,$$

where $a_1\neq 0$. If $f(z_1)=f(z_2)$,

$$\sum_{n=1}^{\infty}a_n(z_1^n-z_2^n)=0,$$

i.e.

$$(z_1-z_2)\Big\{a_1+\sum_{n=2}^{\infty}a_n(z_1^{n-1}+z_1^{n-2}z_2+...+z_2^{n-1})\Big\}=0.$$

If $|z_1|<\rho, |z_2|<\rho$, the modulus of the second factor is greater than

$$|a_1|-\sum_{n=2}^{\infty}n|a_n|\rho^{n-1},$$

which is positive if ρ is small enough. Hence $z_1=z_2$, and the result follows.

6.44. *The limit of a uniformly convergent sequence of simple functions is either simple or constant. More precisely, if $f_n(z)$ is simple in D for each value of n, and $f_n(z)\to f(z)$ uniformly in D, then $f(z)$ is simple in D, or is a constant.*

The possibility of the limit being constant is shown by the example $f_n(z)=z/n$.

In any case, $f(z)$ is analytic and one-valued in D. If it is not simple, there are two points z_1 and z_2 at which $w=f(z)$ takes the same value w_0. Describe, with z_1 and z_2 as centres, two circles which lie in D, do not overlap, and such that $f(z)-w_0$ does not vanish on either circumference (this is possible unless $f(z)$ is a constant). Let m be the lower bound of $|f(z)-w_0|$ on the two circumferences. Then we can choose n so large that $|f(z)-f_n(z)|<m$ on the two circumferences. Hence, by Rouché's theorem, the function

$$f_n(z)-w_0=\{f(z)-w_0\}+\{f_n(z)-f(z)\}$$

has as many zeros in the circles as $f(z)-w_0$, that is, at least two. Hence $f_n(z)$ is not simple, contrary to hypothesis. This proves the theorem.

6.45. *Let C be a simple closed contour in the z-plane, enclosing a region D. Let $w=f(z)$ be an analytic function of z, regular in D and on C, and taking no value more than once on C. Then $f(z)$ is simple in D.*

The contour C corresponds to a contour C' in the w-plane. C' is closed, since $f(z)$ is one-valued; and it has no double points, since $f(z)$ does not take any value twice on C. Let D' be the region enclosed by C'.

We assume that $f(z)$ takes in D values other than those on C, say at z_0. Then, if Δ_C denotes variation round C,

$$\frac{1}{2\pi}\Delta_C \arg\{f(z)-f(z_0)\}$$

is equal to the number of zeros of $f(z)-f(z_0)$ in C; it is therefore a positive integer, since there is at least one such zero. But it is also equal to

$$\frac{1}{2\pi}\Delta_{C'} \arg(w-w_0),$$

where $w_0=f(z_0)$; and this is either 0, if w_0 is outside C', or ± 1, if w_0 is inside C', the sign depending on the direction in which C' is described. Hence it is equal to 1. Hence w_0 lies inside C', C' is described in the positive direction, and $f(z)$ takes the value w_0 just once in D. Thus D is represented simply on D'.

6.46. Extensions. The condition in the above analysis that the function $f(z)$ should be analytic on the contour can be relaxed to a certain extent. The state of affairs is not much altered if $f(z)$ is not analytic, but is continuous, at certain points of C. Suppose that there is a singularity at z_1 on the contour, and that C_1 is C with an indentation round z_1. Then the number of zeros of $f(z)-w_0$ inside C_1 is

$$\frac{1}{2\pi}\Delta_{C_1} \arg\{f(z)-w_0\}=\frac{1}{2\pi i}\int_{C_1}\frac{f'(z)}{f(z)-w_0}\,dz;$$

and, as the indentation is closed up, this tends to the corresponding integral round C, if $f(z)$ is continuous and $f'(z) = O(|z-z_1|^\alpha)$, where $\alpha > -1$. The argument of § 6.45 therefore still applies.

The function $f(z)$ may also have a pole on the contour; the region D' then extends to infinity. The theorem of § 6.45 still holds if the pole is of the first order, and the contour is a fairly ordinary one. Suppose that, by a preliminary change of variable, we take the pole at the origin, and that the direction can be taken so that $\mathbf{R}(z) \geqslant 0$ at all points of D. Let

$$w = f(z) = \frac{1}{z} + g(z),$$

where $g(z)$ is regular in D. Then, for z in D,

$$\mathbf{R}(w) \geqslant \min \mathbf{R}\{g(z)\} = a,$$

say. Let $b < a$. Then $|w-b| \geqslant a-b$ for z in D. Hence

$$\zeta = \frac{1}{w-b} = \frac{z}{1+zg(z)-bz}$$

is regular in D. The theorem applies directly to ζ, and so, since w is a simple function of ζ, to w.

The result is not necessarily true for poles of higher order.

Examples. (i) Let $$w = \frac{1}{i}\frac{z+1}{z-1}.$$

If $z = e^{i\theta}$, then $$w = \frac{1}{i}\frac{e^{i\theta}+1}{e^{i\theta}-1} = -\cot\tfrac{1}{2}\theta.$$

Hence, as z describes the unit circle from 0 to 2π, w describes the real axis from $-\infty$ to ∞. The only singularity on the z-boundary is a simple pole. Hence the unit circle in the z-plane is represented simply on the upper half w-plane. This of course is easily verified.

(ii) Let $$w = -\frac{1}{i}\left(\frac{z+1}{z-1}\right)^3.$$

Then, if $z = e^{i\theta}$, $$w = -\cot^3\tfrac{1}{2}\theta.$$

Hence there is a one-one correspondence between the unit circle in the z-plane and the real w-axis. But since there is a triple pole on the boundary the areas do not necessarily correspond. In fact

$$w = i\left(\frac{x+iy+1}{x+iy-1}\right)^3 = i\frac{(x^2+y^2-1)^3-12(x^2+y^2-1)y^2}{\{(x-1)^2+y^2\}^3}+\ldots,$$

and $v = 0$ corresponds to

$$(x^2+y^2-1)(x^2+y^2-2\sqrt{3}y-1)(x^2+y^2+2\sqrt{3}y-1) = 0,$$

i.e. the equation of three circles. Hence $v > 0$ if z is outside each of these circles, or outside one and inside the other two.

The reader should draw a figure showing the regions of the z-plane which correspond to the upper half of the w-plane.

6.5. The function $w = \int\limits_0^z \frac{dt}{\sqrt{(1-t^2)}}$. We know that this function is equal to $\arcsin z$. Consider, however, what can be deduced from the integral about the representation of the z-plane on the w-plane.

Consider what part of the w-plane corresponds to the first quadrant in the z-plane. If $z = iy$ is purely imaginary, we have

$$w = \int\limits_0^y \frac{i\,ds}{\sqrt{1+s^2}},$$

which is also purely imaginary; and as $y \to \infty$, so does w. The two imaginary axes therefore correspond.

Again, as z increases along the real axis from 0, so does w, until z reaches 1, and w reaches the value

$$\int\limits_0^1 \frac{dt}{\sqrt{(1-t^2)}}.$$

Let us denote the value of this integral by I. Actually $I = \frac{1}{2}\pi$, but we need not assume that this is known.

We must now suppose that the path of integration passes above the point $z = 1$, say by a small semicircle. Then $\arg(1-t)$ decreases from 0 to $-\pi$, and so $\arg(1-t)^{-\frac{1}{2}}$ increases from 0 to $\frac{1}{2}\pi$; thus the integrand becomes purely imaginary, and we have

$$w = I + i\int\limits_1^z \frac{dt}{\sqrt{(t^2-1)}}.$$

Finally, as z tends to infinity along the real axis, w tends to infinity along the line $u = I$.

Hence the boundary of the first quadrant in the z-plane corresponds to the boundary of the half-strip $0 < u < I$, $v > 0$ in the w-plane.

Secondly, the function is simple in this region. We cannot deduce this from § 6.45 without some further argument, since both the regions extend to infinity. But it is easily seen directly. For, if t lies in the first quadrant, the imaginary part of t^2-1

is positive, and $\arg\sqrt{(1-t^2)}$ lies between $-\frac{1}{2}\pi$ and 0. Hence

$$f(z_2)-f(z_1)=\int_{z_1}^{z_2}\frac{dt}{\sqrt{(1-t^2)}},$$

and, taking the integral along the straight line, it is of the form

$$k\int\frac{d\lambda}{\rho e^{i\phi}},$$

where λ is a real variable, $\rho>0$, $-\frac{1}{2}\pi<\phi<0$. Such an integral plainly cannot vanish. Here the function cannot take any value twice.

Also the quadrant of the circle $|z|=R$, where R is large, corresponds to a curve which (by the previous remark) has no double point, and which connects the two sides of the strip and lies entirely at a great distance from the real axis. Hence, by the theorem of § 6.45, the quadrant is represented simply on the whole strip.

The next problem is to continue the function beyond this limited region. This can be done by the method of reflection of § 4.51. In fact all the boundaries in each figure are straight lines.

In the first place, the imaginary axes correspond. Hence, reflecting in these lines, we see that the second quadrant in the z-plane corresponds to the half-strip $-I<u<0$, $v>0$, in the w-plane. Hence the upper half of the z-plane corresponds to the half-strip $-I<u<I$, $v>0$.

Next, reflect with respect to the segment $(0, 1)$ of the real axis in the z-plane. We obtain the lower half of the z-plane. In the w-plane we obtain the half-strip $-I<u<I$, $v<0$.

Hence the whole strip $-I<u<I$ corresponds to the whole z-plane, but there are singularities at $z=\pm 1$ round which we must not pass; we may, for example, suppose the plane cut from $-\infty$ to -1 and from 1 to ∞.

Again, a reflection in the segment $(1,\infty)$ of the real z-axis corresponds to a reflection in the line $u=I$ in the w-plane. Hence the lower half of the z-plane (obtained by continuation to the right of $z=1$) corresponds to the half-strip $I<u<3I$, $v>0$.

It is plain that we can continue this process indefinitely. The whole w-plane is divided up into strips of breadth $2I$, each of which corresponds to the whole z-plane.

If we reflect a point w_0 of the strip $-I < u < I$, first in $u = I$ and then in $u = 3I$, we obtain the point w_0+4I. Meanwhile the corresponding z_0, being reflected twice in the real axis, has returned to its original value. Then w_0 and w_0+4I correspond to the same z_0; i.e. if $z = g(w)$, then $g(w) = g(w+4I)$. The inverse function $g(w)$ is therefore periodic, with period $4I$.

Example. Prove that the function

$$w = \int\limits_0^z \frac{dt}{\sqrt{\{(1-t^2)(1-k^2t^2)\}}} \qquad (0 < k < 1)$$

represents the upper half of the z-plane on the rectangle in the w-plane bounded by the lines $u = -K$, $u = K$, $v = 0$, $v = K'$, where

$$K = \int\limits_0^1 \frac{dt}{\sqrt{\{(1-t^2)(1-k^2t^2)\}}}, \qquad K' = \int\limits_1^{1/k} \frac{dt}{\sqrt{\{(t^2-1)(1-k^2t^2)\}}}.$$

Prove that the inverse function $z = g(w)$ has the two periods $4K$ and $2iK'$. [Hurwitz-Courant, *Funktionentheorie*, pp. 302–3.]

6.6. Representation of a polygon on a half-plane. The functions of the previous section are examples of the representation of a polygon on a half-plane. It is possible to do this with *any* polygon. The complete proof would take us too far, but we can show in a general way how it is to be done.

Consider a polygon in the w-plane with n sides and angles $\alpha_1\pi, \alpha_2\pi,..., \alpha_n\pi$, where $\alpha_1+...+\alpha_n = n-2$. If $\alpha_m < 1$ $(m = 1, 2,...n)$ the polygon is convex. Some of the α's may be greater than 1, but the polygon must never cross itself. Suppose that the vertices of the polygon are to correspond to points $a_1, a_2,..., a_n$ on the real axis in the z-plane. So long as z remains on the real axis without passing any of the points $a_1,...$, w remains on the same side of the polygon; hence the angle between the z-curve and the w-curve is constant, i.e. $\arg(dw/dz)$ is constant (see § 6.1).

If $$\frac{dw}{dz} = C(z-a_1)^{\alpha_1-1}(z-a_2)^{\alpha_2-1}...(z-a_n)^{\alpha_n-1},$$

then dw/dz has this property. When z passes the point a_1 by a small circle above it, $\arg(z-a_1)$ decreases from π to 0, the amplitudes of the other factors returning to their original values. Hence $\arg(dw/dz)$ decreases by $\pi(\alpha_1-1)$. Hence the w-curve turns through $\pi(1-\alpha_1)$ in the positive direction. This corresponds to an angle $\pi\alpha_1$ of the polygon.

The required function is therefore of the form

$$w = C\int\limits_{z_0}^{z} (t-a_1)^{\alpha_1-1}(t-a_2)^{\alpha_2-1}...(t-a_n)^{\alpha_n-1}\,dt.$$

The integrand is $O(1/|t|^2)$ as $|t| \to \infty$; hence the integral converges as $z \to \pm\infty$, and to the same value in each case, since the integral along a large semicircle above the real axis tends to zero. Hence, as z describes the real axis, w describes a closed curve, and in fact, from the construction, a polygon with the prescribed angles. By first considering the real z-axis as closed by a large semicircle above it, we can apply the theorems of §§ 6.45–6, and we find that the interior of the polygon is represented simply on the upper half-plane.

To show that we can choose the constants so that a polygon with given sides, as well as given angles, can be represented, is more difficult. For a triangle, however, the result is easily obtained. Consider, for example, the triangle with vertices $w = i\sqrt{3}$, 0, 1 (and angles $\frac{1}{6}\pi$, $\frac{1}{2}\pi$, $\frac{1}{3}\pi$), and let the vertices correspond to $z = -1$, 0, 1. The above theory gives

$$w = C\int\limits_{z_0}^{z} (t+1)^{-\frac{5}{6}}t^{-\frac{1}{2}}(t-1)^{-\frac{2}{3}}\,dt.$$

The origins correspond if $z_0 = 0$; and if we write the formula as

$$w = C'\int\limits_{0}^{z} (t+1)^{-\frac{5}{6}}t^{-\frac{1}{2}}(1-t)^{-\frac{2}{3}}\,dt,$$

where C' is real and positive, the directions of the real axes correspond. Finally, if

$$1 = C'\int\limits_{0}^{1} (t+1)^{-\frac{5}{6}}t^{-\frac{1}{2}}(1-t)^{-\frac{2}{3}}\,dt,$$

then $z = 1$ corresponds to $w = 1$, and the required representation is obtained.

Examples. (i) Prove that the function

$$w = \int\limits_{0}^{z} \frac{dt}{(1-t^2)^{\frac{2}{3}}}$$

represents a half-plane on an equilateral triangle.

(ii) Prove that the function

$$w = \int_0^z \frac{dt}{\sqrt{(1-t^4)}}$$

represents the unit circle in the z-plane on a square in the w-plane. [Hurwitz-Courant. Put $z = (\zeta - i)/(\zeta + i)$.]

6.7. Representation of any region on a circle. A fundamental theorem of Riemann states that *any region with a suitable boundary can be represented on a circle by a simple analytic function.* It is beyond our scope to inquire exactly what forms of region are suitable. The region may be the interior of a closed curve; or one side of a curve which goes to infinity in both directions (e.g. a half-plane); or any form of strip between two such curves; or even the whole plane cut along a curve (e.g. along the real axis from 0 to infinity).

Let D be a region of one of the above types.

The function which represents any region simply on a bounded region must be simple and bounded. Let us first verify that there are such functions for D. Let a and b be two points on the boundary of D, and let

$$w = \sqrt{\left(\frac{z-a}{z-b}\right)}.$$

In D we can restrict ourselves to one branch of this function; this branch is simple, and the values taken by it cover a part only of the w-plane (since both branches together cover the whole w-plane once). Let w_0 be a point of the region not covered. Then $1/(w-w_0)$ is simple and bounded in D. Also

$$f(z) = \frac{p}{w-w_0} + q$$

is simple and bounded, and we can choose p and q so that, at a given point of D, $f(z) = 0$ and $f'(z) = 1$.

Consider all functions $f(z)$ which are simple and bounded in D, and such that $f(z) = 0$ and $f'(z) = 1$ at a given point P of D. Let $M(f)$ denote the maximum modulus of $f(z)$. Let ρ be the lower bound of $M(f)$ for all such functions.

There is then either a function $\phi(z)$ of the set such that $M(\phi) = \rho$; or a sequence $f_1, f_2, \ldots$ of functions of the set such that

$$\lim M(f_n) = \rho.$$

We shall show that the second alternative reduces to the first. Since the sequence $f_n(z)$ is bounded in D, we can, by § 5.22, select from it a partial sequence which tends uniformly to a limit in any region interior to D. Let $f_{n_1}(z)$, $f_{n_2}(z)$,... be such a sequence, and $\phi(z)$ its limit. Then $\phi(z)$ is also a function of the set; for it is bounded, and $\phi(z) = 0$, $\phi'(z) = 1$ at P, and $\phi(z)$ is simple (§ 6.44), being not constant since $\phi'(z) = 1$. Also

$$M(\phi) \geqslant \rho$$

by definition of ρ; and

$$M(f_{n_\nu}) < \rho + \epsilon \qquad (\nu > \nu_0),$$

i.e.

$$|f_{n_\nu}(z)| < \rho + \epsilon \qquad (\nu > \nu_0).$$

Making $\nu \to \infty$, it follows that

$$|\phi(z)| < \rho + \epsilon,$$

i.e. $M(\phi) \leqslant \rho$.

This proves the existence of a function $\phi(z)$ of the set with $M(\phi) = \rho$; and since $\phi(z)$ is not constant, $\rho > 0$.

We shall show that the function $w = \phi(z)$ represents D simply on the circle $|w| < \rho$. In the proof, we may suppose that $\rho = 1$. Let Δ be the region of the w-plane on which $w = \phi(z)$ represents D. Since $M(\phi) = 1$, Δ is included in $|w| \leqslant 1$, and reaches its circumference at one point at least.

If the theorem is not true, Δ has a boundary point α inside the circle ($|\alpha| < 1$). Then each branch of

$$w_1 = \sqrt{\left(\frac{w-\alpha}{\bar{\alpha}w - 1}\right)}$$

is regular for w in Δ. Also $|w_1| \leqslant 1$ if $|w| \leqslant 1$ (§ 6.24), and $w_1(0) = \sqrt{\alpha}$. Let

$$w_2 = \frac{w_1 - \sqrt{\alpha}}{\sqrt{\bar{\alpha}}w_1 - 1}.$$

Then $|w_2| \leqslant 1$ if $|w_1| \leqslant 1$. Also

$$\frac{dw_2}{dw} = \frac{dw_2}{dw_1}\cdot\frac{dw_1^2}{dw}\cdot\frac{1}{2w_1} = \frac{|\alpha|-1}{(\sqrt{\bar{\alpha}}w_1 - 1)^2}\cdot\frac{|\alpha|^2-1}{(\bar{\alpha}w-1)^2}\cdot\frac{1}{2w_1}$$

$$= \frac{|\alpha|-1}{(|\alpha|-1)^2}\cdot\frac{|\alpha|^2-1}{2\sqrt{\alpha}} = \frac{|\alpha|+1}{2\sqrt{\alpha}}$$

at $w = 0$. The modulus of this is greater than unity. Hence

$$w_3 = \frac{2\sqrt{\alpha}}{|\alpha|+1}w_2$$

is a function of the set considered, and

$$M(w_3) = \frac{2|\sqrt{\alpha}|}{|\alpha|+1} < 1.$$

This gives a contradiction, and the theorem follows.

6.71. Uniqueness theorem. Let D be a region in the z-plane which is the interior of a simple closed contour, or which is of one of the other types considered in § 6.7. *Then there is a uniquely determined function $w = f(z)$ which represents D simply on the interior of the unit circle in the w-plane, and is such that, if z_0 is a given point in D, $f(z_0) = 0$ and $f'(z_0)$ is real and positive.*

It follows from § 6.7 that there is one such function, say $w = f(z)$. Let $z = F(w)$ be the inverse function. Suppose that there is another function $w = g(z)$ with the same properties. Then the function $W = g\{F(w)\}$ represents the unit circle simply on itself, the centre and the direction of the real axis through it remaining unaltered. Hence, by § 6.42, $g\{F(w)\} = w$, i.e. $g(z) = f(z)$.

6.8. Further properties of simple functions. The class of functions $f(z)$ which are simple for $|z| < 1$, and such that $f(0) = 0$, $f'(0) = 1$, has been studied in great detail. The function $w = z$ belongs to the class, and represents the unit circle on itself. For all functions of the class the 'map' of the unit circle is subject to certain limitations. For the details we may refer to Bieberbach, *Funktionentheorie*, ii. 82–94, Landau, *Ergebnisse* (ed. 2), pp. 107–14, or Dienes, *The Taylor Series*, Ch. VIII. We shall, however, obtain the simplest property of the map. *For any function of the class, no boundary point of the map of the unit circle is nearer to the origin than the point* $\frac{1}{4}$.

We deduce this from the two following theorems.

Let
$$w = z + \frac{a_1}{z} + \frac{a_2}{z^2} + \ldots$$

be simple for $|z| > 1$, and regular except for the pole at infinity. Then

$$\sum_{n=1}^{\infty} n|a_n|^2 \leqslant 1.$$

Since the function is simple, any circle $|z| = r > 1$ corresponds to a simple closed curve in the w-plane, which encloses a positive area. If $w = u+iv$, $u = u(\theta)$, $v = v(\theta)$, on the curve, the area enclosed is

$$\int_0^{2\pi} u(\theta)v'(\theta)\,d\theta = \int_0^{2\pi} \frac{w(\theta)+\bar{w}(\theta)}{2}\cdot\frac{w'(\theta)-\bar{w}'(\theta)}{2i}\,d\theta$$

$$= \tfrac{1}{4}\int_0^{2\pi} \left\{re^{i\theta}+re^{-i\theta}+\sum_{n=1}^{\infty}\frac{a_n e^{-in\theta}+\bar{a}_n e^{in\theta}}{r^n}\right\}\times$$

$$\times\left\{re^{i\theta}+re^{-i\theta}-\sum_{n=1}^{\infty}\frac{na_n e^{-in\theta}+n\bar{a}_n e^{in\theta}}{r^n}\right\}d\theta$$

$$= \tfrac{1}{2}\pi\left\{\left(r+\frac{\bar{a}_1}{r}\right)\left(r-\frac{a_1}{r}\right)+\left(r+\frac{a_1}{r}\right)\left(r-\frac{\bar{a}_1}{r}\right)-\sum_{n=2}^{\infty}\frac{2na_n\bar{a}_n}{r^{2n}}\right\}$$

$$= \pi r^2-\pi\sum_{n=1}^{\infty} n|a_n|^2 r^{-2n}.$$

Since this is positive

$$\sum_{n=1}^{\infty} n|a_n|^2 r^{-2n} \leqslant r^2,$$

and making $r \to 1$ the result follows.

If $$w = f(z) = z+a_2 z^2+\ldots$$
is simple in $|z| < 1$, *then* $|a_2| \leqslant 2$.

The function

$$F(z) = \sqrt{\{f(z^2)\}} = z+\tfrac{1}{2}a_2 z^3+\ldots$$

is also simple in $|z| < 1$; for it is regular, since $f(z^2)$ does not vanish except at $z = 0$, where it has a double zero; and if $F(z_1) = F(z_2)$, then $f(z_1^2) = f(z_2^2)$, and hence, since $f(z)$ is simple, $z_1^2 = z_2^2$, i.e. $z_1 = \pm z_2$. But $F(z)$ is an odd function, so that $z_1 = -z_2$ gives $F(z_1) = -F(z_2)$. Hence the only solution of $F(z_1) = F(z_2)$ is $z_1 = z_2$, i.e. $F(z)$ is simple.

It follows that

$$\left\{F\left(\frac{1}{z}\right)\right\}^{-1} = z-\frac{\frac{1}{2}a_2}{z}+\ldots$$

is simple for $|z| > 1$. Hence by the previous theorem

$$\tfrac{1}{4}|a_2|^2+\ldots \leqslant 1,$$

and the result follows.

Now let $$w = f(z) = z+a_2 z^2+\ldots$$

be a function of the class considered in the main theorem. Let c be a value which it does not take in the unit circle, i.e. a point outside the 'map' of the unit circle. Then

$$\frac{c\,f(z)}{c-f(z)} = z+\left(a_2+\frac{1}{c}\right)z^2+\ldots$$

is regular and simple for $|z|<1$. Hence

$$\left|a_2+\frac{1}{c}\right| \leqslant 2,$$

$$\left|\frac{1}{c}\right| \leqslant 2+|a_2| \leqslant 4,$$

$$|c| \geqslant \tfrac{1}{4},$$

and the result follows.

Example. The function $z/(1-z)^2$ belongs to the above class. It has $a_2 = 2$, and it gives a map passing through $w = -\frac{1}{4}$.

[The only solution of

$$\frac{z}{(1-z)^2} = \frac{z'}{(1-z')^2}, \qquad |z|<1, \qquad |z'|<1,$$

is $z = z'$.]

MISCELLANEOUS EXAMPLES

1. In a given linear transformation, the point z_0 is such that there is some circle $|z-z_0| = R$ which transforms into a concentric circle $|w-z_0| = R'$. Show that the locus of z_0 is a rectangular hyperbola; and that to each point z_0 on the locus corresponds just one circle (real or imaginary) which transforms into a concentric circle.

2. Show that, if $$\frac{dz}{dw} = -2i\left(w-\frac{1}{w}\right),$$

and the constant of integration is properly chosen, the whole z-plane cut along the semi-infinite lines $x = \pm\pi$, $y \leqslant 0$, corresponds to the upper half of the w-plane.

3. Show that, if $$\frac{dz}{dw} = \frac{w}{\sqrt{(w^2-a^2)}},$$

and a and the constant of integration and the value of the square root are properly chosen, the upper half of the w-plane corresponds to the upper half of the z-plane, cut along the imaginary axis from $z = 0$ to a point $z = ik$.

4. If $f(z)$ is regular inside and on the unit circle, $|f(z)| \leqslant M$ on the circle, and $f(a) = 0$, where $|a| < 1$, then

$$|f(z)| \leqslant M\left|\frac{z-a}{\bar{a}z-1}\right|$$

inside the circle.

5. If $f(z)$ is regular inside and on the unit circle, $|f(z)| \leqslant M$ on the circle, and $f(0) = a$, where $0 < a < M$, then

$$|f(z)| \leqslant M\frac{M|z|+a}{a|z|+M}$$

inside the circle.

[Consider $F(z) = M\{f(z)-a\}/\{af(z)-M^2\}$.]

6. Either branch of the function

$$\frac{z}{\sqrt{(1-z)}}$$

is simple for $|z| < 1$.

7. Show that the function

$$\frac{z}{(1-z)^3}$$

is simple for $|z| < \frac{1}{2}$, but not in any larger circle with centre at the origin.

8. Show that the function

$$f(z) = z+a_2z^2+a_3z^3+\ldots$$

is simple for $|z| < 1$ if

$$\sum_{n=2}^{\infty} n|a_n| \leqslant 1.$$

CHAPTER VII

POWER SERIES WITH A FINITE RADIUS OF CONVERGENCE

7.1. The circle of convergence. We know that every power series has a circle of convergence, within which it converges, and outside which it diverges. The radius of this circle may, however, be infinite, so that the circle includes the whole plane. In this chapter we shall consider power series which have a finite radius of convergence.

The radius of convergence of a power series is determined by the moduli of the coefficients in the series.

The power series

$$\sum_{n=0}^{\infty} a_n z^n \tag{1}$$

has the radius of convergence

$$R = \varliminf_{n\to\infty} |a_n|^{-1/n}. \tag{2}$$

Suppose that R is defined by (2). If z is a point where the series (1) converges, $a_n z^n \to 0$ as $n \to \infty$. Hence, if n is sufficiently large,

$$|a_n z^n| < 1,$$

i.e.

$$|z| < |a_n|^{-1/n}.$$

Making $n \to \infty$, it follows that $|z| \leqslant R$. Hence the radius of convergence does not exceed R.

On the other hand, for sufficiently large values of n,

$$|a_n|^{-1/n} > R - \epsilon,$$

i.e.

$$|a_n| < (R-\epsilon)^{-n}.$$

Hence the series (1) is convergent if $\sum (R-\epsilon)^{-n}|z|^n$ is convergent, i.e. if $|z| < R-\epsilon$. Since ϵ is arbitrarily small, the series (1) is convergent if $|z| < R$. Thus the radius of convergence is at least equal to R. Putting together the two results, the theorem follows.

Examples. (i) Find the radius of convergence of the series

$$\sum_{n=0}^{\infty} \frac{2n!}{(n!)^2} z^n, \quad \sum_{n=1}^{\infty} \frac{n!}{n^n} z^n, \quad \sum_{n=0}^{\infty} n!\, z^{n^2}.$$

(ii) If $R = 1$, and the only singularities on the unit circle are simple poles, then a_n is bounded. [For

$$f(z) = \frac{a}{1-ze^{-i\alpha}} + \frac{b}{1-ze^{-i\beta}} + \ldots + \frac{k}{1-ze^{-i\kappa}} + g(z),$$

where $g(z)$ is regular for $|z| < 1+\delta$ $(\delta > 0)$. Hence $g(z) = \sum b_n z^n$, where $b_n = o(1)$.]

(iii) If $R = 1$, and the only singularities on the unit circle are poles of order p, then $a_n = O(n^{p-1})$.

7.11. We also know from the Cauchy-Taylor theorem that the circle of convergence of the series passes through the singularity or singularities of the function which are nearest to the origin. *Hence the modulus of the nearest singularity can be determined from the moduli of the coefficients in the series.*

7.2. Position of the singularities. While the modulus of the nearest singularities is determined in quite a simple way, their exact position is not usually so easy to find. There are, however, some special cases in which we can identify a particular point as a singularity.

In the following theorems we shall take the radius of convergence to be unity; we can, of course, pass from this to the general case by a simple transformation.

7.21. *If $a_n \geqslant 0$ for all values of n, then $z = 1$ is a singular point.*

Suppose, on the contrary, that $z = 1$ is regular. Then, if we take a point ρ on the real axis between 0 and 1, there is a circle with centre ρ which includes the point 1, and in which the function is regular. If $f(z)$ is the function, the Taylor's series about ρ is

$$\sum_{\nu=0}^{\infty} \frac{f^{(\nu)}(\rho)}{\nu!}(z-\rho)^{\nu}, \tag{1}$$

and this converges at a point $z = 1+\delta$ $(\delta > 0)$. Now

$$f^{(\nu)}(\rho) = \sum_{n=\nu}^{\infty} n(n-1)\ldots(n-\nu+1)a_n \rho^{n-\nu}, \tag{2}$$

and so the above series is

$$\sum_{\nu=0}^{\infty} \frac{(z-\rho)^{\nu}}{\nu!} \sum_{n=\nu}^{\infty} n(n-1)\ldots(n-\nu+1)a_n \rho^{n-\nu}.$$

This is a double series of positive terms, convergent for $z = 1+\delta$.

Hence we may invert the order of the summations, and we obtain

$$\sum_{n=0}^{\infty} a_n \sum_{\nu=0}^{n} \frac{n(n-1)...(n-\nu+1)}{\nu!}(z-\rho)^\nu \rho^{n-\nu}$$

$$= \sum_{n=0}^{\infty} a_n\{(z-\rho)+\rho\}^n = \sum_{n=0}^{\infty} a_n z^n.$$

Hence the original series is convergent for $z = 1+\delta$, contrary to the hypothesis that the radius of convergence is 1. This proves the theorem.

Another proof, due to Pringsheim, is as follows. There is at least one singularity, say $e^{i\alpha}$, on the unit circle. The Taylor's series about $\rho e^{i\alpha}$, where $0 < \rho < 1$, is

$$\sum_{\nu=0}^{\infty} \frac{f^{(\nu)}(\rho e^{i\alpha})}{\nu!}(z-\rho e^{i\alpha})^\nu,$$

and, since $e^{i\alpha}$ is a singularity, this has the radius of convergence $1-\rho$. But it is clear from (2) that, if $a_n \geqslant 0$ for all values of n,

$$|f^{(\nu)}(\rho e^{i\alpha})| \leqslant f^{(\nu)}(\rho).$$

Hence the radius of convergence of (1) does not exceed $1-\rho$. Hence $z = 1$ is a singularity.

7.22. *If a_n is real for all values of n, and $\sum a_n$ is properly divergent, i.e.*

$$s_n = a_0+a_1+...+a_n \to \infty \text{ (or } \to -\infty),$$

then $z = 1$ is a singular point.

We have, for $|z| < 1$,

$$\frac{f(z)}{1-z} = \sum_{n=0}^{\infty} a_n z^n \sum_{n=0}^{\infty} z^n = \sum_{n=0}^{\infty} s_n z^n. \tag{1}$$

by § 1.65, the series being absolutely convergent. Hence

$$f(z) = (1-z)\sum_{n=0}^{\infty} s_n z^n$$

$$= (1-z)\sum_{n=0}^{N} s_n z^n + (1-z)\sum_{n=N+1}^{\infty} s_n z^n = f_1(z)+f_2(z),$$

say. Suppose that $s_n \to \infty$. Then, given any positive number G, however large, we can choose N so that $s_n > G$ $(n > N)$. Then, if $0 < z < 1$,

$$f_2(z) > (1-z)\sum_{n=N+1}^{\infty} Gz^n = Gz^{N+1}.$$

Having fixed N, we can choose z_0 so near to 1 that

$$z^{N+1} > \tfrac{1}{2}, \qquad |f_1(z)| < \tfrac{1}{4}G \qquad (z > z_0),$$

since $z^{N+1} \to 1$ and $f_1(z) \to 0$. Hence

$$f(z) > \tfrac{1}{4}G \qquad (z > z_0),$$

i.e. $f(z) \to \infty$ as $z \to 1$. This proves the theorem.

If we merely know that $|s_n| \to \infty$, we cannot deduce that $z = 1$ is a singularity. For example,

$$\frac{1}{(1+z)^3} = 1-3z+\ldots+(-1)^n\tfrac{1}{2}(n+1)(n+2)z^n+\ldots,$$

and here $|s_n| \sim \tfrac{1}{4}n^2$, though the function is regular at $z = 1$.

7.23. General tests for singular points. If we consider any particular point on the circle of convergence, we can devise a test to determine whether it is a singularity or not; but it is not one which lends itself to simple calculations.

We may suppose that the radius of convergence is 1, and, by a preliminary transformation, we may bring the point to be considered to $z = 1$.

The principle to be used is that, if we expand $f(z)$ about a point on the real axis between 0 and 1, the circle of convergence includes $z = 1$ if $f(z)$ is regular at this point, and not otherwise. But we can make a transformation which brings the formula into a simpler form than the direct application of the principle would give.

Let
$$F(w) = \frac{1}{1-w} f\left(\frac{w}{1-w}\right).$$

Then $F(w)$ is regular for $\mathbf{R}(w) < \tfrac{1}{2}$, since $\mathbf{R}(w) < \tfrac{1}{2}$ gives $|w| < |1-w|$. Now

$$F(w) = \sum_{m=0}^{\infty} \frac{a_m w^m}{(1-w)^{m+1}} = \sum_{m=0}^{\infty} a_m w^m \sum_{r=0}^{\infty} \frac{(m+r)!}{m!\,r!} w^r$$

$$= \sum_{n=0}^{\infty} w^n \sum_{m=0}^{n} \frac{n!}{m!(n-m)!} a_m.$$

Let
$$b_n = \sum_{m=0}^{n} \frac{n!}{m!(n-m)!} a_m.$$

Then a necessary and sufficient condition that $z = 1$ should be a singularity of $f(z)$, i.e. that $w = \frac{1}{2}$ should be a singularity of $F(w)$, is that

$$\varlimsup_{n \to \infty} |b_n|^{-\frac{1}{n}} = \tfrac{1}{2}.$$

For then $F(w)$ has a singularity on $|w| = \frac{1}{2}$, and every point other than $w = \frac{1}{2}$ is known to be regular.

By using other transformations we obtain a variety of other equivalent conditions.

Example. Prove that every point on the unit circle is a singularity of

$$f(z) = \sum_{n=0}^{\infty} z^{2^n}.$$

[For the point $z = e^{i\theta}$, we have to consider

$$b_n = \sum_{m=0}^{n} \frac{n!}{m!(n-m)!} a_m,$$

where $a_m = e^{i2^r\theta}$ if $m = 2^r$, and $a_m = 0$ otherwise. Clearly

$$|b_n| \leqslant \sum_{m=0}^{n} \frac{n!}{m!(n-m)!} = 2^n.$$

Also
$$b_{2^n} = \sum_{m=0}^{n} \frac{2^n!}{2^m!(2^n-2^m)!} e^{i2^m\theta}.$$

The modulus of the term $m = n-1$ is asymptotic to $A2^{2^n-\frac{1}{2}n}$, by Stirling's theorem. Also, if u_m denotes the general term, and $0 < m \leqslant n-2$,

$$\left|\frac{u_m}{u_{m+1}}\right| = \frac{(2^m+1)\dots 2^{m+1}}{(2^n-2^{m+1}+1)\dots(2^n-2^m)} < \left(\frac{2^{m+1}}{2^n-2^m}\right)^{2^m} < \left(\frac{2}{3}\right)^{2^m} < \frac{4}{9},$$

and the remainder is easily seen to be negligible. Hence

$$\lim |b_{2^n}|^{\frac{1}{2^n}} = 2.]$$

7.3. Convergence of the series and regularity of the function. It will be noticed that we have not used the convergence or divergence of the original series as a test for regularity or singularity of the function. In general no such test is possible, for all possible relations can occur. If

$$f(z) = \sum_{n=1}^{\infty} \frac{(-1)^n z^n}{n} = \log \frac{1}{1+z},$$

the series is convergent, and the function regular, at $z = 1$; and the series is divergent, and the function singular, at $z = -1$.

On the other hand, if

$$f(z)=\sum_{n=0}^{\infty}(-1)^n z^n=\frac{1}{1+z}$$

the series is divergent, but the function is regular, at $z=1$; while if

$$f(z)=\sum_{n=1}^{\infty}\frac{z^n}{n^2}=\int_0^z \frac{1}{w}\log\frac{1}{1-w}\,dw$$

the series is convergent at $z=1$, but $f(z)$ has a singularity.

7.31. There is, however, one case in which divergence of the series indicates a singularity of the function; the case where $a_n \to 0$. This follows from the following theorem.

If

$$f(z)=\sum_{n=0}^{\infty}a_n z^n$$

and $a_n \to 0$, *the series is convergent at every point of the unit circle where the function is regular.*

Two proofs of this theorem have been given. One, due to M. Riesz, is essentially a 'complex variable' method, and is given by Landau, *Ergebnisse*, § 18. The following proof, due to W. H. Young (7), is of Fourier-series type. In some respects it is not so simple as Riesz's, but it can easily be adapted to give more general results.

We may without loss of generality take the point in question to be $z=1$; and we may suppose that $f(1)=0$. We have then to prove that $s_n \to 0$.

It follows from § 7.22 (1) that

$$s_n=\frac{1}{2\pi i}\int \frac{f(z)}{1-z}\frac{dz}{z^{n+1}}.$$

Taking the contour to be the circle $|z|=r<1$, we have

$$s_n=\frac{1}{2\pi r^n}\int_{-\pi}^{\pi}\frac{f(re^{i\theta})}{1-re^{i\theta}}e^{-in\theta}\,d\theta.$$

Let $0<\delta<\pi$, and let $\phi(\theta)=\phi(\theta,\delta,r)$ be such that

(i) $\phi(\theta)=1/(1-re^{i\theta})$ for $-\pi<\theta<-\delta$ and $\delta<\theta<\pi$;

(ii) $\phi(\theta)$ and $\phi'(\theta)$ are continuous for $-\pi<\theta<\pi$;

(iii) $|\phi(\theta)|<K$, $|\phi'(\theta)|<K$, $|\phi''(\theta)|<K$, for $-\pi<\theta<\pi$, where K depends on δ but not on r. For example, if

$$\phi(\theta) = a\theta^3+b\theta^2+c\theta+d \qquad (-\delta \leqslant \theta \leqslant \delta),$$

we can determine the coefficients so that

$$\phi(\pm\delta) = \frac{1}{1-re^{\pm i\delta}}, \quad \phi'(\pm\delta) = \frac{ire^{\pm i\delta}}{(1-re^{\pm i\delta})^2}.$$

Then (ii) is satisfied; and a, b, c, and d are linear functions of $\phi(\pm\delta)$ and $\phi'(\pm\delta)$, the moduli of these not exceeding $\frac{1}{2}\operatorname{cosec}\frac{1}{2}\delta$ and $\frac{1}{4}\operatorname{cosec}^2\frac{1}{2}\delta$ respectively. Hence (iii) is satisfied.

We can then write

$$2\pi r^n s_n = \int_{-\delta}^{\delta} \frac{f(re^{i\theta})}{1-re^{i\theta}} e^{-in\theta}\,d\theta + \int_{-\pi}^{\pi} f(re^{i\theta})\phi(\theta)e^{-in\theta}\,d\theta - {}$$

$$- \int_{-\delta}^{\delta} f(re^{i\theta})\phi(\theta)e^{-in\theta}\,d\theta$$

$$= I_1+I_2-I_3.$$

Since $f(z)$ is regular at $z=1$, and $f(1)=0$, we have

$$f(re^{i\theta}) = O(|1-re^{i\theta}|)$$

in an interval $|\theta| \leqslant \theta_0$, uniformly for $r_0 \leqslant r \leqslant 1$. Hence

$$I_1 = \int_{-\delta}^{\delta} O(1)\,d\theta = O(\delta).$$

Suppose now that δ is fixed. We have

$$I_2 = \sum_{m=1}^{\infty} a_m r^m \int_{-\pi}^{\pi} e^{i(m-n)\theta}\phi(\theta)\,d\theta,$$

by uniform convergence; and integrating by parts twice each integral except the nth,

$$I_2 = a_n r^n \int_{-\pi}^{\pi} \phi(\theta)\,d\theta - \sum_{m\neq n} \frac{a_m r^m}{(m-n)^2} \int_{-\pi}^{\pi} e^{i(m-n)\theta}\phi''(\theta)\,d\theta,$$

all the integrated terms cancelling. Let $\epsilon_\nu = \max_{m\geqslant\nu}(|a_m|)$, so that $\epsilon_\nu \to 0$. Then

$$|I_2| \leqslant 2\pi K\left\{\epsilon_n + \epsilon_0 \sum_{m\leqslant\frac{1}{2}n} \frac{1}{(m-n)^2} + \epsilon_{\frac{1}{2}n} \sum_{m>\frac{1}{2}n} \frac{1}{(m-n)^2}\right\}$$

$$= O(\epsilon_n)+O(1/n)+O(\epsilon_{\frac{1}{2}n}).$$

Finally

$$I_3 = \left[\frac{f\phi e^{-in\theta}}{-in}\right]_{\delta}^{-\delta} + \frac{1}{in}\int_{-\delta}^{\delta} (f'\phi + f\phi')e^{-in\theta}\,d\theta = O\left(\frac{1}{n}\right).$$

Given ϵ, we can choose δ so that $|I_1| < \frac{1}{3}\epsilon$ for all values of n; and, having fixed δ, we can choose $n_0 = n_0(\epsilon)$ so large that $|I_2| < \frac{1}{3}\epsilon$ and $|I_3| < \frac{1}{3}\epsilon$ for $n > n_0$. Hence

$$2\pi r^n |s_n| < \epsilon \qquad (n > n_0).$$

Making $r \to 1$, it follows that $2\pi|s_n| < \epsilon$ $(n > n_0)$, i.e. $s_n \to 0$.

The reader will notice that we have not used the full force of the hypothesis '$f(z)$ is regular at $z = 1$'; and the proof would hold with little change if e.g. $f(z) = O(|1-z|^\alpha)$, where $\alpha > 0$. For the more general form of the theorem we must refer the reader to Young's paper.

7.4. Over-convergence.* We know that, at every point outside the circle of convergence of a power series, the series is divergent. But if, instead of considering the whole sequence of partial sums of the series, we consider particular sequences of these sums, it is sometimes possible to obtain a convergent sequence. This is shown by the following example.

Let
$$f(z) = \sum_{n=1}^{\infty} \frac{\{z(1-z)\}^{4^n}}{p_n},$$

where p_n is the maximum coefficient in the polynomial $\{z(1-z)\}^{4^n}$. Then in each of the polynomials

$$\frac{\{z(1-z)\}^{4^n}}{p_n}$$

the moduli of the coefficients do not exceed 1, and one of them is actually equal to 1. Also the highest term in this polynomial is of degree 2.4^n, whereas the lowest term in the next polynomial is of degree 4^{n+1}. Hence, if we expand $f(z)$ in powers of z, each term is a single term of one of the above polynomials. The radius of convergence of this series is 1, since $|a_n| \leqslant 1$ for all n, while $a_n = 1$ for an infinity of values of n.

In particular, the above series of polynomials is convergent for $|z| < 1$. But, since it is formally unchanged by the substitution $z = 1-w$, it is also convergent for $|w| < 1$, i.e. for $|1-z| < 1$.

* Ostrowski (1), Zygmund (1), Estermann (2).

The special sequence of partial sums obtained by taking each polynomial as a whole is therefore convergent in a region which lies partly outside the unit circle.

A power series which has a sequence of partial sums convergent outside the circle of convergence of the series is said to be 'over-convergent'. Of course a power series can only be over-convergent in the neighbourhood of a point of the circle where the function is regular. We shall next define a class of functions which have this property of over-convergence in the neighbourhood of every point of the circle where the function is regular.

7.41. *Suppose that the power series*

$$f(z)=\sum_{n=0}^{\infty} a_n z^n$$

has the radius of convergence 1, *and that there are an infinite number of gaps in the sequence of coefficients, i.e. there are sequences of suffixes p_k, q_k, such that $a_n=0$ for $p_k<n<q_k$; and $q_k \geqslant (1+\vartheta)p_k$, with a fixed positive ϑ.*

Then the sequence of the corresponding partial sums

$$s_{p_k}(z)=\sum_{n=0}^{p_k} a_n z^n$$

is convergent in a region of which every regular point of $f(z)$ on the circle of convergence is an interior point.

To prove this it is sufficient to consider the point $z=1$. Suppose that $f(z)$ is regular at $z=1$. Then, if δ is small enough, it is regular in and on the circle with centre $\frac{1}{2}$ and radius $\frac{1}{2}+\delta$.

We apply Hadamard's three-circles theorem to the function

$$\phi(z)=f(z)-s_{p_k}(z),$$

and the circles with centre $\frac{1}{2}$ and radii $\frac{1}{2}-\delta$, $\frac{1}{2}+\epsilon$, $\frac{1}{2}+\delta$, where $0<\epsilon<\delta$. If M_1, M_2, M_3 are the maximum moduli of $\phi(z)$ on these circles, then

$$M_2^{\log\frac{1+2\delta}{1-2\delta}} \leqslant M_1^{\log\frac{1+2\delta}{1+2\epsilon}} M_3^{\log\frac{1+2\epsilon}{1-2\delta}}. \tag{1}$$

In order to prove that $s_{p_k}(z)\to f(z)$ in a region including $z=1$, it is sufficient to show that we can take ϵ so small that $M_2\to 0$ when $p_k\to\infty$. The idea of the proof is that, while M_3 is substantially of the order $(1+\delta)^{p_k}$, M_1 behaves like $(1-\delta)^{q_k}$, and so, since q_k is greater than p_k, the right-hand side of (1) is small when p_k is large.

To every positive η (say with $\eta < \frac{1}{2}\delta$) corresponds a K such that

$$|a_n| < K(1-\eta)^{-n}.$$

Hence, as $k \to \infty$,

$$M_1 \leqslant |a_{q_k} z^{q_k}| + |a_{q_k+1} z^{q_k+1}| + \ldots$$

$$< K \frac{\left(\frac{1-\delta}{1-\eta}\right)^{q_k}}{1-\frac{1-\delta}{1-\eta}} = O\left\{\left(\frac{1-\delta}{1-\eta}\right)^{q_k}\right\} = O\left\{\left(\frac{1-\delta}{1-\eta}\right)^{(1+\vartheta)p_k}\right\}.$$

Also, if M is the maximum modulus of $f(z)$ on the outer circle,

$$M_3 \leqslant M + |a_0| + \ldots + |a_{p_k} z^{p_k}|$$

$$\leqslant M + K\left\{1 + \frac{1+\delta}{1-\eta} + \ldots + \left(\frac{1+\delta}{1-\eta}\right)^{p_k}\right\} = O\left\{\left(\frac{1+\delta}{1-\eta}\right)^{p_k}\right\}.$$

Hence the right-hand side of (1) is

$$O\left[\left\{\left(\frac{1-\delta}{1-\eta}\right)^{(1+\vartheta)\log\frac{1+2\delta}{1+2\epsilon}} \left(\frac{1+\delta}{1-\eta}\right)^{\log\frac{1+2\epsilon}{1-2\delta}}\right\}^{p_k}\right].$$

When $\epsilon \to 0$, $\eta \to 0$, the expression in brackets tends to

$$(1-\delta)^{(1+\vartheta)\log(1+2\delta)}(1+\delta)^{-\log(1-2\delta)},$$

which is less than 1 if δ is small enough; for its logarithm $\sim -2\vartheta\delta^2$ as $\delta \to 0$, and so is negative for small δ. Hence we may take ϵ and η so small that the original expression is less than 1; and the result then follows.

7.42. The occurrence of gaps in the series is not merely a useful device for producing over-convergence. It has an essential connexion with it. This is shown by the following theorem, which is a sort of converse of the preceding one.

If a sequence $s_{p_k}(z)$ of partial sums of the series $f(z) = \sum a_n z^n$, with radius of convergence 1, is uniformly convergent in the neighbourhood of a point on the unit circle, then

$$f(z) = g(z) + r(z),$$

where the power series $g(z)$ has an infinite number of gaps p_k, q_k, where $q_k > (1+\vartheta)p_k$, and the radius of convergence of the power series $r(z)$ is greater than 1.

We shall not give the proof, which is more difficult than that of the direct theorem.

7.43. Hadamard's gap theorem. *If, in the power series*

$$f(z) = \sum_{n=0}^{\infty} a_n z^n,$$

$a_n = 0$ *except when* n *belongs to a sequence* n_k *such that* $n_{k+1} > (1+\vartheta)n_k$, *where* $\vartheta > 0$, *then the circle of convergence of the series is a natural boundary of the function.*

This is an almost immediate corollary of the theorem on over-convergence. For, if $f(z)$ were regular at any point of the circle, the series would be over-convergent at that point, i.e. the sequence

$$s_{n_k}(z) = \sum_{n=1}^{n_k} a_n z^n$$

would be convergent at a point outside the circle. But for a series of the given form this sequence of partial sums is the same as the whole sequence of partial sums. Hence over-convergence is impossible, and consequently every point of the circle of convergence is a singularity of $f(z)$.

7.44. Mordell's proof of the theorem.* This is a very simple direct proof. Suppose that the radius of convergence is 1. Let $z = aw^p + bw^{p+1}$, where $0 < a < 1$, $a+b = 1$, and p is a positive integer. Clearly $|z| \leqslant 1$ if $|w| \leqslant 1$; and it is easily seen that $|z| < 1$ if $|w| \leqslant 1$, except that $z = 1$ if $w = 1$. Let

$$\phi(w) = f(z) = \sum a_n(aw^p + bw^{p+1})^n$$
$$= \sum a_n(a^n w^{pn} + \ldots + b^n w^{(p+1)n}) = \sum b_n w^n.$$

Then $\phi(w)$ is regular for $|w| \leqslant 1$, except possibly at $w = 1$. We shall show that the radius of convergence of the power series for $\phi(w)$ is 1, and hence that $w = 1$ is a singularity of $\phi(w)$.

We observe that, in the last expression but one for $\phi(w)$, no power of w occurs twice if

$$(p+1)n_k < pn_{k+1},$$

i.e.
$$p\left(\frac{n_{k+1}}{n_k} - 1\right) > 1$$

throughout the series; and this is true if $p > 1/\vartheta$. The expression $\sum b_n w^n$ is then obtained by simply omitting the brackets in the previous expression.

* Mordell (1).

If the series for $\phi(w)$ had a radius of convergence greater than 1, it would be convergent for a real $w > 1$, and therefore the series for $f(z)$ would be convergent for a real $z > 1$, which is false. This proves the theorem.

There is still another proof,* depending on the criterion of § 7.23.

The theorem of § 7.41 can be proved in a similar way.† Let the series for $f(z)$ satisfy the condition of § 7.41. Then $\phi(w)$ can have no singularity for $|w| \leqslant 1$ except possibly at $w = 1$. Hence if $f(z)$ is regular at $z = 1$, $\phi(w)$ is regular at $w = 1$, and so in $|w| < 1+\delta$ for some positive δ. Hence $\sum b_n w^n$ converges for $|w| < 1+\delta$, and in particular $\sum_{n=0}^{(p+1)p_k} b_n w^n$ converges for $|w| < 1+\delta$. Hence $\sum_{n=0}^{p_k} a_n z^n$ converges in a region of which $z = 1$ is an interior point.

7.5. Asymptotic behaviour near the circle of convergence.

If the coefficients in the power series satisfy a sufficiently simple law as $n \to \infty$, we can deduce an asymptotic expression for the function $f(z)$ as z approaches the circle of convergence along a radius vector. The simplest case of this process is given by the following theorem.

Let $$f(x) = \sum_{n=0}^{\infty} a_n x^n, \qquad g(x) = \sum_{n=0}^{\infty} b_n x^n,$$

where $a_n \geqslant 0$, $b_n \geqslant 0$, *and the series converge for* $0 < x < 1$ *and diverge for* $x = 1$. *If, as* $n \to \infty$,
$$a_n \sim C b_n, \tag{1}$$
then as $x \to 1$
$$f(x) \sim C g(x). \tag{2}$$

Given ϵ, we can find N such that
$$|a_n - C b_n| < \epsilon b_n \qquad (n > N).$$

* See Landau, *Ergebnisse*, § 19.

† Pointed out by Mr. M. M. Crum.

Then

$$|f(x)-Cg(x)| = \left|\sum_{n=0}^{\infty}(a_n-Cb_n)x^n\right|$$
$$\leqslant \left|\sum_{n=0}^{N}(a_n-Cb_n)x^n\right| + \left|\sum_{n=N+1}^{\infty}(a_n-Cb_n)x^n\right|$$
$$\leqslant \sum_{n=0}^{N}|a_n-Cb_n| + \epsilon\sum_{n=N+1}^{\infty}b_n x^n$$
$$\leqslant \sum_{n=0}^{N}|a_n-Cb_n| + \epsilon g(x).$$

Having fixed N, we can, since $g(x)\to\infty$, choose δ so that

$$\sum_{n=0}^{N}|a_n-Cb_n| < \epsilon g(x) \qquad (x>1-\delta).$$

Then $$|f(x)-Cg(x)| < 2\epsilon g(x) \qquad (x>1-\delta),$$

which proves the theorem.

The same result, however, holds under more general conditions. *Let the series converge for* $0<x<1$; *let*

$$s_n = a_0+a_1+\ldots+a_n, \qquad t_n = b_0+b_1+\ldots+b_n,$$

and let s_n *and* t_n *be positive, and* $\sum s_n$ *and* $\sum t_n$ *divergent, and let*

$$s_n \sim Ct_n. \tag{3}$$

Then (2) *is still true.*

For as in § 7.22, for $0<x<1$

$$f(x) = (1-x)\sum_{n=0}^{\infty}s_n x^n, \qquad g(x) = (1-x)\sum_{n=0}^{\infty}t_n x^n,$$

and by the previous theorem

$$\sum_{n=0}^{\infty}s_n x^n \sim C\sum_{n=0}^{\infty}t_n x^n.$$

Hence the result.

In particular, *if* $s_n \sim Cn$, *then*

$$f(x) \sim \frac{C}{1-x}.$$

Examples. (i) If $p<1$, as $x\to 1$

$$\sum_{n=1}^{\infty}\frac{x^n}{n^p} \sim \frac{\Gamma(1-p)}{(1-x)^{1-p}}.$$

[We have $$(1-x)^{p-1} = \frac{1}{\Gamma(1-p)}\sum_{n=0}^{\infty}\frac{\Gamma(n-p+1)}{\Gamma(n+1)}x^n,$$

and we can use the lemma of § 1.87.]

(ii) Show that if

$$F(\alpha,\beta,\gamma,x) = 1+\frac{\alpha.\beta}{1.\gamma}x+\frac{\alpha(\alpha+1)\beta(\beta+1)}{1.2.\gamma(\gamma+1)}x^2+\dots,$$

then, as $x \to 1$,

$$F(\alpha,\beta,\gamma,x) \sim \frac{\Gamma(\gamma)\Gamma(\alpha+\beta-\gamma)}{\Gamma(\alpha)\Gamma(\beta)}\frac{1}{(1-x)^{\alpha+\beta-\gamma}}$$

if $\alpha+\beta > \gamma$; and that

$$F(\alpha,\beta,\alpha+\beta,x) \sim \frac{\Gamma(\alpha+\beta)}{\Gamma(\alpha)\Gamma(\beta)}\log\frac{1}{1-x}.$$

7.51. The converse problem. It is easily seen that there is no general converse of the above theorems; from the asymptotic behaviour of $f(x)$ we cannot deduce that of a_n, or even of s_n. Consider, for example, the function

$$f(x) = \frac{1}{(1+x)^2(1-x)} = (1-x)\sum_{n=0}^{\infty}(n+1)x^{2n}$$

$$= \sum_{n=0}^{\infty}(n+1)(x^{2n}-x^{2n+1}).$$

Here $s_{2m+1} = 0$, while $s_{2m} = m+1$; hence s_n oscillates infinitely, though $f(x) \sim \frac{1}{4}/(1-x)$.

The coefficients in this example are, of course, not all positive; and this is, in a sense, the cause of the failure of the converse theorem. If we assume that all the coefficients are positive, we can state a precise converse of the last result of the previous section.

If $a_n \geqslant 0$ for all values of n, and as $x \to 1$

$$f(x) = \sum_{n=0}^{\infty} a_n x^n \sim \frac{1}{1-x},$$

then as $n \to \infty$

$$s_n = \sum_{\nu=0}^{n} a_\nu \sim n.$$

This theorem is due to Hardy and Littlewood.* We shall give an extremely elegant proof which has recently been obtained by Karamata.†

7.52. In order to appreciate the point of the proof, it may be well to see what can be proved by fairly obvious arguments. In the first place

$$f(x) \geqslant \sum_{\nu=0}^{n} a_\nu x^\nu \geqslant x^n s_n$$

* Hardy and Littlewood (2). † Karamata (1).

for all values of x and n. Taking $x = e^{-1/n}$, we obtain, since $f(x) < A/(1-x)$,

$$e^{-1}s_n \leqslant \frac{A}{1-x} = \frac{A}{1-e^{-1/n}} < An,$$

say $$s_n < A_1 n. \tag{1}$$

On the other hand, using (1), we have

$$\begin{aligned} f(x) &= (1-x)\sum_{m=0}^{\infty} s_m x^m \\ &< (1-x)s_n \sum_{m=0}^{n-1} x^m + A_1(1-x)\sum_{m=n}^{\infty} m x^m \\ &< s_n + A_1 n x^n + \frac{A_1 x^{n+1}}{1-x}. \end{aligned}$$

Taking $x = e^{-\lambda/n}$, we obtain, since $f(x) > A/(1-x) > An/\lambda$ if $n > 2\lambda$,

$$\frac{An}{\lambda} < s_n + Ane^{-\lambda} + \frac{Ane^{-\lambda}}{\lambda}.$$

Hence, if λ is sufficiently large,

$$s_n > A_2 n. \tag{2}$$

What we have to show is that A_1 and A_2 can be replaced by $1+\epsilon$ and $1-\epsilon$ respectively. The above argument is too crude to do this, and the method actually used is far from being an obvious one.

7.53. Karamata's proof. The proof depends on the well-known theorem of Weierstrass, that we can approximate uniformly to any continuous function by a sequence of polynomials.* Let $g(x)$ be continuous in $(0, 1)$, and ϵ a given positive number. Then there are polynomials $p(x)$, $P(x)$, such that

$$p(x) \leqslant g(x) \leqslant P(x), \tag{1}$$

and $$\int_0^1 \{g(x)-p(x)\}\,dx \leqslant \epsilon, \qquad \int_0^1 \{P(x)-g(x)\}\,dx \leqslant \epsilon. \tag{2}$$

This is obviously true if $p(x)$ and $P(x)$ differ by at most $\frac{1}{2}\epsilon$ from $g(x)-\frac{1}{2}\epsilon$ and $g(x)+\frac{1}{2}\epsilon$ respectively.

If $g(x)$ has a discontinuity of the first kind in the interval, say at $x = c$, we can still construct polynomials satisfying (1) and (2). Suppose, for example, that $g(c-0) < g(c+0)$. Let

* A proof is given in § 13.33. For another proof see Goursat, *Cours d'Analyse*, t. 1, § 206.

$\phi(x) = g(x)+\frac{1}{2}\epsilon$ for $x < c-\delta$ and for $x > c$; and, for $c-\delta \leqslant x \leqslant c$, let $\phi(x) = \max\{l(x), g(x)+\frac{1}{4}\epsilon\}$, where $l(x)$ is the linear function of x such that $l(c-\delta) = g(c-\delta)+\frac{1}{2}\epsilon$, $l(c) = g(c+0)+\frac{1}{2}\epsilon$. Then $\phi(x)$ is continuous, and $\phi(x) > g(x)$. It is easily seen that, if δ is small enough, a polynomial $P(x)$ which approximates sufficiently closely to $\phi(x)$ has the required properties. Similarly we may construct $p(x)$.

To prove the theorem of Hardy and Littlewood, we first prove that

$$\lim_{x\to 1}(1-x)\sum_{n=0}^{\infty} a_n x^n P(x^n) = \int_0^1 P(t)\,dt \tag{3}$$

for any polynomial $P(x)$. It is clearly sufficient to consider the case $P(x) = x^k$. Then the left-hand side is

$$(1-x)\sum_{n=0}^{\infty} a_n x^{n+kn} = \frac{1-x}{1-x^{k+1}}\left\{(1-x^{k+1})\sum_{n=0}^{\infty} a_n (x^{k+1})^n\right\}$$

$$\to \frac{1}{k+1} = \int_0^1 x^k\,dx,$$

and the result follows.

Next, we have

$$\lim_{x\to 1}(1-x)\sum_{n=0}^{\infty} a_n x^n g(x^n) = \int_0^1 g(t)\,dt \tag{4}$$

if $g(t)$ is continuous, or has a discontinuity of the first kind. For let $p(x)$ and $P(x)$ be polynomials satisfying (1) and (2). Then, since $g(x) \leqslant P(x)$, and the coefficients are positive,

$$\overline{\lim}(1-x)\sum_{n=0}^{\infty} a_n x^n g(x^n) \leqslant \overline{\lim}(1-x)\sum_{n=0}^{\infty} a_n x^n P(x^n)$$

$$= \int_0^1 P(t)\,dt < \int_0^1 g(t)\,dt+\epsilon.$$

Making $\epsilon \to 0$, it follows that

$$\overline{\lim}(1-x)\sum_{n=0}^{\infty} a_n x^n g(x^n) \leqslant \int_0^1 g(t)\,dt.$$

Similarly, arguing with $p(x)$, we obtain

$$\underline{\lim}(1-x)\sum_{n=0}^{\infty} a_n x^n g(x^n) \geqslant \int_0^1 g(t)\,dt,$$

and (4) follows.

Now let

$$g(t) = 0 \ (0 \leqslant t < e^{-1}), \qquad = 1/t \ (e^{-1} \leqslant t \leqslant 1).$$

Then
$$\int_0^1 g(t)\,dt = \int_{1/e}^1 \frac{dt}{t} = 1. \tag{5}$$

Let $x = e^{-1/N}$. Then

$$\sum_{n=0}^{\infty} a_n x^n g(x^n) = \sum_{n \leqslant 1/\log(1/x)} a_n = \sum_{n=0}^{N} a_n = s_N,$$

and so, by (4) and (5), $s_N \sim 1/(1-x) \sim N$. This proves the theorem.

7.6. Abel's theorem and its converse. In this section we return to a subject already discussed in Chapter I. In § 1.22 we proved Abel's theorem for real power series: *if the series*

$$\sum_{n=0}^{\infty} a_n$$

converges to the sum s, *then*

$$f(x) = \sum_{n=0}^{\infty} a_n x^n \to s$$

as $x \to 1$ *through real values.* In § 1.23 we proved Tauber's theorem, that *the converse deduction holds, provided that* $a_n = o(1/n)$. We shall now consider a number of generalizations of these theorems.*

7.61. *If*
$$\sum_{n=0}^{\infty} a_n = s, \tag{1}$$

then
$$f(z) = \sum_{n=0}^{\infty} a_n z^n \to s \tag{2}$$

as $z \to 1$ *along any path lying between two chords of the unit circle which pass through* $z = 1$.

As in § 1.22, it is sufficient to show that the power series is uniformly convergent, but now we must prove uniform convergence in a region included between two chords through $z = 1$, and a sufficiently small circle with centre at $z = 1$.

We have to adapt the argument used to prove Abel's lemma (§ 1.131) to the present conditions. Let

$$s_{n,p} = a_n + a_{n+1} + \ldots + a_p,$$

* Landau, *Ergebnisse*, Ch. III, and Hardy and Littlewood (1), (2), (3), (4).

so that $|s_{n,p}| < \epsilon$ $(n_0 \leqslant n < p)$. Then

$$\sum_{\nu=n}^{m} a_\nu z^\nu = s_{n,n} z^n + (s_{n,n+1} - s_{n,n}) z^{n+1} + \ldots + (s_{n,m} - s_{n,m-1}) z^m$$
$$= s_{n,n}(z^n - z^{n+1}) + \ldots + s_{n,m-1}(z^{m-1} - z^m) + s_{n,m} z^m.$$

Hence for $n \geqslant n_0$

$$\left|\sum_{\nu=n}^{m} a_\nu z^\nu\right| \leqslant \epsilon\left\{\sum_{\nu=n}^{m-1} |z^\nu - z^{\nu+1}| + |z|^m\right\}$$
$$< \epsilon\left\{|1-z| \sum_{\nu=0}^{\infty} |z|^\nu + 1\right\}$$
$$= \epsilon\left\{\frac{|1-z|}{1-|z|} + 1\right\}.$$

The result now follows as in the previous case, provided that

$$\frac{|1-z|}{1-|z|}$$

is bounded as $z \to 1$ on the path considered. It is this that makes it necessary to restrict the path, for this function can be made large by taking z near to 1, but still nearer to the circumference.

Suppose, then, that

$$|1-z| \leqslant k(1-|z|) \qquad (k > 1). \tag{3}$$

This inequality is satisfied in a region bounded by the curve

$$|1-z| = k(1-|z|).$$

Putting $1-z = \rho e^{i\phi}$, the equation becomes

$$\rho = k - k|1-\rho e^{i\phi}|,$$

i.e.
$$(\rho - k)^2 = k^2(1 - 2\rho\cos\phi + \rho^2),$$

i.e.
$$\rho = 2\frac{k^2\cos\phi - k}{k^2 - 1}.$$

This represents a curve with two branches through $z = 1$, each making an angle $\arccos(1/k)$ with the real axis. By choosing k sufficiently large we can make the curve include any region of the required type. Since (3) is satisfied inside the curve, the theorem now follows.

7.62. We can also obtain a similar extension of Tauber's theorem.

If $f(z) \to s$ as $z \to 1$ along a path satisfying the same conditions as before, and $a_n = o(1/n)$, then $\sum a_n$ converges to the sum s.

In view of the above analysis, the proof given in § 1.23 now requires little modification. We have to prove that

$$S_1 - S_2 = \sum_{n=N+1}^{\infty} a_n z^n - \sum_{n=0}^{N} a_n(1-z^n) \to 0,$$

where $N = [1/(1-|z|)]$. As before, if $|na_n| < \epsilon$ for $n > N$,

$$|S_1| = \left|\sum_{N+1}^{\infty} na_n . \frac{z^n}{n}\right| < \frac{\epsilon}{N+1} \sum_{N+1}^{\infty} |z|^n < \frac{\epsilon}{(N+1)(1-|z|)} < \epsilon.$$

Now $\qquad |1-z^n| = |(1-z)(1+z+...+z^{n-1})| \leqslant |1-z|n.$

Hence, if 7.61 (3) is satisfied,

$$|S_2| \leqslant \sum_{n=1}^{N} |na_n(1-z)| \leqslant k(1-|z|) \sum_{n=1}^{N} n|a_n| \leqslant \frac{k}{N} \sum_{n=1}^{N} n|a_n|,$$

and this tends to zero, by the lemma of § 1.23. This proves the theorem.

7.63. Tauber's theorem for regular paths. It is not possible to extend Abel's theorem, at any rate in its obvious form, to paths which touch the unit circle; for example, it is known* that the series

$$\sum_{n=1}^{\infty} n^{-b} e^{in^a} \qquad (0 < a < 1)$$

is convergent if $b > 1-a$; but, if $b < 1-\frac{1}{2}a$, the function

$$f(z) = \sum_{n=1}^{\infty} n^{-b} e^{in^a} z^n$$

does not tend to a limit as $z \to 1$ along an arc of a circle touching the unit circle at $z = 1$.

On the other hand, we can obtain an extension of Tauber's theorem to paths which touch the circle, provided that they are sufficiently regular.

A path will be called 'regular' if it is defined by equations $x = x(t)$, $y = y(t)$, where $x'(t)$ and $y'(t)$ are continuous and never both 0, so that there is a definite tangent at each point.

If $f(z) \to s$ as $z \to 1$ along a regular path inside the circle, and $a_n = o(1/n)$, then $\sum a_n$ converges to the sum s.

We may suppose without loss of generality that $s = 0$ Let

* See Hardy and Littlewood (3), p. 207.

C be the path in question. Then the integral

$$\int_z^1 f(w)\,dw,$$

taken along C, exists; and it is $o(|1-z|)$ as $z \to 1$. For, given ϵ, we have $|f(w)| \leqslant \epsilon$ for w sufficiently near to 1 on C. Hence

$$\left|\int_z^1 f(w)\,dw\right| \leqslant \epsilon\, l(z),$$

where $l(z)$ is the length of C from z to 1. But $l(z) \sim |1-z|$ as $z \to 1$; for if $t = 0$ corresponds to $z = 1$,

$$\frac{l(z)}{t} = \frac{1}{t}\int_0^t [\{x'(u)\}^2+\{y'(u)\}^2]^{\frac{1}{2}}\,du \to [\{x'(0)\}^2+\{y'(0)\}^2]^{\frac{1}{2}},$$

$x'(u)$ and $y'(u)$ being continuous; and

$$\frac{x-1}{t} \to x'(0), \qquad \frac{y}{t} \to y'(0).$$

Hence
$$\int_z^1 f(w)\,dw = o(|1-z|). \tag{1}$$

Now if z and z' are points on C,

$$\int_z^{z'} f(w)\,dw = \sum_{n=0}^{\infty} \frac{a_n}{n+1}(z'^{n+1}-z^{n+1}).$$

This series converges uniformly with respect to z' for $|z'| \leqslant 1$ (since $a_n = o(1/n)$). Hence, making $z' \to 1$,

$$\int_z^1 f(w)\,dw = \sum_{n=0}^{\infty} \frac{a_n}{n+1}(1-z^{n+1}). \tag{2}$$

Let $N = [1/|1-z|]$. Then

$$\int_z^1 f(w)\,dw = \sum_{n=0}^{N} + \sum_{N+1}^{\infty} \frac{a_n}{n+1}(1-z^{n+1}) = \Sigma_1 + \Sigma_2,$$

and
$$\Sigma_2 = \sum_{N+1}^{\infty} o\left(\frac{1}{n^2}\right) = o\left(\frac{1}{N}\right) = o(|1-z|). \tag{3}$$

Also

$$\begin{aligned}1-z^{n+1} &= (1-z)(1+z+\ldots+z^n)\\ &= (1-z)(n+1)-(1-z)^2\{n+(n-1)z+\ldots+z^{n-1}\}\\ &= (1-z)(n+1)+O(|1-z|^2n^2).\end{aligned}$$

Hence
$$\Sigma_1 = (1-z)\sum_{n=0}^{N} a_n + O\left(|1-z|^2 \sum_{n=1}^{N} n|a_n|\right)$$
$$= (1-z)s_N + o(|1-z|^2 N)$$
$$= (1-z)s_N + o(|1-z|). \tag{4}$$

From (1), (3), and (4) it follows that $s_N = o(1)$, and this proves the theorem.

7.64. Littlewood's extension of Tauber's theorem. We now pass to an extension of quite a different kind. In all the forms of Tauber's theorem so far considered, the condition $a_n = o(1/n)$ has played an apparently essential part. It was, however, discovered by Littlewood that it can be replaced by the more general condition $a_n = O(1/n)$. Here we shall restrict ourselves for the sake of simplicity to the real axis, though it is possible to prove the theorem for complex paths.

7.65. We use the following lemma:

If $f(x)$ is a real function with differential coefficients of the first two orders for $0 \leqslant x < 1$, and, as $x \to 1$,
$$f(x) = o(1), \qquad f''(x) = O\left\{\frac{1}{(1-x)^2}\right\},$$
then
$$f'(x) = o\left(\frac{1}{1-x}\right).$$

Let $x' = x + \delta(1-x)$, where $0 < \delta < \frac{1}{2}$. Then
$$f(x') = f(x) + \delta(1-x)f'(x) + \tfrac{1}{2}\delta^2(1-x)^2 f''(\xi),$$
where $x < \xi < x'$. Hence
$$(1-x)f'(x) = \frac{f(x')-f(x)}{\delta} - \tfrac{1}{2}\delta(1-x)^2 f''(\xi)$$
$$= \frac{f(x')-f(x)}{\delta} + O(\delta), \tag{1}$$
since $f''(\xi) = O\left\{\dfrac{1}{(1-\xi)^2}\right\} = O\left\{\dfrac{1}{(1-x')^2}\right\} = O\left\{\dfrac{1}{(1-x)^2}\right\}$.

By first choosing δ sufficiently small, and then x sufficiently near to 1, the right-hand side of (1) can be made as small as we please. This proves the lemma.

7.66. Littlewood's theorem. *If $f(x) = \sum a_n x^n \to s$ as $x \to 1$, and $a_n = O(1/n)$, then $\sum a_n$ converges to the sum s.*

The proof * depends on the theorem of § 7.51, and in proving

* The original proof, Littlewood (3), was different.

that theorem we have really overcome the most serious difficulties. We may obviously suppose, without loss of generality, that the limit s is zero. Then

$$f(x)=\sum_{n=0}^{\infty}a_n x^n=o(1)$$

as $x \to 1$. Also, since $a_n=O(1/n)$,

$$f''(x)=\sum_{n=2}^{\infty}n(n-1)a_n x^{n-2}=O\left\{\sum_{n=2}^{\infty}(n-1)x^{n-2}\right\}=O\left\{\frac{1}{(1-x)^2}\right\}.$$

Hence, by the lemma,

$$f'(x)=\sum_{n=1}^{\infty}na_n x^{n-1}=o\left(\frac{1}{1-x}\right).$$

Suppose that $|na_n|\leqslant c$. Then

$$\sum_{n=1}^{\infty}\left(1-\frac{na_n}{c}\right)x^{n-1}=\frac{1}{1-x}-\frac{f'(x)}{c}\sim\frac{1}{1-x}.$$

But the coefficients in this series are all positive, and so, by the theorem of § 7.51,

$$\sum_{\nu=1}^{n}\left(1-\frac{\nu a_\nu}{c}\right)\sim n,$$

or

$$\sum_{\nu=1}^{n}\nu a_\nu=o(n). \tag{1}$$

This is an asymptotic formula for a finite sum, and so is a considerable step in the right direction. To get exactly the required result, still another argument is required.

Let w_n denote the left-hand side of (1) if $n>0$, and let $w_0=0$. Then

$$\begin{aligned}f(x)-a_0&=\sum_{n=1}^{\infty}\frac{w_n-w_{n-1}}{n}x^n=\sum_{n=1}^{\infty}w_n\left(\frac{x^n}{n}-\frac{x^{n+1}}{n+1}\right)\\&=\sum_{n=1}^{\infty}w_n\left\{\frac{x^n-x^{n+1}}{n+1}+\frac{x^n}{n(n+1)}\right\}\\&=(1-x)\sum_{n=1}^{\infty}\frac{w_n}{n+1}x^n+\sum_{n=1}^{\infty}\frac{w_n}{n(n+1)}x^n.\end{aligned}$$

Since $w_n=o(n)$, the first term on the right is $o(1)$ as $x\to 1$. Hence, since $f(x)\to 0$,

$$\sum_{n=1}^{\infty}\frac{w_n}{n(n+1)}x^n\to -a_0.$$

But $w_n/\{n(n+1)\} = o(1/n)$, and so, by the ordinary form of Tauber's theorem,

$$\sum_{n=1}^{\infty} \frac{w_n}{n(n+1)} = -a_0.$$

The left-hand side is

$$\lim_{N\to\infty} \sum_{n=1}^{N} \frac{w_n}{n(n+1)} = \lim_{N\to\infty} \sum_{n=1}^{N} w_n \left(\frac{1}{n} - \frac{1}{n+1}\right)$$

$$= \lim_{N\to\infty} \left\{\sum_{n=1}^{N} \frac{w_n - w_{n-1}}{n} - \frac{w_N}{N+1}\right\} = \lim_{N\to\infty} \sum_{n=1}^{N} a_n,$$

and the theorem is therefore proved.

7.7. Partial sums of a power series.* The study of the partial sums of a power series is facilitated by the use of the formulae of the theory of Fourier series. We shall use some of these formulae, and quote them from Chapter XIII; but in each case where they are used here the proof is an immediate consequence of uniform convergence.

Let
$$f(z) = \sum_{n=0}^{\infty} a_n z^n \qquad (|z| < 1),$$
and
$$s_n(z) = a_0 + a_1 z + \ldots + a_n z^n.$$
Let
$$k(r,\theta) = \frac{1-r^2}{2(1-2r\cos\theta+r^2)} = \tfrac{1}{2} + r\cos\theta + r^2\cos 2\theta + \ldots,$$
and
$$k_n(r,\theta) = \frac{1-r^2-2r^{n+1}\{\cos(n+1)\theta - r\cos n\theta\}}{2(1-2r\cos\theta+r^2)}$$
$$= \tfrac{1}{2} + r\cos\theta + \ldots + r^n\cos n\theta.$$
Then
$$s_n(re^{i\theta}) = \frac{1}{\pi}\int_0^{2\pi} f(\rho e^{i(\theta-\phi)})\,k_n\!\left(\frac{r}{\rho},\phi\right)d\phi \qquad (0 < r < \rho < 1). \quad (1)$$

This may be proved directly by term-by-term integration. It is a case of Parseval's formula (§ 13.54).

Also, by Dirichlet's integral (§ 13.2),

$$k_n(r,\theta) = \frac{1}{\pi}\int_0^{2\pi} \frac{\sin(n+\frac{1}{2})(\theta-\phi)}{2\sin\frac{1}{2}(\theta-\phi)}\,k(r,\phi)\,d\phi. \quad (2)$$

* Landau (2), (3), (4), and *Ergebnisse*, Ch. I.

We can thus express s_n as a repeated integral involving f and k,

$$s_n(re^{i\theta}) = \frac{1}{2\pi^2}\int_0^{2\pi} f(\rho e^{i(\theta-\phi)})\,d\phi \int_0^{2\pi} \frac{\sin(n+\frac{1}{2})(\phi-\psi)}{\sin\frac{1}{2}(\phi-\psi)}\,k\left(\frac{r}{\rho},\psi\right)d\psi. \quad (3)$$

We consider also the arithmetic means of the partial sums,

$$\sigma_n(z) = \{s_0(z)+s_1(z)+\ldots+s_{n-1}(z)\}/n.$$

By (1),
$$\sigma_n(re^{i\theta}) = \frac{1}{\pi}\int_0^{2\pi} f(\rho e^{i(\theta-\phi)})K_n\left(\frac{r}{\rho},\phi\right)d\phi, \quad (4)$$

where
$$K_n(r,\theta) = \frac{1}{n}\sum_{\nu=0}^{n-1} k_\nu(r,\theta);$$

and, by Fejér's integral (§ 13.31),

$$K_n(r,\theta) = \frac{1}{2n\pi}\int_0^{2\pi} \frac{\sin^2\frac{1}{2}n(\theta-\phi)}{\sin^2\frac{1}{2}(\theta-\phi)}\,k(r,\phi)\,d\phi. \quad (5)$$

7.71. Bounded power series. Suppose now that $f(z)$ is bounded in the unit circle.

If $|f(z)| \leqslant M$ *for* $|z| < 1$, *then* $|\sigma_n(z)| \leqslant M$ *for all values of* n *and* $|z| < 1$; *and, conversely, if* $|\sigma_n(z)| \leqslant M$ *for all* n *and* $|z| < 1$, *then* $|f(z)| \leqslant M$.

It is clear from the above formulae that $k(r,\theta)$ and $K_n(r,\theta)$ are positive for $r < 1$. Hence, if $|f(z)| \leqslant M$, it follows from (4) that

$$|\sigma_n(re^{i\theta})| \leqslant \frac{1}{\pi}\int_0^{2\pi} MK_n\left(\frac{r}{\rho},\phi\right)d\phi.$$

But the right-hand side is what σ_n reduces to in the case $f(z) = M$, viz. M. This proves the first part.

Again $s_n(z)$, and so also $\sigma_n(z)$, tends to $f(z)$ as $n \to \infty$. The second part follows at once from this.

7.72. The corresponding results for $s_n(z)$ are not so simple. This is due to the fact that k_n, unlike K_n, is not always positive. It is not necessarily true that $|s_n(z)| \leqslant M$ for all values of n and z. In fact it is known* that the upper bound of $|s_n(z)|$, for all functions $f(z)$ such that $|f(z)| \leqslant M$, tends to infinity with n. We have, however, the following result:

* Landau, *Ergebnisse*, § 2.

There is an absolute constant A such that

$$|s_n(z)| < AM \log n$$

for all functions $f(z)$ such that $|f(z)| \leqslant M$.

If $|f(z)| \leqslant M$, by § 7.7 (3),

$$|s_n(re^{i\theta})| \leqslant \frac{M}{2\pi^2} \int_0^{2\pi} k\left(\frac{r}{\rho}, \psi\right) d\psi \int_0^{2\pi} \left| \frac{\sin(n+\frac{1}{2})(\phi-\psi)}{\sin \frac{1}{2}(\phi-\psi)} \right| d\phi.$$

The inner integral is equal to

$$2\int_0^{\pi} \left| \frac{\sin(n+\frac{1}{2})\alpha}{\sin \frac{1}{2}\alpha} \right| d\alpha < 2 \int_0^{1/(n+\frac{1}{2})} \frac{(n+\frac{1}{2})\alpha}{\sin \frac{1}{2}\alpha}\, d\alpha + 2 \int_{1/(n+\frac{1}{2})}^{\pi} \frac{d\alpha}{\sin \frac{1}{2}\alpha}$$

$$= O(1) + O(\log n);$$

and, putting $n = 0$ in 7.7 (2),

$$\frac{1}{2\pi} \int_0^{2\pi} k\left(\frac{r}{\rho}, \psi\right) d\psi = \tfrac{1}{2}.$$

This proves the theorem.

7.73. It is easily seen that $s_n(z)$ is bounded in a circle of radius r' less than 1; for $k_n(r, \theta)$ is obviously bounded in such a circle. The upper bound for $s_n(z)$ depends on M and on r'. What is not so obvious is that we can choose r', independent of M, so that the upper bound is exactly M.

*If $|f(z)| \leqslant M$, then $|s_n(z)| \leqslant M$ for $|z| \leqslant \frac{1}{2}$.**

It is clear that

$$k_n(r, \phi) \geqslant \frac{1-r^2-2r^{n+1}(1+r)}{2(1-r)^2},$$

and if $r \leqslant \frac{1}{2}$, $n \geqslant 1$, the numerator is not less than

$$1-\tfrac{1}{4}-2.\tfrac{1}{4}(1+\tfrac{1}{2}) = 0.$$

Hence $k_n(r, \phi) \geqslant 0$ for $r \leqslant \frac{1}{2}$, and we can now proceed as in § 7.71. We have

$$|s_n(re^{i\theta})| \leqslant \frac{1}{\pi} \int_0^{2\pi} Mk_n\left(\frac{r}{\rho}, \phi\right) d\phi \qquad (r \leqslant \tfrac{1}{2}\rho),$$

and the right-hand side is what $s_n(z)$ reduces to when $f(z) = M$, viz. M. This proves the theorem.

* Fejér (5).

The number $\frac{1}{2}$ is the greatest number with this property. For consider the function

$$f(z) = \frac{z-a}{az-1} \qquad (0 < a < 1).$$

Then $|f(e^{i\theta})| = 1$, so that $|f(z)| \leqslant 1$ for $|z| \leqslant 1$. Also

$$s_1(z) = a+(a^2-1)z,$$

$$s_1\left(-\frac{1}{2a}\right) = \frac{a^2+1}{2a} > 1,$$

and the point $-\frac{1}{2}/a$ where $s_1(z) > 1$ is arbitrarily near to $|z| = \frac{1}{2}$, since a is arbitrarily near to 1.

7.8. The zeros of partial sums.* Let

$$f(z) = a_0+a_1z+\ldots \qquad (a_0 \neq 0),$$

be a power series with radius of convergence 1, and let

$$s_n(z) = a_0+a_1z+\ldots+a_nz^n.$$

Then $s_n(z)$, being a polynomial of degree n, has n zeros.

If $f(z)$ has zeros inside the circle of convergence, then by Hurwitz's theorem (§ 3.45) every such zero is a limit-point of zeros of the polynomials $s_n(z)$.

Now consider the simplest function of the above type,

$$f(z) = \frac{1}{1-z} = 1+z+z^2+\ldots.$$

Here
$$s_n(z) = 1+z+\ldots+z^n = \frac{1-z^{n+1}}{1-z}.$$

Hence $s_n(z)$ has zeros distributed evenly round the circle, and it is plain that every point of the circle is a limit-point of such zeros.

It is somewhat remarkable that the general case is so nearly like this simple case. This was discovered by Jentzsch, who proved that, *for every power series, every point of the circle of convergence is a limit-point of zeros of partial sums.*

We shall deduce this from some quite simple ideas depending on the theory of equations.

Let δ be a given positive number, n a number such that

$$|a_n| > \frac{|a_0|}{(1+\delta)^n}. \tag{1}$$

* Jentzsch (1).

This is true for arbitrarily large values of n, or the radius of convergence would be greater than 1.

Let $z_1, z_2,..., z_n$ be the zeros of the corresponding $s_n(z)$. Then

$$z_1 z_2 ... z_n = (-1)^n a_0/a_n,$$

and so

$$|z_1 z_2 ... z_n| < (1+\delta)^n.$$

Let $z_1,..., z_k$ be the zeros of $s_n(z)$ in the circle $|z| \leqslant 1-\delta$. By Hurwitz's theorem (§ 3.45), k is constant for sufficiently large values of n, and

$$|z_1 z_2 ... z_k| > K,$$

where K depends on δ only.

Let $z_{n-p+1},..., z_n$ be the zeros for which $|z| > 1+\epsilon$. Then

$$(1+\epsilon)^p < |z_{n-p+1}...z_n| = \left|\frac{z_1 z_2 ... z_n}{z_1...z_k . z_{k+1}...z_{n-p}}\right| < \frac{(1+\delta)^n}{K(1-\delta)^{n-p}}.$$

Hence

$$p < \frac{n\{\log(1+\delta)-\log(1-\delta)\}-\log K}{\log(1+\epsilon)-\log(1-\delta)} < \frac{An\delta-\log K}{A\epsilon}.$$

By choosing first ϵ, then δ, and then n, we can make p/n arbitrarily small.

Hence, for given δ, ϵ, and η, the number of zeros in the circle $|z| \leqslant 1+\epsilon$ is greater than $n(1-\eta)$, if n is a sufficiently large integer for which (1) *is true.*

7.81. It is clear from the above result that *the zeros of partial sums have at least one limit-point on the circle of convergence.*

We can obtain a little more information by considering the sum

$$\sum_{\nu=1}^{n} \frac{1}{z_\nu} = -\frac{a_1}{a_0}.$$

Putting $z_\nu = r_\nu e^{i\theta_\nu}$, we have

$$\sum_{\nu=1}^{n} \frac{\cos\theta_\nu}{r_\nu} = -\mathbf{R}\left(\frac{a_1}{a_0}\right). \tag{1}$$

If $\theta_\nu > \frac{1}{2}\pi+\alpha$, or $\theta_\nu < -\frac{1}{2}\pi-\alpha$, where $\alpha > 0$, for every ν, the left-hand side is less than

$$\frac{-n(1-\eta)\sin\alpha}{1+\epsilon}$$

by the above theorem. This is inconsistent with (1), if n is large enough. Hence there must be zeros in any angle including

$(-\frac{1}{2}\pi, \frac{1}{2}\pi)$. Similarly there must be zeros in any angle greater than π.

To prove Jentzsch's theorem we have to replace such an angle by an arbitrarily small one. This is done by using a conformal transformation which magnifies the effect of the zeros in the immediate neighbourhood of the point considered.

7.82. Let

$$w = \frac{\cos\lambda - z}{z\cos\lambda - 1}, \qquad z = \frac{w+\cos\lambda}{1+w\cos\lambda}, \tag{1}$$

where $0 < \lambda < \frac{1}{2}\pi$, and where $f(\cos\lambda) \neq 0$. This transforms the unit circle in the z-plane into the unit circle in the w-plane. The point $z = 1$ becomes $w = 1$. The point $z = e^{i\lambda}$ becomes

$$w = \frac{\frac{1}{2}(e^{-i\lambda} - e^{i\lambda})}{\frac{1}{2}(e^{2i\lambda} - 1)} = -e^{-i\lambda} = e^{i(\pi-\lambda)},$$

and similarly $z = e^{-i\lambda}$ becomes $w = e^{-i(\pi-\lambda)}$. Thus, if $z = re^{i\theta}$, $w = \rho e^{i\phi}$, the arc $-\lambda \leqslant \theta \leqslant \lambda$ of the unit circle is transformed into the arc $-\pi+\lambda \leqslant \phi \leqslant \pi-\lambda$.

The zeros z_ν of $s_n(z)$ are transformed into the zeros $w_\nu = \rho_\nu e^{i\phi_\nu}$ of the function

$$(1+w\cos\lambda)^n s_n\left(\frac{w+\cos\lambda}{1+w\cos\lambda}\right)$$
$$= s_n(\cos\lambda) + w\{n\cos\lambda\, s_n(\cos\lambda) + \sin^2\lambda\, s_n'(\cos\lambda)\} + \dots$$
$$= b_0 + b_1 w + \dots,$$

say; and corresponding to § 7.81 (1) we have

$$\sum_{\nu=1}^{n} \frac{\cos\phi_\nu}{\rho_\nu} = -\mathbf{R}\left(\frac{b_1}{b_0}\right) = -n\cos\lambda - \sin^2\lambda\, \mathbf{R}\left\{\frac{s_n'(\cos\lambda)}{s_n(\cos\lambda)}\right\}. \tag{2}$$

The last term tends to a limit as $n \to \infty$, since $s_n(\cos\lambda) \to f(\cos\lambda)$, which we have supposed is not 0, and $s_n'(\cos\lambda) \to f'(\cos\lambda)$. Hence as $n \to \infty$

$$\sum_{\nu=1}^{n} \frac{\cos\phi_\nu}{\rho_\nu} \sim -n\cos\lambda. \tag{3}$$

Suppose now that the region of the w-plane

$$1-\epsilon < \rho < 1+\epsilon, \qquad -(\pi-\lambda+\alpha) < \phi < \pi-\lambda+\alpha, \tag{4}$$

where $0 < \epsilon < 1$, $0 < \alpha < \lambda$, is free from zeros. Put

$$\sum_{\nu=1}^{n} \frac{\cos\phi_\nu}{\rho_\nu} = \sum_{\rho_\nu \leqslant 1-\epsilon} + \sum_{1-\epsilon<\rho_\nu<1+\epsilon} + \sum_{\rho_\nu \geqslant 1+\epsilon} = \Sigma_1 + \Sigma_2 + \Sigma_3. \tag{5}$$

Since $\rho = 1$ corresponds to $r = 1$, it follows from considerations of continuity that the circles $\rho = 1-\epsilon$, $\rho = 1+\epsilon$ correspond to curves (in fact circles) inside and outside $r = 1$ respectively, which can be made as near to it as we please by taking ϵ small enough.

The number of terms in $\sum_1$ is less than $K = K(\delta, \epsilon, \lambda)$; and ρ_ν has a positive lower bound, since the zeros of $s_n(\cos\lambda)$ in question tend to zeros of $f(\cos\lambda)$. Hence

$$\textstyle\sum_1 < K. \tag{6}$$

The number of terms in $\sum_3$ is, by § 7.8, less than ηn, where $\eta = \eta(n, \delta, \epsilon, \lambda)$ tends to 0 as $n \to \infty$ through a certain sequence of values. Hence

$$\textstyle\sum_3 < \frac{\eta n}{1+\epsilon}. \tag{7}$$

In $\sum_2$ the number of terms exceeds $n(1-\eta)-K$, and by hypothesis $\cos\phi_\nu < -\cos(\lambda-\alpha)$ for each term of this sum. Hence

$$\textstyle\sum_2 < -\frac{n(1-\eta)-K}{1+\epsilon}\cos(\lambda-\alpha). \tag{8}$$

From (5), (6), (7), (8) it follows that

$$\overline{\lim_{n\to\infty}}\frac{1}{n}\sum_{\nu=1}^{n}\frac{\cos\phi_\nu}{\rho_\nu} \leqslant -\frac{\cos(\lambda-\alpha)}{1+\epsilon}.$$

This contradicts (3) if $\alpha > 0$ and ϵ is small enough. There are therefore zeros in the region (4), and hence, since ϵ and α may be as small as we please, in any region containing the arc $\rho = 1$, $-\pi+\lambda < \phi < \pi-\lambda$. Hence, in the z-plane, there are zeros in any region containing the arc $r = 1$, $-\lambda < \theta < \lambda$. Finally, since λ may be as small as we please, it follows that $z = 1$ is a limit-point of zeros. Similarly every point on the unit circle is a limit-point of zeros.

MISCELLANEOUS EXAMPLES

1. If $|a_n/a_{n+1}| \to R$, then the radius of convergence of $\sum a_n z^n$ is R.

2. If
$$\left|\frac{a_n}{a_{n+1}}\right| = \left\{1+\frac{c}{n}+o\left(\frac{1}{n}\right)\right\}R$$
where $c > 1$, then $\sum a_n z^n$ converges absolutely everywhere on its circle of convergence.

3. If $a_n/a_{n+1} \to 1$, then
$$\lim_{n\to\infty}\frac{s_n(z)}{a_n z^n} = \frac{z}{z-1}$$
uniformly for $|z| \geqslant 1+\delta > 1$. Hence show that all the limit-points of the zeros of partial sums are inside or on the unit circle. [S. Izumi (1).]

4. Show that as $x \to 1$

$$\sum_{n=0}^{\infty} x^{n^2} \sim \frac{1}{2}\sqrt{\left(\frac{\pi}{1-x}\right)},$$

and that, as $z \to e^{2i\pi p/q}$ along the radius vector,

$$\sum_{n=0}^{\infty} z^{n^2} \sim \frac{1}{2q}\sqrt{\left(\frac{\pi}{1-|z|}\right)} \sum_{r=0}^{q-1} e^{2i\pi pr^2/q}.$$

5. If $a_n \sim \log n$, then, as $x \to 1$,

$$\sum_{n=0}^{\infty} a_n x^n \sim \frac{1}{1-x}\log\frac{1}{1-x}.$$

$\left[\text{The right-hand side is } \sum\left(1+\frac{1}{2}+\dots+\frac{1}{n}\right)x^n\right].$

6. If $a_n \sim 1/\log n$, then, as $x \to 1$,

$$\sum_{n=0}^{\infty} a_n x^n \sim \frac{1}{(1-x)\log\{1/(1-x)\}}.$$

[If $\sum_p$ denotes a sum over the range $\epsilon p/\log(1/x) < n \leqslant \epsilon(p+1)/\log(1/x)$, then

$$\sum\nolimits_p \frac{x^n}{\log n} < \frac{\epsilon e^{-\epsilon p}}{\log(1/x)\log\{1/\log(1/x)\}}, \text{ etc.]}$$

7. Show that if $a_n \geqslant 0$, and

$$f(x) = \sum_{n=0}^{\infty} a_n x^n \sim \frac{1}{(1-x)^2},$$

then

$$s_n \sim \tfrac{1}{2}n^2.$$

[We have
$$f_1(x) = \int_0^x f(t)\,dt = \sum_{n=0}^{\infty} \frac{a_n x^{n+1}}{n+1} \sim \frac{1}{1-x}.$$

Hence
$$\sum_{\nu=0}^{n} \frac{a_\nu}{\nu+1} \sim n,$$

and the result then follows by partial summation.]

8. Generally, if $a_n \geqslant 0$, and $f(x) \sim (1-x)^{-\alpha}$, where $\alpha > 1$, then

$$s_n \sim \frac{n^\alpha}{\Gamma(\alpha+1)}.$$

[We have

$$f_{\alpha-1}(x) = \frac{1}{\Gamma(\alpha-1)}\int_0^x (x-t)^{\alpha-2} f(t)\,dt = \sum_{n=0}^{\infty} a_n \frac{\Gamma(n+1)}{\Gamma(n+\alpha)} x^{\alpha+n-1} \sim \sum_{n=1}^{\infty} \frac{a_n x^n}{n^{\alpha-1}},$$

and, on the other hand,

$$f_{\alpha-1}(x) \sim \frac{1}{\Gamma(\alpha-1)}\int_0^x (x-t)^{\alpha-2}(1-t)^{-\alpha}\,dt$$

$$= \frac{x^{\alpha-1}}{\Gamma(\alpha)(1-x)} \sim \frac{1}{\Gamma(\alpha)(1-x)}.$$

Hence $$\sum_{\nu=1}^{n} \frac{a_\nu}{\nu^{\alpha-1}} \sim \frac{n}{\Gamma(\alpha)},$$

and the result now follows without difficulty.]

9. If $f(z)$ is regular in a region including the origin, and $f(0) = 1$, then $f(z)$ can be expanded in the form

$$f(z) = (1+a_1 z)(1+a_2 z^2)(1+a_3 z^3)...$$

for sufficiently small values of z.

[Ritt (1): Assuming an expansion of the above form, we write $f'(z)/f(z) = c_1+c_2 z+...$, and determine the numbers a in succession by equating coefficients in the equation

$$c_1+c_2 z+... = \sum_{n=1}^{\infty} \frac{n a_n z^{n-1}}{1+a_n z^n}.$$

If $\mu_n = \max\limits_{\nu \leqslant n} |a_\nu|^{1/\nu}$, we deduce from the recurrence relation that $\mu_n^n \leqslant n\mu_{\frac{1}{2}n}^n + |c_n|$. Hence μ_n is bounded, and the process can be justified.]

10. Show that the circle of convergence of the above product is the same as that of the series $\sum a_n z^n$, but that the power series for $f(z)$ may have a larger circle of convergence.

11. If each of the series

$$\sum_{n=0}^{\infty} a_n z^n, \qquad \sum_{n=0}^{\infty} b_n z^n, \qquad \sum_{n=0}^{\infty} a_n b_n z^n$$

has a radius of convergence equal to 1, then so have the series

$$\sum_{n=0}^{\infty} a_n b_n^2 z^n, \qquad \sum_{n=0}^{\infty} a_n^2 b_n z^n.$$

12. If each of the series

$$f(z) = \sum_{n=0}^{\infty} a_n z^n, \qquad g(z) = \sum_{n=0}^{\infty} b_n z^n, \qquad F(z) = \sum_{n=0}^{\infty} a_n b_n z^n$$

has a radius of convergence equal to 1, if $f(z)$ is regular on its circle of convergence except at $z = 1$, and $b_n \geqslant 0$ for all values of n, then $F(z)$ has a singularity at $z = 1$.

[Bohnenblust (1): the series

$$\phi(z) = \sum_{n=0}^{\infty} |a_n|^2 b_n z^n$$

has the radius of convergence 1, and so by § 7.21 has a singularity at $z = 1$. By Hadamard's multiplication theorem (§ 4.6) the singularities of $\phi(z)$ are products of those of $F(z)$ and of

$$\bar{f}(z) = \sum_{n=0}^{\infty} \bar{a}_n z^n.$$

Thus $1 = \alpha\beta$, where α is a singularity of $F(z)$, β of $\bar{f}(z)$; and β must be 1. Hence $\alpha = 1$.]

13. If $f(z) = \sum a_n z^n$ is regular on its circle of convergence except at z_0, then every series which consists of a selection of terms from $\sum a_n z^n$, and which has the same radius of convergence, has a singularity at z_0.

14. Show that the theorem of § 7.21 is still true if the coefficients a_n are complex, provided that $|\arg a_n| \leqslant \alpha < \frac{1}{2}\pi$ for all values of n.

[We have $|a_n| \leqslant \sec\alpha\, \mathbf{R}a_n$.]

15. The function
$$\sum_{n=1}^{\infty} \frac{z^{2^n}}{n^2}$$
is continuous in and on the unit circle; but every point of the circle is a singularity.

16. If $f(z)$ is bounded in the unit circle, then $\sum |a_n|^2$ is convergent. [See § 2.5.]

The following examples are on the border-line between theory of power series and theory of real functions. It seems most convenient to insert them here, but some of them assume the theory of mean convergence given in § 12.5.

17. If $\sum |a_n|^2$ is convergent, then
$$\frac{1}{2\pi}\int_0^{2\pi} |f(re^{i\theta})-f(r'e^{i\theta})|^2\, d\theta = \sum_{n=0}^{\infty} |a_n|^2(r^n-r'^n)^2.$$
Hence, show that, as $r \to 1$, $f(re^{i\theta})$ converges in mean to a limit-function $F(\theta)$ of the class $L^2(0, 2\pi)$.

18. If $f(z) = u+iv$, $F(\theta) = U+iV$ in the previous example, show that Poisson's formulae
$$u(r,\theta) = \frac{1}{2\pi}\int_0^{2\pi} \frac{1-r^2}{1-2r\cos(\theta-\phi)+r^2}\, U(\phi)\, d\phi,$$
$$v(r,\theta)-v(0) = \frac{1}{2\pi}\int_0^{2\pi} \frac{2r\sin(\theta-\phi)}{1-2r\cos(\theta-\phi)+r^2}\, U(\phi)\, d\phi$$
hold for $r < 1$.

19. Show that, in the above examples, $u(r,\theta) \to U(\theta)$ as $r \to 1$ for every value of θ in the Lebesgue set of $U(\theta)$. Deduce that $f(re^{i\theta}) \to F(\theta)$ as $r \to 1$ for almost all values of θ.

[The analysis is similar to that of § 13.34.]

20. Show that a bounded analytic function tends to a limit radially at almost all points of its circle of convergence.

21. If $U(\theta) \geqslant 0$ for all values of θ, then $u(r,\theta) \geqslant 0$ for all values of r and θ.

22. If $f(z)$ is regular and bounded for $|z| < 1$, and $f(z) \to 0$ as $r \to 1$ throughout an interval of values of θ, then $f(z)$ is identically zero.

[If $0 \leqslant \theta \leqslant 2\pi/p$ is part of the interval, consider the function

$$g(z) = f(z)f(ze^{2i\pi/p})...f(ze^{2(p-1)i\pi/p}).]$$

23. More generally, if $f(z)$ is bounded and tends to zero radially for values of θ in a set of positive measure, then $f(z)$ is identically zero.

[See Bieberbach, ii, p. 156. Let E be the set where $f(z) \to 0$, and let $m(E) = \mu > 0$. Let $u_1(\theta) = \lambda/\mu$ in E, and $= -\lambda/(2\pi-\mu)$ in CE. Let $g(z)$ be the corresponding analytic function defined by the formulae of ex. 18. Then $g(0) = 0$. Let $h(z) = e^{g(z)}$, so that $h(0) = 1$. Then

$$f(0) = f(0)h(0) = \frac{1}{2\pi i} \int\limits_{|z|=1} f(z)h(z)\,\frac{dz}{z} = \frac{1}{2\pi i} \int_{CE} f(z)h(z)\,\frac{dz}{z},$$

$$|f(0)| < Ae^{-\lambda/(2\pi-\mu)}.$$

Since λ may be as large as we please, $f(0) = 0$. Applying the same argument to $f(z)/z$, $f'(0) = 0$, etc.]

24. If $U(\theta)$ is any function integrable in the Lebesgue sense, and

$$f(z) = \frac{1}{2\pi} \int\limits_0^{2\pi} \frac{1+ze^{-i\phi}}{1-ze^{i\phi}}\, U(\phi)\, d\phi \qquad (|z| < 1),$$

then $f(z)$ tends to a limit as $r \to 1$ for almost all values of θ.

[Plessner (1): We may suppose without loss of generality that $U(\phi) \geqslant 0$. Then $\mathbf{R}\{f(z)\} \geqslant 0$. Hence the function $1/\{1+f(z)\}$ is bounded in the unit circle, and so tends to a limit for almost all values of θ. This limit is different from zero almost everywhere.]

CHAPTER VIII

INTEGRAL FUNCTIONS

8.1. Factorization of integral functions. An integral function is an analytic function which has no singularities except at infinity. The simplest such functions are polynomials. A polynomial $f(z)$ which has zeros at the points $z_1, z_2,..., z_n$ can be factorized in the form

$$f(z) = f(0)\left(1-\frac{z}{z_1}\right)\left(1-\frac{z}{z_2}\right)...\left(1-\frac{z}{z_n}\right).$$

The zeros of integral functions in general are equally important. An integral function which is not a polynomial may have an infinity of zeros z_n; and the product

$$\prod\left(1-\frac{z}{z_n}\right)$$

taken over these zeros may be divergent. So we cannot always factorize an integral function in the same way as a polynomial, and we have to consider less simple factors than $1-z/z_n$.

The expressions

$$E(u,0) = 1-u, \qquad E(u,p) = (1-u)e^{u+\frac{u^2}{2}+...+\frac{u^p}{p}} \quad (p = 1, 2,...),$$

are called *primary factors*. Each primary factor vanishes when $u = 1$; but the behaviour of $E(u,p)$ as $u \to 0$ depends on p. For $|u| < 1$,

$$\log E(u,p) = -\frac{u^{p+1}}{p+1}-\frac{u^{p+2}}{p+2}-....$$

Hence, if $k > 1$, and $|u| \leqslant 1/k$,

$$|\log E(u,p)| \leqslant |u|^{p+1}+|u|^{p+2}+...$$

$$\leqslant |u|^{p+1}\left\{1+\frac{1}{k}+\frac{1}{k^2}+...\right\} = \frac{k}{k-1}|u|^{p+1}.$$

It is this inequality which determines the convergence of a product of primary factors.

8.11. The theorem of Weierstrass. If $f(z)$ is an integral function, what can we say about its zeros?

Since $f(z)$ is analytic except at infinity, the zeros can have no limit-point except at infinity. In general, this is all that we can say. This follows from the following theorem of Weierstrass.

Given any sequence of numbers $z_1, z_2,...$ whose sole limiting-point *is at infinity, there is an integral function with zeros at these points, and these points only.*

We may suppose the zeros arranged so that $|z_1| \leqslant |z_2| \leqslant \dots$. Let $|z_n| = r_n$, and let $p_1, p_2, \dots$ be a sequence of positive integers such that the series

$$\sum_{n=1}^{\infty} \left(\frac{r}{r_n}\right)^{p_n}$$

is convergent for all values of r. It is always possible to find such a sequence; for $r_n \to \infty$, since otherwise the zeros would have a limiting-point other than infinity; and we may take $p_n = n$, since

$$\left(\frac{r}{r_n}\right)^n < \frac{1}{2^n}$$

for $r_n > 2r$, and the series is therefore convergent.

Let
$$f(z) = \prod_{n=1}^{\infty} E\left(\frac{z}{z_n}, p_n - 1\right). \tag{1}$$

This function has the required property; for, if $|z_n| > 2|z|$,

$$\left|\log E\left(\frac{z}{z_n}, p_n - 1\right)\right| \leqslant 2\left(\frac{r}{r_n}\right)^{p_n}; \tag{2}$$

hence the series
$$\sum_{|z_n| > 2R} \log E\left(\frac{z}{z_n}, p_n - 1\right)$$

is uniformly convergent for $|z| \leqslant R$, and hence* so is the product

$$\prod_{|z_n| > 2R} E\left(\frac{z}{z_n}, p_n - 1\right).$$

Hence $f(z)$ is regular for $|z| \leqslant R$, and its only zeros in this region are those of

$$\prod_{|z_n| \leqslant 2R} E\left(\frac{z}{z_n}, p_n - 1\right),$$

i.e. the points $z_1, z_2, \dots$. Since R may be as large as we please, this proves the theorem.

The function $f(z)$ is, of course, not uniquely determined by the zeros, since we have a wide choice of the numbers p_n.

8.12. It is possible to factorize any given integral function in the following way.

If $f(z)$ is an integral function, and $f(0) \neq 0$, then

$$f(z) = f(0)\, P(z)\, e^{g(z)}$$

* See § 1.43, end.

where $P(z)$ is a product of primary factors, and $g(z)$ is an integral function.

We form $P(z)$ as in the above theorem from the zeros of $f(z)$. Let

$$\phi(z) = \frac{f'(z)}{f(z)} - \frac{P'(z)}{P(z)}.$$

Then $\phi(z)$ is an integral function, since the poles of one term are cancelled by those of the other. Hence also

$$g(z) = \int_0^z \phi(t)\,dt = \log f(z) - \log f(0) - \log P(z)$$

is an integral function, and the result stated follows on taking exponentials.

If $f(z)$ has a zero of order p at $z = 0$, a factor z^p has to be inserted.

This factorization is not unique.

8.2. Functions of finite order. The general factorization theorem is not precise enough to be of much use; in general the numbers p_n increase indefinitely with n, and we can say little about the function $g(z)$. There is, however, one case in which we can put the theorem into a perfectly definite form, that of functions of finite order.

An integral function $f(z)$ is said to be of finite order if there is a positive number A such that, as $|z| = r \to \infty$,

$$f(z) = O(e^{r^A}).$$

The lower bound ρ of numbers A for which this is true is called the *order* of the function. Thus, if $f(z)$ is of order ρ,

$$f(z) = O(e^{r^{\rho+\epsilon}})$$

for every positive value of ϵ, but not for any negative value. In this, and similar statements throughout the chapter, ϵ is thought of as taking arbitrarily small values, and the constant implied in the O depends in general on ϵ. If it were independent of ϵ, we could replace ϵ by 0 in the formula.

Functions of finite order are, after polynomials, the simplest integral functions. A polynomial is of order zero; some of the properties of functions of small order are similar to those of polynomials.

Many familiar functions are easily seen to be of finite order; e^z is of order 1; so are $\sin z$ and $\cos z$; $\cos \sqrt{z}$ is an integral function of order $\frac{1}{2}$; e^{z^k} is an integral function of order k, if k is a positive integer (if k is not an integer, it is not an integral function). The function e^{e^z} is of infinite order.

In what follows we shall suppose generally that $f(0)$ is not 0. This simplifies the analysis a little, and division by a factor z^k does not affect the order.

8.21. The function $n(r)$. Let $n(r)$ denote the number of zeros $z_1, z_2, \ldots$ of an integral function $f(z)$ for which $|z_n| \leqslant r$. Then $n(r)$ is a non-decreasing function of r which is constant in intervals; it is zero for $r < |z_1|$, if $f(0)$ is not zero.

This function is, as we have seen in § 3.61, connected with $f(z)$ by means of Jensen's formula. In fact

$$\int_0^r \frac{n(x)}{x}\,dx = \frac{1}{2\pi}\int_0^{2\pi} \log|f(re^{i\theta})|\,d\theta - \log|f(0)|. \tag{1}$$

If $f(z)$ is an integral function, this holds for all values of r.

If $f(z)$ is of order ρ, then $n(r) = O(r^{\rho+\epsilon})$. For

$$\log|f(re^{i\theta})| < Kr^{\rho+\epsilon},$$

K depending on ϵ only. Hence, by (1),

$$\int_0^{2r} \frac{n(x)}{x}\,dx < Kr^{\rho+\epsilon}. \tag{2}$$

But, since $n(r)$ is non-decreasing,

$$\int_r^{2r} \frac{n(x)}{x}\,dx \geqslant n(r)\int_r^{2r} \frac{dx}{x} = n(r)\log 2.$$

Hence
$$n(r) \leqslant \frac{1}{\log 2}\int_0^{2r} \frac{n(x)}{x}\,dx < Kr^{\rho+\epsilon}$$

by (2).

We may thus say, roughly, that the higher the order of a function is, the more zeros it may have in a given region.

8.22. *If $r_1, r_2, \ldots$ are the moduli of the zeros of $f(z)$, then the series $\sum r_n^{-\alpha}$ is convergent if $\alpha > \rho$.*

Let β be a number between α and ρ. Then $n(r) < Ar^{\beta}$. Putting $r = r_n$, this gives

$$n < Ar_n^{\beta}.$$

Hence $$r_n^{-\alpha} < An^{-\alpha/\beta},$$
and the result follows.

The lower bound of positive numbers α for which $\sum r_n^{-\alpha}$ is convergent is called the *exponent of convergence of the zeros*, and is denoted by ρ_1. What we have just proved is that $\rho_1 \leqslant \rho$. We may have $\rho_1 < \rho$; for example, if $f(z) = e^z$, $\rho = 1$; but there are no zeros, so that $\rho_1 = 0$.

Notice that $\rho_1 = 0$ for any function with a finite number of zeros; thus $\rho_1 > 0$ implies that there are an infinity of zeros.

8.23. Canonical products. An important consequence of the above theorem is that, if $f(z)$ is of finite order, then there is an integer p, independent of n, such that the product

$$\prod_{n=1}^{\infty} E\left(\frac{z}{z_n}, p\right) \tag{1}$$

is convergent for all values of z; for by 8.11 (1), with $p_n = p+1$, this product is convergent if

$$\sum \left(\frac{r}{r_n}\right)^{p+1} \tag{2}$$

is convergent;* and this is true for all values of r if $p+1 > \rho_1$, and so it is certainly true if $p+1 > \rho$.

If p is the smallest integer for which (2) is convergent, the product (1) is called the *canonical product* formed with the zeros of $f(z)$; and p is called its *genus*.

If ρ_1 is not an integer, then $p = [\rho_1]$; if ρ_1 is an integer, $p = \rho_1$ if $\sum r_n^{-\rho_1}$ is divergent, while $p = \rho_1 - 1$ if it is convergent. In any case
$$p \leqslant \rho_1 \leqslant \rho.$$

8.24. Hadamard's factorization theorem. *If $f(z)$ is an integral function of order ρ, with zeros $z_1, z_2, \ldots$ ($f(0) \neq 0$), then*
$$f(z) = e^{Q(z)}P(z),$$
where $P(z)$ is the canonical product formed with the zeros of $f(z)$, and $Q(z)$ is a polynomial of degree not greater than ρ.

We can now take the $P(z)$ of § 8.12 to be the canonical product. It follows from the factorization theorem of § 8.12 that there is an expression for $f(z)$ in the above form, in which $Q(z)$ is an integral function. What we have to prove is that in this case $Q(z)$ is a polynomial.

* Compare § 1.43, ex. (vii).

Let $\nu = [\rho]$, so that $p \leqslant \nu$. Taking logarithms and differentiating $\nu+1$ times, we obtain

$$\left(\frac{d}{dz}\right)^{\nu}\left\{\frac{f'(z)}{f(z)}\right\} = Q^{(\nu+1)}(z) - \nu! \sum_{n=1}^{\infty} \frac{1}{(z_n - z)^{\nu+1}}.$$

To prove that $Q(z)$ is a polynomial of degree ν at most, we have to prove that $Q^{(\nu+1)}(z) = 0$.

Let
$$g_R(z) = \frac{f(z)}{f(0)} \prod_{|z_n| \leqslant R} \left(1 - \frac{z}{z_n}\right)^{-1}.$$

Since $|1 - z/z_n| \geqslant 1$ for $|z| = 2R$, $|z_n| \leqslant R$, we have

$$|g_R(z)| \leqslant |f(z)/f(0)| = O(e^{(2R)^{\rho+\epsilon}}) \tag{1}$$

for $|z| = 2R$. Since $g_R(z)$ is an integral function, this holds for $|z| < 2R$ also.

Let $h_R(z) = \log g_R(z)$, the logarithm being determined so that $h_R(0) = 0$. Then $h_R(z)$ is regular for $|z| \leqslant R$, and, by (1),

$$\mathbf{R}\{h_R(z)\} < KR^{\rho+\epsilon}. \tag{2}$$

Hence, by § 5.51,

$$|h_R^{(\nu+1)}(z)| \leqslant \frac{2^{\nu+2}(\nu+1)!\,R}{(R-r)^{\nu+2}} KR^{\rho+\epsilon}$$

for $|z| = r < R$; and for $|z| = \frac{1}{2}R$ this gives

$$h_R^{(\nu+1)}(z) = O(R^{\rho+\epsilon-\nu-1}). \tag{3}$$

Hence
$$\begin{aligned} Q^{(\nu+1)}(z) &= h_R^{(\nu+1)}(z) + \nu! \sum_{|z_n| > R} \frac{1}{(z_n - z)^{\nu+1}} \\ &= O(R^{\rho+\epsilon-\nu-1}) + O\Big(\sum_{|z_n| > R} |z_n|^{-\nu-1}\Big) \end{aligned}$$

for $|z| = \frac{1}{2}R$, and so also for $|z| < \frac{1}{2}R$. The first term on the right tends to 0 as $R \to \infty$ if ϵ is small enough, since $\nu+1 > \rho$; and the second term tends to 0 since $\sum |z_n|^{-\nu-1}$ is convergent. Since the left-hand side is independent of R it must be zero, and the theorem follows.*

8.25. *The order of a canonical product is equal to the exponent of convergence of its zeros.*

We know that, for any function, $\rho_1 \leqslant \rho$. Hence we have to prove that, for a canonical product $P(z)$, $\rho \leqslant \rho_1$. Let r_1, r_2,...

* Hadamard (2). This proof is due to Landau (5). For an alternative proof see § 8.72.

be the moduli of the zeros, and k a constant greater than unity. Let

$$\log|P(z)| = \sum_{r_n \leqslant kr} \log\left|E\left(\frac{z}{z_n}, p\right)\right| + \sum_{r_n > kr} \log\left|E\left(\frac{z}{z_n}, p\right)\right| = \sum_1 + \sum_2.$$

In $\sum_2$ we use the inequality 8.11 (2), and obtain

$$\sum_2 = O\left\{\sum_{r_n > kr}\left(\frac{r}{r_n}\right)^{p+1}\right\} = O\left\{r^{p+1}\sum_{r_n > kr}\frac{1}{r_n^{p+1}}\right\}.$$

If $p = \rho_1 - 1$, this is $O(r^{p+1}) = O(r^{\rho_1})$. Otherwise $\rho_1 + \epsilon < p+1$ if ϵ is small enough, and then

$$r^{p+1}\sum_{r_n > kr} r_n^{-p-1} = r^{p+1}\sum_{r_n > kr} r_n^{\rho_1+\epsilon-p-1} r_n^{-\rho_1-\epsilon}$$
$$< r^{p+1}(kr)^{\rho_1+\epsilon-p-1}\sum r_n^{-\rho_1-\epsilon} = O(r^{\rho_1+\epsilon}).$$

Again in $\sum_1$ we have terms involving $E(u, p)$, where $|u| \geqslant 1/k$, so that

$$\log|E(u,p)| \leqslant \log(1+|u|) + |u| + \ldots + \frac{|u|^p}{p} < K|u|^p,$$

where K depends on k only. Hence

$$\sum_1 < O\left(r^p \sum_{r_n \leqslant kr} r_n^{-p}\right) = O\left(r^p \sum_{r_n \leqslant kr} r_n^{\rho_1+\epsilon-p} r_n^{-\rho_1-\epsilon}\right)$$
$$= O\{r^p (kr)^{\rho_1+\epsilon-p}\sum r_n^{-\rho_1-\epsilon}\} = O(r^{\rho_1+\epsilon}).$$

Hence $$\log|P(z)| < O(r^{\rho_1+\epsilon}),$$

and the result follows.

8.26. *If ρ is not an integer, $\rho_1 = \rho$.*

We have in any case $\rho_1 \leqslant \rho$. Suppose that $\rho_1 < \rho$. Then $P(z)$ is of order ρ_1, i.e. of order less than ρ. Also, if $Q(z)$ is of degree q, $e^{Q(z)}$ is of order q; and $q \leqslant \rho$, and in this case $q < \rho$, since q is an integer and ρ is not. Hence $f(z)$ is the product of two functions, each of order less than ρ. Hence $f(z)$ is of order less than ρ, which gives a contradiction. Hence $\rho_1 = \rho$.

In particular, a function of non-integral order must have an infinity of zeros. In fact, if the order is not an integer, the function is dominated by the canonical product $P(z)$; whereas, if the order is an integer, $P(z)$ may reduce to a polynomial or a constant, and the order then depends entirely on the factor $e^{Q(z)}$.

In any case, since $P(z)$ is of order ρ_1, and $e^{Q(z)}$ of order q, we have

$$\rho = \max(q, \rho_1).$$

8.27. Genus. The genus of the integral function $f(z)$ is the greater of the two integers p and q, and is therefore an integer.

Since $p \leqslant \rho$ and $q \leqslant \rho$, the genus does not exceed the order. The actual determination of the genus of a given function is sometimes not easy.

Example. Prove that the genus is not less than $\rho - 1$.

8.3. The coefficients in the expansion of a function of finite order. *A necessary and sufficient condition that*

$$f(z) = \sum_{n=0}^{\infty} a_n z^n \tag{1}$$

should be an integral function of finite order ρ is that

$$\varliminf_{n\to\infty} \frac{\log(1/|a_n|)}{n \log n} = \frac{1}{\rho}.$$

The argument depends on the fact that $\sum |a_n z^n|$ does not differ very much from its greatest term, and that $|f(z)|$ lies between the two. This is further illustrated by the example which follows.

(i) Let
$$\varliminf \frac{\log(1/|a_n|)}{n \log n} = \mu,$$

where μ is 0, positive, or infinite. Then, for every positive ϵ,

$$\log(1/|a_n|) > (\mu - \epsilon) n \log n \qquad (n > n_0),$$

i.e.
$$|a_n| < n^{-n(\mu-\epsilon)}.$$

If $\mu > 0$, it follows that (1) converges for all values of z, so that $f(z)$ is an integral function. Also, if μ is finite,

$$|f(z)| < Ar^{n_0} + \sum_{n=n_0+1}^{\infty} r^n n^{-n(\mu-\epsilon)} \qquad (r > 1).$$

Let $\sum_1$ denote the part of the last series for which $n \leqslant (2r)^{\frac{1}{\mu-\epsilon}}$, $\sum_2$ the remainder. Then in $\sum_1$

$$r^n \leqslant \exp\{(2r)^{\frac{1}{\mu-\epsilon}} \log r\},$$

so that

$$\sum_1 < \exp\{(2r)^{\frac{1}{\mu-\epsilon}} \log r\} \sum n^{-n(\mu-\epsilon)} < K \exp\{(2r)^{\frac{1}{\mu-\epsilon}} \log r\}.$$

In $\sum_2$, $rn^{-(\mu-\epsilon)} < \frac{1}{2}$, so that

$$\sum_2 < \sum (\tfrac{1}{2})^n < 1.$$

Hence
$$|f(z)| < K \exp\{(2r)^{\frac{1}{\mu-\epsilon}} \log r\},$$

i.e. $\rho \leqslant 1/(\mu - \epsilon)$. Making $\epsilon \to 0$, $\rho \leqslant 1/\mu$. In the case $\mu = \infty$, the argument, with an arbitrarily large number instead of μ, shows that $\rho = 0$.

On the other hand, given ϵ, there is a sequence of values of n for which

$$\log(1/|a_n|) < (\mu+\epsilon)n \log n,$$

i.e.
$$|a_n| > n^{-n(\mu+\epsilon)},$$

i.e.
$$|a_n|r^n > \{rn^{-(\mu+\epsilon)}\}^n.$$

Taking $r = (2n)^{\mu+\epsilon}$, this gives

$$|a_n|r^n > 2^{(\mu+\epsilon)n} = \exp\left\{\tfrac{1}{2}(\mu+\epsilon)\log 2 \,.\, r^{\frac{1}{\mu+\epsilon}}\right\}.$$

Since by Cauchy's inequality $M(r) \geqslant |a_n|r^n$, it follows that, for a sequence of values of r tending to infinity,

$$M(r) > \exp\{Ar^{1/(\mu+\epsilon)}\}.$$

Hence $\rho \geqslant 1/(\mu+\epsilon)$, and, making $\epsilon \to 0$, $\rho \geqslant 1/\mu$. If $\mu = 0$, the argument shows that $f(z)$ is of infinite order.

(ii) Let $f(z)$ be a function of finite order ρ. Then $a_n \to 0$, so that μ, defined as before, is not negative. The argument then shows that $\mu = 1/\rho$.

8.4. Examples. (i) Prove that the order of the function

$$f(z) = \sum_{n=0}^{\infty} \frac{z^n}{(n!)^\alpha}$$

is $1/\alpha$.

[We may use the above theorem, or proceed more directly as follows. Suppose z real and positive. The terms of the series increase until n is approximately $z^{1/\alpha}$, and then decrease. Hence, if $z = n^\alpha$, we get a maximum term

$$\frac{n^{n\alpha}}{(n!)^\alpha} \sim \frac{n^{n\alpha}}{(n^{n+\frac{1}{2}}e^{-n}2^{\frac{1}{2}}\pi^{\frac{1}{2}})^\alpha} = \frac{e^{n\alpha}}{n^{\frac{1}{2}\alpha}(2^{\frac{1}{2}}\pi^{\frac{1}{2}})^\alpha} = \frac{e^{\alpha z^{1/\alpha}}}{z^{\frac{1}{2}}(2^{\frac{1}{2}}\pi)^\alpha}.$$

Since $|f(z)|$ is greater than this term, its order is at least $1/\alpha$.

On the other hand, $|f(z)| \leqslant f(|z|)$, and if z is real

$$\begin{aligned} f(z) &= \sum_{n=0}^{N} \frac{z^n}{(n!)^\alpha} + \sum_{N+1}^{\infty} \frac{z^n}{(n!)^\alpha} \\ &< \sum_{n=0}^{N} \frac{z^N}{(n!)^\alpha} + \sum_{N+1}^{\infty} \frac{z^n}{\{(N+1)!\,N^{n-N-1}\}^\alpha} \\ &< Az^N + \frac{z^{N+1}}{\{(N+1)!\}^\alpha(1-z/N^\alpha)}, \end{aligned}$$

provided that $N^\alpha > z$. Taking $N = [(2z)^{1/\alpha}]$, we obtain

$$f(z) = O(z^N) = O\{z^{(2z)^{1/\alpha}}\} = O(e^{z^{1/\alpha+\epsilon}}),$$

so that the order does not exceed $1/\alpha$. Hence $\rho = 1/\alpha$. (See Hardy's *Orders of Infinity*, ed. 1, p. 55.)]

(ii) Discuss in a similar way the function

$$\sum_{n=1}^{\infty} \frac{z^n}{n^{\alpha n}}.$$

(iii) If $\lambda \neq 0$, and $p(z)$ is a polynomial, $e^{\lambda z} - p(z)$ has an infinity of zeros.

[If not, $e^{\lambda z} - p(z) = e^{az+b}P(z)$, where $P(z)$ is a polynomial. By comparing rates of increase in various directions we find that $a = \lambda$, then $e^{\lambda z}$ = rational function.]

(iv) If $f(z)$ is of order ρ, and $g(z)$ of order $\rho' \leqslant \rho$, and the zeros of $g(z)$ are all zeros of $f(z)$, then $f(z)/g(z)$ is of order ρ at most.

[For $f(z) = P_1(z)e^{Q_1(z)}$, $g(z) = P_2(z)e^{Q_2(z)}$, and P_1/P_2 is either the canonical product formed with the zeros of f_1/f_2, or this product multiplied by an exponential factor of order not exceeding ρ. Hence the order of P_1/P_2 does not exceed ρ.]

(v) $\cos z$ and $\sin z$ are of order 1; the product formulae (§ 3.23) are cases of Hadamard's theorem.

(vi) $1/\Gamma(z)$ is of order 1; deduce the product formula (§ 4.41) from Hadamard's theorem.

[With the notation of § 4.41, $f(1-z) = -2i\pi/\Gamma(z)$, and

$$f(z) = O\left\{e^{\pi|z|}\left(1+\int_1^\infty t^{|z|}e^{-t}\,dt\right)\right\} = O\{e^{\pi|z|}(|z|+1)^{|z|+1}\}.\Big]$$

(vii) $\xi(s) = \frac{1}{2}s(s-1)\pi^{-\frac{1}{2}s}\Gamma(\frac{1}{2}s)\zeta(s)$ is an integral function with $\rho = 1$, $\rho_1 = 1$.

[To prove that $\rho \leqslant 1$, use § 4.43 (3) and ex. (iv); and $\rho \geqslant 1$ since $\log\zeta(s) \sim 2^{-s}$, $\log\xi(s) \sim \frac{1}{2}s\log s$ as $s \to \infty$ by real values. Next the functional equation gives $\xi(s) = \xi(1-s)$. Hence $\Xi(z) = \xi(\frac{1}{2}+iz)$ is even, and $\Xi(\sqrt{z})$ is an integral function of order $\frac{1}{2}$, and so has convergence-exponent $\frac{1}{2}$.]

(viii) $z^{-\nu}J_\nu(z)$ is an integral function with $\rho = 1$, $\rho_1 = 1$. Verify the result of § 8.3 in this case. [See p. 60, ex. 5.]

(ix) $F_\alpha(z) = \int\limits_0^\infty e^{-t^\alpha}\cos zt\,dt$ $(\alpha > 1)$ is of order $\alpha/(\alpha-1)$.

[Either directly from the integral, or from the power-series.]

(x) $\vartheta_1(z) = -i\sum\limits_{n=-\infty}^{\infty}(-1)^n q^{(n+\frac{1}{2})^2}e^{(2n+1)iz}$, where $|q| < 1$, is an integral function with $\rho = 2$, $\rho_1 = 2$.

[If $\lambda = (2|z|+\log 2)/\log|1/q|-\frac{1}{2}$,

$$\vartheta_1(z) \leqslant 2\sum_{n\leqslant\lambda}|q|^{(n+\frac{1}{2})^2}e^{(2n+1)|z|}+2\sum_{n>\lambda}(\tfrac{1}{2})^{n+\frac{1}{2}} = O(e^{(2\lambda+1)|z|}) = O(e^{K|z|^2}).$$

$\vartheta_1(z)$ has simple zeros at $z = m\pi+n\pi\tau$, where m and n run through all integers (see e.g. Whittaker and Watson, *Modern Analysis*, § 21.12). Hence $\rho_1 = 2$.]

(xi) $\vartheta_1(z)$ is an integral function of $\sin z$ of order 0.

[If $2\sin z = w$, $\vartheta_1(z) = g(w)$, then

$$g(w) = \sum_{n=0}^{\infty}q^{(n+\frac{1}{2})^2}\{w^{2n+1}-(2n+1)w^{2n-1}+\ldots\}$$

$$= O\left\{\sum_{n=0}^{\infty}|q|^{(n+\frac{1}{2})^2}(|w|+1)^{2n+1}\right\} = O\{e^{K\log^2(|w|+1)}\}.$$

It was proved by Pólya (2) that if g and h are integral functions, and $g\{h(z)\}$ of finite order, then either h is a polynomial and g of finite order, or h is not a polynomial but of finite order, and g of zero order.]

(xii) If $$f(z) = \prod_1^\infty\left(1+\frac{z}{r_n}\right) \qquad (r_n > 0)$$

is of order ρ, $0 < \rho < 1$, then for $\rho < \sigma < 1$

$$\int\limits_0^\infty \frac{\log f(x)}{x^{\sigma+1}}\,dx = \frac{\pi}{\sigma\sin\pi\sigma}\sum_1^\infty \frac{1}{r_n^\sigma}, \qquad \int\limits_0^\infty \frac{\log|f(-x)|}{x^{\sigma+1}}\,dx = \frac{\pi}{\sigma\tan\pi\sigma}\sum_1^\infty \frac{1}{r_n^\sigma}.$$

[We have

$$\int\limits_0^\infty \frac{\log(1+x)}{x^{\sigma+1}}\,dx = \frac{\pi}{\sigma\sin\pi\sigma}, \qquad \int\limits_0^\infty \frac{\log|1-x|}{x^{\sigma+1}}\,dx = \frac{\pi}{\sigma\tan\pi\sigma}.\text{]}$$

8.5. The derived function. Many of the properties of the derived function of an integral function are the same as those of the primitive function. The following theorems are examples of this.

8.51. *The derived function $f'(z)$ is of the same order as $f(z)$.*

Let $M'(r) = \max\limits_{|z|=r}|f'(z)|$. Then

$$\frac{M(r)-|f(0)|}{r} \leqslant M'(r) \leqslant \frac{M(R)}{R-r}. \tag{1}$$

For
$$f(z) = \int\limits_0^z f'(t)\,dt + f(0),$$
the integral being taken along the straight line. Hence

$$M(r) \leqslant rM'(r) + |f(0)|.$$

On the other hand,

$$f'(z) = \frac{1}{2\pi i}\int_C \frac{f(w)}{(w-z)^2}\,dw,$$

where C is the circle $|w-z| = R-r$ $(|z| = r < R)$. Hence, choosing z so that $|f'(z)| = M'(r)$, we have

$$M'(r) \leqslant \frac{M(R)}{R-r}.$$

The result stated now follows on taking, say, $R = 2r$ in (1).

8.52. The well-known theorem, that if $f(z)$ is a polynomial with all its roots real, then $f'(z)$ has the same property, can be

* Pólya (2).

extended to a certain class of integral functions. The result is expressed by the following theorem of Laguerre:

If $f(z)$ is an integral function, real for real z, of order less than 2, with real zeros, then the zeros of $f'(z)$ are also all real, and are separated from each other by the zeros of $f(z)$.

We have
$$f(z) = cz^k e^{az} \prod_{n=1}^{\infty} \left(1-\frac{z}{z_n}\right) e^{\frac{z}{z_n}},$$
where k is zero or a positive integer, and c, a, and z_1, z_2,... are all real. Taking logarithms and differentiating,
$$\frac{f'(z)}{f(z)} = \frac{k}{z} + a + \sum_{n=1}^{\infty} \left(\frac{1}{z-z_n} + \frac{1}{z_n}\right).$$
Hence, if $z = x+iy$,
$$\mathbf{I}\left\{\frac{f'(z)}{f(z)}\right\} = -iy\left\{\frac{k}{x^2+y^2} + \sum_{n=1}^{\infty} \frac{1}{(x-z_n)^2+y^2}\right\},$$
which is zero if $y = 0$ only. Hence $f'(z)$ cannot be zero except on the real axis.

Again,
$$\frac{d}{dz}\left\{\frac{f'(z)}{f(z)}\right\} = -\frac{k}{z^2} - \sum_{n=1}^{\infty} \frac{1}{(z-z_n)^2},$$
which is real and negative if z is real. Hence $f'(z)/f(z)$ decreases steadily as z increases through real values from z_n to z_{n+1}, and so it cannot vanish more than once between z_n and z_{n+1}. Clearly it changes sign, and so vanishes just once in this interval. This proves the theorem.

It is clear from the above result that, if the zeros of $f'(z)$ are z_1', z_2',..., then the series
$$\sum \frac{1}{|z_n|^\alpha}, \quad \sum \frac{1}{|z_n'|^\alpha},$$
converge or diverge together. Hence the zeros of $f'(z)$ have the same exponent of convergence as those of $f(z)$. It may be shown further that $f(z)$ and $f'(z)$ have the same genus, but this is not quite so easy to prove (see ex. 16 at the end of the chapter).

Since $f'(z)$ is of the same order as $f(z)$, and has real zeros only, the theorem may now be applied to it, and we see that $f''(z)$ has real zeros only; and so for $f'''(z)$, etc.

The proof also applies to a function $f(z)$ of order 2, but of genus 1. In this case, however, we cannot extend the result to $f''(z)$,... without considering the problem of the genus of $f'(z)$.

It is easily seen by means of examples that the theorem is not true for functions of genus 2. For example, in the case

$$f(z) = ze^{z^2}, \qquad f'(z) = (2z^2+1)e^{z^2},$$

the zeros of $f'(z)$ are complex; and in the case

$$f(z) = (z^2-4)e^{\frac{1}{3}z^2}, \qquad f'(z) = \tfrac{2}{3}z(z^2-1)e^{\frac{1}{3}z^2},$$

the zeros of $f'(z)$ are real, but are not separated by those of $f(z)$.

Example. The differential equation

$$y\frac{d^2y}{dt^2} = -\sin^2 t$$

has no real solution, other than $y = \pm\sin t$, which is an integral function of finite order.

[Suppose that y is a function of finite order ρ. Then

$$y = e^{Q(t)}P(t),$$

where $P(t)$ is a canonical product, and $Q(t)$ a polynomial of degree not greater than ρ. Since the zeros of $P(t)$ are zeros of $\sin^2 t$, $P(t)$ is of order 1 at most.

Now
$$\frac{dy}{dt} = e^{Q(t)}\{P'(t)+P(t)Q'(t)\},$$

$$\frac{d^2y}{dt^2} = e^{Q(t)}\{P''(t)+2P'(t)Q'(t)+P(t)Q'^2(t)+P(t)Q''(t)\} = e^{Q(t)}f(t),$$

where $f(t)$ is of order 1 at most. Hence

$$f(t) = e^{at+b}P_1(t),$$

where $P_1(t)$ is a canonical product. Hence

$$e^{2Q(t)+at+b}P(t)P_1(t) = -\sin^2 t,$$

i.e.
$$e^{2Q(t)+at+b}P(t) = -\sin^2 t/P_1(t)$$

is of order 1 at most (§ 8.4, ex. (iv)). Hence $P(t)$ is of order 1 and $Q(t)$ is linear.

Hence y is a function of order 1.

We can now use Laguerre's theorem. y is a function of order 1 with real zeros. The zeros of $\frac{dy}{dt}$ are separated by those of y, so that, as y has no triple zeros, all the zeros of $\frac{dy}{dt}$ are simple. So all the zeros of $\frac{d^2y}{dt^2}$ are simple. Hence y has zeros at *all* the zeros of $\sin t$. Suppose y has a double zero at $t = k\pi$. Then $\frac{dy}{dt}$ has a zero between $(k-1)\pi$ and $k\pi$, a zero at $k\pi$, and a zero between $k\pi$ and $(k+1)\pi$. Hence $\frac{d^2y}{dt^2}$ has two zeros between

$(k-1)\pi$ and $(k+1)\pi$, which is impossible. Hence y has all the zeros of $\sin t$ just once. Hence

$$y = e^{\alpha t+\beta}\sin t.$$

Inserting this in the differential equation, we obtain

$$(\alpha^2-1)\sin t+2\alpha\cos t = -e^{-2\alpha t-2\beta}\sin t.$$

Since the left-hand side is bounded for real t, so is the right-hand side, and hence $\alpha = 0$. Then $\beta = 0$ or πi.]

8.6. Functions with real zeros only. A number of important functions have no complex zeros; for example, all the zeros of $1/\Gamma(z)$ are real. On the other hand it is sometimes very difficult to decide whether the zeros are real or not; for example, it was conjectured by Riemann, in 1859, that all the zeros of the function $\Xi(z)$ of § 8.4 (vii) are real, but this has never been proved.

8.61. The theorems of Laguerre. In some cases the question can be decided by the following theorems of Laguerre.*

Let $f(z)$ be a polynomial,

$$f(z) = a_0+a_1z+...+a_pz^p,$$

all of whose zeros are real; and let $\phi(w)$ be an integral function of genus 0 *or* 1, *which is real for real w, and all the zeros of which are real and negative. Then the polynomial*

$$g(z) = a_0\phi(0)+a_1\phi(1)z+...+a_p\phi(p)z^p$$

has all its zeros real, and as many positive, zero and negative zeros as $f(z)$.

Let
$$\phi(w) = ae^{kw}\prod_{n=1}^{\infty}\left(1+\frac{w}{\alpha_n}\right)e^{-\frac{w}{\alpha_n}},$$

where $\alpha_n > 0$ for all values of n. Consider the function

$$\begin{aligned} g_1(z) &= f(z)+\frac{z}{\alpha_1}f'(z) \\ &= \frac{z^{1-\alpha_1}}{\alpha_1}\frac{d}{dz}\{z^{\alpha_1}f(z)\} \qquad (z>0) \\ &= a_0+a_1\left(1+\frac{1}{\alpha_1}\right)z+...+a_p\left(1+\frac{p}{\alpha_1}\right)z^p. \end{aligned}$$

Obviously $g_1(z)$ has as many zeros at $z=0$ as $f(z)$; and the second expression for it shows, by Rolle's theorem, that it has

* *Œuvres*, t. 1, p. 200.

the same number of positive zeros as $f(z)$. Similarly it has the same number of negative zeros.

By repeating the argument, we can obtain the same result for the function

$$g_n(z) = a_0 + a_1\phi_n(1)z + \ldots + a_p\phi_n(p)z^p,$$

where
$$\phi_n(w) = \left(1+\frac{w}{\alpha_1}\right)\ldots\left(1+\frac{w}{\alpha_n}\right).$$

Next, the transformation $z = e^{k_n}z'$, where $k_n = k - \sum_1^n 1/\alpha_\nu$, shows that the same result holds for

$$G_n(z) = a_0\Phi_n(0) + a_1\Phi_n(1)z + \ldots + a_p\Phi_n(p)z^p,$$

where
$$\Phi_n(w) = ae^{kw}\prod_{\nu=1}^{n}\left(1+\frac{w}{\alpha_\nu}\right)e^{-\frac{w}{\alpha_\nu}} = ae^{k_n w}\phi_n(w).$$

Finally, $\Phi_n(w) \to \phi(w)$ uniformly in any finite region. Hence $G_n(z) \to g(z)$ uniformly in any finite region; by Hurwitz's theorem (§ 3.45) the zeros of $g(z)$ are the limits of the zeros of $G_n(z)$; it is clear that $g(z)$ has the same number of zeros at $z = 0$ as $f(z)$; and this completes the proof.

8.62. *Suppose that $\phi(w)$ satisfies the conditions of the previous theorem, and that $f(z)$ is an integral function of the form*

$$f(z) = e^{az+b}\prod_{n=1}^{\infty}\left(1+\frac{z}{z_n}\right),$$

the numbers a and z_n being all positive. Let

$$f(z) = \sum_{n=0}^{\infty} a_n z^n.$$

Then
$$g(z) = \sum_{n=0}^{\infty} a_n\phi(n)z^n$$

is an integral function, all of whose zeros are real and negative.

In the first place, $g(z)$ is an integral function; for, since $(1+x)e^{-x} \leqslant 1$ for $x \geqslant 0$,

$$|\phi(n)| \leqslant |a|e^{kn},$$

and so the series for $g(z)$ is everywhere convergent.

Let
$$f_p(z) = e^b\left(1+\frac{az}{p}\right)^p\prod_{n=1}^{p}\left(1+\frac{z}{z_n}\right)$$
$$= \sum_{n=0}^{2p} a_{n,p}z^n.$$

All the zeros of this are real and negative, and hence, by the previous theorem, so are the zeros of

$$g_p(z) = \sum_{n=0}^{2p} a_{n,p}\,\phi(n) z^n.$$

Finally, $g_p(z) \to g(z)$ uniformly in any finite region. In fact it is clear from the expression

$$\left(1+\frac{az}{p}\right)^p = 1+az+\left(1-\frac{1}{p}\right)\frac{a^2z^2}{2!}+\ldots+\frac{a^pz^p}{p^p}$$

that $a_{n,p} \to a_n$ as $p \to \infty$ for every fixed n, while $|a_{n,p}| \leqslant a_n$ for all values of n and p. Hence, if $N < 2p$,

$$|g(z)-g_p(z)| \leqslant \left|\sum_{n=1}^{N} (a_n - a_{n,p})z^n\right| + \left|\sum_{N+1}^{\infty} a_n z^n\right| + \left|\sum_{N+1}^{2p} a_{n,p} z^n\right|$$

$$\leqslant \left|\sum_{n=1}^{N} (a_n - a_{n,p})z^n\right| + 2\sum_{N+1}^{\infty} |a_n z^n|.$$

We can now choose N so large that the second term is less than any given ϵ, and then, having fixed N, the first term tends to zero. Hence $g_p(z) \to g(z)$.

As in the previous proof, the result now follows from Hurwitz's theorem.

8.63. The simplest case is that of the function $f(z) = e^z$. From this we deduce that *if $\phi(w)$ satisfies the conditions of the previous theorems, then*

$$F(z) = \sum_{n=0}^{\infty} \frac{\phi(n)}{n!} z^n$$

is an integral function, and all of its zeros are real and negative.

Examples. (i) Let

$$\phi(w) = 1/\Gamma(w+\nu+1) \qquad (\nu > -1).$$

This is an integral function of genus 1, with zeros at $w = -\nu-1$, $-\nu-2, \ldots$. These are all real and negative, and hence the zeros of

$$\sum_{n=0}^{\infty} \frac{z^n}{n!\,\Gamma(n+\nu+1)} = \frac{J_\nu(2i\sqrt{z})}{(i\sqrt{z})^\nu}$$

are all real and negative. Hence the zeros of $J_\nu(z)$ are all real.

(ii) The function*

$$F_\alpha(z) = \int_0^\infty e^{-t^\alpha} \cos zt\, dt$$

* Pólya (1).

has no zeros if $\alpha = 2$, and an infinity of real zeros, but no complex zeros, if $\alpha = 4, 6, 8, \ldots$.

[We have, by § 8.47,

$$F_\alpha(z) = \frac{1}{\alpha} \sum_{n=0}^{\infty} (-1)^n \frac{\Gamma\left(\frac{2n+1}{\alpha}\right)}{\Gamma(2n+1)} z^{2n}.$$

If $\alpha = 2$, we have, as in § 8.47,

$$F_2(z) = \tfrac{1}{2}\sqrt{\pi} e^{-\frac{1}{4}z^2},$$

which has no zeros.

If $\alpha = 2k$, where k is a positive integer, let

$$\phi(w) = \Gamma\{(2w+1)/2k\}\Gamma(w+1)/\Gamma(2w+1).$$

Then $\phi(w)$ is an integral function satisfying the conditions of Laguerre's theorems of § 8.6. Hence the zeros of

$$\sum_{n=0}^{\infty} \frac{\phi(n)}{n!} z^n = 2k F_{2k}(i\sqrt{z})$$

are all real and negative, so that the zeros of $F_{2k}(z)$ are all real. Also $\rho = 2k/(2k-1)$, so that $1 < \rho < 2$, and there must be an infinity of zeros.

If α is not an even integer, it can be proved that there are an infinity of complex zeros, and a finite number of real zeros.]

8.64. Functions with real negative zeros. If all the zeros of a function are real and negative, the modulus of the function is related to the distribution of its zeros in a specially simple way.

Suppose that $f(z)$ is such a function, and that its order ρ is less than 1. Then

$$f(z) = \prod_{n=1}^{\infty}\left(1+\frac{z}{z_n}\right).$$

Hence, if z is real,

$$\log f(z) = \sum_{n=1}^{\infty} \log\left(1+\frac{z}{z_n}\right) = \sum_{n=1}^{\infty} n\left\{\log\left(1+\frac{z}{z_n}\right) - \log\left(1+\frac{z}{z_{n+1}}\right)\right\}^*$$

$$= \sum_{n=1}^{\infty} n \int_{z_n}^{z_{n+1}} \frac{z\,dt}{t(z+t)} = z\int_0^{\infty} \frac{n(t)\,dt}{t(z+t)},$$

where $n(t)$ has its usual meaning.

Suppose now that as $t \to \infty$, $n(t) \sim \lambda t^\rho$. Then

$$\log f(x) \sim \pi\lambda \operatorname{cosec} \pi\rho\, x^\rho.$$

For we have $$(\lambda-\epsilon)t^\rho < n(t) < (\lambda+\epsilon)t^\rho$$

* The reader should justify this step, which is a simple example of partial summation.

for $t > t_0(\epsilon)$. Hence

$$\log f(x) < x \int_0^{t_0} \frac{n(t)\,dt}{t(x+t)} + x \int_{t_0}^{\infty} \frac{(\lambda+\epsilon)t^\rho}{t(x+t)}\,dt$$

$$= x \int_0^{t_0} \frac{n(t)-(\lambda+\epsilon)t^\rho}{t(x+t)}\,dt + x \int_0^{\infty} \frac{(\lambda+\epsilon)t^\rho}{t(x+t)}\,dt.$$

The first term is plainly $O(1)$, and, putting $t = xu$ in the second integral, we obtain

$$x^\rho(\lambda+\epsilon) \int_0^{\infty} \frac{u^{\rho-1}\,du}{1+u} = x^\rho(\lambda+\epsilon)\pi \operatorname{cosec} \pi\rho$$

by § 3.123. A similar result holds with $\lambda-\epsilon$, and the theorem follows.

*More generally,**

$$\log f(re^{i\theta}) \sim e^{i\rho\theta}\pi\lambda \operatorname{cosec} \pi\rho\, r^\rho$$

for any fixed θ *in* $(-\pi, \pi)$, $\log f(z)$ *denoting the branch which is real on the positive real axis.*

In fact the above expression for $\log f(z)$ as an integral, obtained for real z, holds by analytic continuation for $-\pi < \arg z < \pi$. Hence we obtain as before

$$\log f(re^{i\theta}) \sim re^{i\theta} \int_0^{\infty} \frac{\lambda t^\rho\,dt}{t(re^{i\theta}+t)}.$$

Turning the line of integration to $t = ue^{i\theta}$, we obtain

$$\lambda re^{\rho i\theta} \int_0^{\infty} \frac{u^\rho\,du}{u(r+u)} = \lambda r^\rho e^{\rho i\theta}\pi \operatorname{cosec} \pi\rho$$

as before.

It is also possible to prove theorems of the converse type, viz. to deduce the asymptotic behaviour of $n(r)$ from that of $\log|f(z)|$. The most interesting is that *if, as* $x \to \infty$ *by real values,* $\log f(x) \sim \pi\lambda \operatorname{cosec} \pi\rho\, x^\rho$, *then* $n(r) \sim \lambda r^\rho$. This theorem† is closely connected with the Tauberian results of §§ 7.41–7.44, but the proof is too complicated to give here.

* Pólya and Szegö, *Aufgaben*, IV Abschn., no. 61.

† See Valiron (1), Titchmarsh (5), (6).

8.7. The minimum modulus. Let $m(r)$ denote the minimum of $|f(z)|$ on the circle $|z| = r$.

The function $m(r)$ cannot be expected to behave in as simple a way as $M(r)$, since it vanishes whenever r is the modulus of a zero of $f(z)$. But we shall see that, if we exclude the immediate neighbourhood of these exceptional points, we can set a lower limit to $m(r)$; and, in general, $m(r)$ tends to zero in somewhat the same way as $1/M(r)$.

8.71. Consider first a canonical product $P(z)$ of order ρ, with zeros $z_1, z_2, \ldots, z_n, \ldots$.

If about each zero z_n ($|z_n| > 1$) we describe a circle of radius $|z_n|^{-h}$, where $h > \rho$, then in the region excluded from these circles

$$|P(z)| > e^{-r^{\rho+\epsilon}} \qquad (r > r_0(\epsilon)).$$

Following the method of § 8.25, it is clear that

$$\log|P(z)| \geqslant \sum_{r_n \leqslant kr} \log\left|1-\frac{z}{z_n}\right| - \sum_{r_n \leqslant kr} O\left\{\left(\frac{r}{r_n}\right)^p\right\} - \sum_{r_n > kr} O\left\{\left(\frac{r}{r_n}\right)^{p+1}\right\}$$

$$= \sum_{r_n \leqslant kr} \log\left|1-\frac{z}{z_n}\right| - O(r^{\rho+\epsilon}).$$

Since $\sum r_n^{-h}$ is convergent, the sum of the radii of the circles is finite, and so there are circles with centre the origin and arbitrarily large radius which lie entirely in the excluded region. Now if z lies outside every circle $|z - z_n| = r_n^{-h}$, and $r_n \leqslant kr$,

$$\left|1-\frac{z}{z_n}\right| > r_n^{-1-h} > (kr)^{-1-h}.$$

Hence
$$\sum_{1 < r_n \leqslant kr} \log\left|1-\frac{z}{z_n}\right| > -(1+h)\log kr \,.\, n(kr)$$

$$> -K \log kr \,.\, r^{\rho+\epsilon} > -r^{\rho+2\epsilon}.$$

Finally
$$\sum_{r_n \leqslant 1} \log\left|1-\frac{z}{z_n}\right| > 0 \qquad (r > 2),$$

and the result follows.

8.711. *If $f(z)$ is a function of order ρ, then*

$$m(r) > e^{-r^{\rho+\epsilon}}$$

on circles of arbitrarily large radius.

For
$$f(z) = P(z)e^{Q(z)},$$

where $Q(z)$ is a polynomial of degree $q \leqslant \rho$; hence

$$|e^{Q(z)}| > e^{-Ar^q} > e^{-Ar^\rho},$$

for sufficiently large values of r; and by the previous result

$$|P(z)| > e^{-r^{\rho_1+\epsilon}} > e^{-r^{\rho+\epsilon}}$$

on circles of arbitrarily large radius. Hence the result.

8.72. Another proof of Hadamard's factorization theorem. The theorem of § 8.71 leads to an alternative proof of Hadamard's factorization theorem. Let

$$f(z) = P(z)e^{Q(z)}$$

where $P(z)$ is the canonical product formed with the zeros of $f(z)$. Then $Q(z)$ is an integral function. Let ρ be the order of $f(z)$, ρ_1 the exponent of convergence. Then $P(z)$ is of order ρ_1, and $\rho_1 \leqslant \rho$. Hence

$$|P(z)| > e^{-r^{\rho_1+\epsilon}} > e^{-r^{\rho+\epsilon}}$$

on circles of arbitrarily large radius. Also

$$f(z) = O(e^{r^{\rho+\epsilon}}).$$

Hence $$\mathbf{R}\{Q(z)\} = \log\left|\frac{f(z)}{P(z)}\right| = \log\{O(e^{r^{\rho+\epsilon}})\} < Kr^{\rho+\epsilon}$$

on circles of arbitrarily large radius. Hence, by the theorem of § 2.54, $Q(z)$ is a polynomial of degree $\leqslant \rho$.

8.73. In special cases it is possible to prove much more precise results than the theorem of § 8.711.

If $\rho < \frac{1}{2}$, there is a sequence of values of r tending to infinity through which $m(r) \to \infty$.

In the first place, there is no line $\arg z = \text{constant}$ on which $f(z)$ is bounded; for the whole plane, bounded by this line, forms an angle 2π, and $2\pi < \pi/\rho$ if $\rho < \frac{1}{2}$. Hence, by § 5.61, if $f(z)$ is bounded on this line it is bounded everywhere, and so reduces to a constant.

Suppose now that

$$f(z) = cz^k \prod_{n=1}^{\infty}\left(1-\frac{z}{z_n}\right),$$

and let $$\phi(z) = cz^k \prod_{n=1}^{\infty}\left(1+\frac{z}{r_n}\right),$$

where $r_n = |z_n|$. Then

$$\min_{|z|=r} |f(z)| \geqslant |\phi(-r)|,$$

since

$$\left|1-\frac{z}{z_n}\right| \geqslant \left|1-\frac{r}{r_n}\right|$$

for every n. Also $\phi(-r)$ is unbounded, since $\phi(z)$ is an integral function of the same order as $f(z)$. This proves the theorem.

8.74. The following result is still more precise.

If $0 < \rho < 1$, there are arbitrarily large values of r for which

$$m(r) > \{M(r)\}^{\cos \pi\rho - \epsilon}.$$

The following proof is due to Pólya (3). Define $f(z)$ and $\phi(z)$ as before, and we may plainly take $c = 1$, $k = 0$. It is sufficient to prove the theorem for $\phi(z)$. If $0 < \rho < \frac{1}{2}$, i.e. $\cos \pi\rho > 0$, this follows at once from the relations $m(r) \geqslant |\phi(-r)|$, $M(r) \leqslant \phi(r)$. In any case, if z' is a point where $|f(z')| = m(r)$, we have

$$|\phi(r)\phi(-r)| = \left|\prod_{n=1}^{\infty}\left(1-\frac{r^2}{r_n^2}\right)\right| \leqslant |f(z')f(-z')| \leqslant m(r)M(r).$$

Hence, if the theorem is true for $\phi(z)$, then

$$m(r)M(r) \geqslant |\phi(r)|^{1+\cos \pi\rho-\epsilon} \geqslant \{M(r)\}^{1+\cos \pi\rho-\epsilon}$$

for arbitrarily large r, and the result for $f(z)$ follows.

If the theorem is false for $\phi(z)$, there are positive constants ϵ and a such that

$$\log|\phi(-x)| < (\cos \pi\rho-\epsilon)\log \phi(x) \qquad (x > a).$$

By § 8.4, ex. (xii), for $\rho < s < 1$, and so also for $\rho < \mathbf{R}(s) < 1$,

$$\int_0^\infty \{\cos \pi s \log \phi(x) - \log|\phi(-x)|\} x^{-s-1}\, dx = 0.$$

Since the integral over $(0, a)$ is regular for $0 < \mathbf{R}(s) < 1$, so is the integral over (a, ∞). Hence

$$F(s) = \int_\alpha^\infty \{\phi_1(e^\xi)+\psi(s)\phi_2(e^\xi)\}e^{-s\xi}\, d\xi,$$

where

$$\phi_1(x) = (\cos \pi\rho-\epsilon)\log \phi(x) - \log|\phi(-x)|, \qquad \phi_2(x) = \log \phi(x),$$

$$\psi(s) = \cos \pi s - \cos \pi\rho + \epsilon, \qquad \alpha = \log a,$$

is regular for $0 < \mathbf{R}(s) < 1$, and in particular at $s = \rho$. Here

ϕ_1 and ϕ_2 are positive for $x > a$, and $\psi(s)$ is positive for s real and sufficiently near to ρ.

Let $h > 0$, $D = d/ds$. Then

$$\left|\left(1-\frac{hD}{m}\right)^m F(s)\right| = \left|\sum_{\mu=0}^{m} \frac{(-h)^\mu}{\mu!} \frac{m(m-1)...(m-\mu+1)}{m^\mu} F^\mu(s)\right|$$
$$\leqslant |F(s)|+\frac{h}{1!}|F'(s)|+\frac{h^2}{2!}|F''(s)|+... = M,$$

say, the series being convergent for sufficiently small positive h and $s-\rho$. Also

$$\left(1-\frac{hD}{m}\right)^m \psi(s)e^{-s\xi} = e^{-s\xi}\left(1+\frac{h\xi-hD}{m}\right)^m \psi(s)$$
$$= e^{-s\xi}\left(1+\frac{h\xi}{m}\right)^m \psi(s)+e^{-s\xi}\sum_{\mu=1}^{m}\binom{m}{\mu}\left(1+\frac{h\xi}{m}\right)^{m-\mu}\left(\frac{-h}{m}\right)^\mu \psi^{(\mu)}(s)$$
$$\geqslant e^{-s\xi}\left(1+\frac{h\xi}{m}\right)^m \psi(s)-e^{-s\xi}\left(1+\frac{h\xi}{m}\right)^m \sum_{\mu=1}^{m}\frac{h^\mu}{\mu!}|\psi^\mu(s)|.$$

Since $|\psi^\mu(s)| \leqslant \pi^\mu$ for real s, this plainly exceeds

$$\tfrac{1}{2}e^{-s\xi}\left(1+\frac{h\xi}{m}\right)^m \psi(s)$$

if h is small enough. Hence

$$\left(1-\frac{hD}{m}\right)^m F(s) \geqslant \frac{1}{2}\int_{\alpha}^{\infty} \{\phi_1(e^\xi)+\psi(s)\phi_2(e^\xi)\}\left(1+\frac{h\xi}{m}\right)^m e^{-s\xi}\, d\xi.$$

In particular for any $\omega > \alpha$

$$\int_{\alpha}^{\omega} \phi_2(e^\xi)\left(1+\frac{h\xi}{m}\right)^m e^{-s\xi}\, d\xi \leqslant 2M/\psi(s).$$

Making $m \to \infty$, then $\omega \to \infty$, it follows that

$$\int_{\alpha}^{\infty} \phi_2(e^\xi)e^{(h-s)\xi}\, d\xi = \int_{a}^{\infty} \frac{\log \phi(x)}{x^{s-h+1}}\, dx$$

is convergent for a value of $s-h$ less than ρ. Hence $\sum r_n^{-s+h}$ is convergent, contrary to § 8.26. This proves the theorem.

8.75. A similar result holds for functions of order 1 and exponential type, i.e. such that $f(z) = O(e^{K|z|})$.

If $f(z) = O(e^{k|z|})$, *then* $m(r) > e^{-(k+\epsilon)r}$ *for some arbitrarily large* r. If $r_1, r_2,...$ are the moduli of the zeros of $f(z)$, we find as in § 8.21 that $n(r) = O(r)$, $1/r_n \leqslant K/n$. Hence

$$\phi(z) = \prod_{n=1}^{\infty}\left(1+\frac{z}{r_n^2}\right) = O\prod_{n=1}^{\infty}\left(1+\frac{K^2r}{n^2}\right) = \frac{\sinh(\pi K\sqrt{r})}{\pi K\sqrt{r}}.$$

Define $h(\theta)$ (§ 5.7) for $\phi(z)$ with $V(r) = \sqrt{r}$. Then $h(\theta) \leqslant \pi K$ for all θ. Since $|\phi(z)| \geqslant 1$ if $\mathbf{R}(z) \geqslant 0$, $h(\theta)$ is finite for $-\frac{1}{2}\pi \leqslant \theta \leqslant \frac{1}{2}\pi$, and so everywhere (§ 5.712). Also $h(-\theta) = h(\theta)$; and § 5.713, with $\theta_1 = -\pi$, $\theta_2 = 0$, $\theta_3 = \pi$, $\rho = \frac{1}{2}$, gives $h(\pi) \geqslant 0$. Hence

$$|f(z)f(-z)| \geqslant |f(0)|^2|\phi(-r^2)| > e^{-\epsilon r}$$

for some arbitrarily large values of r, and the result follows.

8.8. The a-points of an integral function. Our discussion of integral functions has so far centred round the distribution of the zeros of the function. A more general question is that of the distribution of the points where the function takes any given value a—the 'a-points', as we may call them.

There is one case in which we have already obtained fairly precise results, namely, that of functions of finite non-integral order. If $f(z)$ is of order ρ, where ρ is not an integer, then it has an infinity of zeros, and the exponent of convergence of the zeros is ρ. But clearly $f(z)-a$ is also of order ρ, where a is any constant. Hence $f(z)$ has an infinity of a-points, and their exponent of convergence is ρ; i.e. their density is roughly the same for all values of a.

A similar argument may be applied to functions of zero order. Such a function has an infinity of zeros unless it reduces to a polynomial; and $f(z)-a$ is a polynomial for every value of a or for none.

If $f(z)$ is of positive integral order, and $\neq a$, then $f(z)-a = e^{Q(z)}$, where $Q(z)$ is a polynomial. If $b \neq a$, then $Q(z) = \log(b-a)$ for some z, i.e. $f(z) = b$ for some z. Hence $f(z)$ takes every value with one possible exception.

8.81. Picard's theorem. The main theorem of the subject is due to Picard; it is independent of any considerations of order.

An integral function which is not a polynomial takes every value, with one possible exception, an infinity of times.

Picard's proof of the theorem depends on the properties of the elliptic modular function. This function, which we shall denote by $\varpi(z)$, has the following properties: *it is regular everywhere except at* $z=0$, 1, *and* ∞; *and its imaginary part is never negative.*

By means of this function we can easily prove that *an integral function which is not a constant takes every value, with one possible exception, at least once.*

Suppose that $f(z)$ is an integral function which does not take either of the values a or b, where $a \neq b$. Then

$$g(z) = \frac{f(z)-a}{b-a}$$

is an integral function, which does not take either of the values 0 or 1. Consider the function $\varpi\{g(z)\}$. It is regular except at infinity, since $g(z)$ does not take either of the finite values for which ϖ is singular. Also its imaginary part is positive. Hence, by § 2.54, it is a constant. But ϖ is not constant, and so $g(z)$ must be constant. This proves the theorem.

As we have not discussed the construction of the modular function, we shall not complete this proof, but shall give a more direct proof, depending on a theorem of Schottky.*

8.82. The characteristic feature of Picard's theorem is that it admits the possibility of there being an exceptional value. This exceptional value may actually exist; for example, the function e^z never takes the value 0. A value with this property is said to be 'exceptional P'.

There is another sense in which a value may be exceptional. A function may take the value a, but only at points which have a convergence-exponent less than ρ. For example, the function $e^z \cos \sqrt{z}$, of order 1, has zeros, but their convergence exponent is $\frac{1}{2}$. A value with this property is said to be 'exceptional B', i.e. exceptional in the sense of Borel. It is clear that a value which is exceptional P is *a fortiori* exceptional B.

8.83. For functions of positive integral order, Picard's theorem is a consequence of the following theorem of Borel, which shows not merely that there is at most one value 'exceptional P', but at most one 'exceptional B'.

* Another direct proof, depending on a theorem of Bloch, is given by Landau, *Ergebnisse* . . ., ed. 2 (1929), Ch. VII, and by Dienes, *The Taylor Series*, Ch. VIII.

Borel's theorem. *If the order of $f(z)$ is a positive integer, then the exponent of convergence of the a-points of $f(z)$ is equal to the order, except possibly for one value of a.*

Suppose that there are two exceptional values, a and b. Then

$$f(z)-a = z^{k_1}e^{Q_1(z)}P_1(z) \tag{1}$$

and

$$f(z)-b = z^{k_2}e^{Q_2(z)}P_2(z), \tag{2}$$

where $Q_1(z)$ and $Q_2(z)$ are polynomials of degree ρ, and $P_1(z)$ and $P_2(z)$ are canonical products of order less than ρ.

Subtracting, we have

$$b-a = z^{k_1}e^{Q_1(z)}P_1(z)-z^{k_2}e^{Q_2(z)}P_2(z), \tag{3}$$

or

$$z^{k_1}P_1(z)e^{Q_1(z)-Q_2(z)} = z^{k_2}P_2(z)+(b-a)e^{-Q_2(z)}.$$

Since $Q_2(z)$ is of degree ρ, the right-hand side is of order ρ. Hence so is the left-hand side, and so $Q_1(z)-Q_2(z)$ is of degree ρ, since $P_1(z)$ is of order less than ρ.

Differentiating (3), we have

$$(z^{k_1}P_1Q_1'+k_1z^{k_1-1}P_1+z^{k_1}P_1')e^{Q_1}$$
$$= (z^{k_2}P_2Q_2'+k_2z^{k_2-1}P_2+z^{k_2}P_2')e^{Q_2}. \tag{4}$$

Now the order of P_1' is the same as that of P_1, and so is less than ρ. Hence the coefficient of e^{Q_1} is of order less than ρ, and, similarly, so is that of e^{Q_2}. Hence we may write (4) in the form

$$z^{k_3}P_3e^{Q_1+Q_3} = z^{k_4}P_4e^{Q_2+Q_4},$$

where Q_3 and Q_4 are polynomials of degree $\rho-1$ at most, and P_3 and P_4 are canonical products. The two sides must have the same zeros, so that $k_3 = k_4$, $P_3 = P_4$, and so $Q_1+Q_3 = Q_2+Q_4$, i.e. $Q_1-Q_2 = Q_4-Q_3$, which is of degree less than ρ. This contradicts the previous result, that Q_1-Q_2 is of degree ρ, and so proves the theorem.

8.84. For the proof of Schottky's theorem we require the following lemma:

Let $\phi(r)$ be a real function of r for $0 \leqslant r \leqslant R_1$, and let

$$0 \leqslant \phi(r) \leqslant M \qquad (0 < r \leqslant R_1), \tag{1}$$

and also

$$\phi(r) < \frac{C\sqrt{\phi(R)}}{(R-r)^2} \qquad (0<r<R\leqslant R_1). \tag{2}$$

Then

$$\phi(r) < \frac{AC^2}{(R_1-r)^4} \qquad (0<r<R_1). \tag{3}$$

The actual form of the result (3) is not particularly important. What is important is that it depends only on r, R_1, and C, and not on M.

From (1) and (2) we obtain

$$\phi(r) < \frac{C\sqrt{M}}{(R-r)^2} \qquad (0 < r < R \leqslant R_1), \tag{4}$$

so that the upper bound M given in (1) is reduced at once to a multiple of $\sqrt{M}$. If we repeat the process, using first (4) with r_1, r_2 for r, R, and then (2) with r_1 for R, we obtain

$$\phi(r) < \frac{C}{(r_1-r)^2}\left\{\frac{C}{(r_2-r_1)^2}\right\}^{\frac{1}{2}} M^{\frac{1}{4}} \qquad (0 < r < r_1 < r_2 \leqslant R_1).$$

So generally

$$\phi(r) < \frac{C}{(r_1-r)^2}\left\{\frac{C}{(r_2-r_1)^2}\right\}^{\frac{1}{2}} \cdots \left\{\frac{C}{(r_n-r_{n-1})^2}\right\}^{\frac{1}{2^{n-1}}} M^{\frac{1}{2^n}}.$$

Taking $r_1 = \frac{1}{2}(R_1+r)$, $r_2 = \frac{1}{2}(R_1+r_1)$,..., this gives

$$\phi(r) < 4^{1+1+\frac{3}{4}+\ldots+\frac{n}{2^{n-1}}}\left\{\frac{C}{(R_1-r)^2}\right\}^{1+\frac{1}{2}+\ldots+\frac{1}{2^{n-1}}} M^{\frac{1}{2^n}},$$

and, making $n \to \infty$, the result follows.

8.85. Schottky's theorem. *If $f(z)$ is regular and does not take either of the values 0 or 1 for $|z| \leqslant R_1$, then for $|z| \leqslant R < R_1$*

$$|f(z)| < \exp\left\{\frac{KR_1^4}{(R_1-R)^4}\right\},$$

where K depends on $f(0)$ only. For all functions which satisfy the given conditions and are such that $\delta < |f(0)| < 1/\delta$, $|1-f(0)| > \delta$, we can take K to depend on δ only.

We shall not require the actual form of the upper bound for $f(z)$, which could be considerably improved if necessary; what is important is that it depends only on $f(0)$ in the manner stated, and on R/R_1.

Let $\qquad g_1(z) = \log\{f(z)\}, \qquad g_2(z) = \log\{1-f(z)\},$

where each logarithm has its principal value at $z = 0$. Then $g_1(z)$ and $g_2(z)$ are regular for $|z| \leqslant R_1$. Let $M_1(r)$ and $M_2(r)$ be the maxima of $|g_1(z)|$ and $|g_2(z)|$ respectively on $|z| = r$, and let

$$M(r) = \max\{M_1(r), M_2(r)\}.$$

Let $$B_1(r) = -\min_{|z|=r} \mathbf{R}\{g_1(z)\} = \max_{|z|=r} \log\frac{1}{|f(z)|}.$$

Then Carathéodory's theorem (§ 5.5), applied to $g_1(z)$, gives

$$M_1(\rho) \leqslant \frac{2\rho}{r-\rho} B_1(r) + \frac{r+\rho}{r-\rho} |g_1(0)| \quad (0 < \rho < r). \tag{1}$$

There are now two possibilities. Either $B_1(r)$ is not large—say $B_1(r) \leqslant 1$, in which case (1) is a result of the required type; or $B_1(r)$ is large, in which case there is a point z' on $|z| = r$ where $|f(z')|$ is small. But if $|f(z')|$ is small, $g_2(z')$ is (apart from a term $2n\pi i$) approximately equal to $-f(z')$; and then Carathéodory's theorem, applied to $\log g_2$, gives on the left M_1 (not $\log M_1$ as we should in general expect), and on the right $\log M_2 = O(\sqrt{M_2})$. We thus obtain an inequality of the type considered in the elementary lemma. This is the general plan of the proof, and we now proceed to fill in the details.

Suppose that $B_1(r) > 1$, and let z' be a point where

$$B_1(r) = \log 1/|f(z')|.$$

Then
$$|f(z')| = e^{-B_1(r)} < e^{-1} < \tfrac{1}{2}. \tag{2}$$

There is therefore an integer n such that

$$g_2(z') - 2n\pi i = -\sum_{m=1}^{\infty} \frac{\{f(z')\}^m}{m}.$$

Hence
$$|g_2(z') - 2n\pi i| < \sum_{m=1}^{\infty} 2^{-m} = 1,$$

and so
$$2|n|\pi < 1 + |g_2(z')| \leqslant 1 + M_2(r). \tag{3}$$

Let
$$h(z) = \log\{g_2(z) - 2n\pi i\},$$

where the logarithm has its principal value at $z = 0$. Then $h(z)$ is regular for $|z| \leqslant R_1$, since $f(z) \neq 0$ and so $g_2(z) \neq 2n\pi i$; and Carathéodory's theorem gives

$$\max_{|z|=r} |h(z)| \leqslant \frac{2r}{R-r} \max_{|z|=R} \log|g_2(z) - 2n\pi i| + \frac{R+r}{R-r} |h(0)|. \tag{4}$$

The left-hand side is not less than

$$\log\left|\frac{1}{g_2(z') - 2n\pi i}\right| \geqslant \log \frac{1}{|f(z')| + |f(z')|^2 + |f(z')|^3 + \dots}$$
$$\geqslant \log \frac{1}{2|f(z')|} = B_1(r) - \log 2,$$

by (2). On the right we have

$$\max_{|z|=R} \log|g_2(z) - 2n\pi i| \leqslant \log\{M_2(R) + 2|n|\pi\} < \log\{2M_2(R) + 1\},$$

by (3); if $n \neq 0$, $|g_2(0)-2n\pi i| \geqslant \pi > 1$, and so

$$|h(0)| \leqslant \log|g_2(0)-2n\pi i|+\pi \leqslant \log\{|g_2(0)|+1+M_2(r)\}+\pi.$$

If $n = 0$, $|h(0)| \leqslant |\log|g_2(0)||+\pi$. Hence (4) gives

$$B_1(r) \leqslant \frac{2R}{R-r}[2\log\{2M_2(R)+|g_2(0)|+1\}+|\log|g_2(0)||+\pi]+\log 2$$

$$< \frac{4R}{R-r}[\log\{2M_2(R)+|g_2(0)|+1\}+|\log|g_2(0)||+\pi]. \qquad (5)$$

This inequality, proved for $B_1(r) > 1$, is obviously true for $B_1(r) \leqslant 1$. Hence (1) and (5) give in any case

$$M_1(\rho) < \frac{8Rr}{(R-r)(r-\rho)}[\log\{2M_2(R)+|g_2(0)|+1\}+$$
$$+|\log|g_2(0)||+|g_1(0)|+\pi].$$

Since we may interchange $g_1(z)$ and $g_2(z)$ in the whole argument, the inequality is still true if the suffixes 1 and 2 are interchanged. Combining the two results, we have

$$M(\rho) < \frac{8Rr}{(R-r)(r-\rho)}\{\log M(R)+K\},$$

where K depends on $|g_1(0)|$ and $|g_2(0)|$ only. Taking $r = \frac{1}{2}(R+\rho)$, we obtain

$$M(\rho) < \frac{32R_1^2}{(R-\rho)^2}\{\log M(R)+K\} < \frac{KR_1^2\sqrt{M(R)}}{(R-\rho)^2},$$

since $\log M(R) = O\{\sqrt{M(R)}\}$. Hence, by the lemma,

$$M(\rho) < \frac{KR_1^4}{(R-\rho)^4},$$

and

$$|f(z)| \leqslant e^{M(r)} < \exp\left\{\frac{KR_1^4}{(R-r)^4}\right\}.$$

Since K depends on $|g_1(0)|$ and $|g_2(0)|$ only, the last part of the theorem also is true.

8.86. Picard's first theorem. *An integral function which is not constant takes every value, with one possible exception, at least once.*

Suppose that $f(z)$ does not take either of the values a or b $(a \neq b)$. Then $g(z) = \{f(z)-a\}/(b-a)$ does not take either of the values 0 or 1. Hence, by Schottky's theorem,

$$|g(z)| < \exp\left\{\frac{KR_1^4}{(R_1-R)^4}\right\} \qquad (|z| \leqslant R < R_1).$$

Taking $R_1 = 2R$, $|g(z)| < K$. Hence $g(z)$ is a constant.

8.87. We can also prove the following generalization of Picard's theorem.

Landau's theorem.* *If α is any number, and β any number other than 0, there is a number $R = R(\alpha, \beta)$ such that every function*

$$f(z) = \alpha+\beta z+a_2 z^2+a_3 z^3+\dots,$$

regular for $|z| \leqslant R$, takes in this circle one of the values 0 or 1.

We may suppose that $\alpha \neq 0$, $\alpha \neq 1$, for otherwise we have the result at once. If $f(z)$ does not take either of the values 0 or 1, then by Schottky's theorem $|f(z)| < K(\alpha)$ for $|z| \leqslant \frac{1}{2}R$. Hence

$$|\beta| = \left|\frac{1}{2\pi i} \int\limits_{|z|=\frac{1}{2}R} \frac{f(z)}{z^2}\,dz\right| \leqslant \frac{K(\alpha)}{\frac{1}{2}R},$$

$$R \leqslant 2K(\alpha)/|\beta|,$$

and the result follows.

8.88. We have so far stated Picard's theorem in terms of integral functions, i.e. functions with an essential singularity at infinity. But a corresponding theorem holds for any function with an isolated essential singularity.

Picard's second theorem. *In the neighbourhood of an isolated essential singularity, a one-valued function takes every value, with one possible exception, an infinity of times.*

In other words, *if $f(z)$ is regular for $0 < |z-z_0| < \rho$, and there are two unequal numbers a, b, such that $f(z) \neq a$, $f(z) \neq b$, for $|z-z_0| < \rho$, then z_0 is not an essential singularity.*

We may suppose that $z_0 = 0$, $\rho = 1$, $a = 0$, and $b = 1$. We prove that there is a sequence of circles $|z| = r_n$, where $r_n \to 0$, on which $f(z)$ is bounded. By § 2.71 this precludes the existence of a singularity at $z = 0$.

We start from Weierstrass's theorem that, in the neighbourhood of an essential singularity, a function approaches arbitrarily near to any given value an infinity of times. Thus there is a sequence of points $z_1, z_2, \dots$ such that $|z_1| > |z_2| > \dots$, $|z_n| \to 0$, and

$$|f(z_n)-2| < \tfrac{1}{2}. \tag{1}$$

It is clear that Schottky's theorem would enable us to construct a sequence of circles, with these points as centres, in which $f(z)$ is bounded. These circles do not, of course, include

* Landau (1), and *Ergebnisse*, § 25.

the origin; but this is, so to speak, an accident arising from the fact that we have proved Schottky's theorem for a class of convex curves (viz. circles). We can remove the difficulty by making a conformal transformation, which has the effect of replacing a circle by an elongated curve which, though it excludes the origin, passes right round it and overlaps itself on the far side.

Let $z = e^w$ $(w = u+iv)$, and consider the half-strip of the w-plane $u < 0$, $-\pi \leqslant v \leqslant \pi$. This corresponds to the interior of the circle $|z| = 1$. Let $w_n = \log z_n$ $(-\pi < \mathbf{I}(w_n) \leqslant \pi)$, so that $\mathbf{R}(w_n) \to -\infty$; and let $f(z) = g(w)$.

We apply Schottky's theorem to the function

$$h(w') = g(w_n+w').$$

This is regular for $|w'| \leqslant 4\pi$ if n is large enough, and it does not take either of the values 0 or 1. Hence

$$|h(w')| < K = K\{h(0)\} \qquad (|w'| \leqslant 2\pi);$$

and, the numbers $h(0) = g(w_n) = f(z_n)$ satisfying (1), we can replace the right-hand side by an absolute constant. Hence $|g(w)| < A$ for $|w-w_n| \leqslant 2\pi$, and in particular for $u = \mathbf{R}(w_n)$, $-\pi \leqslant v \leqslant \pi$. Hence

$$|f(z)| < A \qquad (|z| = |z_n|),$$

and the result follows.

8.89. Asymptotic values. A number a is said to be an asymptotic value of an integral function $f(z)$ if there is a continuous curve from a given point to infinity, i.e. along which $|z| \to \infty$, and along which $f(z) \to a$ as $z \to \infty$. Thus 0 is an asymptotic value of e^z, since $e^z \to 0$ as $z \to \infty$ along the negative real axis. The function

$$\int_0^z e^{-t^q}\,dt,$$

where q is a positive integer, has the q asymptotic values

$$e^{\frac{2\pi ik}{q}} \int_0^\infty e^{-t^q}\,dt \qquad (k = 0, 1, \ldots, q-1),$$

as $z \to \infty$ along the lines $\arg z = 2\pi k/q$.

We may define the 'asymptotic value ∞' similarly.

We shall now prove the following theorems.

Every function with an isolated essential singularity at infinity, which is not a constant, has the asymptotic value infinity.

By Laurent's theorem, such a function is of the form $f(z)+g(z)$, where $f(z)$ is an integral function, and $g(z)$ tends uniformly to a limit as $|z| \to \infty$. Hence it is sufficient to consider integral functions. For an integral function not a constant, the maximum modulus $M(r)$ tends steadily to infinity. Consider an indefinitely increasing sequence of numbers $X_1 = M(r_1)$, X_2,.... It follows from Liouville's theorem that there is a point outside the circle $|z| = r_1$ at which $|f(z)| > X_1$. The set of points where $|f(z)| > X_1$ constitutes the interior of one or more regions bounded by curves on which $|f(z)| = X_1$; and these regions must be exterior to the circle $|z| = r_1$. Let one such region be D_1. Now D_1 must extend to infinity; for otherwise we should have a finite region with $|f(z)| = X_1$ on the boundary and $|f(z)| > X_1$ inside, contrary to the maximum-modulus theorem. Further, $f(z)$ is unbounded in D_1. For otherwise the principle of Phragmén and Lindelöf would also show that $|f(z)| \leqslant X_1$ at all points inside D_1. In fact, the argument of § 5.6 applies with P at infinity and $\omega(z) = r_1/z$. Hence there is a point of D_1 at which $|f(z)| > X_2$, and consequently a domain D_2, interior to D_1, such that $|f(z)| > X_2$ at all points of D_2. We can now repeat the argument with X_3,.... Hence there is a sequence of infinite regions D_1, D_2,..., each interior to the preceding one and such that $|f(z)| > X_m$ in D_m, and $|f(z)| = X_m$ on its boundary. Now, take a point on the boundary of D_1, and join it to a point on the boundary of D_2 by a continuous curve lying in D_1; then this point to a point of D_3 by a continuous curve lying in D_2, and so on. We clearly obtain a continuous curve along which $f(z) \to \infty$.

If an integral function does not take the value a, then a is an asymptotic value.

For $1/\{f(z)-a\}$ is an integral function, and so has the asymptotic value ∞.

The argument of § 5.64 shows that if an integral function has asymptotic values on two curves, and is bounded between the curves, then these asymptotic values must be the same. Asymptotic values not so connected we should consider as distinct, whether they are equal or not.

It was conjectured by Denjoy that *an integral function of finite order ρ can have at most 2ρ asymptotic values*. The theorem with 5ρ instead of 2ρ was proved by Carleman; and Denjoy's conjecture was finally proved by Ahlfors. The general proof is not easy. It is, however, easy to see that there can be at most 2ρ straight lines from 0 to ∞ along which a function of order ρ has distinct asymptotic values. For by § 5.61 the angle between two such lines must be at least equal to π/ρ.

8.9. Meromorphic functions. We shall now give a short introduction to the theory of meromorphic functions, i.e. functions whose only singularities, except at infinity, are poles.

The theory depends largely on the general Jensen formula (§ 3.61 (4)). Let $f(z)$ be a meromorphic function, with zeros $a_1, a_2,...$ and poles $b_1, b_2,...$ (other than 0) arranged with non-decreasing moduli. Suppose that in the neighbourhood of the origin it is of the form $cz^k+...$, where k may be any integer. Then Jensen's formula for $z^{-k}f(z)$ is

$$\log\left|\frac{b_1...b_n}{a_1...a_m}\right| r^{m-n}+\log|c| = \frac{1}{2\pi}\int\limits_0^{2\pi}\log|f(re^{i\theta})|\,d\theta -k\log r.$$

As in § 3.61

$$\log\frac{r^m}{|a_1...a_m|} = \sum_{\nu=1}^{m-1}\nu\int\limits_{|a_\nu|}^{|a_{\nu+1}|}\frac{dx}{x}+m\int\limits_{|a_m|}^{r}\frac{dx}{x}.$$

Let $n(r,0)$ be the number of zeros of $f(z)$ in $|z|\leqslant r$. If $k>0$, $\nu = n(x,0)-k$ for $|a_\nu|\leqslant x<|a_{\nu+1}|$; hence

$$\log\frac{r^m}{|a_1...a_m|} = \int\limits_0^r\frac{n(x,0)-k}{x}\,dx.$$

If $n(r,\infty)$ is the number of poles of $f(z)$ in $|z|\leqslant r$, we obtain similarly

$$\log\frac{r_n}{|b_1...b_n|} = \int\limits_0^r\frac{n(x,\infty)}{x}\,dx.$$

If k is negative, it appears in the second integral instead of the first. Writing

$$N(r,a) = \int\limits_0^r\frac{n(x,a)-n(0,a)}{x}\,dx +n(0,a)\log r$$

we obtain in any case

$$N(r,0)-N(r,\infty) = \frac{1}{2\pi}\int_0^{2\pi} \log|f(re^{i\theta})|\,d\theta - \log|c|. \tag{1}$$

Let us write $\quad \log^+\alpha = \max(\log\alpha, 0)$

for any positive α. Thus

$$\log\alpha = \log^+\alpha - \log^+1/\alpha.$$

Let $\quad m(r,a) = m\left(r,\frac{1}{f-a}\right) = \frac{1}{2\pi}\int_0^{2\pi} \log^+\left|\frac{1}{f(re^{i\theta})-a}\right| d\theta$

and $\quad m(r,\infty) = m(r,f) = \frac{1}{2\pi}\int_0^{2\pi} \log^+|f(re^{i\theta})|\,d\theta.$

Then (1) may also be written

$$m(r,0)+N(r,0) = m(r,\infty)+N(r,\infty)-\log|c|. \tag{2}$$

Now apply this formula to $f(z)-a$, where a is any number. If $f(z)-a = c_a z^k+\dots$ in the neighbourhood of the origin, we obtain

$$m(r,a)+N(r,a) = m(r,f-a)+N(r,\infty)-\log|c_a|,$$

the term $N(r,\infty)$ being unaltered since the poles of $f(z)-a$ are the same for each a.

Now we have

$$|f|+|a| \leqslant 2|fa|, \quad 2|f|, \quad 2|a| \quad \text{or} \quad 2$$

according as $|f| \geqslant 1$ and $|a| \geqslant 1$, $|f| \geqslant 1$ and $|a| < 1$, $|f| < 1$ and $|a| \geqslant 1$, or $|f| < 1$ and $|a| < 1$. Hence

$$\log(|f|+|a|) \leqslant \log^+|f|+\log^+|a|+\log 2.$$

Hence $\quad \log^+|f-a| \leqslant \log^+|f|+\log^+|a|+\log 2,$

and similarly

$$\log^+|f| \leqslant \log^+|f-a|+\log^+|a|+\log 2.$$

Hence $\quad |m(r,f-a)-m(r,f)| \leqslant \log^+|a|+\log 2.$

It follows that

$$m(r,a)+N(r,a) = m(r,\infty)+N(r,\infty)+\phi(r,a),$$

where $\quad |\phi(r,a)| \leqslant |\log|c_a||+\log^+|a|+\log 2.$

Hence if $f(z)$ is a meromorphic function and not a constant, the

values of the sum
$$m(r,a)+N(r,a)$$
for two given values of a differ by a bounded function of r.

All the sums being to this extent equivalent, we can represent them all, e.g. by the one with $a = \infty$. Thus *if we put*
$$T(r) = m(r,\infty)+N(r,\infty), \tag{3}$$
we have for all values of a
$$m(r,a)+N(r,a) = T(r)+\phi(r,a),$$
where $\phi(r,a)$ *is (for each* a*) bounded as* $r \to \infty$.

$T(r)$ is called the *characteristic function* of $f(z)$.

We shall next show that $T(r)$ *is an increasing convex function of* $\log r$.

Jensen's formula for $f(z)-e^{i\lambda}$ (λ real) is
$$N(r,e^{i\lambda})-N(r,\infty) = \frac{1}{2\pi}\int_0^{2\pi} \log|f(re^{i\theta})-e^{i\lambda}|\,d\theta - \log|f(0)-e^{i\lambda}|, \tag{4}$$
if $f(0) \neq e^{i\lambda}$. Also, for any a,
$$\frac{1}{2\pi}\int_0^{2\pi} \log|e^{i\theta}-a|\,d\theta = \log^+|a|,$$
e.g. by Jensen's theorem with $f(z) = z-a$, $r = 1$. Hence, multiplying (4) by $1/(2\pi)$, and integrating with respect to λ over $(0, 2\pi)$, we obtain
$$\frac{1}{2\pi}\int_0^{2\pi} N(r,e^{i\lambda})\,d\lambda - N(r,\infty) = \frac{1}{2\pi}\int_0^{2\pi} \log^+|f(re^{i\theta})|\,d\theta - \log^+|f(0)|,$$
i.e.
$$T(r) = \frac{1}{2\pi}\int_0^{2\pi} N(r,e^{i\lambda})\,d\lambda + \log^+|f(0)|.$$

Now, for any a, $N(r,a)$ is an increasing convex function of r, since
$$\frac{dN(r,a)}{d\log r} = n(r,a),$$
which is non-negative and non-decreasing. Hence $T(r)$ has the same property.

In the above formulae $N(r,a)$ measures the number of times the function $f(z)$ takes the value a. Since the largest contributions to $m(r,a)$ come from arcs where $f(z)$ is nearly equal to a, $m(r,a)$ measures in a sense the intensity of the approximation of $f(z)$ to a. We could describe $m(r,a)+N(r,a)$ as the total affinity of the function $f(z)$ for the value a.

For a given function, certain values may be exceptional, e.g. in the sense that the function does not take these values. The above theorem shows that there can be no exceptional values in the sense that the total affinity of the function for every value is the same, apart from bounded functions of r.

Examples. (i) Let $f(z)$ be a rational function, say $= P(z)/Q(z)$, where $P(z)$ is of degree μ, $Q(z)$ of degree ν, P and Q having no common factor. If $\mu > \nu$, then

$$m(r,a) = O(1), \qquad N(r,a) = \mu \log r + O(1)$$

for every finite a, while

$$m(r,\infty) = (\mu-\nu)\log r + O(1), \qquad N(r,\infty) = \nu \log r + O(1).$$

If $\mu < \nu$, then

$$m(r,a) = O(1), \qquad N(r,a) = \nu \log r + O(1)$$

for $a \neq 0$, while

$$m(r,0) = (\nu-\mu)\log r + O(1), \qquad N(r,0) = \mu \log r + O(1).$$

If $\mu = \nu$, let a_0 and b_0 be the coefficients of x^μ in P and Q. Then if $a \neq a_0/b_0$,

$$m(r,a) = O(1), \qquad N(r,a) = \mu \log r + O(1),$$

while, if $a_0 Q - b_0 P$ is of degree α,

$$m\left(r,\frac{a_0}{b_0}\right) = (\mu-\alpha)\log r + O(1), \qquad N\left(r,\frac{a_0}{b_0}\right) = \alpha \log r + O(1).$$

In any case $T(r) = O(\log r)$.

(ii) The function e^z does not take the values 0 or ∞; on the other hand these are limiting values of the function as $z \to \infty$. Here

$$N(r,0) = N(r,\infty) = 0, \qquad m(r,0) = m(r,\infty) = \frac{r}{\pi},$$

while for $a \neq 0, \infty$,

$$m(r,a) = O(1), \qquad N(r,a) = \frac{r}{\pi} + O(1).$$

Here $T(r) = r/\pi$.

(iii) Consider similarly $\tan z$ ($\pm i$ are exceptional values).

8.91. Order of a meromorphic function.

The meromorphic function $f(z)$ is said to be of order ρ if

$$\overline{\lim_{r\to\infty}} \frac{\log T(r)}{\log r} = \rho,$$

so that

$$T(r) = O(r^{\rho+\epsilon})$$

for every positive ϵ, but not for $\epsilon < 0$.

To show that this agrees with the definition of order in the case of an integral function, we shall prove:

If $f(z)$ is an integral function,

$$T(r) \leqslant \log^+ M(r) \leqslant \frac{R+r}{R-r} T(R)$$

for $0 < r < R$.

For an integral function, $N(r,\infty) = 0$, and $T(r) = m(r,\infty)$. The left-hand inequality is thus

$$\frac{1}{2\pi}\int_0^{2\pi} \log^+|f(re^{i\theta})|\,d\theta \leqslant \log^+ \max|f(re^{i\theta})|,$$

which is plainly true. Also by the Poisson-Jensen formula

$$\log|f(re^{i\theta})| = \frac{1}{2\pi}\int_0^{2\pi} \frac{R^2-r^2}{R^2-2Rr\cos(\theta-\phi)+r^2}\log|f(Re^{i\phi})|\,d\phi - -\sum_{\mu=1}^{m}\log\left|\frac{R^2-\overline{a_\mu}\,re^{i\theta}}{R(Re^{i\theta}-a_\mu)}\right|.$$

Each term in $\sum$ is negative, and

$$R^2-2Rr\cos(\theta-\phi)+r^2 \geqslant (R-r)^2.$$

Hence, taking θ so that the left-hand side is a maximum,

$$\log|M(r)| \leqslant \frac{R+r}{R-r}\frac{1}{2\pi}\int_0^{2\pi}\log|f(Re^{i\phi})|\,d\phi$$

$$\leqslant \frac{R+r}{R-r}T(R)$$

and the second inequality follows.

Taking $R = 2r$, the identity of the two definitions of the order of an integral function is clear.

Now let $r_n(a)$ be the moduli of the zeros of $f(z)-a$, $r_n(\infty)$ the moduli of the poles of $f(z)$. Then we have the following results.

If $f(z)$ is of order ρ, then for every a

$$m(r,a) = O(r^{\rho+\epsilon}), \qquad N(r,a) = O(r^{\rho+\epsilon}), \qquad n(r,a) = O(r^{\rho+\epsilon})$$

and

$$\sum\left(\frac{1}{r_n(a)}\right)^{\rho+\epsilon}$$

is convergent.

The first two results are immediate since

$$m(r,a) \leqslant T(r)+O(1), \qquad N(r,a) \leqslant T(r)+O(1).$$

The remaining results then follow from that on $N(r,a)$ as in the case of an integral function.

More precise results of the same kind are given by Nevanlinna, *Fonctions Méromorphes*, Ch. II.

8.92. Factorization of meromorphic functions. Let $f(z)$ be a meromorphic function of order ρ, with zeros a_n and poles b_n ($f(0) \neq 0$). Then it follows from the above results that there are integers p_1 and p_2 not exceeding ρ such that

$$P_1(z) = \prod_{n=1}^{\infty} E\left(\frac{z}{a_n}, p_1\right), \qquad P_2(z) = \prod_{n=1}^{\infty} E\left(\frac{z}{b_n}, p_2\right)$$

are convergent for all values of z. Hence $P_1(z)$ and $P_2(z)$ are integral functions of order not exceeding ρ. Also

$$f_1(z) = f(z)P_2(z)$$

is an integral function. Now

$$T(r,f_1) = m(r,\infty,f_1) \leqslant m(r,\infty,f)+m(r,\infty,P_2)$$
$$\leqslant T(r,f)+T(r,P_2) = O(r^{\rho+\epsilon})+O(r^{\rho+\epsilon}).$$

Hence $f_1(z)$ is of order ρ at most, and hence

$$f_1(z) = e^{Q(z)}P_1(z),$$

where $Q(z)$ is a polynomial of degree not exceeding ρ.

We have thus proved that

$$f(z) = e^{Q(z)}P_1(z)/P_2(z),$$

an extension of Hadamard's factorization theorem to meromorphic functions.

A slightly deeper theorem, in which the numerator and denominator do not necessarily converge separately, is proved by Nevanlinna, *Fonctions Méromorphes*, Ch. III.

Further developments of the theory of meromorphic functions are largely concerned with extensions of the theorems of Picard and Borel. For these we must refer to the books of Nevanlinna.

MISCELLANEOUS EXAMPLES

1. Prove that, if a is not a multiple of π,

$$\sin(a-z) = \sin a\, e^{-z\cot a} \prod_{n=-\infty}^{\infty} \left(1-\frac{z}{a+n\pi}\right) e^{\frac{z}{a+n\pi}}.$$

2. Show that the equations

$$\sin z = z^2, \qquad \log z = z^3, \qquad \tan z = az+b,$$

where a and b are any complex numbers, each have an infinity of roots.

3. Find all the zeros of the function

$$e^{e^z}-1,$$

and show that they have no finite exponent of convergence.

4. If $f(z) = \sum_{n=0}^{\infty} a_n z^n$ is a function of non-integral order, show that the coefficients in the polynomial $Q(z) = b_1 z + \ldots + b_q z^q$ of Hadamard's theorem can be expressed in terms of $a_1, a_2, \ldots, a_q$.

[If ρ is not an integer, $q < p+1$, and $P(z) = 1+O(z^{p+1})$ as $z \to 0$. Hence on equating coefficients $P(z)$ is not involved.]

5. If $f(z) = \sum a_n z^n$ is of order ρ, what is the order of $F(z) = \sum |a_n|^p z^n$?

6. The generalized hypergeometric function is defined by the formula

$${}_pF_q(\alpha_1,\ldots,\alpha_p;\beta_1,\ldots,\beta_q;z) = \sum_{n=0}^{\infty} \frac{(\alpha_1)_n\cdots(\alpha_p)_n}{(\beta_1)_n\cdots(\beta_q)_n}\frac{z^n}{n!},$$

where $(\alpha)_n = \alpha(\alpha+1)\ldots(\alpha+n)$, $(\alpha)_0 = 1$. Show that it is an integral function if $q \geqslant p$, and find its order.

7. Show that

$$f(z) = \sum_{n=-\infty}^{\infty} q^{|n|^k} e^{inz}$$

is an integral function if $|q| < 1$, $k > 1$, and find its order.

8. If $\sigma > 1$, the function

$$P_\sigma(z) = \prod_{n=1}^{\infty} \left(1+\frac{z}{n^\sigma}\right)$$

is an integral function of order $1/\sigma$. [For further properties of the function see Hardy (4).]

9. The function

$$\prod_{n=1}^{\infty} \left(1+\frac{z}{e^n}\right)$$

is an integral function of zero order.

10. Show that, if $\alpha > 0$,

$$f_\alpha(z) = \prod_{n=1}^{\infty} \left(1+\frac{e^z}{e^{n^\alpha}}\right)$$

is an integral function of order $1+1/\alpha$; and express it in the standard factor form in the cases $\alpha = 1$, $\alpha = 2$.

11. If $a > 0$, the function

$$E_a(z) = \sum_{n=0}^{\infty} \frac{z^n}{\Gamma(1+an)}$$

is an integral function of order $1/a$. [Several memoirs on this function are to be found in *Acta Math.* 29.]

12. If a is real, all the roots of the equation $\Gamma'(z) = a\Gamma(z)$ are real.

13. Show that

$$\sum_{n=0}^{\infty} \frac{\cosh \sqrt{n}}{n!} z^n$$

is an integral function of order 1, and that it has an infinity of zeros, all of them being real and negative.

14. A function $f(z)$ of order $\frac{1}{2}$ has all its zeros real and negative, and such that $n(r) \sim k\sqrt{r} \log r$. Determine the asymptotic behaviour of $M(r)$.

[Use the method of § 8.64.]

15. Show that, if $f(z)$ is a canonical product with zeros z_n such that $\sum 1/|z_n|$ is convergent, then $f(z) = O(e^{\epsilon|z|})$, and $|f(z)| > e^{-\epsilon|z|}$ on circles of arbitrarily large radius.

[We have

$$|f(z)| \leqslant \prod_{n=1}^{N} \left(1+\left|\frac{z}{z_n}\right|\right) \exp\left(\sum_{N+1}^{\infty} \left|\frac{z}{z_n}\right|\right),$$

whence the first result easily follows. The second part then follows from § 8.75.]

16. In Laguerre's theorem of § 8.52, show that $f'(z)$ has the same genus as $f(z)$.

[The only case in which there is anything to prove is the case $\rho = 1$, when the genus may be 0 or 1. Then we use the fact that the series $\sum 1/|z_n|$ and $\sum 1/|z'_n|$ converge or diverge together, compare $M(r)$ and $M'(r)$ by § 8.51, and apply the previous example.]

17. Show that the genus of a function of exponential type (§ 8.75) is 1.

18. Show that, if $f(z) = \sum_{0}^{\infty} a_n z^n$ is of exponential type, $f^{(n)}(0) = O(e^{An})$, and hence that $\phi(z) = \sum n! a_n z^n$ has a finite radius of convergence.

19. In order that $f(z)$ should be of exponential type, it is necessary and sufficient that it should be expressible in the form

$$f(z) = \frac{1}{2\pi i} \int_C e^{zw} \chi(w)\, dw,$$

where $\chi(w)$ is regular for sufficiently large values of w (including infinity), and C is a circle with centre the origin and sufficiently large radius.

[We have $\chi(w) = 1/w\phi(1/w)$, where ϕ is the function defined in the previous example.]

20. Let $f(z)$ be of exponential type, and let $h(\theta)$, supposed $\geqslant 0$, be the Phragmén-Lindelof function associated with $f(z)$, with $V(r) = r$. Consider the radii vectores of length $h(\theta)$ making angles $-\theta$ with the real axis, and the perpendiculars to these radii vectores at their ends (cf. § 5.72). Then $\chi(w)$ is regular if w lies on the side of one of these perpendiculars opposite to the origin.

[We have

$$\phi(z) = \int_0^\infty e^{-t}f(zt)\,dt$$

by term-by-term integration, if $|z|$ is small enough; turning the contour through an angle λ,

$$\phi(z) = \int_0^\infty e^{-te^{i\lambda}}f(zte^{i\lambda})e^{i\lambda}\,dt.$$

Here the integrand is $\quad O(e^{-t\cos\lambda+rt\{h(\theta+\lambda)+\epsilon\}})$,

and the integral is convergent if

$$rh(\theta+\lambda) < \cos\lambda. \tag{1}$$

Hence $\phi(z)$ is regular at $z = re^{i\theta}$ if (1) holds for some value of λ. If $w = r'e^{i\theta'}$, then $\chi(w)$ is regular if $r' > h(\lambda-\theta')\sec\lambda$ for some λ. This is equivalent to the above statement.

For a detailed discussion see Pólya (4).]

21. The function

$$\sum_{n=0}^{\infty} \frac{z^n}{(n+a)^s n!}$$

is of exponential type. [For further properties of the function see Hardy (2).]

22. Show that the function

$$f(z) = \int_a^b e^{izt}g(t)\,dt,$$

where $g(t)$ is continuous, is of exponential type, and that the corresponding function $\chi(w)$ is regular except in the interval (ia, ib) of the imaginary axis.

23. Show that the function $f(z)$ of the above example tends to zero as $z \to \infty$ in either direction along the real axis, and deduce that $f(z)$ has an infinity of zeros.

24. A function $f(z)$ is said to be of zero type if $f(z) = O(e^{\epsilon r})$.

In order that $f(z)$ should be of zero type, it is necessary and sufficient that

$$f(z) = \frac{1}{2\pi i}\int_C e^{zw}\chi(w)\,dw,$$

where $\chi(w)$ is an integral function of $\dfrac{1}{w}$.

[The situation is similar to that of examples 18–19, except that here $f^{(n)}(0) = O(e^{\epsilon n})$.]

25. A necessary and sufficient condition that $f(z) = \sum_{n=0}^{\infty} a_n z^n$ should be an integral function of $1/(1-z)$ is that there should be an integral function $g(z)$ of zero type such that $a_n = g(n)$ for $n = 1, 2,\ldots$.

[Carlson (1), Wigert (1), Hardy (14). If there is such a function $g(z)$, let

$$g(z) = \frac{1}{2\pi i}\int_C e^{zw}\chi(w)\,dw.$$

Then $$f(z)-a_0 = \sum_{n=1}^{\infty} a_n z^n = \frac{1}{2\pi i}\int_C \frac{ze^w}{1-ze^w}\chi(w)\,dw,$$

if C is a contour enclosing the origin, and on which $\mathbf{R}(w) < \log|1/z|$. This is an integral of the type used in § 4.6, and by deforming it we can show that any branch of $f(z)$ is regular except at $z = 1$ (where the contour is nipped between 0 and $\log 1/z$). Also

$$f(z)-a_0 = \chi\left(\log\frac{1}{z}\right)+\frac{1}{2\pi i}\int_{C'} \frac{ze^w}{1-ze^w}\chi(w)\,dw,$$

where C' is a contour enclosing $w = \log 1/z$. This shows that $f(z)$ is one-valued near $z = 1$, and so an integral function of $1/(1-z)$.

Conversely, if $f(z)$ is of the type prescribed, we have

$$a_n = \frac{1}{2\pi i}\int \frac{f(z)}{z^{n+1}}\,dz,$$

and we can put $z = e^{-w}$, and deform the resulting contour into any simple closed contour which encloses the origin but lies entirely inside the circle $|w| = 2\pi$. Finally $f(e^{-w})$ is regular except for $w = 0$ and $w = \pm 2k\pi i$ $(k = 1, 2,\ldots)$, and so $f(e^{-w}) = g(w)+\psi(w)$, where $g(w)$ is an integral function of $1/w$, and $\psi(w)$ is regular for $|w| < 2\pi$. Hence the result.]

CHAPTER IX

DIRICHLET SERIES

9.1. Introduction. By a Dirichlet series we mean, in this chapter, a series of the form

$$\sum_{n=1}^{\infty} \frac{a_n}{n^s}, \tag{1}$$

where the coefficients a_n are any given numbers, and s is a complex variable. The more general type of series

$$\sum a_n e^{-\lambda_n s}$$

is also known as a Dirichlet series. The special type is obtained by putting $\lambda_n = \log n$. For the theory of the general type we must refer to Hardy and Riesz's *General Theory of Dirichlet's Series.*

Throughout the chapter we shall write $s = \sigma + it$, where σ and t are real. If the Dirichlet series is convergent, we shall denote its sum by $f(s)$. We have already had one important example of a Dirichlet series, the zeta-function of Riemann,

$$\zeta(s) = \sum_{n=1}^{\infty} \frac{1}{n^s}. \tag{2}$$

Dirichlet series are not of such importance in general analysis as power series because they only represent a very special class of analytic functions. They are, however, of great importance in applications of analysis to theory of numbers. In several ways their theory is more complicated than that of power series. For example, the circle of convergence, circle of absolute convergence, and circle of regularity of sum-function are all the same for a power series. In the theory of Dirichlet series, in which 'circle' must be replaced by 'half-plane', the three corresponding half-planes may be all different.

9.11. The association of half-planes with a Dirichlet series depends on the following theorem:

If the Dirichlet series is convergent for $s = s_0$, then it is uniformly convergent throughout the angular region in the s-plane defined by the inequality

$$|\arg(s - s_0)| \leqslant \tfrac{1}{2}\pi - \delta,$$

where δ is any positive number less than $\tfrac{1}{2}\pi$.

It is sufficient to consider the case where $s_0 = 0$; for

$$\sum \frac{a_n}{n^s} = \sum \frac{a'_n}{n^{s'}},$$

where $$a'_n = a_n n^{-s_0}, \qquad s' = s - s_0,$$

and the latter series is convergent for $s' = 0$.

We suppose, then, that $\sum a_n$ is convergent. Let

$$r_n = a_{n+1} + a_{n+2} + \dots$$

so that $r_n \to 0$. Then

$$\sum_{n=M}^{N} \frac{a_n}{n^s} = \sum_{M}^{N} \frac{r_{n-1} - r_n}{n^s} = \sum_{M}^{N} r_n \left\{\frac{1}{(n+1)^s} - \frac{1}{n^s}\right\} + \frac{r_{M-1}}{M^s} - \frac{r_N}{(N+1)^s}. \tag{1}$$

Now

$$\left|\frac{1}{(n+1)^s} - \frac{1}{n^s}\right| = \left| s \int_n^{n+1} \frac{du}{u^{s+1}} \right| \leqslant |s| \int_n^{n+1} \frac{du}{u^{\sigma+1}} = \frac{|s|}{\sigma} \left\{\frac{1}{n^\sigma} - \frac{1}{(n+1)^\sigma}\right\}, \tag{2}$$

and $|r_n| < \epsilon$ for $n \geqslant n_0 = n_0(\epsilon)$, n_0 being independent of s. Hence for $M > n_0$

$$\left|\sum_{n=M}^{N} \frac{a_n}{n^s}\right| < \frac{\epsilon|s|}{\sigma} \sum_{n=M}^{N} \left\{\frac{1}{n^\sigma} - \frac{1}{(n+1)^\sigma}\right\} + \frac{\epsilon}{M^\sigma} + \frac{\epsilon}{(N+1)^\sigma}$$

$$= \frac{\epsilon|s|}{\sigma} \left\{\frac{1}{M^\sigma} - \frac{1}{(N+1)^\sigma}\right\} + \frac{\epsilon}{M^\sigma} + \frac{\epsilon}{(N+1)^\sigma}.$$

$$< 2\epsilon|s|/\sigma + 2\epsilon.$$

If $|\arg s| \leqslant \frac{1}{2}\pi - \delta$, i.e. $t/\sigma \leqslant \tan(\frac{1}{2}\pi - \delta) = \cot\delta$, we have

$$|s|/\sigma = \sqrt{(1 + t^2/\sigma^2)} \leqslant \operatorname{cosec}\delta.$$

Hence $$\sum_{n=M}^{N} \frac{a_n}{n^s} < 2\epsilon(\operatorname{cosec}\delta + 1).$$

The right-hand side is independent of s, and tends to 0 with ϵ; and this establishes uniform convergence.

In particular, *if the series is convergent for $s_0 = \sigma_0 + it_0$, it is convergent for $s = \sigma + it$, provided that $\sigma > \sigma_0$.* For we can choose the δ of the above proof so small that $|\arg(s - s_0)| < \frac{1}{2}\pi - \delta$.

9.12. *The region of convergence of the series is a half-plane.* For we can divide values of σ' into two classes, those for which the series is convergent for $\sigma > \sigma'$, and other values of σ'. By the above theorem, every member of the first class lies to the right of every member of the second class. Let σ_0 be the real number defined by this section. Then the series is convergent for $\sigma > \sigma_0$, divergent for $\sigma < \sigma_0$.

The number σ_0 is called *the abscissa of convergence* of the series.

The series may converge for all values of s (e.g. $a_n = 1/n!$), or for no values of s (e.g. $a_n = n!$).

The sum $f(s)$ of the series is an analytic function of s, regular for $\sigma > \sigma_0$. For every term of the series is analytic, and any point s with $\sigma > \sigma_0$ is included in a domain of uniform convergence.

The questions of the convergence of the series, and the regularity of the function, *on* the line $\sigma = \sigma_0$, remain open; and (as in the case of power series) various different cases are possible.

We have, however, the following analogue of Abel's theorem for power series:

If the series is convergent for $s = s_0$, and has the sum $f(s_0)$, then $f(s) \to f(s_0)$ when $s \to s_0$ along any path lying entirely inside the region $|\arg(s-s_0)| \leqslant \frac{1}{2}\pi-\delta$.

This follows at once from the theorem of uniform convergence.

9.13. Absolute convergence. *The region of absolute convergence of the Dirichlet series is a half-plane.*

For the series is absolutely convergent if the series

$$\sum_1^\infty \frac{|a_n|}{n^\sigma}$$

is convergent. If this is convergent for a particular value of σ, it is clearly convergent for any greater value. Hence, as in the case of convergence, there is a number $\bar{\sigma}$ such that it is convergent for $\sigma > \bar{\sigma}$, and divergent for $\sigma < \bar{\sigma}$.

Hence the original series is absolutely convergent for $\sigma > \bar{\sigma}$, and not absolutely convergent for $\sigma < \bar{\sigma}$.

The number $\bar{\sigma}$ is called the *abscissa of absolute convergence.*

The numbers σ_0 and $\bar{\sigma}$ are not necessarily equal, i.e. there may be a strip of the plane in which the series is convergent, but not absolutely convergent.

This is shown by the following example. If $\sigma > 1$, we have

$$\left(1-\frac{1}{2^{s-1}}\right)\zeta(s) = \left(\frac{1}{1^s}+\frac{1}{2^s}+\frac{1}{3^s}+\ldots\right)-2\left(\frac{1}{2^s}+\frac{1}{4^s}+\ldots\right)$$
$$= \frac{1}{1^s}-\frac{1}{2^s}+\frac{1}{3^s}-\frac{1}{4^s}+\ldots.$$

The last series, as it is arranged here, is convergent for $\sigma > 0$ (and uniformly convergent in any finite region to the right of $\sigma = 0$). For, by a well-known theorem (*P.M.* § 188), it is convergent if s is real and positive. Hence, by the theory of analytic continuation, the formula holds for $\sigma > 0$.

In this case $\sigma_0 = 0$, $\bar{\sigma} = 1$.

In any case $\bar{\sigma} - \sigma_0 \leqslant 1$.

For if $\sum a_n n^{-s}$ is convergent, $|a_n| n^{-\sigma}$ is bounded as $n \to \infty$, and hence

$$\sum \frac{a_n}{n^{s+1+\delta}}$$

is absolutely convergent if $\delta > 0$, which gives the result.

In the above example, $\bar{\sigma} - \sigma_0 = 1$, so that the strip of non-absolute convergence may be as wide as 1, though it can be no wider.

9.14. The abscissa of convergence. The formula for σ_0, analogous to the formula (§ 7.1) for the radius of convergence of a power series, takes slightly different forms according to whether $\sum a_n$ is convergent or not. Let

$$s_n = a_1 + a_2 + \ldots + a_n,$$

and, if $\sum a_n$ is convergent, let $r_n = a_{n+1} + a_{n+2} + \ldots$. Let

$$\alpha = \overline{\lim_{n\to\infty}} \frac{\log|s_n|}{\log n}, \qquad \beta = \overline{\lim_{n\to\infty}} \frac{\log|r_n|}{\log n},$$

β being defined only if $\sum a_n$ is convergent.

Then $\sigma_0 = \alpha$ *if* $\sum a_n$ *is divergent, and otherwise* $\sigma_0 = \beta$.

In the former case $\sigma_0 \geqslant 0$, and in the latter case $\sigma_0 \leqslant 0$; for $\sum a_n$ is convergent if $\sigma_0 < 0$.

(i) Let $\sum a_n$ be divergent, and let s have a real positive value for which the Dirichlet series is convergent. Let

$$b_n = a_n n^{-s}, \qquad B_n = b_1 + b_2 + \ldots + b_n, \qquad B_0 = 0,$$

so that B_n is bounded, say $|B_n| \leqslant B$. Then

$$s_N = \sum_{n=1}^{N} b_n n^s = \sum_{n=1}^{N} (B_n - B_{n-1}) n^s$$

$$= \sum_{n=1}^{N-1} B_n \{n^s - (n+1)^s\} + B_N N^s.$$

Hence $$|s_N| \leqslant B\sum_{n=1}^{N-1}\{(n+1)^s-n^s\}+BN^s < 2BN^s,$$
$$\log|s_N| < s\log N+\log 2B,$$
and so $\alpha \leqslant s$. Hence $\alpha \leqslant \sigma_0$.

A similar argument holds if $\sum a_n$ is convergent. If s has a real negative value for which the Dirichlet series is convergent,

$$r_N = \sum_{n=N+1}^{\infty} b_n n^s = \sum_{n=N+1}^{\infty}(B_n-B_{n-1})n^s$$
$$= \sum_{n=N}^{\infty} B_n\{n^s-(n+1)^s\}-B_N N^s,$$

so that $$|r_N| \leqslant B\sum_{n=N}^{\infty}\{n^s-(n+1)^s\}+BN^s = 2BN^s.$$

Hence, as in the other case, $\beta \leqslant \sigma_0$.

(ii) Since $s_n = O(n^{\alpha+\epsilon})$, and, if s is real,

$$\frac{1}{n^s}-\frac{1}{(n+1)^s} = s\int_n^{n+1}\frac{du}{u^{s+1}} = O(n^{-s-1}),$$

we have

$$\sum_{n=M+1}^{N}\frac{a_n}{n^s} = \sum_{n=M+1}^{N}\frac{s_n-s_{n-1}}{n^s}$$
$$= \sum_{n=M+1}^{N} s_n\left\{\frac{1}{n^s}-\frac{1}{(n+1)^s}\right\}+\frac{s_N}{(N+1)^s}-\frac{s_M}{(M+1)^s} \tag{1}$$
$$= \sum_{M+1}^{N} O(n^{\alpha+\epsilon-s-1})+O(N^{\alpha+\epsilon-s})+O(M^{\alpha+\epsilon-s})$$
$$= O(M^{\alpha+\epsilon-s}) = o(1)$$

if $s > \alpha$ and ϵ is small enough. Hence the Dirichlet series is convergent if $s > \alpha$; hence $\sigma_0 \leqslant \alpha$. Since $a_n = r_{n-1}-r_n$ if $\sum a_n$ is convergent, we find similarly that $\sigma_0 \leqslant \beta$. This proves the theorem.

If $\alpha = \infty$, the series is nowhere convergent, and if $\beta = -\infty$ it is everywhere convergent. This easily follows from the above argument.

9.15. The abscissa of absolute convergence. We have

$$\bar{\sigma} = \overline{\lim_{n\to\infty}}\frac{\log(|a_1|+|a_2|+\ldots+|a_n|)}{\log n},$$

or $$\bar{\sigma} = \overline{\lim_{n\to\infty}}\frac{\log(|a_{n+1}|+|a_{n+2}|+\ldots)}{\log n},$$

according to whether $\sum |a_n|$ is divergent or convergent. This is a particular case of the previous result.

Example. Determine σ_0 and $\bar{\sigma}$ for the series in which $a_n = 1, (-1)^n, n^{-\frac{1}{2}}, (-1)^n n^{-\frac{1}{2}}, a^n\,(0 < a < 1), \log n, 1/\log n$, respectively; and for the series in which $a_n = 1$ (n a perfect square), $a_n = 0$ otherwise.

9.2. Convergence of the series and regularity of the function. The region of convergence of a power series is determined in a perfectly simple way by the analytic character of the function which it represents—the circle of convergence passes through the singularity nearest to the centre. There is no such simple relation in the case of Dirichlet series. There is not necessarily any singularity on the line of convergence, and in fact $f(s)$ may be an integral function even though the abscissa of convergence of the series is finite. This is shown by the above example of the series for $(1-2^{1-s})\zeta(s)$. This is an integral function, since the pole of $\zeta(s)$ at $s = 1$ is cancelled by the zero of $1-2^{1-s}$. But the corresponding series converges for $\sigma > 0$ only.

On the other hand, the series for $\zeta(s)$, § 9.1 (2), has a singularity on its line of convergence. This is a particular case of the following theorem:

If $a_n \geqslant 0$ for all values of n, then the real point of the line of convergence is a singularity of $f(s)$.

The proof is similar to that of the corresponding theorem for power series (§ 7.21).

We may suppose without loss of generality that $\sigma_0 = 0$. Then if $s = 0$ is a regular point, the Taylor's series for $f(s)$, at the point $s = 1$, has a radius of convergence greater than 1. Hence we can find a negative value of s for which

$$f(s) = \sum_{\nu=0}^{\infty} \frac{(s-1)^\nu}{\nu!} f^{(\nu)}(1) = \sum_{\nu=0}^{\infty} \frac{(1-s)^\nu}{\nu!} \sum_{n=1}^{\infty} \frac{(\log n)^\nu a_n}{n}.$$

But every term in this repeated series is positive. Hence the order of summation may be inverted, and we obtain

$$f(s) = \sum_{n=1}^{\infty} \frac{a_n}{n} \sum_{\nu=0}^{\infty} \frac{(1-s)^\nu (\log n)^\nu}{\nu!}$$

$$= \sum_{n=1}^{\infty} \frac{a_n}{n} e^{(1-s)\log n} = \sum_{n=1}^{\infty} \frac{a_n}{n^s}.$$

Thus the Dirichlet series is convergent for a negative value of s, contrary to hypothesis.

9.3. Asymptotic behaviour of the function as $t \to \infty$. *The function $f(s)$ is bounded in any half-plane included in the half-plane of absolute convergence.*

For
$$|f(s)| \leqslant \sum_{n=1}^{\infty} \frac{|a_n|}{n^\sigma} \leqslant \sum_{n=1}^{\infty} \frac{|a_n|}{n^\alpha}$$
for $\sigma \geqslant \alpha > \bar{\sigma}$, and all values of t.

If the series
$$\sum_{n=1}^{\infty} \frac{|a_n|}{n^{\bar{\sigma}}}$$
is convergent, we can take $\alpha = \bar{\sigma}$, and the function is bounded in the half-plane of absolute convergence. This is true, for example, of the function
$$f(s) = \sum_{n=2}^{\infty} \frac{1}{n^s \log^2 n}.$$
But in general the half-plane of absolute convergence is not a region where $f(s)$ is bounded, even if we exclude the neighbourhood of singularities on the line $\sigma = \bar{\sigma}$ (see § 9.32).

Even in the half-plane of absolute convergence, the behaviour of $f(\sigma+it)$ as $t \to \infty$ is, in general, rather complicated. Take, for example, a series with real positive coefficients in which
$$\sum_{n=3}^{\infty} a_n n^{-\sigma} < a_2 2^{-\sigma}$$
for a certain value of σ. Then
$$\mathbf{R}f(s) = \sum_{n=1}^{\infty} \frac{a_n \cos(t \log n)}{n^\sigma} > a_1 + \frac{a_2}{2^\sigma} - \sum_{n=3}^{\infty} \frac{a_n}{n^\sigma}$$
for $t = 2m\pi/\log 2$, $m = 1, 2,\ldots$. Also
$$\mathbf{R}f(s) < a_1 - \frac{a_2}{2^\sigma} + \sum_{n=3}^{\infty} \frac{a_n}{n^\sigma}$$
for $t = (2m+1)\pi/\log 2$, $m = 1, 2,\ldots$. Hence $\mathbf{R}f(s)$ oscillates as $t \to \infty$.

9.31. The following theorem is due to Dirichlet: *Given N real numbers $c_1, c_2,\ldots, c_N$, a positive integer q, and a positive number τ,* *we can find a number t in the range $\tau \leqslant t \leqslant \tau q^N$, and integers $x_1,\dots, x_N$, such that*

$$|tc_n - x_n| \leqslant \frac{1}{q} \qquad (n = 1, 2,\dots, N).$$

The proof is based on the argument that if there are $m+1$ points in m regions, there must be at least one region which contains at least two points.

Consider the N-dimensional unit cube with a vertex at the origin and edges along the coordinate axes. Divide each edge into q equal parts, and thus the cube into q^N equal compartments. Consider the q^N+1 points in the cube

$$(\lambda c_1 - [\lambda c_1], \lambda c_2 - [\lambda c_2],\dots, \lambda c_N - [\lambda c_N]),$$

where λ takes the values 0, $\tau,\dots, \tau q^N$. At least two of these points must lie in the same compartment. If they are given by $\lambda = \lambda_1$, $\lambda = \lambda_2$ $(\lambda_1 < \lambda_2)$, then there are integers $x_1, x_2,\dots$ such that

$$|(\lambda_2 - \lambda_1)c_n - x_n| \leqslant 1/q \qquad (n = 1, 2,\dots, N),$$

and so $t = \lambda_2 - \lambda_1$ gives the required result.

9.32. We can now deduce the following theorem.

If $f(s) = \sum a_n n^{-s}$, where $a_n \geqslant 0$ for every value of n, and where $\sum a_n n^{-\bar{\sigma}}$ is divergent, the function $f(s)$ is not bounded in the region $\sigma > \bar{\sigma}$, $|t| \geqslant t_0 > 0$.

That there may be no singularities on the boundary of this region is shown by the function $\zeta(s) = \sum n^{-s}$, which is regular except at $s = 1$. We make $|t| \geqslant t_0$ to exclude the neighbourhood of the point $s = \bar{\sigma}$, where there is a singularity.

We have, for every value of N, and $\sigma > \bar{\sigma}$,

$$f(s) = \sum_{n=1}^{N} \frac{a_n}{n^\sigma} e^{-it\log n} + \sum_{N+1}^{\infty} \frac{a_n}{n^s},$$

and so

$$|f(s)| \geqslant \mathbf{R} \sum_{n=1}^{N} \frac{a_n}{n^\sigma} e^{-it\log n} - \left|\sum_{N+1}^{\infty} \frac{a_n}{n^s}\right|$$

$$\geqslant \sum_{n=1}^{N} \frac{a_n}{n^\sigma} \cos(t\log n) - \sum_{N+1}^{\infty} \frac{a_n}{n^\sigma}.$$

By Dirichlet's theorem there is, for given N and q, a number t $(t_0 \leqslant t \leqslant t_0 q^N)$ and integers $x_1,\dots, x_N$, such that

$$\left|\frac{t\log n}{2\pi} - x_n\right| \leqslant \frac{1}{q} \qquad (n = 1, 2,\dots, N).$$

Hence $\cos(t\log n) \geqslant \cos(2\pi/q)$ for these values of n, and so

$$\sum_{n=1}^{N} \frac{a_n}{n^\sigma}\cos(t\log n) \geqslant \cos\frac{2\pi}{q}\sum_{n=1}^{N}\frac{a_n}{n^\sigma}$$

$$> \cos\frac{2\pi}{q} f(\sigma) - \sum_{N+1}^{\infty}\frac{a_n}{n^\sigma}.$$

Hence, taking $q = 6$, say, so that $\cos 2\pi/q = \frac{1}{2}$,

$$|f(s)| > \tfrac{1}{2}f(\sigma) - 2\sum_{N+1}^{\infty}\frac{a_n}{n^\sigma}.$$

Given any positive number H, we can choose $\sigma - \bar{\sigma}$ so small that $f(\sigma) > 4H$ (since $f(\sigma) \to \infty$ as $\sigma \to \bar{\sigma}$). Having fixed σ, we can choose N so large that

$$\sum_{N+1}^{\infty}\frac{a_n}{n^\sigma} < \tfrac{1}{2}H.$$

Then $|f(s)| > H$, and the result follows.

9.33. In the half-plane of convergence, the function may, for certain values of t, become as large as a power of t. For example, the function $f(s) = (1-2^{1-s})\zeta(s)$, referred to in § 9.13, satisfies the inequality

$$|f(s)| > At^{\frac{1}{2}-\sigma}$$

for some arbitrarily large values of t, and values of σ between 0 and $\frac{1}{2}$.*

On the other hand, the function cannot have values greater than every power of t. This is shown by the following theorem.

We have

$$f(s) = O(|t|^{1-(\sigma-\sigma_0)+\epsilon})$$

as $|t| \to \infty$, *for any value of* σ *between* σ_0 *and* $\sigma_0 + 1$; *and also uniformly in the half-plane to the right of any such line.*

Suppose first that $\sum a_n$ is convergent. Then a_n and s_n are bounded. Now (§ 9.14 (1))

$$\sum_{1}^{N}\frac{a_n}{n^s} = \sum_{1}^{M}\frac{a_n}{n^s} + \sum_{M+1}^{N} s_n\left\{\frac{1}{n^s} - \frac{1}{(n+1)^s}\right\} - \frac{s_M}{(M+1)^s} + \frac{s_N}{(N+1)^s}.$$

If $\sigma > 0$, the last term tends to zero as $N \to \infty$, and we obtain

$$f(s) = \sum_{1}^{M}\frac{a_n}{n^s} + \sum_{M+1}^{\infty} s_n\left\{\frac{1}{n^s} - \frac{1}{(n+1)^s}\right\} - \frac{s_M}{(M+1)^s}.$$

* See Miscellaneous Examples, no. 18.

Hence, by § 9.11 (2), if $0 < \sigma < 1$,

$$|f(s)| < A\sum_{1}^{M}\frac{1}{n^\sigma}+A\frac{|s|}{\sigma}\sum_{M+1}^{\infty}\left\{\frac{1}{n^\sigma}-\frac{1}{(n+1)^\sigma}\right\}+\frac{A}{(M+1)^\sigma}$$

$$< AM^{1-\sigma}+AtM^{-\sigma}+A.$$

Taking $M = [t]$, we obtain

$$f(s) = O(t^{1-\sigma}) \qquad (0 < \sigma < 1),$$

and similarly $f(s) = O(t^{1-\alpha})$ $(0 < \alpha < 1,\ \sigma \geqslant \alpha)$. In the general case, the series $\sum a_n n^{-s}$ is convergent for $s = \sigma_0+\epsilon$, and we obtain the above case by changing the origin to this point. Hence the general result follows.

9.4. Functions of finite order. At this point we adopt a slightly different point of view. We have so far considered $f(s)$ as being defined by the series $\sum a_n n^{-s}$, and we have confined our attention to the half-plane of convergence of the series. It may, however, be possible to continue the function outside this half-plane. The function, so defined, may be regular in a wider half-plane; or it may be regular in a wider half-plane except for a certain finite region. We shall now consider the relations between a function defined in this way, and the Dirichlet series from which it originated.

The theorem of § 9.33 suggests that it will be particularly interesting to consider functions which satisfy the condition

$$f(s) = O(|t|^A)$$

for some positive value of A. A function which satisfies this condition for a particular value of σ is said to be of finite order for that value; if the condition is satisfied uniformly for $\sigma_1 \leqslant \sigma \leqslant \sigma_2$, we say that the function is of finite order in this strip. Similarly we can define a function of finite order in a half-plane $\sigma \geqslant \sigma_1$.

We have seen that any function defined by a Dirichlet series is of finite order in a half-plane included in the half-plane of convergence. It may be of finite order outside this half-plane; for $\zeta(s)$, for example, $\sigma_0 = 1$; but (§ 9.13)

$$\zeta(s) = (1-2^{1-s})^{-1}\sum_{n=1}^{\infty}\frac{(-1)^n}{n^s} \qquad (\sigma > 0),$$

and hence, by § 9.33,

$$\zeta(s) = O(|t|^{1-\sigma+\epsilon}) \qquad (0 < \sigma \leqslant 1).$$

9.41. The function $\mu(\sigma)$. The lower bound μ of numbers ξ such that $f(s) = O(|t|^{\xi})$ is called the *order* of $f(s)$ for that particular value of σ. Thus μ is a function of σ.

The main properties of the function $\mu(\sigma)$ follow from the Phragmén-Lindelöf theorem proved in § 5.65. Suppose that $f(s)$ is regular and of finite order for $\sigma_1 \leqslant \sigma \leqslant \sigma_2$, $t \geqslant t_0$, and let $\mu(\sigma_1) = \mu_1$, $\mu(\sigma_2) = \mu_2$. Then, if ϵ is any positive number,

$$f(\sigma_1+it) = O(t^{\mu_1+\epsilon}), \qquad f(\sigma_2+it) = O(t^{\mu_2+\epsilon}).$$

Hence, by the theorem referred to,

$$f(s) = O(t^{k(\sigma)}) \qquad (\sigma_1 \leqslant \sigma \leqslant \sigma_2),$$

where $$k(\sigma) = \frac{(\sigma_2-\sigma)(\mu_1+\epsilon)+(\sigma-\sigma_1)(\mu_2+\epsilon)}{\sigma_2-\sigma_1}.$$

Making $\epsilon \to 0$, it follows that

$$\mu(\sigma) \leqslant \frac{(\sigma_2-\sigma)\mu_1+(\sigma-\sigma_1)\mu_2}{\sigma_2-\sigma_1} \qquad (\sigma_1 \leqslant \sigma \leqslant \sigma_2). \tag{1}$$

Hence *the function* $\mu(\sigma)$ *is convex downwards.*

It follows also that $\mu(\sigma)$ *is continuous* (§ 5.31).

Secondly, $\mu(\sigma) = 0$ for sufficiently large values of σ; for since $f(s)$ is bounded for $\sigma > \bar{\sigma}$, $\mu(\sigma) \leqslant 0$ for $\sigma > \bar{\sigma}$; on the other hand, if a_m is the first coefficient in the Dirichlet series which does not vanish, and $\bar{\sigma} < \alpha < \sigma$,

$$|f(s)| \geqslant \frac{|a_m|}{m^\sigma} - \sum_{n=m+1}^{\infty} \frac{|a_n|}{n^\sigma} \geqslant \frac{|a_m|}{m^\sigma} - (m+1)^{\alpha-\sigma} \sum_{n=m+1}^{\infty} \frac{|a_n|}{n^\alpha},$$

which can be made positive by taking σ large enough. Thus $|f(s)|$, considered as a function of t, has a positive lower bound if σ is large enough. Hence $\mu(\sigma) \geqslant 0$, so that in fact $\mu(\sigma) = 0$.

If now $\mu(\sigma)$ were negative for any value σ_1 in the region where $f(s)$ is of finite order, it would follow from (1), with σ_2 so large that $\mu_2 = 0$, that $\mu(\sigma) < 0$ for $\sigma_1 < \sigma < \sigma_2$; and we have shown that this is impossible if σ_2 is large enough. *Hence* $\mu(\sigma)$ *is never negative.*

In particular, $\mu(\sigma) = 0$ for $\sigma > \bar{\sigma}$; for we have already shown that $\mu(\sigma) \leqslant 0$ for $\sigma > \bar{\sigma}$.

Again, take $\sigma_2 > \bar{\sigma}$ in (1), so that $\mu_2 = 0$. Then if $\mu_1 > 0$,

$$\mu(\sigma) \leqslant \frac{\sigma_2 - \sigma}{\sigma_2 - \sigma_1} \mu_1 < \mu_1 \qquad (\sigma > \sigma_1).$$

Hence $\mu(\sigma)$ is a steadily decreasing function of σ.

9.42. Perron's formula. We next require an expression for the sum s_n as an integral. This is a particular case of the following theorem.

If x is not an integer, c is any positive number, and $\sigma > \sigma_0 - c$, then

$$\sum_{n<x} \frac{a_n}{n^s} = \frac{1}{2\pi i} \int_{c-i\infty}^{c+i\infty} f(s+w) \frac{x^w}{w}\, dw. \tag{1}$$

Suppose first that $\sigma > \bar{\sigma} - c$. Then the series for $f(s+w)$ is absolutely and uniformly convergent, and we have

$$\frac{1}{2\pi i} \int_{c-iU}^{c+iT} f(s+w) \frac{x^w}{w}\, dw = \frac{1}{2\pi i} \int_{c-iU}^{c+iT} \sum_{1}^{\infty} \frac{a_n}{n^{s+w}} \frac{x^w}{w}\, dw$$

$$= \frac{1}{2\pi i} \sum_{1}^{\infty} \frac{a_n}{n^s} \int_{c-iU}^{c+iT} \left(\frac{x}{n}\right)^w \frac{dw}{w}. \tag{2}$$

Now by § 3.126

$$\frac{1}{2\pi i} \int_{c-i\infty}^{c+i\infty} \left(\frac{x}{n}\right)^w \frac{dw}{w} = \begin{matrix} 1 \\ 0 \end{matrix} \qquad \begin{matrix} (n < x) \\ (n > x). \end{matrix}$$

It is therefore sufficient to prove that we can replace U and T in (2) by ∞; that is, we want

$$\lim_{T\to\infty} \sum_{1}^{\infty} \frac{a_n}{n^s} \int_{c+iT}^{\infty+iT} \left(\frac{x}{n}\right)^w \frac{dw}{w} = 0,$$

with a similar result for U.

Now for a fixed x,

$$\int_{c+iT}^{c+i\infty} \left(\frac{x}{n}\right)^w \frac{dw}{w} = -\left(\frac{x}{n}\right)^{c+iT} \frac{1}{(\log x/n)(c+iT)} + \frac{1}{\log x/n} \int_{c+iT}^{c+i\infty} \left(\frac{x}{n}\right)^w \frac{dw}{w^2}$$

$$= O\left(\frac{1}{n^c T}\right) + O\left(\frac{1}{n^c} \int_{T}^{\infty} \frac{dv}{c^2+v^2}\right) = O\left(\frac{1}{n^c T}\right).$$

Hence
$$\sum_{n=1}^{\infty}\frac{a_n}{n^s}\int\limits_{c+iT}^{c+i\infty}\left(\frac{x}{n}\right)^w\frac{dw}{w}=O\left(\frac{1}{T}\sum_{n=1}^{\infty}\frac{|a_n|}{n^{\sigma+c}}\right),$$
and the result (for $\sigma>\bar{\sigma}-c$) follows.

Suppose next that $\sigma_0-c<\sigma\leqslant\bar{\sigma}-c$. Let $\alpha>\bar{\sigma}-\sigma$, and consider the integral
$$\int f(s+w)\frac{x^w}{w}\,dw$$
taken round the rectangle formed by the lines $\mathbf{R}(w)=c$, $\mathbf{R}(w)=\alpha$, $\mathbf{I}(w)=-U$, $\mathbf{I}(w)=T$. By the theorem of § 9.33, the integrand is
$$O(t^{-(\sigma+c-\sigma_0)+\epsilon}),$$
and so the integrals along the horizontal sides tend to zero as $U\to\infty$, $T\to\infty$. The integrand is regular inside the rectangle, and so, by Cauchy's theorem,
$$\frac{1}{2\pi i}\int\limits_{c-i\infty}^{c+i\infty}f(s+w)\frac{x^w}{w}\,dw=\frac{1}{2\pi i}\int\limits_{\alpha-i\infty}^{\alpha+i\infty}f(s+w)\frac{x^w}{w}\,dw.$$
Since $\sigma>\bar{\sigma}-\alpha$, the right-hand side is equal to $\sum_{n<x}a_n n^{-s}$, by the first part. This completes the proof.

The particular case $s=0$ is
$$\sum_{n<x}a_n=\frac{1}{2\pi i}\int\limits_{c-i\infty}^{c+i\infty}f(w)\frac{x^w}{w}\,dw\qquad(c>\sigma_0).\tag{3}$$
This is Perron's formula.

9.43. There are several other formulae of the same type as Perron's. One which we shall use later is
$$\sum_{n=1}^{\infty}\frac{a_n}{n^s}e^{-(n\delta)^\lambda}=\frac{1}{2\pi i\lambda}\int\limits_{c-i\infty}^{c+i\infty}\Gamma\left(\frac{w}{\lambda}\right)f(s+w)\delta^{-w}\,dw,$$
where $\delta>0$, $\lambda>0$, and $c>0$, $c>\bar{\sigma}-\sigma$. To prove this, write the right-hand side as
$$\frac{1}{2\pi i\lambda}\int\limits_{c-i\infty}^{c+i\infty}\Gamma\left(\frac{w}{\lambda}\right)\sum_{n=1}^{\infty}\frac{a_n}{n^{s+w}}\delta^{-w}\,dw,$$
and observe that we can invert the order of summation and

integration by 'absolute convergence'. We therefore obtain

$$\sum_{n=1}^{\infty} \frac{a_n}{n^s} \frac{1}{2\pi i\lambda} \int_{c-i\infty}^{c+i\infty} \Gamma\left(\frac{w}{\lambda}\right) (n\delta)^{-w}\, dw,$$

and the result follows by the calculus of residues.

9.44. The theorem of § 9.42 enables us to obtain a result of the opposite type to the previous ones—we can pass from the order of the function to the convergence of the series.

The Dirichlet series is convergent in the half-plane where $f(s)$ is regular and $\mu(\sigma) = 0$.

Let s be a point in the interior of this half-plane, and let δ be a positive number so small that $\sigma-\delta$ is still in the same half-plane. Let $c > \bar{\sigma}-\sigma+1$ (so that the simpler case of the theorem of § 9.42 can be used). We deform the contour of § 9.42 (1) into the form

$$c-i\infty, \quad c-iT, \quad -\delta-iT, \quad -\delta+iT, \quad c+iT, \quad c+i\infty,$$

where $T > |t|$. In doing so we pass over a pole at $w = 0$, with residue $f(s)$. Hence

$$\sum_{n<x} \frac{a_n}{n^s} - f(s)$$

$$= \frac{1}{2\pi i}\left\{\int_{c-i\infty}^{c-iT} + \int_{c-iT}^{-\delta-iT} + \int_{-\delta-iT}^{-\delta+iT} + \int_{-\delta+iT}^{c+iT} + \int_{c+iT}^{c+i\infty}\right\} f(s+w)\frac{x^w}{w}\, dw.$$

Since we are in the half-plane where $\mu(\sigma) = 0$, we have $f(s) = O(|t|^\epsilon)$ for every positive ϵ. Hence

$$\int_{-\delta-iT}^{-\delta+iT} f(s+w)\frac{x^w}{w}\, dw = \int_{-T}^{T} O\{(|t|+|v|)^\epsilon\} \frac{x^{-\delta}\, dv}{\sqrt{(\delta^2+v^2)}} = O(x^{-\delta}T^\epsilon),$$

and $$\int_{-\delta+iT}^{c+iT} f(s+w)\frac{x^w}{w}\, dw = \int_{-\delta}^{c} O(T^\epsilon)\frac{x^c}{T}\, du = O(x^c T^{\epsilon-1}).$$

A similar result holds for the integral over $(c-iT, -\delta-iT)$. Finally, as in § 9.42,

$$\int_{c+iT}^{c+i\infty} f(s+w)\frac{x^w}{w}\, dw = \sum_{n=1}^{\infty} \frac{a_n}{n^s} \int_{c+iT}^{c+i\infty} \left(\frac{x}{n}\right)^w \frac{dw}{w}$$

$$= O\left(\frac{x^c}{T}\sum_{n=1}^{\infty} \frac{|a_n|}{n^{\sigma+c}|\log x/n|}\right).$$

We may suppose without loss of generality that x is half an odd integer. Then

$$|\log x/n| > \log\{(n+\tfrac{1}{2})/n\} > A/n.$$

Hence the above expression is

$$O\left(\frac{x^c}{T}\sum_{n=1}^{\infty}\frac{|a_n|}{n^{\sigma+c-1}}\right) = O\left(\frac{x^c}{T}\right),$$

since $\sigma+c-1 > \bar{\sigma}$.

A similar result holds for the integral over $(c-i\infty, c-iT)$; and adding, we obtain

$$\sum_{n<x} a_n n^{-s} - f(s) = O(x^{-\delta}T^{\epsilon}) + O(x^c T^{\epsilon-1}).$$

Taking $T = x^{2c}$, this is

$$O(x^{-\delta+2c\epsilon}) + O(x^{-c+2c\epsilon}),$$

which tends to zero as $x \to \infty$ if $\epsilon < \delta/2c$ and $\epsilon < \frac{1}{2}$. This proves the theorem.

A more general theorem of the same type is given in Landau's *Handbuch*, § 238, Satz 57.

9.45. Let σ_ϵ be the abscissa limiting the half-plane where $f(s)$ is regular and of the form $O(t^\epsilon)$. Then we have proved that

$$\sigma_0 \leqslant \sigma_\epsilon \leqslant \bar{\sigma}.$$

It is not easy to give an example where these numbers are all different. There is some reason to suppose that, for the function $(1-2^{1-s})\zeta(s)$, we have $\sigma_\epsilon = \frac{1}{2}$, so that the three numbers are 0, $\frac{1}{2}$, and 1 respectively. But this has not been proved.

9.5. The mean-value formula. *If* $\sigma > \bar{\sigma}$,

$$\lim_{T\to\infty}\frac{1}{2T}\int_{-T}^{T}|f(s)|^2\,dt = \sum_{n=1}^{\infty}\frac{|a_n|^2}{n^{2\sigma}}.$$

For
$$|f(s)|^2 = \sum_{m=1}^{\infty}\frac{a_m}{m^{\sigma+it}}\sum_{n=1}^{\infty}\frac{\bar{a}_n}{n^{\sigma-it}}$$
$$= \sum_{n=1}^{\infty}\frac{|a_n|^2}{n^{2\sigma}} + \sum_{m\neq n}\sum\frac{a_m\bar{a}_n}{m^\sigma n^\sigma}\left(\frac{n}{m}\right)^{it},$$

the series being absolutely convergent, and uniformly convergent in any finite t-range. Hence we may integrate term by term, and obtain

$$\frac{1}{2T}\int_{-T}^{T}|f(s)|^2\,dt=\sum_{n=1}^{\infty}\frac{|a_n|^2}{n^{2\sigma}}+\sum_{m\neq n}\sum\frac{a_m\bar{a}_n}{m^\sigma n^\sigma}\frac{2\sin(T\log n/m)}{2T\log n/m}.$$

The factor involving T is bounded for all T, m, and n, so that the double series converges uniformly with respect to T; and each term tends to zero as $T\to\infty$. Hence the sum tends to zero as $T\to\infty$, and the result follows.

9.51. The mean-value half-plane. Let σ_m be the least number such that $f(s)$ is regular and of finite order, and the mean-value formula holds, for every σ greater than σ_m. We shall call the half-plane $\sigma>\sigma_m$ the mean-value half-plane. This expression is justified by the following theorem:*

If $f(s)$ is regular and of finite order for $\sigma\geqslant\alpha$, and

$$\frac{1}{2T}\int_{-T}^{T}|f(\alpha+it)|^2\,dt \tag{1}$$

is bounded as $T\to\infty$, then

$$\lim_{T\to\infty}\frac{1}{2T}\int_{-T}^{T}|f(\sigma+it)|^2\,dt=\sum_{n=1}^{\infty}\frac{|a_n|^2}{n^{2\sigma}} \tag{2}$$

for $\sigma>\alpha$, and uniformly in any strip $\alpha<\sigma_1\leqslant\sigma\leqslant\sigma_2$.

Starting with the formula of § 9.43, and moving the contour to $\mathbf{R}(w)=\alpha-\sigma$, where $\sigma>\alpha$, we pass a pole at $w=0$, with residue $\lambda f(s)$; and if $\lambda>\sigma-\alpha$, no other pole is passed. Hence

$$\sum_{n=1}^{\infty}\frac{a_n}{n^s}e^{-(n\delta)^\lambda}-f(s)=\frac{1}{2\pi i\lambda}\int_{\alpha-\sigma-i\infty}^{\alpha-\sigma+i\infty}\Gamma\left(\frac{w}{\lambda}\right)f(s+w)\delta^{-w}\,dw$$

$$=O\left\{\delta^{\sigma-\alpha}\int_{-\infty}^{\infty}e^{-A|v|}|f\{\alpha+i(t+v)\}|\,dv\right\}$$

by the asymptotic formula for the Γ-function (§ 4.42, ex. (ii)). Now if $|t|<T$,

$$\int_{2T}^{\infty}e^{-Av}|f\{\alpha+i(t+v)\}|\,dv=O\left(\int_{2T}^{\infty}e^{-Av}v^A\,dv\right)=O(e^{-AT}),$$

* Carlson (1). The theorem is analogous to Parseval's theorem for Fourier series (§ 13.54).

and a similar result holds for the integral over $(-\infty, -2T)$. Also, by Schwarz's inequality,*

$$\left\{\int_{-2T}^{2T} e^{-A|v|}|f\{\alpha+i(t+v)\}|\,dv\right\}^2$$

$$\leqslant \int_{-2T}^{2T} e^{-A|v|}|f\{\alpha+i(t+v)\}|^2\,dv \int_{-2T}^{2T} e^{-A|v|}\,dv$$

$$< A\int_{-2T}^{2T} e^{-A|v|}|f\{\alpha+i(t+v)\}|^2\,dv.$$

Hence

$$\left|\sum \frac{a_n}{n^s} e^{-(n\delta)^\lambda} - f(s)\right|^2$$

$$< A\delta^{2\sigma-2\alpha}\int_{-2T}^{2T} e^{-A|v|}|f\{\alpha+i(t+v)\}|^2\,dv + A\delta^{2\sigma-2\alpha}e^{-AT},$$

and, integrating with respect to t over $(-T, T)$,

$$\int_{-T}^{T}\left|\sum \frac{a_n}{n^s} e^{-(n\delta)^\lambda} - f(s)\right|^2 dt$$

$$< A\delta^{2\sigma-2\alpha}\int_{-2T}^{2T} e^{-A|v|}\,dv \int_{-T}^{T} |f\{\alpha+i(t+v)\}|^2\,dt + O(\delta^{2\sigma-2\alpha}).$$

Now

$$\int_{-T}^{T} |f\{\alpha+i(t+v)\}|^2\,dt = \int_{-T+v}^{T+v} |f(\alpha+it)|^2\,dt = O(T)$$

uniformly for $|v| < 2T$. Hence

$$\frac{1}{2T}\int_{-T}^{T}\left|\sum \frac{a_n}{n^s} e^{-(n\delta)^\lambda} - f(s)\right|^2 dt = O(\delta^{2\sigma-2\alpha})$$

uniformly with respect to T. Hence†

$$\left\{\frac{1}{2T}\int_{-T}^{T}\left|\sum \frac{a_n}{n^s} e^{-(n\delta)^\lambda}\right|^2 dt\right\}^{\frac{1}{2}} - \left\{\frac{1}{2T}\int_{-T}^{T} |f(s)|^2\,dt\right\}^{\frac{1}{2}} = O(\delta^{\sigma-\alpha}) \quad (3)$$

uniformly with respect to T.

* See *P.M.* Ch. VII, ex. 42; or § 12.41 below.

† By Minkowski's inequality (§ 12.43), but only in the case where $p = 2$ and the functions are continuous.

If $\delta > 0$, the series $\sum a_n n^{-s} e^{-(n\delta)^\lambda}$ is absolutely convergent, and so, by § 9.5,

$$\lim_{T\to\infty} \frac{1}{2T} \int_{-T}^{T} \left| \sum \frac{a_n}{n^s} e^{-(n\delta)^\lambda} \right|^2 dt = \sum \frac{|a_n|^2}{n^{2\sigma}} e^{-2(n\delta)^\lambda} \tag{4}$$

Taking, say, $\delta = 1$, it follows from (3) and (4) that

$$\frac{1}{2T} \int_{-T}^{T} |f(s)|^2 \, dt < A.$$

Hence by (3), with any positive δ,

$$\sum \frac{|a_n|^2}{n^{2\sigma}} e^{-2(n\delta)^\lambda} < A,$$

and so, since δ may be as small as we please,

$$\sum \frac{|a_n|^2}{n^{2\sigma}}$$

is convergent, and

$$\lim_{\delta\to 0} \sum \frac{|a_n|^2}{n^{2\sigma}} e^{-2(n\delta)^\lambda} = \sum \frac{|a_n|^2}{n^{2\sigma}}.$$

Given ϵ, we can choose δ so that the absolute value of the left-hand side of (3) is less than ϵ, for all values of T, and so that

$$\left| \left\{ \sum \frac{|a_n|^2}{n^{2\sigma}} e^{-2(n\delta)^\lambda} \right\}^{\frac{1}{2}} - \left\{ \sum \frac{|a_n|^2}{n^{2\sigma}} \right\}^{\frac{1}{2}} \right| < \epsilon.$$

Having fixed δ, we can, by (4), choose T_0 so large that

$$\left| \left\{ \frac{1}{2T} \int_{-T}^{T} \left| \sum \frac{a_n}{n^s} e^{-(n\delta)^\lambda} \right|^2 dt \right\}^{\frac{1}{2}} - \left\{ \sum \frac{|a_n|^2}{n^{2\sigma}} e^{-2(n\delta)^\lambda} \right\}^{\frac{1}{2}} \right| < \epsilon.$$

for $T > T_0$. Then

$$\left| \left\{ \frac{1}{2T} \int_{-T}^{T} |f(s)|^2 \, dt \right\}^{\frac{1}{2}} - \left\{ \sum \frac{|a_n|^2}{n^{2\sigma}} \right\}^{\frac{1}{2}} \right| < 3\epsilon$$

for $T > T_0$, and the theorem is proved.

9.52. *If $|f(s)|^2$ has a mean-value for $\sigma = \alpha$, then the Dirichlet series is absolutely convergent for $\sigma > \alpha + \frac{1}{2}$. In symbols,*

$$\bar{\sigma} \leqslant \sigma_m + \tfrac{1}{2}.$$

It follows from the above theorem that

$$\sum \frac{|a_n|^2}{n^{2\alpha+2\epsilon}}$$

is convergent for every positive ϵ. Now

$$\left\{\sum_{n=1}^{N}\left|\frac{a_n}{n^s}\right|\right\}^2 \leqslant \sum_{n=1}^{N}\frac{|a_n|^2}{n^{2\alpha+2\epsilon}}\sum_{n=1}^{N}\frac{1}{n^{2\sigma-2\alpha-2\epsilon}}$$

and this is bounded if ϵ is small enough and $\sigma-\alpha > \frac{1}{2}$. This result was obtained in another way by Hardy (10).

9.53. *If $f(s)$ is bounded for $\sigma > \alpha$, then $\sum |a_n|^2 n^{-2\alpha}$ is convergent; if $|f(s)| \leqslant M$, then*

$$\sum \frac{|a_n|^2}{n^{2\alpha}} \leqslant M^2.$$

This also follows from the theorem of § 9.51. For

$$\sum \frac{|a_n|^2}{n^{2\sigma}} = \lim \frac{1}{2T}\int_{-T}^{T} |f(s)|^2\,dt \leqslant M^2$$

for every $\sigma > \alpha$; and making $\sigma \to \alpha$ the result follows.

If we assume that $f(s)$ is bounded, the analysis of § 9.51 can, of course, be very much simplified.

9.54. Another consequence of these theorems is that a strip in which $f(s)$ is bounded, but in which the Dirichlet series is not absolutely convergent, can be at most of breadth $\frac{1}{2}$.

9.55. *The Dirichlet series converges in the half-plane in which $f(s)$ is regular and of finite order, and*

$$\lim_{T\to\infty}\frac{1}{2T}\int_{-T}^{T} |f(s)|^2\,dt$$

exists. That is, $\sigma_0 \leqslant \sigma_m$.

We have first to deduce an 'order' result for $f(s)$ from the given mean-value result.

We have $f(s) = O(|t|^{\frac{1}{2}})$ uniformly in any strip $\alpha \leqslant \sigma \leqslant \beta$, where $\alpha > \sigma_m$.

Let s be a point of the strip (α, β), R a number less than 1 and less than $\alpha-\sigma_m$, independent of t. Then, if $0 < \rho < R$,

$$f(s) = \frac{1}{2\pi i}\int_{|z-s|=\rho} \frac{f(z)}{z-s}\,dz = \frac{1}{2\pi}\int_{0}^{2\pi} f(s+\rho e^{i\phi})\,d\phi.$$

Hence, by Schwarz's inequality,

$$|f(s)|^2 \leqslant \frac{1}{4\pi^2}\int_0^{2\pi} d\phi . \int_0^{2\pi} |f(s+\rho e^{i\phi})|^2\, d\phi = \frac{1}{2\pi}\int_0^{2\pi} |f(s+\rho e^{i\phi})|^2\, d\phi.$$

Multiplying by ρ, and integrating with respect to ρ from 0 to R, we have

$$\tfrac{1}{2}R^2|f(s)|^2 \leqslant \frac{1}{2\pi}\int_0^R\int_0^{2\pi} |f(s+\rho e^{i\phi})|^2\rho\, d\rho d\phi$$

$$< \frac{1}{2\pi}\int_{\sigma-R}^{\sigma+R}\int_{t-R}^{t+R} |f(x+iy)|^2\, dxdy < \frac{1}{2\pi}\int_{\sigma-R}^{\sigma+R} dx \int_{-|t|-1}^{|t|+1} |f(x+iy)|^2\, dy.$$

Now
$$\int_{-|t|-1}^{|t|+1} |f(x+iy)|^2\, dy = O(t)$$

uniformly in x; hence

$$\tfrac{1}{2}R^2|f(s)|^2 = O(t),$$

which gives the required result.

We use the same contour integral as in § 9.44, s and $\sigma-\delta$ now being in the half-plane $\sigma > \sigma_m$. Then

$$\left|\int_{-\delta-iT}^{-\delta+iT} f(s+w)\frac{x^w}{w}\, dw\right| \leqslant x^{-\delta}\left\{\int_{-T}^{T} \frac{|f(s+w)|^2}{\sqrt{(\delta^2+v^2)}}\, dv \int_{-T}^{T} \frac{dv}{\sqrt{(\delta^2+v^2)}}\right\}^{\frac{1}{2}}.$$

Let
$$\phi(v) = \int_0^v |f(\sigma+u+iy)|^2\, dy = O(v).$$

Then

$$\int_0^T \frac{|f|^2}{\sqrt{(\delta^2+v^2)}}\, dv = \frac{\phi(T)}{\sqrt{(\delta^2+T^2)}} + \int_0^T \frac{v\phi(v)}{(\delta^2+v^2)^{\frac{3}{2}}}\, dv$$

$$= O(1) + \int_{-T}^{T} \frac{O(v^2)}{(\delta^2+v^2)^{\frac{3}{2}}}\, dv = O(\log T).$$

Similarly the integral over $(-T, 0)$ is $O(\log T)$. Hence

$$\int_{-\delta-iT}^{-\delta+iT} f(s+w)\frac{x^w}{w}\, dw = O(x^{-\delta}\log T).$$

Also, by the lemma,

$$\int_{-\delta+iT}^{c+iT} f(s+w)\frac{x^w}{w}\,dw = \int_{-\delta}^{c} O(\sqrt{T})\frac{x^c}{T}\,du = O(x^c T^{-\frac{1}{2}}),$$

with a similar result for the integral over $(c-iT, -\delta-iT)$. As in § 9.44, the remaining integrals are $O(x^c T^{-1})$. Hence

$$\sum_{n<x}\frac{a_n}{n^s} - f(s) = O(x^{-\delta}\log T) + O(x^c T^{-\frac{1}{2}}),$$

and, taking $T = x^{2c+1}$, this tends to 0 as $x \to \infty$, so that the result follows.

9.6. The uniqueness theorem. A function $f(s)$ can have at most one representation as a Dirichlet series. More precisely, *if*

$$\sum_1^\infty \frac{a_n}{n^s} = \sum_1^\infty \frac{b_n}{n^s}$$

in any region of values of s, then $a_n = b_n$ for all values of n.

For the series $\sum (a_n - b_n)n^{-s}$ is uniformly convergent in a region including part of the given region and extending arbitrarily far to the right; and so its sum is the same analytic function, viz. 0, in the whole region. But, if m is the first value of n for which $a_n \neq b_n$,

$$|\sum (a_n-b_n)n^{-s}| \geqslant |a_m - b_m| m^{-\sigma} - \sum_{m+1}^{\infty} |a_n - b_n| n^{-\sigma},$$

and, as in § 9.41, this is positive if σ is large enough. This leads to a contradiction, and so proves the theorem.

9.61. The zeros of $f(s)$. The above argument shows that *$f(s)$ always has a half-plane free from zeros.*

The problem of the distribution of the zeros of any given $f(s)$ is usually a very difficult one, and the results for different functions may be very different. For example, it is supposed that all the complex zeros of the function

$$\frac{1}{1^s} - \frac{1}{2^s} + \frac{1}{3^s} - \ldots = (1-2^{1-s})\zeta(s)$$

lie on the lines $\sigma = 1$ (zeros of $1-2^{1-s}$) and $\sigma = \frac{1}{2}$ (zeros of $\zeta(s)$); the zeros on $\sigma = 1$ are easily identified, but the remaining statement has never been proved.

On the other hand, it is known that the function $\zeta'(s)/\zeta(s)$, which is represented by an absolutely convergent Dirichlet series for $\sigma > 1$, has no zeros in a certain half-plane $\sigma > E$ $(E > 1)$, and zeros on lines $\sigma = \sigma'$ which are dense everywhere in the interval $1 < \sigma < E$.

It is interesting to compare the general problem with the particular case where $a_n = 0$ except when n is a power of 2. Then the function is of the form

$$f(s) = \sum_{n=0}^{\infty} \frac{b_n}{2^{ns}} = \sum_{n=0}^{\infty} b_n z^n,$$

where $z = 2^{-s}$. The series is a power series as well as a Dirichlet series. To each zero z_ν of the power series corresponds a sequence of zeros

$$s_{\mu,\nu} = -\frac{\log z_\nu + 2\mu\pi i}{\log 2} \qquad (\mu = 0, \pm 1, \pm 2, \ldots)$$

of $f(s)$. If z_0 is the zero of smallest modulus (other than 0), $f(s)$ has no zero to the right of the line

$$\sigma = \frac{\log 1/|z_0|}{\log 2},$$

there being an infinity of zeros on this line.

9.62. The function $N(\sigma, T)$. Let t_0 be a positive number such that $f(s)$ is regular for $t \geqslant t_0$ and σ sufficiently large, and let $N(\sigma, T)$ be the number of zeros $\sigma' + it'$ of $f(s)$ such that $\sigma' > \sigma$, $t_0 < t' < T$. Then we have the following theorems:

9.621. *If $f(s)$ is of finite order for $\sigma \geqslant \alpha$, then*

$$N(\sigma, T) = O(T \log T) \qquad (\sigma > \alpha).$$

We can find a number β so large that $|f(s)|$ has positive lower and upper bounds on the line $\sigma = \beta$. Let $0 < \delta < \frac{1}{2}(\beta - \alpha)$. We apply Jensen's theorem to the circle with centre $\beta + in\delta$ and radius $\beta - \alpha$. If $n(r)$ is the number of zeros of $f(s)$ in the circle $|s - (\beta + in\delta)| \leqslant r$, Jensen's theorem gives

$$\int_0^{\beta-\alpha} \frac{n(r)}{r}\, dr = \frac{1}{2\pi} \int_0^{2\pi} \log|f\{\beta + in\delta + (\beta - \alpha)e^{i\theta}\}|\, d\theta - \log|f(\beta + in\delta)|.$$

Now $f(s) = O(t^A)$ for $\sigma \geqslant \alpha$, and so

$$\log|f\{\beta + in\delta + (\beta - \alpha)e^{i\theta}\}| = \log|O\{(n\delta + \beta - \alpha)^A\}| < K \log n,$$

where K depends on α, β, δ,... only. Also $\log|f(\beta+in\delta)| = O(1)$; hence

$$\int_0^{\beta-\alpha} \frac{n(r)}{r}\,dr < K\log n.$$

But

$$\int_0^{\beta-\alpha} \frac{n(r)}{r}\,dr \geqslant n(\beta-\alpha-\delta) \int_{\beta-\alpha-\delta}^{\beta-\alpha} \frac{dr}{r} > Kn(\beta-\alpha-\delta),$$

and $n(\beta-\alpha-\delta)$ is, if δ is small enough, greater than the number of zeros in the strip

$$\sigma \geqslant \alpha+2\delta, \qquad (n-\tfrac{1}{2})\delta \leqslant t < (n+\tfrac{1}{2})\delta.$$

Denoting this number by ν_n, we have therefore

$$\nu_n < K\log n.$$

Hence $$N(\alpha+2\delta, T) \leqslant \sum_{t_0/\delta < n < T/\delta} \nu_n < KT\log T,$$

and the theorem follows.

9.622. *If $f(s)$ is bounded for $\sigma \geqslant \alpha$,*

$$N(\sigma, T) = O(T) \qquad (\sigma > \alpha).$$

The proof is similar to the previous one, but here the factor $\log T$ obviously does not occur. The example at the end of § 9.61 shows that we may have $N(\sigma, T) > AT$.

9.623. *If $f(s)$ has a mean value for $\sigma = \alpha$, and is of finite order for $\sigma \geqslant \alpha$, then*

$$N(\sigma, T) = O(T) \qquad (\sigma > \alpha).$$

We use the following lemma:

If $\phi(t)$ is a positive continuous function in (a, b),

$$\frac{1}{b-a}\int_a^b \log\phi(t)\,dt \leqslant \log\left\{\frac{1}{b-a}\int_a^b \phi(t)\,dt\right\}.$$

Divide the interval (a, b) into n equal parts by the points $a = x_0, x_1,..., x_n = b$. We have

$$\{\phi(x_1)\phi(x_2)...\phi(x_n)\}^{1/n} \leqslant \{\phi(x_1)+\phi(x_2)+...+\phi(x_n)\}/n.$$

Hence $$\frac{1}{n}\sum \log\phi(x_\nu) \leqslant \log\left\{\frac{1}{n}\sum \phi(x_\nu)\right\},$$

i.e.

$$\frac{1}{b-a}\sum(x_\nu-x_{\nu-1})\log\phi(x_\nu)\leqslant\log\left\{\frac{1}{b-a}\sum(x_\nu-x_{\nu-1})\phi(x_\nu)\right\}.$$

Making $n\to\infty$, the result follows.

The theorem may be deduced from Jensen's theorem by an elaboration of the argument of § 9.621, but it is more convenient to use the theorem of § 3.8. Applying it to the function $f(s)$ and the rectangle $(\alpha,\beta;t_0,T)$, we have, on taking real parts,

$$2\pi\int_\alpha^\beta N(\sigma,T)\,d\sigma=\int_{t_0}^T\log|f(\alpha+it)|\,dt-$$

$$-\int_{t_0}^T\log|f(\beta+it)|\,dt+\int_\alpha^\beta\arg f(\sigma+iT)\,d\sigma-\int_\alpha^\beta\arg f(\sigma+it_0)\,d\sigma. \quad (1)$$

Applying the lemma to the first term on the right of (1), we have

$$\frac{1}{T-t_0}\int_{t_0}^T\log|f(\alpha+it)|\,dt=\frac{1}{2}\frac{1}{T-t_0}\int_{t_0}^T\log|f(\alpha+it)|^2\,dt$$

$$\leqslant\tfrac{1}{2}\log\left\{\frac{1}{T-t_0}\int_{T_0}^T|f(\alpha+it)|^2\,dt\right\}<A$$

by hypothesis. Thus the term in question is less than AT.

Secondly, as in § 9.621, $\log|f(\beta+it)|$ is bounded if β is large enough. Hence, if β is suitably chosen, the second term on the right of (1) is $O(T)$.

To deal with the third term, suppose first that $f(s)$ is real for real s. We can take β so large that $\mathbf{R}\{f(s)\}$ does not vanish on $\sigma=\beta$. Then, as in § 3.56, $\arg f(s)$ is bounded on $\sigma=\beta$, and, on $t=T$, $\arg f(s)=O(q)$, where q is the number of times $\mathbf{R}\{f(s)\}$ vanishes on $t=T$, $\alpha\leqslant\sigma<\beta$. Now on $t=T$

$$\mathbf{R}\{f(s)\}=\tfrac{1}{2}\{f(\sigma+iT)+f(\sigma-iT)\}=g(\sigma),$$

say, and q is the number of zeros of $g(z)$ on the real z-axis such that $\alpha\leqslant z\leqslant\beta$. Since $g(z)=O(T^A)$, it follows from Jensen's theorem as in § 9.621 that $q=O(\log T)$. Hence the third term on the right of (1) is $O(\log T)$.

If $f(s)$ is not real on the real axis, we can consider instead the function

$$f_1(s) = \sum \frac{a_n}{n^s} \sum \frac{\bar{a}_n}{n^s} = f(s)\bar{f}(s)$$

and apply the same proof to this.

Finally, the last term on the right of (1) is a constant. Hence

$$\int_{\alpha}^{\beta} N(\sigma, T)\, d\sigma = O(T).$$

But $$\int_{\alpha}^{\beta} N(\sigma, T)\, d\sigma \geqslant \int_{\alpha}^{\alpha+\delta} N(\sigma, T)\, d\sigma \geqslant \delta N(\alpha+\delta, T),$$

and the result now follows.

9.7. Representation of functions by Dirichlet series. What sort of function can be represented by a Dirichlet series?

It would take us much too far to give anything like an adequate answer to this question, but we can give some indications. It is not difficult to see that a Dirichlet series can only represent functions of a very special kind.

If $f(s)$ is representable by a Dirichlet series, it must, in the first place, be regular and bounded in a certain half-plane (viz. $\sigma \geqslant \bar{\sigma}+\epsilon$). Further, it must have a mean-value

$$\lim_{T\to\infty} \frac{1}{2T} \int_{-T}^{T} |f(\sigma+it)|^2\, dt$$

for all sufficiently large values of σ, and the value of the limit must decrease steadily as σ increases.

Again, if $f(s) = \sum a_n n^{-s}$, and x is real, then for $\sigma > \bar{\sigma}$

$$\frac{1}{2T} \int_{-T}^{T} f(s)x^s\, dt = \frac{1}{2T} \int_{-T}^{T} \left(\sum_{n=1}^{\infty} \frac{a_n}{n^s}\right) x^s\, dt = \frac{x^\sigma}{2T} \sum_{n=1}^{\infty} \frac{a_n}{n^\sigma} \int_{-T}^{T} \left(\frac{x}{n}\right)^{it} dt$$

$$= a_x + x^\sigma \sum_{n\neq x} \frac{a_n}{n^\sigma} \frac{2\sin(T\log x/n)}{2T\log x/n},$$

the term a_x occurring if x is a positive integer only. The last series, being uniformly convergent in T, tends to zero as $T \to \infty$. Hence

$$\lim_{T\to\infty} \frac{1}{2T} \int_{-T}^{T} f(s)x^s\, dt = \begin{matrix} a_x & \text{(x a positive integer),} \\ 0 & \text{(otherwise).} \end{matrix}$$

This, therefore, is a necessary condition for $f(s)$ to have the form $\sum a_n n^{-s}$ (the formulae are due to Hadamard). It is, however, not sufficient. But it shows what special properties a function representable by a Dirichlet series must have.

If the Dirichlet series reduces to a single term, say $f(s) = ak^{-s}$, then $f(s)$ is periodic, with period $2\pi i/\log k$. The general Dirichlet series with period $2\pi i/\log k$ is $\sum_{n=0}^{\infty} b_n k^{-ns}$. If we insert other terms, the property of periodicity disappears; but $f(s)$ always retains a certain more general property, which resembles that of periodicity, and any such function is said to be 'almost periodic' It is in the study of almost periodic functions that answers to the question which we have raised are to be found. We have no space to go into this question further. But we may say roughly that, if an almost periodic function takes a certain value, it repeats this value, not *exactly*, but *approximately*, an infinity of times; and the points where it does this are distributed in much the same way as the periods $(a, 2a, 3a,...)$ of a periodic function.

The theory of almost periodic functions is due to H. Bohr (1), (2), (3).

MISCELLANEOUS EXAMPLES

1. Prove that, if $\phi(x) = \sum_{n=1}^{\infty} a_n e^{-nx}$, then

$$f(s) = \frac{1}{\Gamma(s)} \int_0^{\infty} x^{s-1}\phi(x)\, dx$$

(i) for $\sigma > 0$, $\sigma > \bar{\sigma}$, (ii) for $\sigma > 0$, $\sigma > \sigma_0$.

2. If $0 < \theta < 2\pi$, the function $f(s)$, defined for $\sigma > 0$ by the equation

$$f(s) = \sum_{n=1}^{\infty} \frac{e^{in\theta}}{n^s},$$

is an integral function.

[Use ex. 1 and proceed as in the case of $\zeta(s)$.]

3. The functions defined for $\sigma > 1$ by the series

$$\sum_{n=1}^{\infty} \frac{e^{ain^b}}{n^s} \quad (a > 0,\ 0 < b < 1), \qquad \sum_{n=1}^{\infty} \frac{e^{ai(\log n)^2}}{n^s} \quad (a > 0),$$

are both integral functions. [Hardy (7), (10).]

4. A function represented by a Dirichlet series cannot tend to a limit (in the half-plane of absolute convergence) as $t \to \infty$, unless it is a constant.

[If $f(s) = \sum_1^\infty a_n n^{-s}$, then for $\sigma > \bar{\sigma}$

$$\lim \frac{1}{2T}\int_{-T}^{T} f(s)\,dt = a_1, \qquad \lim \frac{1}{2T}\int_{-T}^{T} |f(s)|^2\,dt = \sum a_n^2.$$

Hence, if $f(s) \to a$, $\quad a_1 = a, \quad \sum |a_n|^2 = |a|^2.$

Hence $\quad |a_2|^2+|a_3|^2+\ldots = 0,$

i.e. $a_2 = 0$, $a_3 = 0,\ldots$. Hence $f(s) = a_1$.]

5. Show that

$$\frac{1}{\zeta(s)} = \sum_{n=1}^{\infty} \frac{\mu(n)}{n^s} \qquad (\sigma > 1),$$

where $\mu(1) = 1$, $\mu(n) = (-1)^r$ if n is the product of r different primes, and otherwise $\mu(n) = 0$. Show also that

$$\frac{\zeta(s)}{\zeta(2s)} = \sum_{n=1}^{\infty} \frac{|\mu(n)|}{n^s}.$$

[The infinite product for $\zeta(s)$ is given in § 1.44, ex. 1.]

6. Verify the formulae*

$$\{\zeta(s)\}^2 = \sum_{n=1}^{\infty} \frac{d(n)}{n^s}, \qquad \frac{\{\zeta(s)\}^3}{\zeta(2s)} = \sum_{n=1}^{\infty} \frac{d(n^2)}{n^s}, \qquad \frac{\{\zeta(s)\}^4}{\zeta(2s)} = \sum_{n=1}^{\infty} \frac{\{d(n)\}^2}{n^s},$$

where $d(n)$ denotes the number of divisors of n, and $\sigma > 1$.

[If the expression of n in prime factors is

$$n = p_1^{m_1} p_2^{m_2} \ldots p_r^{m_r},$$

then $\quad d(n) = (m_1+1)(m_2+1)\ldots(m_r+1).$

Hence

$$\sum \frac{d(n^2)}{n^s} = \prod_p \sum_{m=0}^{\infty} \frac{(2m+1)}{p^{ms}}$$

and

$$\sum \frac{\{d(n)\}^2}{n^s} = \prod_p \sum_{m=0}^{\infty} \frac{(m+1)^2}{p^{ms}}.]$$

7. Verify the formulae

$$\zeta(s)\zeta(s-a) = \sum_{n=1}^{\infty} \frac{\sigma_a(n)}{n^s} \qquad (\sigma > 1,\ \sigma > a+1),$$

* A number of other formulae of this kind are given by Pólya and Szegö, *Aufgaben*, VIII. Abschn., nos. 49–64.

and*

$$\frac{\zeta(s)\zeta(s-a)\zeta(s-b)\zeta(s-a-b)}{\zeta(2s-a-b)} = \sum_{n=1}^{\infty} \frac{\sigma_a(n)\sigma_b(n)}{n^s}$$

$$(\sigma > 1,\ \sigma > a+1,\ \sigma > b+1,\ \sigma > a+b+1),$$

where $\sigma_a(n)$ denotes the sum of the ath powers of the divisors of n.

[The second formula follows from the identity

$$\frac{1-p^{a+b-2s}}{(1-p^{-s})(1-p^{a-s})(1-p^{b-s})(1-p^{a+b-s})}$$

$$= \frac{1}{(1-p^a)(1-p^b)} \sum_{m=0}^{\infty} \frac{(1-p^{(m+1)a})(1-p^{(m+1)b})}{p^{ms}}.\Bigg]$$

8. Let $d_k(n)$, where $k = 2, 3,\ldots$, denote the number of ways of expressing n as a product of k factors, the order of the factors being taken into account. Then

$$\sum_{n=1}^{\infty} \frac{d_k(n)}{n^s} = \{\zeta(s)\}^k \qquad (\sigma > 1);$$

and

$$\sum_{n=1}^{\infty} \frac{\{d_k(n)\}^2}{n^s} = \{\zeta(s)\}^k \prod_p \left\{P_{k-1}\left(\frac{1+p^{-s}}{1-p^{-s}}\right)\right\} \qquad (\sigma > 1),$$

where p runs through all prime numbers, and $P_n(z)$ is the Legendre polynomial of degree n. [Titchmarsh (8).]

9. Show that, if $f(s)$ has the period $2\pi i/\log k$, Hadamard's formulae for the coefficients a_n (§ 9.7) are equivalent to Laurent's formulae for the coefficients in a power series.

10. A necessary and sufficient condition that a function $f(s)$ should be of the form

$$\sum_{n=0}^{\infty} \frac{b_n}{k^{ns}}$$

is that $f(s)$ should be regular and bounded for sufficiently large values of σ, and have the period $2\pi i/\log k$.

11. If $a_n = 0$ unless n is a power of k, then

$$\sigma_0 = \bar{\sigma} = \sigma_\epsilon = \sigma_m.$$

12. The function $f(s) = \sum_{m=0}^{\infty} 2^{-m!s}$ has the line $\sigma = 0$ as a natural boundary. [See § 4.71.]

13. The function $f(s) = \sum p^{-s}$, where p runs through all prime numbers, has the line $\sigma = 0$ as a natural boundary.

[This is a more recondite example than the previous one; see Landau and Walfisz (1).]

* Ramanujan (1). B. M. Wilson (1).

14. The function

$$f(s) = \sum_{n=1}^{\infty} \frac{\{d_k(n)\}^r}{n^s}$$

is meromorphic if $r = 1$, or if $r = 2$, $k = 2$; for other values of r and k it has the line $\sigma = 0$ as a natural boundary. [Estermann (1).]

15. Show that, for the function $\zeta(s)$,

$$\mu(\sigma) = 0 \ (\sigma \geqslant 1), \qquad = 1-2\sigma \ (\sigma \leqslant 0),$$

and that $\mu(\sigma) \leqslant 1-\sigma$ for $0 < \sigma < 1$.

[The result for $\sigma < 0$ follows from the functional equation for $\zeta(s)$. The actual value of $\mu(\sigma)$ for $0 < \sigma < 1$ is not known.]

16. Calculate the mean value

$$\lim_{T\to\infty} \frac{1}{2T} \int_{-T}^{T} |f(s)|^2 \, dt \qquad (\sigma > 1)$$

for the functions $f(s) = \zeta(s)$, $1/\zeta(s)$, $\{\zeta(s)\}^2$.

17. Show that, if $f(s)$ is unbounded on any line $\sigma = \alpha$ in the half-plane where it is of finite order, it is also unbounded on every line $\sigma = \beta < \alpha$ in the same half-plane.

18. Show that the function $f(s) = (1-2^{1-s})\zeta(s)$ is unbounded on every line $\sigma = \alpha \leqslant 1$; and that $t^{\sigma-\frac{1}{2}}f(s)$ is unbounded on every line $\sigma = \alpha$, where $0 < \alpha < \frac{1}{2}$.

[The theorem of § 9.32 shows that $\zeta(s)$, and so also $(1-2^{1-s})\zeta(s)$, is unbounded for $\sigma > 1$, $|t| > 1$. The theorem of § 9.41 then gives the first result, and the second result then follows from the functional equation for $\zeta(s)$, § 4.44, and the asymptotic formula for the Γ-function, § 4.42.]

19. If $s_n = \sum_{\nu=1}^{n} a_\nu$ is bounded, then $f(s) = \sum a_n n^{-s}$ is regular for $\sigma > 0$; and, if $f(s)$ has a pole on $\sigma = 0$, it is at most of the first order.

[If $\phi(u) = \sum_{\nu \leqslant u} a_\nu$, we have

$$f(s) = s\int_{1}^{\infty} \frac{\phi(u)}{u^{s+1}}\,du = O\left(s\int_{1}^{\infty} \frac{du}{u^{\sigma+1}}\right) = O\left(\frac{s}{\sigma}\right).]$$

20. If $s_n \sim n$, then $f(s) \sim 1/(s-1)$ as $s \to 1$ by real values greater than 1.

If $s_n \sim n \log^k n$, where k is a positive integer, then

$$f(s) \sim \frac{k!}{(s-1)^{k+1}}.$$

CHAPTER X

THE THEORY OF MEASURE AND THE LEBESGUE INTEGRAL

10.1. Riemann integration. In the theory of analytic functions we have used the familiar definition of an integral due to Riemann. In the theory of functions of a real variable, however, Riemann's definition has been almost entirely superseded by a more general one, due to Lebesgue.

Lebesgue's definition enables us to integrate functions for which Riemann's method fails; but this is only one of its advantages. The new theory gives us a command over the whole subject which was previously lacking. It deals, so to speak, automatically with many of the limiting processes which present difficulties in the Riemann theory. At this early stage it is difficult to say anything more precise.

Let us begin by recalling the definition of the Riemann integral of a bounded function. Suppose that $f(x)$ is bounded in the interval (a,b); we subdivide this interval by means of the points x_0, x_1,..., x_n, so that

$$a = x_0 < x_1 < ... < x_{n-1} < x_n = b.$$

Let m_ν, M_ν be the lower and upper bounds of $f(x)$ in the interval $x_\nu < x \leqslant x_{\nu+1}$, and let

$$s = \sum_{\nu=0}^{n-1} m_\nu(x_{\nu+1} - x_\nu), \qquad S = \sum_{\nu=0}^{n-1} M_\nu(x_{\nu+1} - x_\nu).$$

When the number of division-points is increased indefinitely so that the greatest interval $x_{\nu+1} - x_\nu$ tends to zero, each of the sums s and S tends to a limit. If the limits are the same, their common value is the Riemann integral

$$\int_a^b f(x)\,dx.$$

In certain cases, e.g. if $f(x)$ is continuous, we can say definitely that this integral exists.

Suppose in particular that $f(x)$ takes the values 0 and 1 only, say $f(x) = 1$ in a set E, and $f(x) = 0$ elsewhere. Then it is easily seen that s is equal to the sum of the lengths of those intervals throughout which $f(x) = 1$, i.e. intervals consisting entirely of points of E; while S is the sum of lengths of intervals which include any point of E. If the set E consists of a finite number of intervals, there is no difficulty in proving that s and S tend to the same limit, viz. the sum of the lengths of the intervals of E.

The Riemann integral of such a function ($f(x) = 1$ in E, 0 elsewhere) may be called the *extent* of the set E. *Extent* is thus a generalization of the *length* of an interval. The extent of E, if it exists, is written $e(E)$, so that

$$e(E) = \int_a^b f(x)\,dx.$$

Whether the extent exists or not, the limits of s and S exist. These limits are called the interior and exterior extents* of E, and are written $e_i(E)$, $e_e(E)$.

The function $f(x)$ is called the *characteristic function* of the set E.

It is easy to define a set which has no extent. Let E be the set of all rational values of x in (a, b). Since every interval contains both rational and irrational numbers, we have $m_\nu = 0$, $M_\nu = 1$, for all modes of division and all values of ν. Hence $s = 0$, $S = b-a$, and consequently

$$e_i(E) = 0, \qquad e_e(E) = b-a.$$

The extent of this set is therefore undefined, and the characteristic function $f(x)$ has no Riemann integral.

In the general case we may say that the definition of the extent of E depends on the consideration of certain sets of intervals related to E, the number of such intervals being always finite.

Lebesgue's generalization is in the first place a generalization of extent; and it consists fundamentally in removing the restriction that our sets of intervals must be finite. Before we can introduce it formally we must make some further remarks about sets of points.

10.2. Sets of points. For the fundamental ideas concerning sets of points we refer to Hardy's *Pure Mathematics*, Chapter I.

We usually denote sets of points by E, E_1,... and suppose

* The exterior extent is sometimes called the content.

them all to lie within a finite interval (a, b). We denote by CE the complement of E, i.e. the set of all points of the interval (a, b) which do not belong to E.

If E_1 and E_2 are two sets, we denote by E_1+E_2 the set of all points belonging to E_1 or E_2, and by E_1E_2 the set of all points belonging to both E_1 and E_2. The notation is suggested by the fact that, if $f_1(x)$, $f_2(x)$ are the characteristic functions of E_1 and E_2, then $f_1(x)f_2(x)$ is the characteristic function of E_1E_2; while, if E_1 and E_2 have no common points, $f_1(x)+f_2(x)$ is the characteristic function of E_1+E_2.

Note that $$C(E_1+E_2) = CE_1 . CE_2.$$

The notation extends in an obvious way to any finite number of sets; also, if there are an infinity of given sets $E_1, E_2,\ldots$, then $E_1+E_2+\ldots$ denotes the set of points belonging to any of the given sets, and $E_1E_2\ldots$ denotes the set of points belonging to each of the given sets.

By $E_1 < E_2$ we mean that every point of E_1 is a point of E_2. Two sets are said to 'overlap' if they have common points.

An infinite set of points is said to be *enumerable* if it is possible to define a one-to-one correspondence between the points of the set and the integers 1, 2, 3,...; that is, we must be able to arrange the points in a sequence $x_1, x_2, x_3,\ldots$ such that every point occupies a definite place in the sequence. For example, the set of numbers $1, \frac{1}{2}, \frac{1}{3}, \frac{1}{4},\ldots$ is enumerable; so is the set $\frac{1}{2}, \frac{1}{4}, \frac{1}{8},\ldots$.

The set of all proper rational fractions is enumerable; for we can arrange them as follows:

$$\tfrac{1}{2}, \tfrac{1}{3}, \tfrac{2}{3}, \tfrac{1}{4}, \tfrac{3}{4},\ldots$$

taking the denominators in order of magnitude, then the numerators.

The 'sum' of two enumerable sets is enumerable; for if E_1 consists of the points $x_1, x_2,\ldots$, and E_2 of $\xi_1, \xi_2,\ldots$, then all points of E_1+E_2 are given by the sequence

$$x_1, \xi_1, x_2, \xi_2,\ldots.$$

A similar argument applies to any finite number of enumerable sets. Further, *the sum of an enumerable infinity of enumerable sets is enumerable*; for let the sets be $E_1, E_2,\ldots$, and let E_n consist of the points

$$x_{1,n}, x_{2,n},\ldots, x_{m,n},\ldots.$$

We can arrange the double infinity of points $x_{m.n}$ as a single infinity in various ways, e.g. by taking together points for which $m+n = k$ $(k = 2, 3,...)$, and in each such group taking m increasing; thus

$$x_{1,1}, x_{1,2}, x_{2,1}, x_{1,3}, x_{2,2}, x_{3,1}, x_{1,4},....$$

This proves the theorem.

Finally *a sub-set of an enumerable set is enumerable.* For any sub-set of x_1, x_2, x_3,... clearly has a first member, a second member, a third member, and so on, and this gives the required enumeration.

10.201. The reader might begin to suspect that all sets were enumerable; but this is not the case. *The set of all numbers between* 0 *and* 1 *is not enumerable.*

To prove this, suppose on the contrary it were possible to arrange all such numbers in a sequence x_1, x_2,.... Suppose each such number expressed as an infinite decimal ('terminating' decimals end with an infinity of 0's; we exclude a recurring 9). We then form a new decimal ξ, such that, for every value of n, the nth term in the decimal for ξ exceeds by 1 the nth term in the decimal for x_n, if it is 0, 1,..., 7, and is 0 if it is 8 or 9. This rule defines ξ completely, and ξ does not end with a recurring 9. But ξ is a number between 0 and 1, and is different from any of the numbers x_n. This contradicts the assumption that the sequence x_n contains *all* the numbers between 0 and 1.

A similar argument applies to any interval. We call all the points of an interval a *continuum*. Our result is that *a continuum is not enumerable.*

10.202. A point ξ is called a 'limit-point' of a set E if, however small δ may be, there are points of E, other than ξ, in the interval $(\xi-\delta, \xi+\delta)$. (See *P.M.*, p. 30, where a limit-point is called a 'point of accumulation'.)

A set which contains all its limit-points is called a *closed* set. Thus an interval together with its end-points is a closed set. Such an interval is called a closed interval.

An *open* interval is an interval without its end-points. An *open* set is the complement of a closed set with respect to an open interval.

An open set consists of an enumerable set of non-overlapping open intervals. For let E be an open set, and let x be a point of E. Then, for sufficiently small values of δ, the interval $(x, x+\delta)$ consists entirely of points of E; for otherwise x would be a limit-point of CE, so that CE would not be closed. Let δ_1 be the upper bound of values of δ with this property. Then ξ belongs to E for $x \leqslant \xi < x+\delta_1$; but $x+\delta_1$ is not a point of E, since, if it were, the interval of points of E would extend beyond it, by the above argument.

Similarly there is a number δ_2 such that ξ is in E for $x-\delta_2 < \xi \leqslant x$, while $x-\delta_2$ is not in E.

Thus x is a point of an open interval $(x-\delta_2, x+\delta_1)$ of points of E.

Similarly all points of E fall into open intervals. To arrange these intervals as an enumerable sequence, take first the interval, if there is one, greater than $\frac{1}{2}(b-a)$; next, in the order in which they occur on the line, those whose length is $\leqslant \frac{1}{2}(b-a)$ and $> \frac{1}{3}(b-a)$; and so on. Every interval of E has a definite place in this enumeration.

The 'sum' of two open sets is an open set. For if $E = E_1+E_2$, and E_1 and E_2 are open, every point of E is an interior point of an interval of points of E.

The same argument shows that *the sum of any finite number, or of an enumerable infinity, of open sets is open.* In particular (the converse of the above theorem), *the sum of an infinity of open intervals is an open set.*

Also *if E_1 and E_2 are open sets, then E_1E_2 is open.* For a point of E_1E_2 is an interior point of intervals both of E_1 and of E_2; and so it is not a limit-point of $C(E_1E_2)$, which consists of points of either CE_1 or CE_2.

This argument cannot be extended to an infinity of sets; e.g. if E_n is the open interval $-1/n < x < 1/n$, then $E_1E_2...$ is the single point $x = 0$.

10.21. The measure of a set of points. We are now in a position to define a new generalization of 'length'. Instead of starting from a finite number of intervals, we start from an open set, which may contain an infinity of intervals.

The measure of an open set is defined to be the sum of the lengths of its intervals. This sum is, in general, the sum of an infinite series. It is always convergent, since the sum of any finite number of terms is the sum of the lengths of a finite number of non-overlapping intervals, all contained in an interval (a,b), and so is not greater than $b-a$. Hence the measure of any open set contained in (a,b) does not exceed $b-a$.

The exterior measure of a set E is the lower bound of the measures of all open sets which contain E. It is denoted by $m_e(E)$. It is clear that

$$0 \leqslant m_e(E) \leqslant b-a,$$

and that, if $E_1 < E_2$, then $m_e(E_1) \leqslant m_e(E_2)$.

The interior measure, $m_i(E)$, is defined by the formula

$$m_i(E) = b-a-m_e(CE).$$

If $m_i(E) = m_e(E)$, then the set E is said to be measurable, and the common value of $m_i(E)$ and $m_e(E)$ is called its measure, and is denoted by $m(E)$.

We have also

$$m_i(CE) = b-a-m_e(E).$$

If E is measurable, so that $m_i(E) = m_e(E)$, it follows that $m_i(CE) = m_e(CE)$. Hence CE is measurable, and

$$m(E)+m(CE) = b-a.$$

Notice that we have given two definitions of the measure of an open set, one direct and one indirect. It will appear before long that they are equivalent. Meanwhile, in arguments involving open sets, we use the direct definition.

10.22. *For any set E we have*

$$m_i(E) \leqslant m_e(E).$$

For, by the definition of exterior measure, there are open sets O and O', including E and CE respectively, and such that

$$m(O) < m_e(E)+\epsilon,$$

$$m(O') < m_e(CE)+\epsilon.$$

If $\epsilon > 0$, every point of the interval $(a+\epsilon, b-\epsilon)$ is an interior point of an interval of O or of O'; and so, by the Heine-Borel theorem,* we can select from these intervals a finite set, say Q, which together include $(a+\epsilon, b-\epsilon)$. Then plainly

$$m(Q) \geqslant b-a-2\epsilon$$

and

$$m(Q) \leqslant m(O)+m(O').$$

* *P.M.* § 105. In the proof there given we start with an interval ending at a, whereas here there is an interval including a. This does not affect the proof.

Combining these inequalities we have

$$b-a < m_e(E)+m_e(CE)+4\epsilon.$$

Making $\epsilon \to 0$, it follows that

$$b-a \leqslant m_e(E)+m_e(CE),$$

which is equivalent to the result stated.

If $m_e(E) = 0$, it follows that $m_i(E) = 0$. Hence E is measurable, and its measure is zero.

10.23. We now come to the two fundamental theorems in the theory of measure.

First fundamental theorem. *If E_1, E_2,..., E_n,... are measurable sets, then the set $E = E_1+E_2+E_3+...$ is measurable, and*

$$m(E) \leqslant m(E_1)+m(E_2)+....$$

If E_1, E_2,... do not overlap, then the equality holds. (*Otherwise the series may diverge.*)

Second fundamental theorem. *If E_1, E_2,... are measurable sets, then the set $E_1 E_2 E_3$... is measurable.*

That is, the set of points belonging to any of the sets E_1, E_2,... is measurable, and so is the set of points belonging to all of them.

We shall begin by proving two lemmas on open sets, the first of which is the first fundamental theorem for open sets. We next prove a general theorem on exterior measure, and deduce from it the first fundamental theorem for the case where the sets do not overlap. Then we obtain the second theorem for two sets, and use it to complete the first theorem. Finally we use this result to complete the second theorem.

10.24. *If O_1, O_2,... are open sets (overlapping or not), and*

$$O = O_1+O_2+O_3+...,$$

then

$$m(O) \leqslant m(O_1)+m(O_2)+.... \tag{1}$$

We assume the convergence of the series on the right, since otherwise the theorem is meaningless.

Let the intervals of O_n be $(a_{m,n}, b_{m,n})$ $(m = 1, 2,...)$, and let those of O be (A_k, B_k) $(k = 1, 2,...)$. Let ϵ be a positive number less than $\frac{1}{2}(B_k-A_k)$. Then every point of the interval $(A_k+\epsilon, B_k-\epsilon)$ is an interior point of one of the intervals $(a_{m,n}, b_{m,n})$ which make up (A_k, B_k). If $\sum_k$ denotes a summation over these intervals, it follows from the Heine-Borel theorem, as in the previous proof, that

$$B_k - A_k - 2\epsilon \leqslant \sum_k (b_{m,n} - a_{m,n}).$$

Making $\epsilon \to 0$, we obtain

$$B_k - A_k \leqslant \sum_k (b_{m,n} - a_{m,n}), \tag{2}$$

and, summing with respect to k,

$$m(O) \leqslant \sum_{k=1}^{\infty} \sum_k (b_{m,n} - a_{m,n}). \tag{3}$$

Since a convergent double series of positive terms can be summed in any manner, the right-hand side of (3) can be rearranged in the form

$$\sum_{n=1}^{\infty} \sum_{m=1}^{\infty} (b_{m,n} - a_{m,n}) = \sum_{n=1}^{\infty} m(O_n).$$

This proves the theorem.

If none of the sets overlap, each interval (A_k, B_k) coincides with one interval $(a_{m,n}, b_{m,n})$, and the inequalities (2) and (3), and so also (1), become equalities.

An enumerable set is measurable, and its measure is zero. For let the set be $x_1, x_2, \ldots$. Include x_1 in an open interval of length ϵ. If this does not include x_2, we can include x_2 in an interval of length $\frac{1}{2}\epsilon$; and so generally x_n in an interval of length $\epsilon/2^n$. Thus the given set can be included in an open set of measure not greater than 2ϵ. Since ϵ may be as small as we please, the exterior measure of the set is zero. Hence its measure is zero.

10.241. *If O and O' are open sets which together include all points of the interval (a, b), then*

$$m(OO') \leqslant m(O) + m(O') - (b-a).$$

By the Heine-Borel theorem we can select finite sets of the intervals of O and O', say Q from O and Q' from O', such that Q and Q' together include the whole interval $(a+\epsilon, b-\epsilon)$; and we may, by adding further intervals if necessary, suppose that

$$O = Q + R, \qquad O' = Q' + R',$$

where $m(R) < \epsilon$, $m(R') < \epsilon$. Now

$$OO' \prec QQ' + R + R',$$

so that by the previous lemma

$$m(OO') \leqslant m(QQ') + m(R) + m(R') < m(QQ') + 2\epsilon.$$

But $m(Q)+m(Q')-m(QQ') \geqslant b-a-2\epsilon$, from elementary considerations, and $m(O) \geqslant m(Q)$, $m(O') \geqslant m(Q')$. Making $\epsilon \to 0$, the result follows.*

10.25. *If E_1, E_2,... are any sets, and*

$$E = E_1+E_2+...,$$

then $$m_e(E) \leqslant m_e(E_1)+m_e(E_2)+....$$

We can enclose E_n in an open set O_n such that

$$m(O_n) < m_e(E_n)+\frac{\epsilon}{2^n}.$$

Summing with respect to n, and using the result of § 10.24,

$$m(O) \leqslant m(O_1)+m(O_2)+... < m_e(E_1)+m_e(E_2)+...+\epsilon.$$

But O is an open set which includes E. Hence

$$m_e(E) \leqslant m(O).$$

Hence $$m_e(E) < m_e(E_1)+m_e(E_2)+...+\epsilon,$$

and, making $\epsilon \to 0$, the result follows.

10.26. *If E_1, E_2,... are non-overlapping measurable sets, and*

$$E = E_1+E_2+...,$$

then E is measurable, and

$$m(E) = m(E_1)+m(E_2)+....$$

We may suppose that all the sets are included in (a, b).

(i) Consider first the case of two sets, $E = E_1+E_2$. We know already that

$$m_e(E) \leqslant m_e(E_1)+m_e(E_2) = m(E_1)+m(E_2).$$

Hence it is sufficient to prove that

$$m_i(E) \geqslant m(E_1)+m(E_2),$$

i.e. that $$m_e(CE) \leqslant m(CE_1)+m(CE_2)-(b-a).$$

Now we can include CE_1, CE_2, in open sets O_1, O_2, such that

$$m(O_1) < m(CE_1)+\epsilon, \qquad m(O_2) < m(CE_2)+\epsilon.$$

Since E_1 and E_2 have no common points, CE_1 and CE_2 together include the whole interval, and hence so do O_1 and O_2. Hence

$$m(O_1O_2) \leqslant m(O_1)+m(O_2)-(b-a).$$

* Actually the two sides are equal. This follows in due course from the first fundamental theorem.

But $O_1 O_2$ includes CE. Hence

$$m_e(CE) \leqslant m(O_1 O_2) \leqslant m(O_1)+m(O_2)-(b-a)$$
$$< m(CE_1)+m(CE_2)+2\epsilon-(b-a),$$

and, making $\epsilon \to 0$, the result follows.

(ii) The theorem for any finite number of sets follows by repeated application of (i).

(iii) In the case of an infinity of sets, we have, for all values of n,

$$m(E_1)+m(E_2)+...+m(E_n) = m(E_1+...+E_n) \leqslant b-a.$$

Hence $\sum m(E_n)$ is convergent.

Let $S_n = E_1+...+E_n$. Then $CE < CS_n$, so that

$$m_e(CE) \leqslant m_e(CS_n) = m(CS_n) = b-a-m(E_1)-...-m(E_n).$$

Making $n \to \infty$ we obtain

$$m_e(CE) \leqslant b-a-\sum m(E_n),$$

i.e.

$$m_i(E) \geqslant \sum m(E_n).$$

Combining this with § 10.25, the result follows.

In particular, taking E_1, E_2,... to be open intervals, it follows that any open set is measurable in the general sense, and that the two definitions of the measure of an open set agree. Also any closed set, as the complement of an open set, is measurable.

If E_1 and E_2 are measurable sets, E_1 being included in E_2, then E_2-E_1 is measurable.

For

$$C(E_2-E_1) = E_1+CE_2.$$

10.27. *If E and F are measurable sets, so is EF.*

Let both sets be included in (a, b), and suppose first that F is an interval (α, β). Let E_1 be the part of E in (α, β), E_2 the remainder. Similarly, if O is an open set containing E, let $O = O_1+O_2$. O_1 and O_2 are open sets containing respectively E_1 and E_2, if we neglect the points α and β, as we obviously may; and clearly

$$m(O) = m(O_1)+m(O_2).$$

Taking lower bounds,

$$m_e(E) = m_e(E_1)+m_e(E_2). \tag{1}$$

Similarly, if $e = CE = e_1+e_2$,

$$m_e(e) = m_e(e_1)+m_e(e_2). \tag{2}$$

But, since E is measurable,

$$m_e(E)+m_e(e) = b-a, \tag{3}$$

and by § 10.25

$$m_e(E_2)+m_e(e_2) \geqslant m_e(E_2+e_2) = b-a-(\beta-\alpha). \tag{4}$$

From (1), (2), (3) and (4) it follows that

$$m_e(E_1)+m_e(e_1) \leqslant \beta-\alpha,$$

and hence E_1 is measurable.

The result is therefore proved if F is an interval, and so, by the previous theorem, if F is an open set. In the general case we can include F in an open set O, and CF in O', so that $m(O)+m(O') < b-a+\epsilon$. Then

$$EF < EO, \qquad C(EF) = CF+F.CE < O'+O.CE,$$

so that

$$\begin{aligned} m_e(EF)+m_e\{C(EF)\} &\leqslant m(EO)+m(O')+m(O.CE) \\ &= m(O)+m(O') < b-a+\epsilon. \end{aligned}$$

Making $\epsilon \to 0$, $m_e(EF)+m_e\{C(EF)\} \leqslant b-a$, whence the result.

If E_1 and E_2 are measurable, the set E of points belonging to E_2 but not to E_1 is measurable.

For $E = E_2 . CE_1$.

10.28. We can now complete the proofs of the fundamental theorems. Let E_1, E_2,... be any sets, overlapping or not, and let E be their sum. Let

$$E_2' = E_2 . CE_1, \qquad E_3' = E_3 . C(E_1+E_2'),$$
$$E_4' = E_4 . C(E_1+E_2'+E_3'),$$

and so on. Then E_1, E_2', E_3',... are non-overlapping measurable sets, and $E = E_1+E_2'+E_3'+\dots$. Hence E is measurable, by § 10.26, and the proof of the first fundamental theorem is completed, the inequality then stated following from § 10.25.

Again, if $F = E_1E_2E_3\dots$, then

$$CF = CE_1+CE_2+\dots.$$

Hence, by what has just been proved, CF is measurable, and so F is measurable. This proves the second fundamental theorem.

10.29. Limiting sets. *If E_1, E_2,... are measurable sets, each contained in the following one, and E is their sum, then*

$$\lim_{n\to\infty} m(E_n) = m(E).$$

For the sets E_2-E_1, E_3-E_2,... are measurable and non-overlapping, and

$$E = E_1+(E_2-E_1)+(E_3-E_2)+...,$$

so that

$$m(E) = m(E_1)+m(E_2-E_1)+...$$
$$= \lim\{m(E_1)+m(E_2-E_1)+...+m(E_n-E_{n-1})\} = \lim m(E_n).$$

The set E is called the *outer limiting set* of the sets E_1, E_2,... .

If each of the sets E_1, E_2,... *contains the next, and* $E = E_1E_2...$, *then*

$$\lim_{n\to\infty} m(E_n) = m(E).$$

This follows by complementary sets from the previous theorem. In this case the set E is called the *inner limiting set.*

Unlike most of the theorems on the measure of sets, the first of these results holds if 'measure' is replaced by 'exterior measure', whether the sets are measurable or not. This remark will be useful in the next chapter, where it happens to be inconvenient to verify that certain sets are measurable.

If E is the outer limiting set of a sequence E_n, *then*

$$\lim_{n\to\infty} m_e(E_n) = m_e(E).$$

Let E_n be included in an open set O_n such that

$$m(O_n) < m_e(E_n)+\epsilon.$$

Let $S_n = O_nO_{n+1}O_{n+2}...$, and let $S = S_1+S_2+...$. Then $E_n < S_n < O_n$, $E < S$, and $S_n < S_{n+1}$, so that S is the outer limiting set of the sets S_n (this is not necessarily true for O_n, which is why we introduce S_n). Hence

$$m_e(E) \leqslant m(S) = \lim m(S_n) < \lim m_e(E_n)+\epsilon,$$

and, making $\epsilon \to 0$, $m_e(E) \leqslant \lim m_e(E_n)$. But since a set which includes E also includes E_n, $m_e(E) \geqslant m_e(E_n)$ for every n. This proves the theorem.

10.291. Cantor's ternary set. The following set of points, defined by Cantor, has many interesting properties.

Divide the interval (0, 1) into three equal parts, and remove the interior of the middle part. Next subdivide each of the two remaining parts into three equal parts, and remove the interiors of the middle parts of each of them; and repeat this process indefinitely. Thus at the pth step we remove 2^{p-1} intervals.

We denote these intervals, from left to right, by $\delta_{p,k}$, where k runs from 1 to 2^{p-1}. For each k the length of $\delta_{p,k}$ is 3^{-p}.

Let E be the set of points which remain. Then E is the set of points represented by the infinite decimals

$$\cdot a_1 a_2 \ldots a_n \ldots (3)$$

in the scale of 3 (indicated by the final figure), where the numbers $a_1, a_2, \ldots$ take the values 0 or 2 only, never the value 1; for example, E includes $\frac{2}{3} = \cdot 200\ldots$, and also $\frac{1}{3}$, which can be represented as $\cdot 0222\ldots$. In fact the first step described above removes from the interval all points for which the first figure is a 1 (except $\cdot 100\ldots = \cdot 022\ldots$); the second step removes all remaining points for which the second figure is a 1 (except $\cdot 010\ldots = \cdot 0022\ldots$, and $\cdot 210\ldots = \cdot 2022\ldots$); and so on. Notice also that the end-points of the intervals $\delta_{p,k}$ consist of all decimals $\cdot a_1 a_2 \ldots (3)$, where the digits after a certain point are all 0's or all 2's. This is obviously true for $\delta_{1,1}$; then $\delta_{2,1}$, $\delta_{2,2}$ are obtained by taking the first decimal as 0 or 2 and then the rest as the decimals corresponding to the ends of $\delta_{1,1}$; and so on. Thus the general form of the end-points of a $\delta_{p,k}$ is

$$\cdot a_1 \ldots a_m 0222\ldots (3), \qquad \cdot a_1 \ldots a_m 2000\ldots (3).$$

The set E is not enumerable; this may be proved in the same way that it was proved that the continuum was not enumerable. On the other hand, the measure of E is zero; for

$$m(E) = 1 - \sum m(\delta_{p,k}) = 1 - \sum_{p=1}^{\infty} \frac{2^{p-1}}{3^p} = 0.$$

We shall refer to this set again in § 11.72.

Example. Prove that the measure of the set of points in the interval (0, 1) representing numbers whose expressions as infinite decimals do not contain some particular digit (say 7) is zero.

10.3. Measurable functions. Let $f(x)$ be a bounded function of x in the interval $a \leqslant x \leqslant b$. We denote by $E(f > c)$ the set of points in (a, b) where $f(x) > c$; and similarly with other inequalities.

The function $f(x)$ is said to be measurable if any one of the sets

$$E(f \geqslant c), \qquad E(f < c), \qquad E(f > c), \qquad E(f \leqslant c)$$

is measurable for all values of c.

Any one of these four conditions implies the other three.

Suppose, for example, that the first holds. The second follows by complementary sets. Hence also the sets

$$E_n = E\left(f < c + \frac{1}{n}\right) \qquad (n = 1, 2, \ldots)$$

are all measurable. Hence the set

$$(E_1 - E_2) + (E_2 - E_3) + \ldots = E(c < f < c+1)$$

is measurable. Hence

$$E(f = c) = E(f \geqslant c) - E(f \geqslant c+1) - E(c < f < c+1)$$

is measurable, and the result clearly follows from this.

10.31. General properties of measurable functions.

(i) *Let f be a measurable function, k a constant. Then $k+f$, kf, and in particular $-f$, are measurable.*

This is obvious.

(ii) *If f and ϕ are measurable functions, the set $E(f > \phi)$ is measurable.*

If $f > \phi$, there is a rational number r such that $f > r > \phi$. Hence

$$E(f > \phi) = \sum_r E(f > r) \,.\, E(\phi < r)$$

where r runs through all rational numbers. Hence the result.

(iii) *If f and ϕ are measurable, so are $f+\phi$ and $f-\phi$.*

For

$$E(f+\phi > c) = E(f > c - \phi)$$

and the result follows from (ii). Similarly for $f-\phi$.

(iv) *If f and ϕ are measurable, so is $f\phi$.*

The function $\{f(x)\}^2$ is measurable, for, if $c > 0$,

$$E(f^2 > c) = E(f > \surd c) + E(f < -\surd c).$$

The general theorem then follows from the fact that

$$f\phi = \tfrac{1}{4}(f+\phi)^2 - \tfrac{1}{4}(f-\phi)^2.$$

(v) *If $f_n(x)$ is a sequence of measurable functions, then*

$$\varliminf_{n\to\infty} f_n(x), \qquad \varlimsup_{n\to\infty} f_n(x),$$

supposed finite, are measurable. In particular, if the sequence tends to a limit, the limit is measurable.

Let $f(x) = \varlimsup f_n(x)$. Let c be any real number, let

$$E_{m,n} = E\left(f_n > c + \frac{1}{m}\right) + E\left(f_{n+1} > c + \frac{1}{m}\right) + \ldots,$$

and let $E_m = E_{m,1} E_{m,2} E_{m,3} \ldots$. By the fundamental theorems

$E_{m,n}$ and E_m are measurable. Now E_m, the set of points common to all the sets $E_{m,n}$, is the set where $f_\nu > c+1/m$ for arbitrarily large values of ν. Hence

$$f = \lim f_\nu \geqslant c+\frac{1}{m} > c$$

in E_m. Let $E = E_1+E_2+E_3+\ldots$. Then E is measurable, and $f > c$ at all points of E. Conversely, if $f(x) > c$, then there is an integer m such that $f_\nu(x) > c+1/m$ for arbitrarily large values of ν, and so x belongs to one of the sets E_m. Hence $E = E(f > c)$, which proves the theorem.

(vi) *A continuous function is measurable.* For if $f(x)$ is continuous, it is easily seen that $E(f \leqslant c)$ is closed. Hence $E(f > c)$ is open, and so measurable.

All the ordinary functions of analysis may be obtained by limiting processes from continuous functions, and so are measurable. The same thing is true of some of the more artificial functions. For example,

$$\lim_{n\to\infty}\{\cos m!\pi x\}^{2n}$$

is the limit of a continuous function, and is equal to 1 if $m!x$ is an integer, and otherwise is zero. If x is rational, $m!x$ is an integer if m is large enough. Hence

$$f(x) = \lim_{m\to\infty}\lim_{n\to\infty}\{\cos m!\pi x\}^{2n}$$

is equal to 1 if x is rational, and to 0 otherwise. The fact that this function is measurable has, of course, been proved more directly (§ 10.22).

10.4. The Lebesgue integral of a bounded function. We are now in a position to define the Lebesgue integral of any bounded measurable function.

If $f(x)$ is the characteristic function of a set E, i.e. $f(x) = 1$ in E and 0 elsewhere, a natural definition of the integral is

$$\int_a^b f(x)\,dx = m(E).$$

If $f(x) = k$ in E and 0 elsewhere, then we take

$$\int_a^b f(x)\,dx = km(E).$$

In the general case, let α and β be the lower and upper bounds of $f(x)$. As in the case of Riemann integration, the integral is defined as the limit of the sum; but this time the sum is obtained by dividing up the interval of variation of $f(x)$. We take numbers $y_0, y_1,..., y_{n+1}$ such that

$$\alpha = y_0 < y_1 < y_2 < ... < y_{n-1} < y_n = \beta.$$

Let e_ν be the set where $y_\nu \leqslant f(x) < y_{\nu+1}$ $(\nu = 0,..., n-1)$, and e_n the set where $f(x) = \beta$. Since $f(x)$ is measurable, all the sets e_ν are measurable. Putting $y_{n+1} = \beta$, let

$$s = \sum_{\nu=0}^{n} y_\nu m(e_\nu), \qquad S = \sum_{\nu=0}^{n} y_{\nu+1} m(e_\nu).$$

The Lebesgue integral of $f(x)$ over (a, b) is the common limit of the sums s and S when the number of division-points y_ν is increased indefinitely, so that the greatest value of $y_{\nu+1} - y_\nu$ tends to zero.

To justify the definition we have to prove that the two limits exist and are equal.

Suppose the interval (α, β) divided up in two different ways, each difference $y_{\nu+1} - y_\nu$ in each way being less than ϵ. Let the sums formed in these two ways be s, S and s', S'. Then

$$S - s = \sum_{\nu=0}^{n} (y_{\nu+1} - y_\nu) m(e_\nu) \leqslant \epsilon \sum_{\nu=0}^{n} m(e_\nu) = \epsilon(b-a),$$

and similarly $S' - s' \leqslant \epsilon(b-a)$.

We now divide up the interval (α, β) by taking all the division-points of the first two ways at once. This gives two more sums, s'' and S''. Now the insertion of a new division-point does not decrease a lower sum or increase an upper sum; for example, if we insert a point η between y_ν and $y_{\nu+1}$ we have

$$y_\nu m(e_\nu) \leqslant y_\nu m\{E(y_\nu \leqslant f < \eta)\} + \eta m\{E(\eta \leqslant f < y_{\nu+1})\},$$

so that the lower sum is not decreased. Applying this principle repeatedly, we obtain

$$s \leqslant s'', \qquad s' \leqslant s'',$$

and similarly $\qquad S'' \leqslant S, \qquad S'' \leqslant S'.$

It follows that the intervals (s, S) and (s', S') have points in common, e.g. all points of the interval (s'', S''). Hence the numbers s, s', S, S' all lie within an interval of length $2\epsilon(b-a)$. The existence and equality of the limits then follow from the general principle of convergence.

10.41. Comparison with Riemann's definition. Perhaps the most obvious difference to the beginner is that, in Lebesgue's definition, we divide up the interval of variation of the function instead of the interval of integration. This, however, is comparatively unimportant. What is essential is that we use the general theory of 'measure' of sets instead of the more limited theory of 'extent'. It would be possible to build up an integral from integrals of characteristic functions, but using extent instead of measure. This would be substantially equivalent to Riemann's definition. On the other hand, it is possible to define an integral equivalent to Lebesgue's by dividing up the interval of integration in a suitable way.

In both Riemann's and Lebesgue's definitions we have upper and lower sums which tend to limits. In the Riemann case the two limits are not necessarily the same, and the function is only integrable if they are the same. In the Lebesgue case the two limits are necessarily the same, their equality being a consequence of the assumption that the function is measurable.

Lebesgue's definition is more general than Riemann's. For the characteristic function of the set of rational points has a Lebesgue integral, but not a Riemann integral; and we shall see later that, if a function has a Riemann integral, then it also has a Lebesgue integral, and the two are equal.

We use the same notation

$$\int_a^b f(x)\,dx$$

for a Lebesgue integral as we have done for a Riemann integral. When it is necessary to distinguish a Riemann integral from a Lebesgue integral, we shall denote the former by

$$R\int_a^b f(x)\,dx.$$

10.42. Integral over any measurable set. Let E be any measurable set contained in an interval (a, b). The integral of $f(x)$ over E may be defined in the same way as the integral over an interval. The sets e_ν of § 10.4 are now the sub-sets of E where $y_\nu \leqslant f(x) < y_{\nu+1}$; the proof of the existence of the integral is practically unchanged. The integral is written

$$\int_E f(x)\,dx.$$

Any integral over a set of measure zero is zero. For all the sets e_ν are of measure zero, and so the sums s and S are always 0.

We might also define the integral by putting $f(x)=0$ in CE, and then using the definition of the integral over an interval. It is easily seen that the two definitions are equivalent.

10.43. Henceforward we shall assume that all sets and functions introduced are measurable, without always saying so explicitly.

10.44. Elementary properties of the integral of a bounded function.

(i) **The mean-value theorem.** *If* $\alpha \leqslant f(x) \leqslant \beta$, *then*

$$\alpha m(E) \leqslant \int_E f(x)\,dx \leqslant \beta m(E).$$

For it is easily seen that $\alpha m(E) \leqslant s \leqslant \beta m(E)$, and the result follows in the limit.

(ii) *The integral is additive for a finite number or for an enumerable infinity of non-overlapping sets included in a finite interval. That is, if*

$$E = E_1 + E_2 + \ldots,$$

then
$$\int_E f(x)\,dx = \int_{E_1} f(x)\,dx + \int_{E_2} f(x)\,dx + \ldots.$$

Suppose first that there are two sets, E_1 and E_2. Inserting division-points y_ν, the sets E, E_1, E_2 are divided into sub-sets e_ν, e_ν^1, e_ν^2, such that

$$m(e_\nu) = m(e_\nu^1) + m(e_\nu^2).$$

Hence
$$\int_{E_1} + \int_{E_2} = \lim \sum y_\nu m(e_\nu^1) + \lim \sum y_\nu m(e_\nu^2)$$
$$= \lim \sum y_\nu m(e_\nu) = \int_E.$$

Similarly for any finite number of sets.

If there are an infinity of sets, let S_n be the sum of the first n, R_n the remainder. Then

$$\int_E = \int_{S_n} + \int_{R_n}.$$

But, by the mean-value theorem, if $|f(x)| \leqslant M$, then

$$\left|\int_{R_n} f(x)\,dx\right| \leqslant Mm(R_n),$$

and this tends to zero as $n \to \infty$, since the series $\sum m(E_n)$ is convergent. Hence

$$\int_E = \lim \int_{S_n} = \int_{E_1} + \int_{E_2} + \dots .$$

(iii) *If, in a set* E, $f(x) \leqslant \phi(x)$, *then*

$$\int_E f(x)\,dx \leqslant \int_E \phi(x)\,dx.$$

Take division-points y_ν, and define the sets e_ν by means of $f(x)$. Then, in e_ν, $\phi(x) \geqslant f(x) \geqslant y_\nu$. Hence

$$\int_E \phi(x)\,dx = \sum \int_{e_\nu} \phi(x)\,dx \geqslant \sum y_\nu m(e_\nu).$$

The right-hand side tends to $\int_E f(x)\,dx$, whence the result follows.

(iv) *The integral of the sum of a finite number of bounded measurable functions is the sum of the integrals of the separate functions.*

In the first place, if k is a constant,

$$\int_E (f+k)\,dx = \int_E f\,dx + \int_E k\,dx = \int_E f\,dx + km(E).$$

For calculate the sum s relative to $f(x)$ with the scale $y_0, y_1, \dots$, and the sum s' relative to $f(x)+k$ with the scale $y_0+k, y_1+k, \dots$. Then

$$s' = \sum (y_\nu + k)m(e_\nu) = s + km(E),$$

and the result follows in the limit.

Now consider any two functions $f(x)$ and $\phi(x)$. We have

$$\begin{aligned}\int_E \{f(x)+\phi(x)\}\,dx &= \sum \int_{e_\nu} (f+\phi)\,dx \\ &\geqslant \sum \int_{e_\nu} (y_\nu + \phi)\,dx \\ &= s + \int_E \phi\,dx\end{aligned}$$

by what has just been proved. Similarly, replacing y_ν by $y_{\nu+1}$, we obtain

$$\int_E (f+\phi)\,dx \leqslant S + \int_E \phi\,dx.$$

The result now follows in the limit.

The result for any finite number of functions is obtained by repeated application of the result for two functions.

(v) *If k is a constant,*

$$\int_E k\,f(x)\,dx = k\int_E f(x)\,dx.$$

This is obvious if $k = 0$. If $k > 0$, calculate the second integral with the scale y_ν, and the first with the scale ky_ν. Then the sets e_ν are the same in each case, and $s = ks'$, whence the result.

(vi) *We have*

$$\left|\int_E f(x)\,dx\right| \leqslant \int_E |f(x)|\,dx.$$

Let E_1 be the set where $f(x) \geqslant 0$, E_2 the set where $f(x) < 0$.

Then
$$\int_E f\,dx = \int_{E_1} f\,dx - \int_{E_2} |f|\,dx,$$

$$\int_E |f|\,dx = \int_{E_1} f\,dx + \int_{E_2} |f|\,dx,$$

and the result is obvious.

(vii) A relation which holds except in a set of measure zero is said to hold *almost everywhere.*

Two functions which are equal almost everywhere have the same integral.

Let $f(x) = \phi(x)$ at all points of E, except in a set e of measure zero. Then

$$\int_E (f-\phi)\,dx = \int_e (f-\phi)\,dx + \int_{E.Ce} (f-\phi)\,dx.$$

The first term is zero because $m(e) = 0$, and the second because the integrand is everywhere zero. Hence

$$\int_E f\,dx = \int_E \phi\,dx.$$

(viii) *If* $f(x) \geqslant 0$ *and* $\int_E f(x)\,dx = 0$, *then* $f(x) = 0$ *almost everywhere in* E.

Let $E_0 = E(f = 0)$, and

$$E_n = E\big(M/(n+1) < f \leqslant M/n\big), \qquad n = 1, 2, \ldots,$$

where M is the upper bound of f. Then $E = E_0 + E_1 + E_2 + \ldots$; and

$$m(E_n) \leqslant \frac{n+1}{M} \int_{E_n} f\,dx \leqslant \frac{n+1}{M} \int_E f\,dx = 0.$$

Thus $m(E_n) = 0$ for $n = 1, 2, \ldots$, and the result follows.

10.5. Lebesgue's convergence theorem (theorem of bounded convergence). *Let $f_n(x)$ be a sequence of measurable functions such that $|f_n(x)| \leqslant M$ for all values of n, when x is in a set E, and let*

$$\lim_{n\to\infty} f_n(x) = f(x)$$

for all values of x in E. Then

$$\lim_{n\to\infty}\int_E f_n(x)\,dx=\int_E f(x)\,dx.$$

Since sets of measure zero can be omitted from the integrals, it is sufficient that the conditions should hold *almost everywhere.*

Since $|f_n(x)|\leqslant M$ for each n, $|f(x)|\leqslant M$. Hence $f(x)$ is integrable, and we have to prove that

$$\lim\int_E\{f(x)-f_n(x)\}\,dx=0.$$

Let $g_n=|f-f_n|$, let ϵ be any positive number, and let

$$E_1=E(\epsilon>g_1,g_2,\ldots),\qquad E_2=E(g_1\geqslant\epsilon>g_2,g_3,\ldots),$$
$$E_3=E(g_2\geqslant\epsilon>g_3,g_4,\ldots),$$

and so on. Then the sets E_k are measurable; they are non-overlapping, since $g_k\geqslant\epsilon$ in E_{k+1}, but not in $E_1,\ldots,E_k$, so that E_{k+1} has no point in common with $E_1,\ldots,E_k$; and every point of E belongs to some E_k; for $g_n(x)\to 0$ for every x, so that to every x corresponds a first number k such that g_k, $g_{k+1},\ldots$ are all less than ϵ, and then x belongs to E_k.

It follows that

$$\int_E g_n\,dx=\int_{E_1}g_n\,dx+\int_{E_2}g_n\,dx+\ldots.$$

Now $g_n<\epsilon$ in $E_1,\ldots,E_n$, and $g_n\leqslant 2M$ everywhere. Hence

$$\int_E g_n\,dx\leqslant\epsilon\{m(E_1)+\ldots+m(E_n)\}+2M\{m(E_{n+1})+\ldots\}.$$

Making $n\to\infty$, it follows that

$$\overline{\lim}\int_E g_n\,dx\leqslant\epsilon m(E).$$

Hence, making $\epsilon\to 0$, it follows that

$$\lim\int_E g_n\,dx=0,$$

and the theorem follows.

The theorem is not true for Riemann integrals, because the function $f(x)$ is not necessarily integrable in Riemann's sense, even if each $f_n(x)$ is. For example, let r_1, $r_2,\ldots$ be the rational points in $(0,1)$, and let $f_n(x)=1$ if $x=r_1,r_2,\ldots$ or r_n, and $f_n(x)=0$ elsewhere. Then

$$R\int_0^1 f_n(x)\,dx=0$$

for every n; but $f(x) = 1$ for every rational x, and $f(x) = 0$ for irrational x, so that $f(x)$ is not integrable in Riemann's sense.

10.51. The theorem of bounded convergence may be stated as a theorem on term-by-term integration of series. *If the series*

$$u_1(x)+u_2(x)+\ldots$$

converges in a set E to $s(x)$, and its partial sums

$$s_n(x) = u_1(x)+\ldots+u_n(x)$$

are bounded for all values of n, when x is in E, then

$$\int_E s(x)\,dx = \int_E u_1(x)\,dx + \int_E u_2(x)\,dx + \ldots.$$

This is the final form of the theorem of bounded convergence proved for Riemann integrals in § 1.76.

10.52. Egoroff's theorem.* *If a sequence of functions converges to a finite limit almost everywhere in a set E, then, given δ, we can find a set of measure greater than $m(E)-\delta$ in which the sequence converges uniformly.*

Let $f_n(x)$ be the sequence, let E' be the set where $f_n(x)$ converges, say to $f(x)$, and let $g_n = |f-f_n|$.

Let $\epsilon_1, \ldots, \epsilon_r, \ldots$ be a sequence of positive numbers tending to zero. Let $S_{n,r}$ be the sub-set of E' where $g_\nu < \epsilon_r$ for $\nu \geqslant n$. Then each of the sets $S_{1,r}, S_{2,r}, \ldots$ is contained in the next, and their outer limiting set (§ 10.29) is E', since $g_\nu \to 0$ everywhere in E'. Hence we can determine $n(r)$ so that

$$m(E' - S_{n(r),r}) < \frac{\delta}{2^r}.$$

Let $$S = S_{n(1),1} S_{n(2),2} \ldots S_{n(r),r} \ldots.$$

Then, in S, $g_n < \epsilon_r$ $(n \geqslant n(r))$ for all values of r, i.e. $g_n \to 0$ uniformly in S; and

$$m(E-S) = m(E'-S) \leqslant \sum_{r=1}^{\infty} m(E' - S_{n(r),r}) < \sum_{r=1}^{\infty} \frac{\delta}{2^r} = \delta.$$

This proves the theorem.

Example. Use Egoroff's theorem to prove Lebesgue's convergence theorem.

10.6. *If $f(x)$ has a Riemann integral over (a, b), then it has a Lebesgue integral over the same interval, and the two are equal.*

The result is easily proved if we assume that $f(x)$ is measurable;

* Egoroff (1).

for then it certainly has a Lebesgue integral. Dividing up the interval (a, b) by the points $x_0, x_1, \ldots, x_n$, and denoting by m_ν, M_ν the lower and upper bounds of $f(x)$ in $x_\nu < x \leqslant x_{\nu+1}$, we have

$$\sum_{\nu=0}^{n-1} m_\nu(x_{\nu+1}-x_\nu) \leqslant \sum_{\nu=0}^{n-1} \int_{x_\nu}^{x_{\nu+1}} f(x)\,dx \leqslant \sum_{\nu=0}^{n-1} M_\nu(x_{\nu+1}-x_\nu).$$

The middle term is the Lebesgue integral, while each of the extreme terms tends to the Riemann integral. Hence they are equal.

To prove that $f(x)$ is necessarily measurable if it has a Riemann integral, let

$$\phi(x) = m_\nu \quad (x_\nu < x \leqslant x_{\nu+1}), \qquad \Phi(x) = M_\nu \quad (x_\nu < x \leqslant x_{\nu+1}).$$

Then

$$\sum_{\nu=0}^{n-1} m_\nu(x_{\nu+1}-x_\nu) = \int_a^b \phi(x)\,dx, \qquad \sum_{\nu=0}^{n-1} M_\nu(x_{\nu+1}-x_\nu) = \int_a^b \Phi(x)\,dx.$$

Consider now an enumerable infinity of modes of division of the interval (a, b) such that $\max(x_{\nu+1}-x_\nu) \to 0$; and let each set of division-points contain the previous set. Let E be the set of all the division-points. E is enumerable and so of measure zero, and so may be neglected in integration. At any point x not in E, $\phi(x)$ does not decrease, and $\Phi(x)$ does not increase, as we insert division points. Hence $\phi(x) \to m(x)$, $\Phi(x) \to M(x)$, where $m(x)$ and $M(x)$ are the 'lower and upper bounds of $f(x)$ at x', i.e. the limits of the lower and upper bounds in indefinitely small intervals containing x. Also $\phi(x)$ and $\Phi(x)$ are measurable, and hence so are $m(x)$ and $M(x)$; and, by Lebesgue's convergence theorem,

$$\lim \int_a^b \phi(x)\,dx = \int_a^b m(x)\,dx, \qquad \lim \int_a^b \Phi(x)\,dx = \int_a^b M(x)\,dx.$$

But if $f(x)$ has a Riemann integral, each of these limits is equal to it. Hence

$$\int_a^b \{M(x)-m(x)\}\,dx = 0.$$

Since $M(x) \geqslant m(x)$ it follows by § 10.44 (viii) that $M(x) = m(x)$ almost everywhere; and since $M(x) \geqslant f(x) \geqslant m(x)$ it follows that $f(x) = m(x)$ almost everywhere. Hence $f(x)$ is measurable.

10.7. The Lebesgue integral of an unbounded function. Let $f(x)$ be an unbounded measurable function, and suppose first that $f(x) \geqslant 0$. Let $\{f(x)\}_n$, or simply $(f)_n$, denote $f(x)$ at points where $f(x) \leqslant n$, but n where $f(x) > n$. Then $\{f(x)\}_n$ is bounded and measurable, and so integrable. We define the integral of $f(x)$ over the set E to be the limit, if it exists, of the integral of $\{f(x)\}_n$,

$$\int_E f(x)\,dx = \lim_{n\to\infty} \int_E \{f(x)\}_n\,dx.$$

For a positive function $f(x)$ to be integrable over E, it is clearly necessary and sufficient that

$$\int_E \{f(x)\}_n\,dx$$

should be bounded.

The integral of a negative function may be defined in a similar way. In the general case, let $f(x) \geqslant 0$ in E_1, $f(x) < 0$ in E_2. Then we define the integral of $f(x)$ by the equation

$$\int_E f(x)\,dx = \int_{E_1} f(x)\,dx + \int_{E_2} f(x)\,dx.$$

A function which is integrable in this sense is 'absolutely integrable', i.e. $|f(x)|$ is also integrable. In fact it is clear that

$$\int_E |f(x)|\,dx = \int_{E_1} f(x)\,dx - \int_{E_2} f(x)\,dx.$$

It would of course be possible to define integrals which are not absolutely convergent; but we shall see that integrals of the above kind preserve all the characteristic properties of integrals of bounded functions, whereas this would not be true of non-absolutely convergent integrals.

We shall henceforth use the word 'integrable' to describe any function, bounded or unbounded, which has an integral in the above sense.

The use of the expression 'infinity', introduced in § 5.701, is also very convenient here. For example, if

$$\int_E \{f(x)\}_n\,dx$$

tends to infinity with n, we write

$$\int_E f(x)\,dx = \infty.$$

Examples. (i) Show that $\int_0^1 x^{-a}\,dx$ exists as a Lebesgue integral, and is equal to $1/(1-a)$, if $0 < a < 1$; but is infinite if $a \geqslant 1$.

[The Lebesgue definition of the integral is

$$\lim_{n\to\infty}\left\{\int_0^{n^{-1/a}} n\,dx + \int_{n^{-1/a}}^1 x^{-a}\,dx\right\},$$

and the results are the same as in the elementary theory.]

(ii) More generally, let $f(x)$ be positive, and bounded in $(\epsilon, 1)$ for every positive ϵ. Then

$$\int_0^1 f(x)\,dx = \lim_{\epsilon\to 0}\int_\epsilon^1 f(x)\,dx$$

in the sense that both sides are finite and equal, or both infinite.

(iii) The function

$$f(x) = \frac{d}{dx}\left(x^2\sin\frac{1}{x^2}\right) = 2x\sin\frac{1}{x^2} - \frac{2}{x}\cos\frac{1}{x^2}$$

is not integrable in Lebesgue's sense over (0, 1).

[The function is continuous over $(\epsilon, 1)$, and $\lim_{\epsilon\to 0}\int_\epsilon^1 f(x)\,dx$ exists. But

$$\int_0^1 |f(x)|\,dx = \infty;$$

for

$$|f(x)| \geqslant \frac{2}{x}\left|\cos\frac{1}{x^2}\right| - 2x \geqslant \frac{1}{x} - 2x$$

in each of the intervals $\{(2n+\frac{1}{3})\pi\}^{-\frac{1}{2}} \leqslant x \leqslant \{(2n-\frac{1}{3})\pi\}^{-\frac{1}{2}}$, and it is easily seen from this that

$$\int_0^1 \{|f(x)|\}_n\,dx > A\log n.]$$

(iv) Let $f(x)$ be any measurable function in E, and let e_n be the sub-set of E where $n-1 \leqslant f(x) < n$. Then the necessary and sufficient condition that $f(x)$ should be integrable over E is that $\sum_{n=-\infty}^{\infty} |n|m(e_n)$ should be convergent.

(v) We might define the integral of a positive unbounded function $f(x)$ by taking $\{f(x)\}^n = f(x)$ if $f(x) \leqslant n$, and otherwise $\{f(x)\}^n = 0$, and substituting $\{f(x)\}^n$ for $\{f(x)\}_n$ in Lebesgue's definition. Show that this definition is equivalent to that of Lebesgue.

(vi) If $|f(x)| \leqslant \phi(x)$, and $\phi(x)$ is integrable over E, then $f(x)$ is integrable over E.

(vii) If $f(x)$ is integrable over E, and E_n is the part of E where $|f(x)| \geqslant n$, then $m(E_n) = o(1/n)$.

(viii) If $f(x) = 0$ at every point of Cantor's ternary set, and $f(x) = p$ in each of the complementary intervals of length 3^{-p}, then

$$\int_0^1 f(x)\,dx$$

exists in Lebesgue's sense and is equal to 3.

10.71. Elementary properties of integrals. *The integral is additive, i.e. if* $E_1, E_2, \ldots$ *are non-overlapping sets, and* $E = E_1 + E_2 + \ldots$, *then*

$$\int_E f\,dx = \int_{E_1} f\,dx + \int_{E_2} f\,dx + \ldots.$$

We may suppose without loss of generality that $f \geqslant 0$; for if the result is true for positive functions it is true similarly for negative functions, and so by addition in the general case. This remark simplifies many of our proofs.

We define $(f)_n$ as before. The integral of $(f)_n$ is additive, so that

$$\int_E (f)_n\,dx = \sum_{k=1}^{\infty} \int_{E_k} (f)_n\,dx \leqslant \sum_{k=1}^{\infty} \int_{E_k} f\,dx.$$

Now make $n \to \infty$. If there are only a finite number of sets, the result follows (from the equality). If there are an infinity of sets we obtain (from the inequality)

$$\int_E f\,dx \leqslant \sum \int_{E_k} f\,dx.$$

But for any value of K

$$\int_E (f)_n\,dx \geqslant \sum_{k=1}^{K} \int_{E_k} (f)_n\,dx.$$

Making $n \to \infty$ first, and then $K \to \infty$, we obtain

$$\int_E f\,dx \geqslant \sum \int_{E_k} f\,dx.$$

Hence the result. (Notice the analogy with the proof given in § 1.62 that a double series of positive terms may be summed by rows or by columns to the same sum.)

10.72. *The sum of a finite number of integrable functions is integrable, and the integral of the sum is the sum of the integrals of the separate functions.*

It is sufficient to consider two functions, say $f(x)$ and $g(x)$. Suppose first that they are both positive, and let $\phi = f + g$. Then

$$(\phi)_n \leqslant (f)_n + (g)_n \leqslant (\phi)_{2n}.$$

Hence

$$\int_E (\phi)_n\,dx \leqslant \int_E (f)_n\,dx + \int_E (g)_n\,dx \leqslant \int_E (\phi)_{2n}\,dx,$$

and making $n \to \infty$

$$\int_E \phi\,dx \leqslant \int_E f\,dx + \int_E g\,dx \leqslant \int_E \phi\,dx.$$

which gives the required result.

If $f \geqslant 0$, $g < 0$, consider the set where $\phi \geqslant 0$. Here

$$f = \phi + (-g),$$

and the result follows from the previous case. Similarly where $\phi < 0$ we consider $-g = f + (-\phi)$.

Having proved the result for the sum and difference of positive functions, the general result now follows.

10.73. The following results can easily be deduced from the corresponding results for bounded functions:

(i) *If k is a constant,*

$$\int_E kf\,dx = k\int_E f\,dx.$$

(ii)
$$\left|\int_E f\,dx\right| \leqslant \int_E |f|\,dx.$$

(iii) *Two functions which are equal almost everywhere have the same integral.*

(iv) *If $f(x) \geqslant 0$, $\int_E f(x) = 0$, then $f(x) = 0$ almost everywhere in E.*

(v) *If $f(x)$ is integrable over E, and $E_1, E_2,\ldots$ is a sequence of sets contained in E such that $m(E_k) \to 0$, then $\int_{E_k} f(x)\,dx \to 0$, and indeed uniformly for all such sequences of sets.*

For, supposing, as we may, that $f(x) \geqslant 0$, choose n so that

$$\int_E [f(x) - \{f(x)\}_n]\,dx < \epsilon.$$

Having fixed n, we have

$$\int_{E_k} \{f(x)\}_n\,dx \leqslant nm(E_k) < \epsilon \qquad (k > k_0).$$

Hence

$$\begin{aligned}\int_{E_k} f(x)\,dx &= \int_{E_k} \{f(x)\}_n\,dx + \int_{E_k} [f(x) - \{f(x)\}_n]\,dx \\ &\leqslant \int_{E_k} \{f(x)\}_n\,dx + \int_E [f(x) - \{f(x)\}_n]\,dx \\ &< 2\epsilon \qquad (k > k_0),\end{aligned}$$

and the result follows.

Example. Let $f(x)$ be integrable, and $\phi(x)$ integrable in Riemann's sense, over (a, b). Dividing up the interval (a, b) by points x_ν as in § 10.1, prove that, as $\max(x_{\nu+1}-x_\nu) \to 0$,

$$\lim \sum_{\nu=0}^{n-1} \phi(x_\nu) \int_{x_\nu}^{x_{\nu+1}} f(x)\, dx = \int_a^b \phi(x) f(x)\, dx.$$

[Titchmarsh (1).]

10.8. The general convergence theorem of Lebesgue.

If $f_n(x)$ is a sequence of functions such that $|f_n(x)| \leqslant F(x)$, where $F(x)$ is integrable over E, for all values of n and all values of x in E, and

$$\lim_{n\to\infty} f_n(x) = f(x)$$

for all values of x in E, then

$$\lim_{n\to\infty} \int_E f_n(x)\, dx = \int_E f(x)\, dx.$$

As usual, it is sufficient that the conditions should hold almost everywhere. The proof is almost the same as that of the theorem of bounded convergence. We define the sets E_n as before; by § 10.71 the series

$$\sum \int_{E_n} F(x)\, dx$$

is convergent, and we have

$$\int_E g_n\, dx \leqslant \epsilon\{m(E_1)+...+m(E_n)\}+$$
$$+2\int_{E_{n+1}} F(x)\, dx + 2\int_{E_{n+2}} F(x)\, dx + \dots.$$

Making $n \to \infty$ it follows that

$$\lim \int_E g_n\, dx \leqslant \epsilon m(E),$$

and the result now follows as before.

The above theorem enables us to prove a new theorem on term-by-term integration of series. *We may multiply a boundedly convergent series by any integrable function, and integrate term by term.* For if $s_n(x)$ is the nth partial sum of the series, and $|s_n(x)| \leqslant M$, and $\phi(x)$ is the integrable function, we have

$$|\phi(x)s_n(x)| \leqslant M|\phi(x)|,$$

which is integrable, and may be taken as the $F(x)$ of the above proof.

10.81. The following theorem is often useful. Its original form is due to Fatou.*

If $f_n(x) \geqslant 0$ for all values of n, and x in E, and $f_n(x) \to f(x)$ as $n \to \infty$, then

$$\int_E f(x)\,dx \leqslant \varliminf_{n\to\infty} \int_E f_n(x)\,dx.$$

The statement implies that, if the right-hand side is finite, then $f(x)$ is finite almost everywhere and integrable; while, if $f(x)$ is not integrable, or is infinite in a set of positive measure, then

$$\lim_{n\to\infty} \int_E f_n(x)\,dx = \infty.$$

It is easily seen that, with the usual notation,

$$\lim_{n\to\infty} \{f_n(x)\}_k = \{f(x)\}_k.$$

Hence, by the theorem of bounded convergence,

$$\lim_{n\to\infty} \int_E \{f_n(x)\}_k\,dx = \int_E \{f(x)\}_k\,dx.$$

But
$$\int_E \{f_n(x)\}_k\,dx \leqslant \int_E f_n(x)\,dx,$$

and hence
$$\varliminf_{n\to\infty} \int_E f_n(x)\,dx \geqslant \int_E \{f(x)\}_k\,dx.$$

Making $k \to \infty$, the result follows at once if $f(x)$ is finite almost everywhere, the set where $f(x)$ is infinite being omitted from the integral. If $f(x) = \infty$ in a set e of positive measure, then

$$\int_E \{f(x)\}_k\,dx \geqslant km(e)$$

for all values of k, and the result follows.

10.82. A convergence theorem for monotonic sequences. *Let $f_1(x), f_2(x), \ldots$ be a sequence of positive integrable functions, non-decreasing for every value of x in E. Let $f(x)$ be the limit, finite or infinite, of the sequence. Then*

$$\lim_{n\to\infty} \int_E f_n(x)\,dx = \int_E f(x)\,dx$$

in the following sense:

(i) *if the left-hand side is finite, then $f(x)$ is finite almost everywhere and integrable, and the equality holds;*

* Fatou (1), p. 375.

(ii) *if the right-hand side is finite, so is the left-hand side, and equality holds*;

(iii) *if the left-hand side is infinite, then $f(x)$ is not integrable or is infinite in a set of positive measure*;

(iv) *the converse of* (iii) *holds.*

If the left-hand side is finite, so is the right-hand side, by Fatou's theorem; and equality in cases (i) and (ii) follows from Lebesgue's convergence theorem, since $f_n(x) \leqslant f(x)$. Then (iii) follows from (ii) and (iv) from (i).

10.83. We can now put the theorem of § 1.77 on integration of series into a more satisfactory form.

If $u_n(x) \geqslant 0$ for all values of n and x, then

$$\int_a^b \{\sum u_n(x)\}\,dx = \sum \int_a^b u_n(x)\,dx,$$

provided that either side is convergent.

For the partial sum $s_n(x) = u_1(x)+...+u_n(x)$ is positive, and non-decreasing for every value of x.

In particular, the convergence of the right-hand side implies the convergence of $\sum u_n(x)$ for almost all values of x.

We have still to consider the case where the range of integration is infinite; but as we have not yet discussed infinite Lebesgue integrals of this kind, we must postpone the complete result until the end of the next section.

10.9. Integrals over an infinite range. Let $f(x)$ be a function which is integrable over the interval (a, b), for all finite values of b. Let $f_1(x) = f(x)$ where $f(x) \geqslant 0$, and $f_1(x) = 0$ elsewhere; and let $f_2(x) = -f(x)$ where $f(x) < 0$, and $f_2(x) = 0$ elsewhere. Then

$$\int_a^b f(x)\,dx = \int_a^b f_1(x)\,dx - \int_a^b f_2(x)\,dx.$$

Each integral on the right is a non-decreasing function of b, and so tends to a finite limit or to positive infinity as $b \to \infty$. We write

$$\int_a^\infty f_1(x)\,dx = \lim_{b\to\infty} \int_a^b f_1(x)\,dx, \qquad \int_a^\infty f_2(x)\,dx = \lim_{b\to\infty} \int_a^b f_2(x)\,dx,$$

if both the limits are finite; and we then define the integral of $f(x)$ over (a, ∞) by the equation

$$\int_a^\infty f(x)\,dx = \int_a^\infty f_1(x)\,dx - \int_a^\infty f_2(x)\,dx.$$

It is clear from the definition that a convergent integral of this kind is absolutely convergent; for

$$\int_a^\infty |f(x)|\,dx = \int_a^\infty f_1(x)\,dx + \int_a^\infty f_2(x)\,dx.$$

Thus
$$\int_0^\infty \frac{\sin x}{x}\,dx$$

is not, in the strict sense, a Lebesgue integral, because it is not absolutely convergent.

Naturally many of the properties of finite integrals can be extended to infinite integrals. It is usually quite easy to see when this can be done, and we leave the details to the reader.

The theorem of § 10.83 has an immediate extension: *if* $u_n(x) \geqslant 0$, *then*

$$\int_a^\infty \{\sum u_n(x)\}\,dx = \sum \int_a^\infty u_n(x)\,dx,$$

provided that either side is convergent.

The convergence of either side implies the convergence of the corresponding expression in which the upper limit is replaced by a finite b. Hence, by § 10.83, the equation with upper limit b on both sides holds; and the required result now follows as in § 1.77.

It may be well to remark finally that the examples given in § 1.75 and § 1.78, where

$$\sum \int \neq \int \sum,$$

are just as cogent with the Lebesgue as with the Riemann integral. The same sort of restrictions still have to be made, though the theorem as a whole takes a simpler form.

CHAPTER XI

DIFFERENTIATION AND INTEGRATION

11.1. Introduction. The 'fundamental theorem of the integral calculus' is that differentiation and integration are inverse processes. This general principle may be interpreted in two different ways. If $f(x)$ is integrable, the function

$$F(x) = \int_a^x f(t)\,dt \tag{1}$$

is called the indefinite integral of $f(x)$; and the principle asserts that

$$F'(x) = f(x). \tag{2}$$

On the other hand, if $F(x)$ is a given function, and $f(x)$ is defined by (2), the principle asserts that

$$\int_a^x f(t)\,dt = F(x) - F(a). \tag{3}$$

The main object of this chapter is to consider in what sense these theorems are true.

As in elementary theory, (2) follows from (1) for every value of x for which $f(x)$ is continuous. For we can choose h_0 so small that $|f(t)-f(x)| < \epsilon$ for $|t-x| \leqslant h_0$; and then

$$\left|\frac{F(x+h)-F(x)}{h} - f(x)\right| = \left|\frac{1}{h}\int_x^{x+h} \{f(t)-f(x)\}\,dt\right| \leqslant \epsilon \quad (|h| < h_0),$$

by the mean-value theorem. This proves (2).

However, in the Lebesgue theory we consider functions which are in general discontinuous, so that the above argument does not apply to them. Actually the interesting question is, not whether (2) holds at particular points, but whether it is true in general; and to this we can give a satisfactory answer.

If $f(x)$ is any integrable function, its indefinite integral $F(x)$ has almost everywhere a finite differential coefficient equal to $f(x)$.

The problem of deducing (3) from (2) is much more difficult. We require in the first place that $F'(x)$ should exist at any rate almost everywhere, and, as we shall see in § 11.22, this is not necessarily so. Secondly, if $F'(x)$ exists we require that it should be integrable. If we were relying on the Riemann theory, we should find a fundamental difficulty here; for Volterra has shown by an example* that $F'(x)$ may exist everywhere and be bounded, and yet not be integrable in Riemann's sense. In the Lebesgue theory, a differential coefficient is measurable, and so integrable if it is bounded. But, if it is unbounded, it is not necessarily integrable in the Lebesgue sense. The problem has received a satisfactory answer, but it requires a more general process, known as totalization, or Denjoy integration, which we have not space to consider here. The result is that if $F'(x)$ is finite everywhere, then (3) follows from (2) if the integral is taken in the Denjoy sense.

11.2. Differentiation throughout an interval. The ordinary functions of analysis are differentiable in general, i.e. for most values of the variable, though there may be special points at which they are not differentiable. The exceptional points are usually isolated. This seems to have created the impression at one time that a continuous function necessarily has a differential coefficient in general. It was, however, shown by Weierstrass that this is quite untrue. *There is a continuous function which has no differential coefficient anywhere.*

Nevertheless, the idea that an 'ordinary function' has a differential coefficient in general is correct, if we attach this vague expression to a different class of functions. We shall see that it is true in the sense that *a monotonic function has a finite differential coefficient almost everywhere.*

We shall first consider non-differentiable functions, and then proceed to the constructive side of the theory.

11.21. Continuous non-differentiable functions. There are many simple examples of continuous functions which are not differentiable at particular points; for example, if $f(x) = |x|$, the ratio

$$\frac{f(h)-f(0)}{h}$$

tends to different limits, 1 and -1, as $h \to 0$ by positive or negative values; and if $f(x) = x \sin 1/x$ $(x \neq 0)$, $f(0) = 0$, the ratio does not tend to any definite limit.

We can next, by a method known as the condensation of

* Hobson, vol. i, p. 461.

singularities, construct continuous functions which are not differentiable in a set which is everywhere dense, for example in the set of rational points. Let r_1, r_2,... denote the rational numbers between 0 and 1, and let

$$F(x)=\sum_{n=1}^{\infty} a_n f(x-r_n),$$

where $f(x)$ has an assigned singularity at $x=0$, and the coefficients a_n tend to zero sufficiently rapidly. Then $F(x)$ will have the assigned singularity at every rational point. For example,

$$F(x)=\sum_{n=1}^{\infty}\frac{|x-r_n|}{3^n}$$

is continuous, since the series is uniformly convergent; but it is not differentiable at any rational point; for

$$\frac{F(r_k+h)-F(r_k)}{h}=\sum_{n=1}^{k-1}\frac{|r_k+h-r_n|-|r_k-r_n|}{h.3^n}+\frac{|h|}{h.3^k}+$$

$$+\sum_{k+1}^{\infty}\frac{|r_k+h-r_n|-|r_k-r_n|}{h.3^n};$$

and as $h\to 0$ the first term tends to a limit, the second term tends to $\pm 1/3^k$ according as $h>0$ or $h<0$, and, if $|h|<1$, the third term does not exceed

$$\sum_{k+1}^{\infty}\frac{1}{3^n}=\frac{1}{2.3^k}$$

in absolute value. Hence $F'(r_k)$ does not exist.

To obtain functions which are *everywhere* non-differentiable we have to use quite different methods. The first example of such a function was given by Weierstrass.

11.22. Weierstrass's non-differentiable function. This function is defined by the series

$$f(x)=\sum_{n=0}^{\infty} b^n\cos(a^n\pi x),$$

where $0<b<1$, and a is an odd positive integer. The series is uniformly convergent in any interval, so that $f(x)$ is everywhere continuous. On the other hand, if $ab>1$, the series obtained by term-by-term differentiation is divergent. This in itself does not prove that $f(x)$ is not differentiable, but it suggests possibilities in this direction. We shall prove that *if* $ab > 1+\frac{3}{2}\pi$, *the function has no finite differential coefficient for any value of* x.

We have

$$\frac{f(x+h)-f(x)}{h} = \sum_{n=0}^{\infty} b^n \frac{\cos\{a^n\pi(x+h)\}-\cos(a^n\pi x)}{h}$$

$$= \sum_{n=0}^{m-1} + \sum_{m}^{\infty} = S_m + R_m,$$

say. Now

$$|\cos\{a^n\pi(x+h)\}-\cos(a^n\pi x)| = |a^n\pi h \sin\{a^n\pi(x+\theta h)\}| \leqslant a^n\pi|h|,$$

so that

$$|S_m| \leqslant \sum_{n=0}^{m-1} \pi a^n b^n = \pi\frac{a^m b^m - 1}{ab-1} < \pi\frac{a^m b^m}{ab-1}.$$

We next obtain a lower limit for R_m, giving h a particular value. We can write

$$a^m x = \alpha_m + \xi_m,$$

where α_m is an integer, and $-\frac{1}{2} \leqslant \xi_m < \frac{1}{2}$. Let

$$h = \frac{1-\xi_m}{a^m}.$$

Then

$$0 < h \leqslant \frac{3}{2a^m}.$$

Also $\quad a^n\pi(x+h) = a^{n-m}\,.\,a^m\pi(x+h) = a^{n-m}\pi(\alpha_m+1).$

Since a is odd, it follows that

$$\cos\{a^n\pi(x+h)\} = (-1)^{a^{n-m}(\alpha_m+1)} = (-1)^{\alpha_m+1}.$$

Again

$$\cos(a^n\pi x) = \cos\{a^{n-m}\pi(\alpha_m+\xi_m)\} = \cos(a^{n-m}\pi\alpha_m)\cos(a^{n-m}\pi\xi_m)$$
$$= (-1)^{\alpha_m}\cos(a^{n-m}\pi\xi_m).$$

Hence

$$R_m = \frac{(-1)^{\alpha_m+1}}{h}\sum_{n=m}^{\infty} b^n\{1+\cos(a^{n-m}\pi\xi_m)\}.$$

All the terms of this series are positive, and hence, taking the first term only,

$$|R_m| > \frac{b^m}{|h|} > \frac{2}{3}a^m b^m.$$

Hence

$$\left|\frac{f(x+h)-f(x)}{h}\right| \geqslant |R_m|-|S_m| > \left(\frac{2}{3}-\frac{\pi}{ab-1}\right)a^m b^m.$$

If $ab > 1+\frac{3}{2}\pi$, the factor in brackets is positive; and when $m \to \infty$, $h \to 0$, and the expression on the right tends to infinity. Hence $\{f(x+h)-f(x)\}/h$ takes arbitrarily large values, so that $f'(x)$ does not exist or is not finite.

The graph of the function may be said to consist of an infinity of infinitesimal crinkles; but it is almost impossible to form any definite picture of it which does not obscure its essential feature.*

11.23. The following example of a continuous non-differentiable function is due to van der Waerden.† The function is similar to Weierstrass's, but the result is obtained in quite a different way.

Let $f_n(x)$ denote the distance between x and the nearest number of the form $m/10^n$, where m is an integer. Then the function

$$f(x) = \sum_{n=1}^{\infty} f_n(x)$$

is a continuous non-differentiable function.

Each $f_n(x)$ is continuous; and $|f_n(x)| < 10^{-n}$, so that the series is uniformly convergent. Hence $f(x)$ is continuous.

Let x be any number in the interval $(0, 1)$, and suppose it expressed as a decimal. If the qth figure is 4 or 9, let $x' = x-10^{-q}$; otherwise let $x' = x+10^{-q}$. Then if $n < q$, the nearest number $m/10^n$ is the same for x and x', and x and x' lie on the same side of it; while if $n \geqslant q$, the numbers $m/10^n$ and $m'/10^n$ corresponding to x and x' differ by $x-x'$. These rules may be verified by considering simple examples, such as $q = 2$, $x = \cdot 326$, $\cdot 346$, or $\cdot 396$.

It follows that

$$\begin{aligned} f_n(x')-f_n(x) &= \pm(x'-x) \qquad (n < q) \\ &= 0 \qquad (n \geqslant q). \end{aligned}$$

Hence $$f(x')-f(x) = \sum_{n=1}^{q-1} \pm(x'-x) = p(x'-x),$$

* For further properties of this function see Hardy (7), where the same result is obtained for $ab > 1$. A general method of constructing continuous non-differentiable functions is given by Knopp (2).

† Van der Waerden (1).

where p is an integer, and is odd or even with $q-1$. Hence $\{f(x')-f(x)\}/(x'-x)$ cannot tend to a finite limit as $x' \to x$.

11.3. The four derivates of a function. Whether the differential coefficient

$$f'(x) = \lim_{h\to 0} \frac{f(x+h)-f(x)}{h}$$

exists or not, the four expressions

$$\overline{\lim_{h\to +0}} \frac{f(x+h)-f(x)}{h}, \qquad \underline{\lim_{h\to +0}} \frac{f(x+h)-f(x)}{h},$$

$$\overline{\lim_{h\to -0}} \frac{f(x+h)-f(x)}{h}, \qquad \underline{\lim_{h\to -0}} \frac{f(x+h)-f(x)}{h}$$

always have a meaning, being either finite, or positive or negative infinity. They are called the upper and lower derivates on the right, and the upper and lower derivates on the left, respectively. We shall denote them by

$$D^+f(x), \qquad D_+f(x), \qquad D^-f(x), \qquad D_-f(x)$$

respectively, the sign referring to that of h in the above ratio, and its position corresponding to the 'lower' or 'upper' limit. If $D^+f = D_+f$, the function is said to have a right-hand derivative, if $D^-f = D_-f$, a left-hand derivative. The necessary and sufficient condition for the existence of the ordinary differential coefficient is that all the derivates should be equal.

We denote the left-hand and right-hand derivatives, when they exist, by $f'_-(x)$ and $f'_+(x)$.

Examples. (i) The function $\sqrt{x^2}$, where the positive value of the square root is always taken, has different left-hand and right-hand derivatives at $x = 0$.

(ii) Let $f(x) = x \sin 1/x$ $(x \neq 0)$, 0 $(x = 0)$. Then at $x = 0$

$$D_+f = -1, \qquad D^+f = 1, \qquad D_-f = -1, \qquad D^-f = 1.$$

(iii) Let
$$f(x) = \begin{cases} ax\sin^2 1/x + bx\cos^2 1/x & (x > 0) \\ 0 & (x = 0) \\ a'x\sin^2 1/x + b'x\cos^2 1/x & (x < 0), \end{cases}$$

where $a < b$, $a' < b'$. Then at $x = 0$

$$D_+f = a, \qquad D^+f = b, \qquad D_-f = a', \qquad D^-f = b'.$$

(iv) If $f(x)$ is continuous in (a, b), and one of its derivates is non-negative in the interval, then $f(a) \leqslant f(b)$.

[Let $D^+f \geqslant 0$, for example. Suppose that $f(b)-f(a) < -\epsilon(b-a)$, and let $\phi(x) = f(x)-f(a)+\epsilon(x-a)$. Then $\phi(b) < 0$. Also $\phi(x) > 0$ for some sufficiently small values of $x-a$, since $D^+f(a) \geqslant 0$. Hence $\phi(x) = 0$ for some values of x between a and b. Let ξ be the greatest such value. Then $D^+\phi(\xi) \leqslant 0$, $D^+f(\xi)+\epsilon \leqslant 0$, contrary to hypothesis. Hence $f(b)-f(a) \geqslant -\epsilon(b-a)$ for every positive ϵ, and the result follows.]

(v) The derivates and incrementary ratios of a continuous function have the same bounds in any interval; i.e. if any one of the derivates satisfies $\alpha \leqslant Df \leqslant \beta$, then $\alpha \leqslant \{f(x_2)-f(x_1)\}/(x_2-x_1) \leqslant \beta$, and conversely.

[Consider $\phi(x) = f(x)-\alpha x$, and use the previous example.]

(vi) If one of the derivates of a continuous function $f(x)$ is continuous at a certain point, then $f(x)$ has a differential coefficient at the point.

11.4. Functions of bounded variation. We say that $f(x)$ is of bounded variation in (a,b) if, in this interval, it can be expressed in the form $\phi(x)-\psi(x)$, where ϕ and ψ are non-decreasing bounded functions.

It is easily seen that the sum, difference, or product of two functions of bounded variation is also of bounded variation.

An alternative definition is obtained by assuming that, if the interval (a,b) is divided up by points $a = x_0 < x_1 < \ldots < x_n = b$, then

$$\sum_{\nu=0}^{n-1} |f(x_{\nu+1})-f(x_\nu)|$$

is less than a constant independent of the mode of division. The upper bound of these sums is called the total variation.

It is easily seen that, if the first condition holds, then so does the second. For

$$|f(x_{\nu+1})-f(x_\nu)| \leqslant \phi(x_{\nu+1})-\phi(x_\nu)+\psi(x_{\nu+1})-\psi(x_\nu),$$

so that

$$\sum_{\nu=0}^{n-1} |f(x_{\nu+1})-f(x_\nu)| \leqslant \phi(b)-\phi(a)+\psi(b)-\psi(a).$$

To prove the converse, let p be the sum of those differences $f(x_{\nu+1})-f(x_\nu)$ which are positive, $-n$ the sum of those which are negative. Then, if v is the sum $\sum |f(x_{\nu+1})-f(x_\nu)|$, we have

$$v = p+n, \qquad f(b)-f(a) = p-n,$$

and so $$v = 2p+f(a)-f(b), \qquad v = 2n+f(b)-f(a).$$

Hence, if v is bounded for all modes of division, so are p and n. Let V, P, and N be the upper bounds of v, p, and n. Then

$$V = 2P+f(a)-f(b), \qquad V = 2N+f(b)-f(a).$$

Let $V(x)$, $P(x)$, and $N(x)$ be the corresponding numbers for the interval (a, x). They are obviously bounded non-decreasing functions of x; and

$$V(x) = 2P(x) + f(a) - f(x), \qquad V(x) = 2N(x) + f(x) - f(a),$$

so that
$$f(x) = f(a) + P(x) - N(x).$$

This is the required expression for $f(x)$.

The functions $V(x)$, $P(x)$, and $N(x)$ are called the total variation and the positive and negative variations of $f(x)$ in (a, x).

If $f(x)$ is continuous and of bounded variation, its variation $V(x)$ is continuous. We can find a mode of division of the interval (a, x), with a point of division x' as near x as we please, such that

$$v > V(x) - \epsilon$$

and also
$$|f(x) - f(x')| < \epsilon.$$

Let
$$v' = v - |f(x) - f(x')|.$$

Then v' is a sum corresponding to the interval (a, x'), and so

$$V(x') \geqslant v' > V(x) - 2\epsilon.$$

Since $V(x')$ is non-decreasing, it follows that $V(x') \to V(x)$ as $x' \to x$ from below. Similarly $V(x') \to V(x)$ as $x' \to x$ from above. Hence $V(x)$ is continuous.

A continuous function of bounded variation is the difference between two continuous non-decreasing functions. For if $f(x)$ is continuous, so are $P(x)$ and $N(x)$.

11.41. The differential coefficient of a function of bounded variation. The object of the next three sections is to prove that *a function of bounded variation has a finite differential coefficient almost everywhere.*

Our proof depends on the following lemmas, due to Sierpinski.* They are of the same type as the Heine-Borel theorem, but apply to sets which need not even be measurable.

LEMMA 1. *Suppose that each point x of a set E in (a, b) is the left-hand end-point of one or more intervals $(x, x+h_x)$ of a family H. Then there is a finite non-overlapping set S of intervals of H which includes a sub-set E' of E such that $m_e(E') > m_e(E) - \epsilon$.*

Let E_n be the set of points of E which are associated with some $h_x > 1/n$. Then E is the outer limiting set of the sets E_n, we have $\lim m_e(E_n) = m_e(E)$ (§ 10.29), and we can take n so large that $m_e(E_n) > m_e(E) - \frac{1}{2}\epsilon$.

* Sierpinski (1). A similar lemma is given by W. H. and G. C. Young (1).

Let a_1 be the lower bound of E_n, b_1 its upper bound, and let $l = b_1 - a_1$. Let $\eta = \frac{1}{2}\epsilon/(nl+1)$. Then there is a point x_1 of E_n such that $a_1 \leqslant x_1 < a_1 + \eta$. Let (x_1, x_1+h_1) be an associated interval for which $h_1 > 1/n$.

If there are points of E_n to the right of x_1+h_1, let a_2 be their lower bound. Then there is a point x_2 of E_n in $(a_2, a_2+\eta)$. Let (x_2, x_2+h_2) be an associated interval with $h_2 > 1/n$.

Continuing the process, we reach b_1 in a finite number of steps, since each step takes us at least $1/n$ nearer to it. In fact, if there are N steps, then $(N-1)/n < l$, i.e. $N < nl+1$.

Let S denote the set of intervals $(x_\nu, x_\nu+h_\nu)$ so constructed, and T the set of intervals $(x_\nu-\eta, x_\nu)$. Then $E_n < S+T$, and $m(T) < N\eta < \frac{1}{2}\epsilon$. Hence

$$m_e(E) - \tfrac{1}{2}\epsilon < m_e(E_n) \leqslant m_e(E_n S) + m_e(E_n T) < m_e(E_n S) + \tfrac{1}{2}\epsilon,$$

and the set $E' = E_n S$ has the required property.

LEMMA 2. *Suppose in addition that for every x there are arbitrarily small intervals $(x, x+h_x)$. Then we may conclude in addition that*

$$m(S) < m_e(E) + \epsilon.$$

The additional condition is necessary; we might, for example, take E to be a single point x, and associate with it the interval $(x, x+1)$. Then Lemma 1 would hold, but not Lemma 2.

Let O be an open set containing E, such that

$$m(O) < m_e(E) + \epsilon.$$

Let H_1 be the sub-class of the family of intervals H consisting of those intervals that lie in O. In view of the additional condition imposed in Lemma 2, every point of E is the left-hand end-point of one or more intervals of H_1. We can now apply Lemma 1 with H replaced by H_1. We obtain a new set of intervals S which has the same property as that constructed in the proof of Lemma 1. But now S is a set of non-overlapping intervals included in O. Hence

$$m(S) \leqslant m(O) < m_e(E) + \epsilon.$$

This proves the lemma.

In these lemmas the intervals of which S consists may be regarded as either open or closed, whichever is most convenient in any particular case. For if the result has been obtained with S consisting of closed intervals, we can replace them by open intervals by removing a finite number of points, i.e. a set of measure zero. This clearly does not affect the result.

LEMMA 3. *We may suppose S in the above construction to be included in any given set of intervals G which contains E.*

For we may replace O by OG in the construction.

11.42. *If* $f(x)$ is non-decreasing in (a, b), it has almost everywhere in (a, b) a differential coefficient $f'(x)$.*

Let E be the set where $D_+ f < D^+ f$. We shall first prove that $m_e(E) = 0$.

Now E is the sum of the sets $E(u, v)$ where

$$D_+ f < u < v < D^+ f,$$

u and v running through all rational numbers $(u < v)$. Hence it is sufficient to prove that $m_e\{E(u, v)\} = 0$ for every pair of such numbers.

Suppose on the contrary that one of these sets $E(u, v)$ has a positive exterior measure, say μ. Every point x of it is the left-hand end-point of arbitrarily small intervals $(x, x+h)$ for which

$$f(x+h) - f(x) < hu.$$

Hence by Lemma 2 there is a finite set S of such intervals, containing a part E' of $E(u, v)$ such that $m_e(E') > \mu - \epsilon$, and such that $\sum_1 h < \mu + \epsilon$, where $\sum_1$ denotes a summation over S. Hence

$$\sum_1 \{f(x+h) - f(x)\} < u \sum_1 h < u(\mu + \epsilon).$$

Again, every point of E' is the left-hand end-point of intervals $(x, x+k)$ such that

$$f(x+k) - f(x) > kv,$$

and by Lemma 3 there is a finite set of these intervals, included in S and of measure greater than $m_e(E') - \epsilon > \mu - 2\epsilon$. If $\sum_2$ denotes a summation over these intervals,

$$\sum_2 \{f(x+k) - f(x)\} > v \sum_2 k > v(\mu - 2\epsilon).$$

But since $f(x)$ is non-decreasing, and the k-intervals are included in the h-intervals,

$$\sum_2 \{f(x+k) - f(x)\} \leqslant \sum_1 \{f(x+h) - f(x)\}.$$

Hence $v(\mu - 2\epsilon) < u(\mu + \epsilon)$, which is false if ϵ is small enough. Hence $f'_+(x)$ (and similarly $f'_-(x)$) exists almost everywhere.

* This proof is due to Rajchman and Saks (1).

Further, we can argue in the above way with D^+ replaced by D^-; every point of E' is then the right-hand end-point of arbitrarily small intervals $(x-k, x)$ such that $f(x)-f(x-k) > kv$, and the conclusion follows as before. Hence almost everywhere $D_+f \geqslant D^-f$, i.e. almost everywhere $f'_+(x) \geqslant f'_-(x)$. Similarly we can prove the reversed inequality, and the result follows.

11.43. There is a more general theorem on the possible sets where $f'_-(x) \neq f'_+(x)$, and the result has nothing to do with monotony.

The set of points where the right-hand and left-hand derivatives of any function exist and are different is enumerable.

Let E be the set where $f'_-(x) < f'_+(x)$, and let all rational numbers be arranged in a sequence $r_1, r_2, \ldots$. Then if x is a point of E, there is a smallest integer k such that

$$f'_-(x) < r_k < f'_+(x).$$

There is then a smallest integer m such that $r_m < x$, and such that

$$\{f(\xi)-f(x)\}/(\xi-x) < r_k$$

for $r_m < \xi < x$; and a smallest integer n such that $r_n > x$, and

$$\{f(\xi)-f(x)\}/(\xi-x) > r_k$$

for $x < \xi < r_n$. The two inequalities together give

$$f(\xi)-f(x) > r_k(\xi-x) \qquad (r_m < \xi < r_n,\ \xi \neq x). \tag{1}$$

Thus to every x corresponds a unique triad of numbers (k, m, n); and no two values of x correspond to the same triad; for if x_1 and x_2 correspond to the same triad, we have, on putting $x = x_1$, $\xi = x_2$ in (1), $f(x_2)-f(x_1) > r_k(x_2-x_1)$, and, on putting $x = x_2$, $\xi = x_1$, the same inequality reversed.

Since the set of triads (k, m, n) is enumerable, it follows that E is enumerable or finite. This is the required result. Since the measure of an enumerable set is zero, this theorem can be used to give an alternative ending to the proof of the theorem of the previous section.

11.5. Integrals. A function which is the Lebesgue indefinite integral of another function is called *an integral.*

An integral is continuous. For if $F(x)$ is the integral of $f(x)$, then

$$F(x+h)-F(x) = \int\limits_{x}^{x+h} f(t)\, dt,$$

which tends to 0 with h, by § 10.73 (v).

The integral of a positive function is a non-decreasing function. For if $f(x) \geqslant 0$, $h > 0$,

$$F(x+h)-F(x) = \int_x^{x+h} f(t)\,dt \geqslant 0.$$

An integral is a function of bounded variation. For let

$$F(x) = F(a) + \int_a^x f(t)\,dt,$$

and let $f_1(x) = f(x)$ where $f(x) \geqslant 0$, and $f_1(x) = 0$ elsewhere, and $-f_2(x) = f(x) - f_1(x)$. Then $f_1(x) \geqslant 0$, $f_2(x) \geqslant 0$, and

$$\begin{aligned} F(x) &= F(a) + \int_a^x f_1(t)\,dt - \int_a^x f_2(t)\,dt \\ &= F(a) + F_1(x) - F_2(x), \end{aligned}$$

where $F_1(x)$ and $F_2(x)$ are bounded non-decreasing functions.

11.51. Differentiation of the indefinite integral. Let $f(x)$ be integrable over (a, b), and let

$$F(x) = \int_a^x f(t)\,dt.$$

Since $F(x)$ is a function of bounded variation, it has a finite differential coefficient $F'(x)$ almost everywhere. Our next object is to prove that $F'(x) = f(x)$ almost everywhere.

11.52. The proof depends on the following lemma.

If $\phi(x)$ is integrable, and $\int_a^x \phi(t)\,dt = 0$ for all values of x in (a, b), then $\phi(x) = 0$ for almost all values of x in (a, b).

If this is not so, then either $\phi(x) > 0$ in a set of positive measure, or $\phi(x) < 0$ in a set of positive measure—suppose, for example, the former. Any set of positive measure contains a closed set of positive measure, since its complement can be included in an open set less than the whole interval. Hence $\phi(x) > 0$ in a closed set of positive measure—say E.

Now the integral of ϕ over any interval is zero; hence, by § 10.71, the integral over any open set is zero. Hence the integral over any closed set is zero, and in particular

$$\int_E \phi(x)\,dx = 0.$$

Hence, by § 10.73, $\phi(x) = 0$ almost everywhere in E, contrary to hypothesis. This proves the lemma.

11.53. *If $f(x)$ is bounded, and $F(x)$ is its integral, then $F'(x) = f(x)$ almost everywhere.*

Let $|f(x)| \leqslant M$. Then

$$\left|\frac{F(x+h)-F(x)}{h}\right| = \left|\frac{1}{h}\int_x^{x+h} f(t)\,dt\right| \leqslant M,$$

and
$$\lim_{h\to 0} \frac{F(x+h)-F(x)}{h} = F'(x)$$

almost everywhere. Hence, by the theorem of bounded convergence,* as $h \to 0$,

$$\int_a^x \frac{F(t+h)-F(t)}{h}\,dt \to \int_a^x F'(t)\,dt.$$

But the left-hand side is equal to

$$\frac{1}{h}\int_{a+h}^{x+h} F(t)\,dt - \frac{1}{h}\int_a^x F(t)\,dt = \frac{1}{h}\int_x^{x+h} F(t)\,dt - \frac{1}{h}\int_a^{a+h} F(t)\,dt,$$

which tends to $F(x)-F(a)$, since F is continuous. Hence

$$\int_a^x F'(t)\,dt = F(x)-F(a), \tag{1}$$

i.e.
$$\int_a^x \{F'(t)-f(t)\}\,dt = 0, \tag{2}$$

for all values of x. The result now follows from the lemma.

11.54. To extend the theorem to unbounded functions, we require another lemma.

If $\phi(x)$ is continuous and non-decreasing in (a, b), then $\phi'(x)$ is integrable, and

$$\int_a^b \phi'(x)\,dx \leqslant \phi(b)-\phi(a).$$

For $\{\phi(x+h)-\phi(x)\}/h \geqslant 0$, and $\{\phi(x+h)-\phi(x)\}/h$ tends to

* To apply the theorem as given in § 10.5, we make $h \to 0$ through an enumerable sequence; so also in the next section.

$\phi'(x)$ almost everywhere as $h \to 0$. Hence, by Fatou's theorem (§ 10.81),

$$\varliminf_{h\to 0} \int_a^b \frac{\phi(x+h)-\phi(x)}{h}\,dx \geqslant \int_a^b \phi'(x)\,dx.$$

Also, since ϕ is continuous, the left-hand side is equal to $\phi(b)-\phi(a)$, as in the above proof. Hence the result.

11.55. *If $f(x)$ is any integrable function, $F'(x) = f(x)$ almost everywhere.*

We may as usual suppose that $f(x) \geqslant 0$. We define $\{f(x)\}_n$ as in § 10.7. Since $f(t)-\{f(t)\}_n \geqslant 0$, the function

$$\int_a^x [f(t)-\{f(t)\}_n]\,dt$$

is non-decreasing, so that its differential coefficient is never negative. Hence

$$\frac{d}{dx}\left\{\int_a^x f(t)\,dt\right\} \geqslant \frac{d}{dx}\left\{\int_a^x \{f(t)\}_n\,dt\right\}$$

wherever these differential coefficients exist. Hence, by the theorem for bounded functions, $F'(x) \geqslant \{f(x)\}_n$ almost everywhere. Making $n \to \infty$ we see that $F'(x) \geqslant f(x)$ almost everywhere. Hence

$$\int_a^b F'(x)\,dx \geqslant \int_a^b f(x)\,dx.$$

The above lemma, however, gives this inequality reversed. Hence in fact the two sides are equal, i.e.

$$\int_a^b \{F'(x)-f(x)\}\,dx = 0.$$

Since the integrand is never negative, it must be zero almost everywhere. This is the required result.

11.6. The Lebesgue set. The theorem that $F'(x) = f(x)$ almost everywhere was extended by Lebesgue as follows.

If $f(x)$ is integrable,

$$\lim_{h\to 0} \frac{1}{h} \int_x^{x+h} |f(t)-\alpha|\,dt = |f(x)-\alpha|$$

for all values of α, except when x belongs to a set of measure zero; that is, $|f(x)-\alpha|$ is the derivative of its indefinite integral for all values of α and almost all values of x.

If α were fixed there would be nothing to prove, since $|f(x)-\alpha|$ is integrable, and the result follows from the above fundamental theorem.

Consider next all rational values of α, say $\alpha_1, \alpha_2, \ldots$. The sets in which the theorem is false for $\alpha_1, \alpha_2, \ldots$ are all of measure zero, and so their aggregate is of measure zero. Hence $|f(x)-\alpha|$ is the derivative of its integral for all rational values of α, except when x belongs to a set E of measure zero.

Now let x be a point not in E, α an irrational number, and β a rational number near to α. Since

$$\big||f(t)-\alpha|-|f(t)-\beta|\big| \leqslant |\beta-\alpha|.$$

we have

$$\left|\frac{1}{h}\int_x^{x+h} |f(t)-\alpha|\, dt - \frac{1}{h}\int_x^{x+h} |f(t)-\beta|\, dt\right| \leqslant |\beta-\alpha|.$$

But
$$\left|\frac{1}{h}\int_x^{x+h} |f(t)-\beta|\, dt - |f(x)-\beta|\right| \leqslant \epsilon$$

if $|h| < h_0(\beta, \epsilon)$. Hence

$$\left|\frac{1}{h}\int_x^{x+h} |f(t)-\alpha|\, dt - |f(x)-\alpha|\right|$$
$$\leqslant \left|\frac{1}{h}\int_x^{x+h} |f(t)-\alpha|\, dt - \frac{1}{h}\int_x^{x+h} |f(t)-\beta|\, dt\right| +$$
$$+\left|\frac{1}{h}\int_x^{x+h} |f(t)-\beta|\, dt - |f(x)-\beta|\right| + \big||f(x)-\beta|-|f(x)-\alpha|\big|$$
$$\leqslant |\beta-\alpha|+\epsilon+|\beta-\alpha|,$$

which may be made as small as we please, by choice first of β and then of ϵ. Hence $|f(x)-\alpha|$ is also the derivative of its indefinite integral for all irrational α, if x is not a point of E. This proves the theorem.

We may, in particular, take $\alpha = f(x)$. Hence

$$\int_0^h |f(x+t)-f(x)|\, dt = o\,(h)$$

as $h \to 0$, for almost all values of x. The set where this holds is called the *Lebesgue set.*

All points of continuity are of course included in the Lebesgue set.

The interest of the Lebesgue set lies in the fact that many theorems which hold at all points of continuity are also found to hold at all points of the Lebesgue set, and so almost everywhere. We shall have examples of this in the chapter on Fourier series.

We note finally that if the modulus sign is omitted from the formula, the α disappears, and the result reduces to the previous theorem.

11.7. Absolutely continuous functions. *A function $f(x)$ is said to be absolutely continuous in an interval (a, b) if, given ϵ, we can find δ such that*

$$\sum_{\nu=1}^{n} |f(x_\nu+h_\nu)-f(x_\nu)| \leqslant \epsilon$$

for every set of non-overlapping intervals $(x_\nu, x_\nu+h_\nu)$ such that $\sum h_\nu \leqslant \delta$.

An absolutely continuous function is continuous, since we can take the above sum to consist of one term only.

An absolutely continuous function is of bounded variation, since its total variation over an interval of length δ is at most ϵ, and consequently its total variation over (a, b) is at most $(b-a)\epsilon/\delta$.

On the other hand, there are continuous functions of bounded variation which are not absolutely continuous. An example of such a function will be given in § 11.72.

11.71. *A necessary and sufficient condition that a function should be an integral is that it should be absolutely continuous.*

If $F(x)$ is the integral of $f(x)$,

$$\sum_{\nu=1}^{n} |F(x_\nu+h_\nu)-F(x_\nu)| \leqslant \sum_{\nu=1}^{n} \int_{x_\nu}^{x_\nu+h_\nu} |f(x)|\, dx = \int_E |f(x)|\, dx,$$

where E denotes the set of intervals $(x_\nu, x_\nu+h_\nu)$. The right-hand side tends to zero with $\sum h_\nu$, in the sense of the above definition, by § 10.73 (v). Hence $F(x)$ is absolutely continuous.

To prove the converse we require the following lemma.

If $\phi(x)$ is absolutely continuous in (a,b), and $\phi'(x)=0$ almost everywhere, then $\phi(x)$ is a constant.

Let E be the set where $\phi'(x)=0$. Every point x of E is the left-hand end-point of arbitrarily small intervals $(x, x+h)$, such that

$$|\phi(x+h)-\phi(x)|<\epsilon h.$$

By the lemmas of § 11.41, we can select a finite set S of these intervals which do not overlap, and which contain all E except a set of measure δ, and so all (a,b) except a set of measure δ.

Let x_1, x_2,... be the end-points of the intervals of S, and let $\sum_1$ denote a summation over the intervals of S, and $\sum_2$ over the complementary intervals. Then

$$|\phi(b)-\phi(a)|\leqslant \sum\nolimits_1 |\phi(x_{\nu+1})-\phi(x_\nu)|+\sum\nolimits_2 |\phi(x_{\nu+1})-\phi(x_\nu)|.$$

Now $\quad \sum_1 |\phi(x_{\nu+1})-\phi(x_\nu)|<\epsilon\sum_1 (x_{\nu+1}-x_\nu)<\epsilon(b-a).$

Also $\sum_2 (x_{\nu+1}-x_\nu)<\delta$, and so, by the property of absolute continuity,

$$\sum\nolimits_2 |\phi(x_{\nu+1})-\phi(x_\nu)|$$

tends to zero with δ. Hence, making $\delta\to 0$,

$$|\phi(b)-\phi(a)|\leqslant\epsilon(b-a).$$

Making $\epsilon\to 0$, it follows that $\phi(b)=\phi(a)$; and similarly $\phi(x)=\phi(a)$ for every value of x.

Suppose now that $F(x)$ is any absolutely continuous function. Then it is continuous and of bounded variation, and we may write

$$F(x)=F_1(x)-F_2(x),$$

where F_1 and F_2 are continuous non-decreasing functions. By the lemma of § 11.54, $F_1'(x)$ and $F_2'(x)$ are integrable, and hence so is $F'(x)$. Hence

$$\int_a^x F'(t)\,dt$$

is absolutely continuous, and so also is

$$\phi(x)=F(x)-\int_a^x F'(t)\,dt.$$

But $\phi'(x) = 0$ almost everywhere. Hence, by the lemma, $\phi(x)$ is a constant, i.e.

$$F(x) - \int_a^x F'(t)\,dt = F(a).$$

Thus $F(x)$ is the integral of $F'(x)$.

11.72. A continuous increasing function which is not an integral.* We can define a function of this type by means of Cantor's ternary set (§ 10.291).

Let a_n always take the values 0 or 2, and let $b_n = \frac{1}{2}a_n$, so that b_n is always 0 or 1. If

$$x = \cdot a_1 a_2 a_3 \ldots (3)$$

is a point of Cantor's set E, we define

$$f(x) = \cdot b_1 b_2 b_3 \ldots (2)$$

(in the scale of 2).

At the ends of an interval δ_{pk}, $f(x)$ therefore has the values

$$\cdot b_1 \ldots b_m 0111 \ldots (2), \qquad \cdot b_1 \ldots b_m 1000 \ldots (2),$$

and these are equal. We define $f(x)$ throughout the interval δ_{pk} to be equal to its value at the end-points.

The function $f(x)$ is non-decreasing. In proving this it is sufficient to consider points x of E, since $f(x)$ is constant in the intervals of CE. Let

$$x' = \cdot a_1' a_2' \ldots (3), \qquad x'' = \cdot a_1'' a_2'' \ldots (3)$$

be points of E, $x'' > x'$. Then there is a suffix n such that $a_m' = a_m''$ $(m < n)$, $a_n' < a_n''$. Hence

$$f(x') = \cdot b_1' \ldots b_{n-1}' b_n' \ldots (2) \leqslant \cdot b_1'' \ldots b_{n-1}'' b_n'' \ldots (2) = f(x'').$$

The function $f(x)$ is continuous. We have to prove that $f(x') \to f(x)$ as $x' \to x$, and again it is sufficient to consider points x, x' of E. Let

$$x = \cdot a_1 a_2 \ldots (3), \qquad x' = \cdot a_1' a_2' \ldots (3).$$

If now $x' \to x$, there will be a value of n, which tends to infinity as $x' \to x$, such that $a_m = a_m'$ $(m < n)$. Hence

$$f(x) - f(x') = \cdot 00 \ldots 0 b_n \ldots - \cdot 00 \ldots 0 b_n' \ldots \to 0.$$

On the other hand,

$$\int_0^1 f'(x)\,dx \neq f(1) - f(0).$$

* A detailed discussion of this function is given by Hille and Tamarkin (1),

For the right-hand side is 1, since $f(1) = \cdot 111...(2) = 1$, $f(0) = 0$; but $f(x)$ is constant in the intervals δ_{pk}, so that $f'(x) = 0$ in the interior of any of these intervals. Hence $f'(x) = 0$ almost everywhere, and the left-hand side is 0.

It follows that $f(x)$ is not the integral of its differential coefficient, and so is not absolutely continuous. It is easy to see this directly. Consider the sum

$$\sum |f(\beta_k)-f(\alpha_k)|$$

taken over the intervals (α_k, β_k) which remain after the pth step of removing intervals δ_{pk}. It is equal to

$$\sum \{f(\beta_k)-f(\alpha_k)\} = f(1)-f(0) = 1.$$

But
$$\sum (\beta_k-\alpha_k) = 1-\frac{1}{3}-\frac{2}{9}-...-\frac{2^{p-1}}{3^p} = \left(\frac{2}{3}\right)^p,$$

which tends to zero as $p \to \infty$. Hence $f(x)$ is not absolutely continuous.

11.8. Integration of a differential coefficient. If $f(x)$ has a differential coefficient almost everywhere, or even everywhere, in an interval (a, b), the formula

$$\int\limits_a^x f'(t)\,dt = f(x)-f(a) \qquad (a \leqslant x \leqslant b) \tag{1}$$

is not necessarily true. It may fail in one or other of two ways. Consider, for example, the function

$$f(x) = x^2 \sin\frac{1}{x^2} \quad (x > 0), \qquad f(0) = 0,$$

already referred to in § 10.7. Here

$$f'(x) = 2x\sin\frac{1}{x^2}-\frac{2}{x}\cos\frac{1}{x^2} \quad (x > 0), \qquad f'(0) = 0,$$

so that $f'(x)$ exists everywhere; but, as we saw in § 10.7, it is not integrable in the Lebesgue sense, so that (1), on the Lebesgue theory, has no meaning.

If we can imagine a function with this kind of singularity distributed everywhere in an interval, we shall obtain some idea of the nature of the problem of integrating a differential coefficient. The problem has been solved by means of the Denjoy integral. This is a highly general type of non-absolutely convergent integral, and it would take us too far to discuss its properties here. The result is that, *if $f'(x)$ exists everywhere, the formula* (1) *is true, the integral being a Denjoy integral.*

If we do not assume that $f'(x)$ exists everywhere, but merely almost everywhere, the formula (1) may break down still more completely. The integral on the left may exist as a Lebesgue integral, but be unequal to the right-hand side. We have already had an example of this in § 11.72—in fact an example where $f'(x) = 0$ almost everywhere, without $f(x)$ being a constant.

In order to obtain the formula (1), the integral being a Lebesgue integral, we have therefore to impose further conditions on $f(x)$ or on $f'(x)$. There are several theorems, varying in difficulty according to what is assumed. Their common feature is that we suppose that $f'(x)$ exists everywhere. The example of § 11.72 shows that no set of conditions which is merely given almost everywhere is sufficient.

11.81. *If $f'(x)$ exists everywhere and is bounded, then* 11.8 (1) *is true.*

If $|f'(x)| \leqslant M$, then (*P.M.* § 125) there is a number θ between 0 and 1 such that

$$\left|\frac{f(x+h)-f(x)}{h}\right| = |f'(x+\theta h)| \leqslant M. \tag{1}$$

Hence $\{f(x+h)-f(x)\}/h$ converges boundedly to $f'(x)$, and the proof is now the same as that of 11.53 (1) (with $f(x)$ instead of $F(x)$).

Alternatively, we may observe that it follows from (1) that

$$\sum_{\nu=1}^{n} |f(x_\nu+h_\nu)-f(x_\nu)| \leqslant M \sum_{\nu=1}^{n} h_\nu.$$

Hence $f(x)$ is absolutely continuous, and the required result follows from § 11.71.

11.82. *If $f(x)$ is any function such that $f'(x)$ is finite everywhere and is integrable, then* 11.8 (1) *is true.*

This evidently shows in particular that 11.8 (1) holds if $f(x)$ is of bounded variation and $f'(x)$ is finite everywhere; for if $f(x)$ is of bounded variation, $f'(x)$ is integrable (see § 11.54 and example 12 below).

The following proof is substantially that given by Schlesinger and Plessner.† It depends on the two following lemmas.

LEMMA 1. *Let E be any set in (a,b) of measure zero, ϵ a given positive number. Then there is a non-decreasing absolutely continuous function $\chi(x)$ such that $\chi'(x) = +\infty$ in E, and $\chi(b)-\chi(a) < \epsilon$.*

We can include E in a sequence of open sets $O_1 > O_2 > \ldots$ such that $m(O_n) < \epsilon_n$, $\epsilon_1+\epsilon_2+\ldots = \epsilon$. Let $f_n(x)$ be the characteristic function of the set O_n. Then

$$\int_a^b f_n(t)\,dt = m(O_n) < \epsilon_n.$$

Let $$\phi_n(x) = f_1(x)+f_2(x)+\ldots+f_n(x).$$

Then $\phi_n(t)$ is non-decreasing as $n \to \infty$ for every t, and

$$\int_a^x \phi_n(t)\,dt < \epsilon_1+\epsilon_2+\ldots+\epsilon_n \leqslant \epsilon.$$

Hence by § 10.82 $\phi_n(t)$ tends to a finite limit $\phi(t)$ almost everywhere, and

$$\lim_{n\to\infty} \int_a^x \phi_n(t)\,dt = \int_a^x \phi(t)\,dt = \chi(x),$$

say.

This function $\chi(x)$ has the required properties. Since it is the integral of a non-negative function it is non-decreasing and absolutely continuous, and

$$\chi(b)-\chi(a) = \int_a^b \phi(t)\,dt < \epsilon.$$

Also $$\frac{d}{dx}\int_a^x f_\nu(t)\,dt = 1$$

in O_ν, and so, if $\chi_n(x) = \int_a^x \phi_n(t)\,dt$,

$$\chi_n'(x) = \sum_{\nu=1}^{n} \frac{d}{dx}\int_a^x f_\nu(t)\,dt = n$$

† *Lebesguesche Integrale*, pp. 166–74.

in O_n. Hence

$$\frac{\chi(x+h)-\chi(x)}{h} \geqslant \frac{\chi_n(x+h)-\chi_n(x)}{h} > n-\delta$$

for $|h| < h_0(\delta)$ and x in O_n. Hence $D\chi \geqslant n$ for each of the four derivates and x in O_n. Since a point of E belongs to O_n for every n, it follows that $\chi'(x) = +\infty$ in E.

LEMMA 2. *If $f(x)$ is continuous in (a, b), and $D^+f \geqslant 0$ almost everywhere in the interval, and D^+f is nowhere $-\infty$, then $f(x)$ is a non-decreasing function.*

It is sufficient to prove that $f(b) \geqslant f(a)$, since the general result then follows by a similar argument.

Let E be the set of measure zero where $D^+f < 0$. By Lemma 1 there is an absolutely continuous function $\chi(x)$ such that $\chi'(x) = +\infty$ in E, and $\chi(b)-\chi(a) < \epsilon$.

Let $$g(x) = f(x)+\chi(x).$$

Then in E, $D^+g = +\infty$, since $D_+\chi = +\infty$ and D^+f is finite, and $D^+g \geqslant D_+\chi+D^+f$. Also in CE

$$D^+g \geqslant D^+f \geqslant 0$$

since χ is non-decreasing. Hence $D^+g \geqslant 0$ everywhere, and so, by § 11.3, ex. (iv), $g(b) \geqslant g(a)$. Hence

$$f(b)-f(a) \geqslant -\{\chi(b)-\chi(a)\} > -\epsilon,$$

and, making $\epsilon \to 0$, the result follows.

11.83. We can now prove the theorem stated in § 11.82. Let n be any positive number, and let

$$g_n(x) = \min\{f'(x), n\}, \qquad G_n(x) = \max\{f'(x), -n\}.$$

Then $g_n(x) \leqslant f'(x) \leqslant G_n(x)$, and, since $f'(x)$ is integrable, so are $g_n(x)$ and $G_n(x)$. Let

$$f_n(x) = \int_a^x g_n(t)\,dt, \qquad F_n(x) = \int_a^x G_n(t)\,dt.$$

Then $$\lim_{n\to\infty} f_n(x) = \lim_{n\to\infty} F_n(x) = \int_a^x f'(t)\,dt = \phi(x),$$

say. Now $$D^+\{F_n(x)-f(x)\} \geqslant D^+F_n - D^+f.$$

This is almost everywhere equal to $G_n(x)-f'(x)$, i.e.

$$D^+\{F_n(x)-f(x)\} \geqslant 0$$

almost everywhere. Also

$$\frac{F_n(x+h)-F_n(x)}{h} \geqslant \frac{1}{h}\int\limits_x^{x+h} (-n)\,dt = -n,$$

so that $D^+F_n \geqslant -n$, and so $D^+(F_n-f)$ is nowhere $-\infty$. Hence, by Lemma 2, $F_n(x)-f(x)$ is non-decreasing, i.e.

$$F_n(x)-f(x) \geqslant F_n(a)-f(a) = -f(a).$$

Making $n \to \infty$, we obtain

$$\phi(x) \geqslant f(x)-f(a).$$

A similar argument with $f_n(x)$ gives the reversed inequality, and this proves the theorem.

MISCELLANEOUS EXAMPLES

1. For $x = \frac{1}{3}$ the function $f(x)$ of § 11.23 has the derivative $+\infty$.

2. The density of a set E at a point x may be defined as

$$\lim_{h\to 0} \frac{m(EH)}{2h},$$

where H is the interval $(x-h, x+h)$.

Prove that the density of a set is 1 almost everywhere in the set, and 0 almost everywhere outside it.

[Consider the integral of the characteristic function of E.]

3. A set E in (0, 1) is such that, if (α, β) is any interval, then

$$m\{E(\alpha, \beta)\} \geqslant \delta(\beta-\alpha)$$

where $\delta > 0$. Show that $m(E) = 1$.

4. If, as $h \to 0$,

$$\int\limits_a^b |f(x+h)-f(x)|\,dx = o(h),$$

then $f(x)$ is almost everywhere equal to a constant.

[Consider $\int\limits_{x_1}^{x_2} \{f(x+h)-f(x)\}\,dx$. See Titchmarsh (7), where, however, the proof is unnecessarily complicated.]

5. Let α and β be positive numbers, $f(x) = x^\alpha \sin x^{-\beta}$ $(0 < x \leqslant 1)$, and $f(0) = 0$. Then $f(x)$ is of bounded variation in (0, 1) if $\alpha > \beta$, but not if $\alpha \leqslant \beta$.

6. A function $f(x)$, defined for $0 \leqslant x < 1$, is absolutely continuous in every interval $(0, \xi)$, where $\xi < 1$, and its total variation in $(0, \xi)$ is bounded as $\xi \to 1$. Show that $f(x)$ tends to a limit as $\xi \to 1$, and that, if we define $f(1)$ to be equal to this limit, then $f(x)$ is absolutely continuous in the whole interval (0, 1).

[The point of this example is that the difference between 'continuity plus bounded variation' and absolute continuity is a property of a whole interval, and cannot be traced to the behaviour of the function in the neighbourhood of any one point.]

7. The theorem of § 11.82 remains true if $f'(x) = +\infty$ in an enumerable set.

8. A necessary and sufficient condition that a function should be convex in an interval (a, b), in the sense of § 5.31, is that it should be the integral of a bounded increasing function over any interval interior to (a, b).

9. If $f(x)$ is absolutely continuous, so is $|f(x)|^p$, where $p \geqslant 1$.

10. A necessary and sufficient condition that $f(x)$ should be almost everywhere equal to a function of bounded variation in (a, b) is that as $h \to 0$

$$\int_a^b |f(x+h)-f(x)|\,dx = O(h)$$

[where $f(x) = 0$, say, outside (a, b)].*

[If $f(x)$ is of bounded variation, we have $f(x) = \phi(x)-\psi(x)$, where ϕ and ψ are positive, non-decreasing and bounded in (a, b). Then, if $h > 0$,

$$\int_a^b |f(x+h)-f(x)|\,dx \leqslant \int_a^b \{\phi(x+h)-\phi(x)\}\,dx + \int_a^b \{\psi(x+h)-\psi(x)\}\,dx$$

$$= \int_b^{b+h} \phi(t)\,dt - \int_a^{a+h} \phi(t)\,dt + \int_b^{b+h} \psi(t)\,dt - \int_a^{a+h} \psi(t)\,dt = O(h),$$

so that the condition is necessary.

Suppose now that the condition is satisfied. Let

$$\phi_n(x) = n \int_x^{x+1/n} f(t)\,dt.$$

Then

$$\int_a^b |\phi_n(x+h)-\phi_n(x)|\,dx = n\int_a^b dx \left| \int_{x+h}^{x+h+1/n} - \int_x^{x+1/n} f(t)\,dt \right|$$

$$= n\int_a^b dx \left| \int_0^{1/n} \{f(x+t+h)-f(x+t)\}\,dt \right|$$

$$\leqslant n\int_a^b dx \int_0^{1/n} |f(x+t+h)-f(x+t)|\,dt$$

$$= n\int_0^{1/n} dt \int_a^b |f(x+t+h)-f(x+t)|\,dx = O(h).$$

* Hardy and Littlewood (5), pp. 599–601, and (6), p. 619.

If $(x_\nu, x_\nu+h_\nu)$ is any set of non-overlapping intervals,

$$\sum|\phi_n(x_\nu+h_\nu)-\phi_n(x_\nu)| = \sum\left|\int\limits_{x_\nu}^{x_\nu+h_\nu} \phi_n'(x)\,dx\right|$$

$$\leqslant \sum \int\limits_{x_\nu}^{x_\nu+h_\nu} |\phi_n'(x)|\,dx \leqslant \int\limits_a^b |\phi_n'(x)|\,dx,$$

and, by Fatou's lemma and the above result,

$$\int\limits_a^b |\phi_n'(x)|\,dx \leqslant \varliminf \int\limits_a^b \left|\frac{\phi_n(x+h)-\phi_n(x)}{h}\right| dx = O(1).$$

Hence $$\sum|\phi_n(x_\nu+h_\nu)-\phi_n(x_\nu)| = O(1).$$

But $\phi_n(x) \to f(x)$ almost everywhere. Hence

$$\sum|f(x_\nu+h_\nu)-f(x_\nu)| < A$$

if none of the points x_ν, $x_\nu+h_\nu$ belong to a certain set E of measure zero. If a does not belong to E, it follows as in § 11.4 that $f(x) = f(a)+P(x)-N(x)$ in CE, where $P(x)$ and $N(x)$ are bounded and non-decreasing in CE. In E we can define $P(x)$ as $\lim P(x')$, where $x' \to x$ from below through CE. The result follows without difficulty from this.]

11. In § 11.4 the existence of $f'(x)$ at a point does not imply that of $V'(x)$.

[Consider $f(x) = x^2 \cos x^{-\alpha}$ $(0 < x \leqslant 1)$, $f(0) = 0$, $1 < \alpha < 2$.]

12. In § 11.54 the condition that $\phi(x)$ is continuous can be omitted.

[The proof shows that if α and β are any two points of continuity

$$\int\limits_\alpha^\beta \phi'(x)\,dx \leqslant \phi(\beta)-\phi(\alpha).$$

But for any non-decreasing function points of continuity are everywhere dense. Hence, making $\alpha \to a+0$, $\beta \to b-0$, through such points, we obtain

$$\int\limits_a^b \phi'(x)\,dx \leqslant \phi(b-0)-\phi(a+0).]$$

13. The set consisting of the intervals $\left(\frac{1}{2n+1}, \frac{1}{2n}\right)$, $n = 1, 2,\ldots$, has density $\frac{1}{2}$ at $x = 0$.

14. A convergent series of non-decreasing functions can be differentiated term by term almost everywhere.

Fubini: see Rajchman and Saks (1).

[Let $$u_1(x)+u_2(x)+\ldots+u_n(x) = s_n(x) \to s(x) \quad (a \leqslant x \leqslant b).$$

Then $s(x)$ is non-decreasing; and

$$\frac{s(x+h)-s(x)}{h} = \sum_{n=1}^{\infty} \frac{u_n(x+h)-u_n(x)}{h} \geqslant \sum_{n=1}^{N} \frac{u_n(x+h)-u_n(x)}{h}$$

for every N. Making $h \to 0$, it follows that

$$s'(x) \geqslant \sum_{n=1}^{N} u_n'(x)$$

almost everywhere. Hence $\sum u_n'(x)$ converges almost everywhere, to $\phi(x)$ say, and $\phi(x) \leqslant s'(x)$.

Suppose that the set $E(u,v)$, where $\phi(x) < u < v < s'(x)$, has positive measure μ. Almost everywhere in $E(u,v)$

$$s_n'(x) < u < v < s'(x),$$

so that $$s_n(x+h)-s_n(x) < hu < hv < s(x+h)-s(x)$$

for sufficiently small h. This holds over a finite non-overlapping set of intervals of total length $l > \frac{1}{2}\mu > 0$. Summing over these intervals

$$l(v-u) < \sum \{s(x+h)-s_n(x+h)\}-\{s(x)-s_n(x)\}$$
$$\leqslant \{s(b)-s_n(b)\}-\{s(a)-s_n(a)\}$$

since $s(x)-s_n(x)$ is non-decreasing. Making $n \to \infty$, $l \leqslant 0$, a contradiction. Hence $\phi(x) = s'(x)$ almost everywhere.]

15. If $f'(x)$ is finite everywhere, and equal to a continuous function almost everywhere, it is equal to it everywhere.

16. Show that $$\lim_{\delta \to 0} \int_{\delta}^{1} \frac{f(x+t)-f(x-t)}{t}\,dt$$

exists for every x if $f(x)$ is Weierstrass's non-differentiable function.

Show that the limit does not exist at $x = 0$ if $f(x)$ is the continuous function 0 $(x \leqslant 0)$, $1/\log(1/x)$ $(x > 0)$.

[This limit exists almost everywhere if $f(x)$ is any integrable function. See Titchmarsh, *Fourier Integrals*, Theorem 105.]

CHAPTER XII

FURTHER THEOREMS ON LEBESGUE INTEGRATION

12.1. In this chapter we adopt a slightly more practical point of view than in the two preceding ones. We have carried the general theory of definite and indefinite integrals as far as we shall require it, and we shall now prove a number of theorems which are useful in the manipulation of integrals.

12.11. Integration by parts. The formula of integration by parts in the Lebesgue theory is, of course, the same as the ordinary one: if $G(x)$ is an indefinite integral of $g(x)$, then

$$\int_a^b f(x)g(x)\,dx = [f(x)G(x)]_a^b - \int_a^b f'(x)G(x)\,dx.$$

The formula holds if $g(x)$ is any integrable function, and $f(x)$ is an integral.

The proof depends on the fact that *the product of two absolutely continuous functions is absolutely continuous.* For let $\phi(x)$ and $\psi(x)$ be absolutely continuous in (a,b), and let M and M' be the upper bounds of $|\phi(x)|$ and $|\psi(x)|$. Let $(x_\nu, x_\nu+h_\nu)$ be a set of non-overlapping intervals in (a,b). Then

$$\begin{aligned}\sum |\phi(x_\nu+h_\nu)\psi(x_\nu+h_\nu)-\phi(x_\nu)\psi(x_\nu)| \\ = \sum |\phi(x_\nu+h_\nu)\{\psi(x_\nu+h_\nu)-\psi(x_\nu)\}+\psi(x_\nu)\{\phi(x_\nu+h_\nu)-\phi(x_\nu)\}| \\ \leqslant M\sum |\psi(x_\nu+h_\nu)-\psi(x_\nu)|+M'\sum |\phi(x_\nu+h_\nu)-\phi(x_\nu)|.\end{aligned}$$

The last two sums tend to zero with $\sum h_\nu$, and so $\phi(x)\psi(x)$ is absolutely continuous.

In the given formula, $f(x)$ and $G(x)$ are absolutely continuous, and hence so is $f(x)G(x)$; and

$$\int_a^b \frac{d}{dx}\{f(x)G(x)\}\,dx = [f(x)G(x)]_a^b.$$

But $$\frac{d}{dx}\{f(x)G(x)\} = f'(x)G(x)+f(x)g(x)$$

wherever $f'(x)$ and $G'(x)$ exist, and $G'(x)=g(x)$. Since this is true almost everywhere the result follows.

12.2. Approximation to an integrable function. The following theorem is often useful.

If $f(x)$ is measurable over a finite interval, then, given two positive numbers δ and ϵ, we can define an absolutely continuous function $\phi(x)$ such that $|f-\phi| < \delta$ except in a set of measure less than ϵ.

Suppose first that $f(x)$ is bounded. We may suppose without loss of generality that $f(x) \geqslant 0$. Divide up the interval of variation of $f(x)$ by the scale

$$0,\ \delta,\ 2\delta,\ldots,\ n\delta.$$

Let e_ν be the set where $\nu\delta \leqslant f(x) < (\nu+1)\delta$. Let $\psi_\nu(x) = \nu\delta$ in e_ν, and zero elsewhere. Then the function

$$\psi(x) = \psi_0(x)+\ldots+\psi_{n-1}(x)$$

differs from $f(x)$ by less than δ.

Let E_ν be an open set, including e_ν, of measure less than $m(e_\nu)+\epsilon/3n$. Let S_ν be the sum of a finite number of the intervals of E_ν, such that $m(E_\nu-S_\nu) < \epsilon/3n$. Let $\phi_\nu(x) = \nu\delta$ in S_ν, and zero elsewhere. Then $\phi_\nu = \psi_\nu$ except in a set of measure less than $2\epsilon/3n$; also ϕ_ν is discontinuous at a finite number of points, viz. the ends of the intervals of S_ν. To remove these discontinuities, we join the graph of the function to zero at the end of each interval by a straight line inclined so that the modifications all occur in a set of measure less than $\epsilon/3n$. Thus if ϕ'_ν is the modified function, ϕ'_ν is absolutely continuous, and $\phi'_\nu = \psi_\nu$ except in a set of measure ϵ/n.

Let
$$\phi(x) = \phi'_0+\phi'_1+\ldots+\phi'_{n-1}.$$

Then $\phi(x)$ is absolutely continuous, and $\phi(x) = \psi(x)$ except in a set of measure ϵ. Hence $\phi(x)$ has the required property.

If $f(x)$ is not bounded, let $\{f(x)\}_k = f(x)$ where $|f(x)| \leqslant k$, and $\{f(x)\}_k = 0$ elsewhere. We can take k so large that $\{f(x)\}_k = f(x)$ except in a set of measure $\frac{1}{2}\epsilon$. By the first part, we can determine $\phi(x)$ so that $|\{f(x)\}_k-\phi(x)| < \delta$ except in a set of measure $\frac{1}{2}\epsilon$. Then $\phi(x)$ has the required property.

Notice that, if $f(x)$ is bounded, $\phi(x)$ can be constructed to lie between the same bounds as $f(x)$.

If $f(x)$ is integrable, we can construct $\phi(x)$ so that, in addition to the above properties,

$$\int_a^b |f(x)-\phi(x)|\,dx < \eta, \tag{1}$$

where η is arbitrarily small. If $f(x)$ is bounded, say $|f(x)| \leqslant M$, then $|\phi(x)| \leqslant M$, and

$$\int_a^b |f(x)-\phi(x)|\, dx \leqslant \delta(b-a)+2\epsilon M,$$

giving the required result. If $f(x)$ is unbounded, we define $\{f(x)\}_k$ as above, and then determine $\phi(x)$ so that $|\{f(x)\}_k-\phi(x)| < \delta$ except in a set of measure $\frac{1}{2}\epsilon/k$. Then

$$\int_a^b |f(x)-\phi(x)|\, dx \leqslant \int_a^b |f(x)-f\{(x)\}_k|\, dx + \int_a^b |\{f(x)\}_k-\phi(x)|\, dx.$$

The first term tends to zero as $k \to \infty$, and the second term does not exceed $\delta(b-a)+\epsilon$. Hence the result.

Example. If $f(x)$ is integrable over $(a-\epsilon, b+\epsilon)$, then

$$\lim_{h\to 0}\int_a^b |f(x+h)-f(x)|\, dx = 0.$$

12.21. Change of the independent variable. Here again the formula is familiar, but the conditions under which it holds are novel.

If $f(x)$ and $g(x)$ are integrable, $g(x) \geqslant 0$, and $G(x)$ is an indefinite integral of $g(x)$, $a = G(\alpha)$, $b = G(\beta)$, then

$$\int_a^b f(t)\, dt = \int_\alpha^\beta f\{G(x)\}g(x)\, dx,$$

where $f\{G(x)\}g(x)$ is defined as 0 if $g(x) = 0$.

The inverse function of $t = G(x)$, of which α and β are values, is not necessarily one-valued, since $G(x)$ may be constant in some intervals. But if more than one value of x corresponds to a given value of t, these values of x form a closed interval, and we can make the inverse function one-valued by taking x to be, say, the left-hand end-point of the interval.

We next observe that *if $F(x)$ and $G(x)$ are absolutely continuous functions, and $G(x)$ is monotonic, then $F\{G(x)\}$ is absolutely continuous.* For, since F is absolutely continuous,

$$\sum |F\{G(x_\nu+h_\nu)\}-F\{G(x_\nu)\}|$$

tends to zero with

$$\sum |G(x_\nu+h_\nu)-G(x_\nu)|,$$

and, since $G(x)$ is absolutely continuous, this tends to zero with $\sum h_\nu$.

It follows that, if $F(x)$ and $G(x)$ are integrals of $f(x)$ and $g(x)$, then $F\{G(x)\}$ has a finite differential coefficient for almost all values of x, and

$$\int_\alpha^\beta \frac{d}{dx}[F\{G(x)\}]\,dx = F\{G(\beta)\}-F\{G(\alpha)\} = \int_a^b f(t)\,dt.$$

The result will now follow if

$$\frac{d}{dx}[F\{G(x)\}] = f\{G(x)\}g(x) \tag{1}$$

for almost all values of x. But this is not obviously true. For

$$\frac{F\{G(x+h)\}-F\{G(x)\}}{h} = \frac{F\{G(x+h)\}-F\{G(x)\}}{G(x+h)-G(x)}\cdot\frac{G(x+h)-G(x)}{h},$$

and the second factor on the right tends to $g(x)$ for almost all values of x, while the first factor tends to $f\{G(x)\}$ for almost all values of $G(x)$; and the difficulty is that the exceptional set of values of $G(x)$, of measure zero, does not necessarily correspond to a set of values of x of measure zero.

Let $f(x)$ be bounded, say $|f(x)| \leqslant M$. Divide the interval (α, β) into sets $E_1,\ldots, E_4$ as follows. In E_1, $G'(x) = g(x) > 0$, and the first factor on the right tends to $f\{G(x)\}$; in E_2,

$$G'(x) = g(x) > 0$$

but the other condition is negatived; in E_3, $G'(x) = g(x) = 0$; in E_4, $G'(x) \neq g(x)$. Clearly (1) holds in E_1; and it holds in E_3, since there

$$\left|\frac{F\{G(x+h)\}-F\{G(x)\}}{h}\right|$$

$$= \left|\frac{1}{h}\int_{G(x)}^{G(x+h)} f(t)\,dt\right| \leqslant M\left|\frac{G(x+h)-G(x)}{h}\right| \to 0$$

and each side of (1) is zero; $m(E_4) = 0$; and we have to prove that $m(E_2) = 0$.

Let $E_{2,n}$ be the part of E_2 in which $G'(x) > 1/n$. Enclose the corresponding t-set in an open set O of measure less than a given ϵ. With each x of $E_{2,n}$ associate an interval $(x, x+h_x)$ such that $G(x+h_x)-G(x) > h_x/n$, and such that the interval $G(x)$, $G(x+h_x)$ is in O. By Lemma 1 of § 11.41, there is a finite non-overlapping set S of the intervals $(x, x+h_x)$ such that

$$m_e(E_{2,n}) < m(S)+\epsilon = \sum_S h_x+\epsilon.$$

This is less than

$$n\sum_S \{G(x+h_x)-G(x)\}+\epsilon \leqslant nm(O)+\epsilon < (n+1)\epsilon.$$

Hence $m(E_{2,n}) = 0$, and, since E_2 is the outer limiting set of the sets $E_{2,n}$, $m(E_2) = 0$.

Lastly let $f(x)$ be any integrable function. We may suppose without loss of generality that it is positive. Defining $\{f(x)\}_n$ in the usual way, the theorem holds for $\{f(x)\}_n$, and it is sufficient to prove that

$$\lim \int_\alpha^\beta [f\{G(x)\}]_n\, g(x)\, dx = \int_\alpha^\beta f\{G(x)\}g(x)\, dx.$$

But $$\int_\alpha^\beta [f\{G(x)\}]_n\, g(x)\, dx = \int_a^b \{f(t)\}_n\, dt \leqslant \int_a^b f(t)\, dt.$$

The result therefore follows from the convergence theorem of § 10.82 (regarding $f\{G(x)\}g(x)$ as 0 if $f\{G(x)\} = \infty$, $g(x) = 0$).

12.3. The second mean-value theorem. *If $f(x)$ is integrable over (a,b), and $\phi(x)$ is positive, bounded, and non-increasing, then*

$$\int_a^b f(x)\phi(x)\, dx = \phi(a+0)\int_a^\xi f(x)\, dx,$$

where ξ is some number between a and b.

Let ϵ be a positive number less than $\phi(a+0)-\phi(b-0)$. Then there is a point x_1 such that

$$\begin{aligned} \phi(a+0)-\phi(x) &< \epsilon \qquad (a < x < x_1) \\ &\geqslant \epsilon \qquad (x > x_1). \end{aligned}$$

Similarly there are points $x_2, x_3, \ldots$ such that

$$\begin{aligned} \phi(x_{\nu-1}+0)-\phi(x) &< \epsilon \qquad (x_{\nu-1} < x < x_\nu) \\ &\geqslant \epsilon \qquad (x > x_\nu), \end{aligned}$$

so long as $\phi(x_{\nu-1}+0)-\phi(b-0) > \epsilon$. Otherwise we take $x_n = b$. The point b is thus reached in a finite number of steps, since the variation of $\phi(x)$ in each interval $(x_{\nu-1}, x_\nu)$ is at least ϵ.

Let $\psi(x) = \phi(x_\nu+0)$ in each interval $x_\nu \leqslant x < x_{\nu+1}$. Then $0 \leqslant \psi(x)-\phi(x) < \epsilon$ except possibly at the points $a = x_0$, x_1, $x_2, \ldots, b$, and

$$\int_a^b \psi(x)f(x)\,dx = \sum_{\nu=0}^{n-1} \phi(x_\nu+0) \int_{x_\nu}^{x_{\nu+1}} f(x)\,dx.$$

Let $F(x) = \int_a^x f(t)\,dt$; then, if m and M are the lower and upper bounds of $F(x)$, it follows from Abel's lemma (§ 1.131) that

$$m\phi(a+0) \leqslant \int_a^b \psi(x)f(x)\,dx \leqslant M\phi(a+0).$$

But

$$\left| \int_a^b \psi(x)f(x)\,dx - \int_a^b \phi(x)f(x)\,dx \right| \leqslant \epsilon \int_a^b |f(x)|\,dx,$$

which tends to zero with ϵ. Hence, making $\epsilon \to 0$, it follows that

$$m\phi(a+0) \leqslant \int_a^b \phi(x)f(x)\,dx \leqslant M\phi(a+0).$$

Since $F(x)$ is continuous, it takes every value between m and M, and so, at $x = \xi$ say, the value

$$\frac{1}{\phi(a+0)} \int_a^b \phi(x)f(x)\,dx.$$

This proves the theorem.

If $\phi(x)$ is positive and non-decreasing, the corresponding formula is

$$\int_a^b f(x)\phi(x)\,dx = \phi(b-0) \int_\xi^b f(x)\,dx,$$

where $a < \xi < b$.

If $\phi(x)$ is any monotonic function, there is a number ξ between a and b such that

$$\int_a^b f(x)\phi(x)\,dx = \phi(a+0)\int_a^\xi f(x)\,dx + \phi(b-0)\int_\xi^b f(x)\,dx.$$

This is obtained from the previous results by considering $\phi(x)-\phi(a+0)$ or $\phi(x)-\phi(b-0)$.

12.4. The Lebesgue class* L^p. We denote by $L^p(a,b)$ the class of functions $f(x)$ such that $f(x)$ is measurable, and $|f(x)|^p$, where $p>0$, is integrable over (a,b). If it is not necessary to specify the interval, we denote the class by L^p simply. The class L^1 is the class of functions integrable over (a,b), and is denoted simply by L.

We may classify functions defined over any set, or over an infinite interval, in the same way; for example, the function $(1+x)^{-\frac{1}{2}}$ belongs to $L^p(0,\infty)$ if $p>2$.

If $f(x)$ belongs to L^p, and $|g(x)| \leqslant |f(x)|$, then clearly $g(x)$ also belongs to L^p.

Examples. (i) A bounded function belongs to $L^p(a, b)$, where (a, b) is a finite interval, for all values of p.

(ii) If $f(x)$ belongs to $L^p(a, b)$, where (a, b) is a finite interval, then it also belongs to $L^q(a, b)$ for $q < p$.

(iii) If $f(x)$ belongs to $L^p(0, \infty)$ and to $L^q(0, \infty)$, where $p < q$, then it also belongs to $L^r(0, \infty)$ if $p < r < q$.

[Consider separately the sets where $|f(x)| \leqslant 1$ and $|f(x)| > 1$.]

(iv) The sum of two functions of L^p also belongs to L^p.

[For $|f(x)+g(x)|^p \leqslant \max\{2^p|f(x)|^p, 2^p|g(x)|^p\}$.]

(v) The function $\{x \log^2 1/x\}^{-1}$ belongs to $L(0, \frac{1}{2})$, but not to any $L^p(0, \frac{1}{2})$ for $p > 1$.

(vi) The function $\{x^{\frac{1}{2}}(1+|\log x|)\}^{-1}$ belongs to $L^2(0, \infty)$, but not to $L^p(0, \infty)$ for any other value of p.

12.41. Schwarz's inequality. *If $f(x)$ and $g(x)$ belong to L^2, then $f(x)g(x)$ belongs to L, and*

$$\left|\int f(x)g(x)\,dx\right| \leqslant \left\{\int |f(x)|^2\,dx \int |g(x)|^2\,dx\right\}^{\frac{1}{2}}.$$

The interval of integration may be finite or infinite.

* See in particular F. Riesz (2).

Since $2|fg| \leqslant f^2+g^2$, fg belongs to L. Hence the integral

$$\int \{\lambda f(x)+\mu g(x)\}^2\,dx$$
$$=\lambda^2 \int \{f(x)\}^2\,dx+2\lambda\mu \int f(x)g(x)\,dx+\mu^2 \int \{g(x)\}^2\,dx$$

exists for all values of λ and μ. It is evidently never negative. But the necessary and sufficient condition that $a\lambda^2+2h\lambda\mu+b\mu^2$ should be never negative is that $h^2 \leqslant ab$, $a \geqslant 0$, $b \geqslant 0$; and this gives the inequality stated.

Examples. (i) The case of equality in the above theorem occurs only if $f(x)/g(x)$ is almost everywhere equal to a constant.

(ii) If $f(x)$ and $g(x)$ belong to L^p, where $p > 2$, then $f(x)g(x)$ belongs to $L^{\frac{1}{2}p}$.

12.42. Hölder's inequality. This is a generalization of Schwarz's inequality.

If $f(x)$ belongs to L^p, and $g(x)$ to $L^{p/(p-1)}$, where $p > 1$, then $f(x)g(x)$ belongs to L, and

$$\left|\int f(x)g(x)\,dx\right| \leqslant \left\{\int |f(x)|^p\,dx\right\}^{1/p}\left\{\int |g(x)|^{p/(p-1)}\,dx\right\}^{1-1/p}. \tag{1}$$

The interval of integration may be finite or infinite.

Let E be the set where $|g(x)| \leqslant |f(x)|^{p-1}$. Then

$$|f(x)g(x)| \leqslant |f(x)|^p$$

in E; hence $f(x)g(x)$ is integrable over E. In the complementary set CE, $|f(x)| < |g(x)|^{1/(p-1)}$. Hence

$$|f(x)g(x)| < |g(x)|^{p/(p-1)}$$

in CE; hence $f(x)g(x)$ is integrable over CE, and so over the whole interval considered.

This argument can be used to obtain an inequality similar to (1), but with a factor 2 on the right-hand side. Let

$$I=\int_a^b |f(x)|^p\,dx, \qquad J=\int_a^b |g(x)|^{p/(p-1)}\,dx.$$

Then

$$\left|\int_a^b fg\,dx\right| \leqslant \int_E |fg|\,dx+\int_{CE} |fg|\,dx$$
$$\leqslant \int_E |f|^p\,dx+\int_{CE} |g|^{p/(p-1)}\,dx \leqslant I+J. \tag{2}$$

If we replace $f(x)$ and $g(x)$ in this inequality by

$$(J/I)^{(p-1)/p^2}f(x), \qquad (I/J)^{(p-1)/p^2}g(x),$$

respectively, the left-hand side is unchanged, and each term on the right-hand side is replaced by $I^{1/p}J^{1-1/p}$. Hence

$$\left|\int fg\,dx\right| \leqslant 2I^{1/p}J^{1-1/p}. \tag{3}$$

The inequality (1) can be deduced from the well-known inequality

$$x^m - 1 < m(x-1) \qquad (x > 1,\ 0 < m < 1). \tag{4}$$

Putting $x = a/b$ $(a > b)$, and multiplying by b,

$$a^m b^{1-m} < b + m(a-b).$$

Putting $m = \alpha$, $1-m = \beta$, so that $\alpha+\beta = 1$, this takes the form

$$a^\alpha b^\beta < a\alpha + b\beta, \tag{5}$$

and since this is symmetrical it holds if a and b are any unequal positive numbers. If $a = b$ it becomes an equality.

Using (5), we have, if $F(x) \geqslant 0$, $G(x) \geqslant 0$,

$$\int_a^b \left(\frac{F(x)}{\int_a^b F(t)\,dt}\right)^\alpha \left(\frac{G(x)}{\int_a^b G(t)\,dt}\right)^\beta dx \leqslant \int_a^b \left(\frac{\alpha F(x)}{\int_a^b F(t)\,dt} + \frac{\beta G(x)}{\int_a^b G(t)\,dt}\right) dx$$

$$= \alpha + \beta = 1,$$

i.e.
$$\int_a^b \{F(x)\}^\alpha \{G(x)\}^\beta\,dx \leqslant \left\{\int_a^b F(x)\,dx\right\}^\alpha \left\{\int_a^b G(x)\,dx\right\}^\beta.$$

Finally, putting $\alpha = 1/p$, $F(x) = |f(x)|^p$, and $G(x) = |g(x)|^{p/(p-1)}$, the result (1) follows.*

Example. The case of equality occurs only if $|f(x)|^p/|g(x)|^{p/(p-1)}$ is almost everywhere equal to a constant.

12.421. Hölder's inequality for sums. This is

$$\left|\sum a_n b_n\right| \leqslant \left(\sum |a_n|^p\right)^{1/p} \left(\sum |b_n|^{p/(p-1)}\right)^{1-1/p}.$$

The proof is similar to that of the integral inequality. We have

$$\sum \left\{\left(\frac{A_n}{\sum A_n}\right)^\alpha \left(\frac{B_n}{\sum B_n}\right)^\beta\right\} \leqslant \sum \left(\alpha\frac{A_n}{\sum A_n} + \beta\frac{B_n}{\sum B_n}\right)$$

$$= \alpha + \beta = 1,$$

i.e.
$$\sum A_n^\alpha B_n^\beta \leqslant \left(\sum A_n\right)^\alpha \left(\sum B_n\right)^\beta,$$

and writing $\alpha = 1/p$, $A_n = |a_n|^p$, $B_n = |b_n|^{p/(p-1)}$, the result follows.

* This proof is given by Hardy (20).

12.43. Minkowski's inequality. *If $f(x)$ and $g(x)$ belong to L^p, where $p > 1$, then*

$$\left\{\int |f(x)+g(x)|^p \, dx\right\}^{1/p} \leqslant \left\{\int |f(x)|^p \, dx\right\}^{1/p} + \left\{\int |g(x)|^p \, dx\right\}^{1/p}. \quad (1)$$

For

$$\int |f+g|^p \, dx \leqslant \int |f| \cdot |f+g|^{p-1} \, dx + \int |g| \cdot |f+g|^{p-1} \, dx$$
$$\leqslant \left\{\int |f|^p \, dx\right\}^{1/p} \left\{\int |f+g|^p \, dx\right\}^{1-1/p} +$$
$$+ \left\{\int |g|^p \, dx\right\}^{1/p} \left\{\int |f+g|^p \, dx\right\}^{1-1/p}$$

by Hölder's inequality. Dividing each side by

$$\left\{\int |f+g|^p \, dx\right\}^{1-1/p},$$

the result follows.

The corresponding inequality for sums

$$\left(\sum |a_n+b_n|^p\right)^{1/p} \leqslant \left(\sum |a_n|^p\right)^{1/p} + \left(\sum |b_n|^p\right)^{1/p} \quad (2)$$

can be proved in a similar way.

12.44. The integral of a function of L^p. We have seen in the previous chapter that a necessary and sufficient condition that a function should be an integral is that it should be absolutely continuous. There is a corresponding condition that a function should be an integral of a function of the class L^p.

A necessary and sufficient condition that a function $F(x)$ should be the integral of a function of the class L^p, where $p > 1$, is that the sum

$$\sum |F(x_\nu+h_\nu)-F(x_\nu)|^p h_\nu^{1-p},$$

taken over any system of non-overlapping intervals $(x_\nu, x_\nu+h_\nu)$, should be bounded.

If instead of 'should be bounded' we say 'should be bounded and tend to zero with $\sum h_\nu$', the theorem is still true, and in this form it is true for $p = 1$ also, and so includes the theorem on absolute continuity as a particular case. For $p > 1$ the two conditions, one of which appears to be more restrictive than the other, turn out to be equivalent.

To prove that the condition is necessary, suppose that

$$F(x) = F(a) + \int_a^x f(t) \, dt,$$

where $f(t)$ belongs to L^p. Then

$$|F(x_\nu+h_\nu)-F(x_\nu)| = \left|\int\limits_{x_\nu}^{x_\nu+h_\nu} f(t)\,dt\right|$$

$$\leqslant \left\{\int\limits_{x_\nu}^{x_\nu+h_\nu} |f(t)|^p\,dt\right\}^{1/p}\left\{\int\limits_{x_\nu}^{x_\nu+h_\nu} dt\right\}^{1-1/p} = h_\nu^{1-1/p}\left\{\int\limits_{x_\nu}^{x_\nu+h_\nu} |f(t)|^p\,dt\right\}^{1/p}.$$

Hence

$$\sum |F(x_\nu+h_\nu)-F(x_\nu)|^p h_\nu^{1-p} \leqslant \sum \int\limits_{x_\nu}^{x_\nu+h_\nu} |f(t)|^p\,dt \leqslant \int\limits_a^b |f(t)|^p\,dt,$$

so that the condition is necessary. Since

$$\sum \int\limits_{x_\nu}^{x_\nu+h_\nu} |f(t)|^p\,dt$$

tends to zero with $\sum h_\nu$, the alternative condition is also necessary.

Suppose now that the condition is satisfied, and let M be the upper bound of the given sums. Then, by Hölder's inequality for sums,

$$\sum |F(x_\nu+h_\nu)-F(x_\nu)| = \sum |F(x_\nu+h_\nu)-F(x_\nu)|h_\nu^{1/p-1}.h_\nu^{1-1/p}$$

$$\leqslant \{\sum |F(x_\nu+h_\nu)-F(x_\nu)|^p h_\nu^{1-p}\}^{1/p}(\sum h_\nu)^{1-1/p} \leqslant M^{1/p}(\sum h_\nu)^{1-1/p},$$

which tends to zero with $\sum h_\nu$. Hence $F(x)$ is absolutely continuous, and so is an integral, say

$$F(x) = F(a)+\int\limits_a^x f(t)\,dt.$$

It remains to prove that $f(t)$ belongs to L^p. Consider a sequence of finite sets of points in the interval, the mth set being $x_{m,1}, x_{m,2},\ldots, x_{m,n}$, such that

$$\lim_{m\to\infty} \max_\nu(x_{m,\nu+1}-x_{m,\nu}) = 0.$$

For example, if the interval is $(0, 1)$ we may take $x_{m,\nu} = \nu/2^m$.

Let
$$f_m(x) = \frac{F(x_{m,\nu+1})-F(x_{m,\nu})}{x_{m,\nu+1}-x_{m,\nu}}$$

in each interval $x_{m,\nu} \leqslant x < x_{m,\nu+1}$. If x is not one of the points

$x_{m,\nu}$, and $F'(x)$ exists, and $x_{m,\nu} < x < x_{m,\nu+1}$, then

$$f_m(x) = \frac{F(x_{m,\nu+1})-F(x)}{x_{m,\nu+1}-x}\,\frac{x_{m,\nu+1}-x}{x_{m,\nu+1}-x_{m,\nu}}+$$
$$+\frac{F(x_{m,\nu})-F(x)}{x_{m,\nu}-x}\,\frac{x-x_{m,\nu}}{x_{m,\nu+1}-x_{m,\nu}}$$
$$= \{F'(x)+\delta_1\}\frac{x_{m,\nu+1}-x}{x_{m,\nu+1}-x_{m,\nu}}+\{F'(x)+\delta_2\}\frac{x-x_{m,\nu}}{x_{m,\nu+1}-x_{m,\nu}}$$
$$= F'(x)+\delta_3,$$

where $|\delta_3| \leqslant |\delta_1|+|\delta_2|$, and δ_1 and δ_2 tend to zero as

$$x_{m,\nu+1}-x_{m,\nu} \to 0.$$

Hence
$$\lim_{m\to\infty} f_m(x) = F'(x) = f(x)$$

almost everywhere. Also

$$\int_a^b |f_m(x)|^p\,dx = \sum |F(x_{m,\nu+1})-F(x_{m,\nu})|^p|x_{m,\nu+1}-x_{m,\nu}|^{1-p} \leqslant M.$$

Hence, by Fatou's theorem (§ 10.81), $f(x)$ belongs to L^p, and

$$\int_a^b |f(x)|^p\,dx \leqslant \underline{\lim}\int_a^b |f_m(x)|^p\,dx \leqslant M.$$

12.5. Mean convergence. If we are given a sequence of numbers, say s_n, we have usually to consider the behaviour of the difference s_n-s between s_n and a given number s. In dealing with a sequence of functions, say $f_n(x)$, and a given function $f(x)$, it is often not the difference but the mean or average value of the difference which is important. This can be defined in various ways. If the functions belong to the class L^p, where $p \geqslant 1$, we consider the integral

$$\int_a^b |f_n(x)-f(x)|^p\,dx. \tag{1}$$

If this integral tends to zero as $n \to \infty$, we say that $f_n(x)$ converges in mean (*en moyenne*, *im Mittel*), to $f(x)$, with index p.

If
$$\int_a^b |f_m(x)-f_n(x)|^p\,dx \tag{2}$$

tends to zero as m and n tend independently to infinity, we say that the sequence $f_n(x)$ converges in mean, with index p. Here the function $f(x)$ is not involved explicitly.

The fundamental theorem* of the subject is that *if the sequence $f_n(x)$ converges in mean, with index p, then there is a function $f(x)$ of the class L^p, defined uniquely apart from sets of measure zero, to which $f_n(x)$ converges in mean.*

The theorem is analogous to the 'general principle of convergence', that if $s_m - s_n \to 0$, then there is a number s to which s_n tends.

A word of explanation is necessary with regard to the 'uniqueness' of the limit-function $f(x)$. Suppose that we have found a function $f(x)$ which satisfies the given conditions. Then obviously any other function $g(x)$ which is equal to $f(x)$ almost everywhere also satisfies the conditions. So at any particular point the value of $f(x)$ is undetermined, though its general aggregate of values is in a sense determined. The function $f(x)$ should be regarded as a representative of a class of functions, any two of which are equal almost everywhere, and so all of which behave in the same way in integration.

The theorem for a finite interval and $p > 1$ may be proved as follows. To every integer ν corresponds a smallest positive integer n_ν such that

$$\int_a^b |f_m(x) - f_n(x)|^p \, dx < \frac{1}{3^\nu} \qquad (m \geqslant n_\nu,\ n \geqslant n_\nu).$$

In particular,

$$\int_a^b |f_{n_{\nu+1}}(x) - f_{n_\nu}(x)|^p \, dx < \frac{1}{3^\nu} \qquad (\nu = 1, 2, 3, \ldots).$$

If E_ν is the set where $|f_{n_{\nu+1}}(x) - f_{n_\nu}(x)| > 2^{-\nu/p}$, it follows that $m(E_\nu) < (\frac{2}{3})^\nu$. Hence the series

$$\sum_{\nu=1}^{\infty} |f_{n_{\nu+1}}(x) - f_{n_\nu}(x)|$$

is convergent, by comparison with $\sum 2^{-\nu/p}$, if, for some value of N, x does not belong to the set $E_{N+1} + E_{N+2} + \ldots$. Since the measure of this set tends to 0 as $N \to \infty$, it follows that the

* Fischer (1), F. Riesz (1), (2); W. H. and G. C. Young (2), where several alternative proofs are given; and Hobson (1).

above series is convergent for almost all values of x; hence so is

$$\sum_{\nu=1}^{\infty} \{f_{n_{\nu+1}}(x) - f_{n_\nu}(x)\},$$

i.e. there is a function $f(x)$ (defined almost everywhere) such that

$$\lim_{\nu\to\infty} f_{n_\nu}(x) = f(x)$$

almost everywhere.

This function $f(x)$ has the required property. For by Fatou's theorem

$$\varliminf_{\nu\to\infty} \int_a^b |f_m(x) - f_{n_\nu}(x)|^p \, dx \geqslant \int_a^b |f_m(x) - f(x)|^p \, dx;$$

but
$$\int_a^b |f_m(x) - f_{n_\nu}(x)|^p \, dx < \epsilon \qquad (m > m_0,\ \nu > \nu_0).$$

Hence
$$\int_a^b |f_m(x) - f(x)|^p \, dx \leqslant \epsilon \qquad (m > m_0),$$

i.e.
$$\lim \int_a^b |f_m(x) - f(x)|^p \, dx = 0,$$

i.e. the sequence $f_m(x)$ converges in mean to $f(x)$.

Finally, suppose that $f_n(x)$ converges in mean to $f(x)$ and also to $g(x)$. Then, by Minkowski's inequality,

$$\left(\int_a^b |f-g|^p \, dx\right)^{1/p} \leqslant \left(\int_a^b |f-f_n|^p \, dx\right)^{1/p} + \left(\int_a^b |g-f_n|^p \, dx\right)^{1/p} \to 0.$$

Hence the left-hand side is 0, and so $f(x) = g(x)$ almost everywhere.

If $p = 1$ the proof is simpler, since it is not necessary to use Minkowski's inequality. We have

$$|f-g| = |f-f_n-g+f_n| \leqslant |f-f_n| + |g-f_n|,$$

and hence

$$\int_a^b |f-g| \, dx \leqslant \int_a^b |f-f_n| \, dx + \int_a^b |g-f_n| \, dx.$$

The same result therefore follows.

12.51. The proof applies almost unchanged to an infinite interval. We find that the above series are convergent almost everywhere in (a, b) for every b, i.e. almost everywhere in (a, ∞); and Fatou's theorem holds for an infinite interval; for, taking the set E of § 10.81 to be the interval (a, b),

$$\int_a^b f(x)\,dx \leqslant \underline{\lim} \int_a^b f_n(x)\,dx \leqslant \underline{\lim} \int_a^\infty f_n(x)\,dx,$$

and, making $b \to \infty$, we obtain the required extension of Fatou's theorem. The proof for an infinite interval now follows.

12.52. *We have also (for a finite or infinite interval)*

$$\lim_{n\to\infty} \int |f_n(x)|^p\,dx = \int |f(x)|^p\,dx.$$

For by Minkowski's inequality

$$\left\{\int |f_n(x)|^p\,dx\right\}^{1/p} \leqslant \left\{\int |f(x)|^p\,dx\right\}^{1/p} + \left\{\int |f(x)-f_n(x)|^p\,dx\right\}^{1/p},$$

and also

$$\left\{\int |f(x)|^p\,dx\right\}^{1/p} \leqslant \left\{\int |f_n(x)|^p\,dx\right\}^{1/p} + \left\{\int |f(x)-f_n(x)|^p\,dx\right\}^{1/p}.$$

Hence $$\lim_{n\to\infty} \left\{\int |f_n(x)|^p\,dx\right\}^{1/p} = \left\{\int |f(x)|^p\,dx\right\}^{1/p},$$

and the result follows.

12.53. *If $f_n(x)$ converges in mean to $f(x)$ with index p, and $g(x)$ belongs to $L^{p/(p-1)}$, then*

$$\lim_{n\to\infty} \int f_n(x)g(x)\,dx = \int f(x)g(x)\,dx. \tag{1}$$

For

$$\left|\int \{f_n(x)-f(x)\}g(x)\,dx\right| \leqslant \left\{\int |f_n(x)-f(x)|^p\,dx\right\}^{1/p} \left\{\int |g(x)|^{p/(p-1)}\,dx\right\}^{1-1/p},$$

which tends to zero.

In particular

$$\lim_{n\to\infty} \int_a^x f_n(t)\,dt = \int_a^x f(t)\,dt \tag{2}$$

for all values of x in the interval considered.

For the function $g(t) = 1$ $(a < t < x)$, $= 0$ $(t > x)$, belongs to $L^{p/(p-1)}$.

Examples. (i) If $f_n(x) \to f(x)$ boundedly over a finite interval, then $f_n(x)$ converges in mean to $f(x)$ with any index.

(ii) Consider the closed intervals $(0, \frac{1}{2})$, $(\frac{1}{2}, 1)$, $(0, \frac{1}{3})$, $(\frac{1}{3}, \frac{2}{3})$, $(\frac{2}{3}, 1)$, $(0, \frac{1}{4})$, etc. Let $f_n(x) = 1$ in the nth interval, and $f_n(x) = 0$ in the remainder of the interval $(0, 1)$. Then $f_n(x)$ converges in mean to zero in $(0, 1)$, with any index; but $f_n(x)$ does not tend to zero for any value of x.

(iii) If $f_n(x)$ converges in mean to $f(x)$, and $f_n(x) \to g(x)$ almost everywhere, then $f(x) = g(x)$ almost everywhere. [Use Egoroff's theorem.]

(iv) If $f_n(x)$ converges in mean to $f(x)$ with index p, and $g_n(x)$ to $g(x)$ with index $p/(p-1)$, then $\int f_n g_n\, dx \to \int fg\, dx$.

12.6. Repeated integrals. As in the elementary cases considered in § 1.8, the equation

$$\int_a^b dx \int_\alpha^\beta f(x,y)\, dy = \int_\alpha^\beta dy \int_a^b f(x,y)\, dx \tag{1}$$

is in general true in the Lebesgue theory. The general discussion of this, however, depends on the theory of the double integral

$$\int\int f(x,y)\, dxdy,$$

which in turn depends on the theory of two-dimensional sets of points. It would take us too far to carry this out in detail.

There is, however, a particular kind of repeated integral which includes many cases of interest, and which can be dealt with by the theory already developed.

Let $f(x)$ be integrable in the Lebesgue sense over (a, b), and $g(y)$ over (α, β), and let $k(x, y)$ be a continuous function of both variables, or, if it has discontinuities, let them be of the type described in § 1.82. *Then*

$$\int_a^b f(x)\, dx \int_\alpha^\beta g(y)k(x,y)\, dy = \int_\alpha^\beta g(y)\, dy \int_a^b f(x)k(x,y)\, dx. \tag{2}$$

Suppose first that $f(x)$ and $g(y)$ are bounded, say $|f(x)| \leqslant M$, $|g(y)| \leqslant M$. Let $|k(x,y)| \leqslant K$.

Let $\phi(x)$ be a continuous function satisfying 12.2 (1); and let $|\phi(x)| \leqslant M$. Let $\psi(y)$ be a continuous function related in the same way to $g(y)$.

Call the left-hand side of (2) I, and let

$$I' = \int_a^b \phi(x)\, dx \int_\alpha^\beta \psi(y)k(x,y)\, dy.$$

Then

$$I-I' = \int_a^b \{f(x)-\phi(x)\}\,dx \int_\alpha^\beta g(y)k(x,y)\,dy +$$

$$+\int_a^b \phi(x)\,dx \int_\alpha^\beta \{g(y)-\psi(y)\}k(x,y)\,dy,$$

and hence

$$|I-I'| \leqslant \int_a^b |f(x)-\phi(x)|(\beta-\alpha)MK\,dx + (b-a)M\eta$$

$$\leqslant MK(\beta-\alpha+b-a)\eta.$$

Similarly, if the right-hand side of (2) is J, and

$$J' = \int_\alpha^\beta \psi(y)\,dy \int_a^b \phi(x)k(x,y)\,dx,$$

then $|J-J'|$ tends to 0 with η.

But, by the theorem of § 1.81, $I' = J'$, since $\phi(x)\psi(y)k(x,y)$ is continuous, or has discontinuities of the restricted type.

Hence $|I-J|$ tends to 0 with η, and so $I = J$.

The extension to unbounded functions may be left to the reader; we suppose first that $f(x)$ and $g(x)$ are positive, and argue with $\{f(x)\}_n$ and $\{g(x)\}_n$ in the usual manner.

12.61. *If $f(x)$ is integrable over $(0, 1)$, and $g(x)$ over $(0, 2)$, then the integral*

$$\int_0^1 f(x)g(x+t)\,dx$$

exists for almost all values of t in $(0, 1)$, and represents an integrable function of t.

It is sufficient to consider the case where f and g are positive. Define $\{f(x)\}_n$ as usual, and let

$$F_n(t) = \int_0^1 \{f(x)\}_n\, g(x+t)\,dx.$$

This integral exists for all values of t, and, for a given n, $F_n(t)$ is bounded, and for each value of t it is a non-decreasing function of n. Also

$$\int_0^1 F_n(t)\,dt = \int_0^1 dt \int_0^1 \{f(x)\}_n\, g(x+t)\,dx = \int_0^1 \{f(x)\}_n\,dx \int_0^1 g(x+t)\,dt, \tag{1}$$

if we may invert the order of integration.

To justify this, approximate to g by a continuous function ψ, as in the previous proof, and let

$$\chi(t) = \int_0^1 \{f(x)\}_n \psi(x+t)\, dx.$$

Then
$$\int_0^1 \chi(t)\, dt = \int_0^1 \{f(x)\}_n\, dx \int_0^1 \psi(x+t)\, dt, \tag{2}$$

this inversion being justified by the above theorem. Now

$$|F_n(t) - \chi(t)| \leqslant \int_0^1 \{f(x)\}_n |g(x+t) - \psi(x+t)|\, dx < n\eta,$$

so that the left-hand side of (1) differs from that of (2) by less than $n\eta$. Similarly the right-hand sides differ by less than $n\eta$. Hence, making $\eta \to 0$, we obtain (1).

Hence
$$\int_0^1 F_n(t)\, dt \leqslant \int_0^1 f(x)\, dx \int_0^2 g(y)\, dy.$$

Hence, as $n \to \infty$, $F_n(t)$ tends to a finite limit for almost all values of t (§ 10.82). The result now follows from the theorem of § 10.82.

12.62. Repeated infinite integrals. *If $f(x)$, $g(y)$, and $k(x,y)$ are positive, and the conditions of* § 12.6 *are satisfied for all values of $b > a$ and $\beta > \alpha$, then*

$$\int_a^\infty f(x)\, dx \int_\alpha^\infty g(y)k(x,y)\, dy = \int_\alpha^\infty g(y)\, dy \int_a^\infty f(x)k(x,y)\, dx \tag{1}$$

provided that either side is convergent.

The theorem is similar to that of § 1.85, but the supplementary conditions which appear there are now a consequence of the main hypothesis.

Suppose that the right-hand side of (1) is convergent. Since

$$\int_a^X f(x)k(x,y)\, dx \leqslant \int_a^\infty f(x)k(x,y)\, dx, \tag{2}$$

and the left-hand side of (2) is a measurable (in fact a continuous) function of y, it follows that

$$\int_\alpha^\infty g(y)\, dy \int_a^X f(x)k(x,y)\, dx$$

is convergent. Hence

$$\lim_{n\to\infty}\int_\alpha^n g(y)\,dy\int_a^X f(x)k(x,y)\,dx = \lim_{n\to\infty}\int_a^X f(x)\,dx\int_\alpha^n g(y)k(x,y)\,dy$$

is finite. Also

$$F_n(x) = f(x)\int_\alpha^n g(y)k(x,y)\,dy$$

is a non-decreasing function of n for each value of x. It therefore follows from § 10.82 that $F_n(x)$ tends to a finite limit, as $n \to \infty$, for almost all values of x in (a, X); i.e.

$$\int_\alpha^\infty g(y)k(x,y)\,dy$$

is convergent for almost all values of x in (a, X); and by § 10.82

$$\int_a^X f(x)\,dx\int_\alpha^\infty g(y)k(x,y)\,dy = \lim_{n\to\infty}\int_a^X f(x)\,dx\int_\alpha^n g(y)k(x,y)\,dy$$

$$= \lim_{n\to\infty}\int_\alpha^n g(y)\,dy\int_a^X f(x)k(x,y)\,dx = \int_\alpha^\infty g(y)\,dy\int_a^X f(x)k(x,y)\,dx. \quad (3)$$

By (2), the right-hand side of (3) is bounded as $X \to \infty$; hence so is the left-hand side, and therefore the left-hand side of (1) is convergent.

We can now prove in a similar way that the order of integration in

$$\int_\alpha^Y g(y)\,dy\int_a^\infty f(x)k(x,y)\,dx$$

may be inverted. The final result then follows as in § 1.85.

MISCELLANEOUS EXAMPLES.

1. If $f(x)$ is integrable over (a, b), and $a = x_0 < x_1 < x_2 < \ldots < x_n = b$, then

$$\sum_{\nu=0}^{n-1}\left|\int_{x_\nu}^{x_{\nu+1}} f(t)\,dt\right| \to \int_a^b |f(t)|\,dt$$

as the greatest partial interval tends to zero.

[The proof is elementary for continuous functions; and then the general result may be deduced by means of the theorem of § 12.2.]

2. If $F(x)$ is absolutely continuous in (a, b), its total variation in the interval is

$$\int_a^b |F'(x)|\,dx.$$

[Use the result of the previous example.]

3. Show that, if $f(x)$ and $g(x)$ belong to L^2,

$$\int \{f(x)\}^2\,dx \int \{g(x)\}^2\,dx - \left\{\int f(x)g(x)\,dx\right\}^2$$

$$= \tfrac{1}{2}\int dy \int \{f(x)g(y)-f(y)g(x)\}^2\,dx,$$

and hence obtain another proof of Schwarz's inequality.

4. We use $\log' x$ to denote $\log x$ if $x \geqslant e$, and $\log' x = 1$ if $x < e$.

Show that if $\{f(x)\}^2 \log' f(x)$ and $\{g(x)\}^2/\log' g(x)$ are integrable over (a, b), then $f(x)g(x)$ is integrable over (a, b).

[Let E be the set where $f \leqslant g/\log' g$. Then

$$\int_E fg\,dx \leqslant \int_E \frac{\{g(x)\}^2}{\log' g(x)}\,dx.$$

In CE, $g \leqslant f \log' g$. If $g \leqslant e$ this gives

$$g \leqslant f \leqslant f \log' f.$$

If $g > e$, $\sqrt{g} < Ag/\log' g < Af$, $\log g < A \log f$, and hence again, $g < Af \log' f$. Hence

$$\int_{CE} fg\,dx \leqslant A \int_{CE} \{f(x)\}^2 \log' f(x)\,dx.$$

5. If $f^p(\log' f)^q$ and $g^{p/(p-1)}(\log' g)^{-q/(p-1)}$ are integrable, then $f(x)g(x)$ is integrable.

6. Prove that $$uv \leqslant u \log u + e^{v-1} \qquad (u > 1,\ v > 1).$$

Deduce that if $f(x)\log' f(x)$ and $e^{g(x)}$ are integrable, so is $f(x)g(x)$.

[W. H. Young (4). The inequality may be verified by putting $u = e^x$, $v = y+1$.]

7. If $\alpha > 0$, $\beta > 0$, $\gamma > 0$, $\alpha+\beta+\gamma = 1$,

$$\left|\int fgh\,dx\right| \leqslant \left(\int |f|^{1/\alpha}\,dx\right)^{\alpha}\left(\int |g|^{1/\beta}\,dx\right)^{\beta}\left(\int |h|^{1/\gamma}\,dx\right)^{\gamma}.$$

8. If $\lambda > 0$, $\mu > 0$, $\lambda\mu < 1$, and $f(x)$ and $g(x)$ belong to suitable L-classes, then

$$\left|\int fg\,dx\right|^{(1+\lambda)(1+\mu)/(1-\lambda\mu)}$$

$$\leqslant \int |f|^{1+\lambda}|g|^{1+\mu}\,dx \left(\int |f|^{1+\lambda}\,dx\right)^{\mu(1+\lambda)/(1-\lambda\mu)}\left(\int |g|^{1+\mu}\,dx\right)^{\lambda(1+\mu)/(1-\lambda\mu)}$$

[W. H. Young (2); the result may be obtained by suitable substitutions in ex. 7.]

9. If $F(x)$ is the integral of a function of the class L^p, where $p > 1$, then as $h \to 0$ $$F(x+h)-F(x) = o\,(h^{1-1/p}).$$

[If $F(x)$ is the integral of $f(x)$,

$$|F(x+h)-F(x)| = \left|\int_x^{x+h} f(t)\,dt\right|$$

$$\leqslant \left\{\int_x^{x+h} |f(t)|^p\,dt\right\}^{1/p}\left\{\int_x^{x+h} dt\right\}^{1-1/p} = h^{1-1/p}\left\{\int_x^{x+h} |f(t)|^p\,dt\right\}^{1-1/p},$$

and the last factor tends to zero with h.]

10. If $f(x)$ belongs to $L^p(0, \infty)$, where $p > 1$, the integral

$$\int_0^\infty f(x)\,\frac{\sin xy}{x}\,dx$$

is uniformly convergent in any finite interval.

11. If $f(x)$ belongs to L^p, where $p > 1$, and $\phi(y)$ is the integral defined in the previous example, then as $h \to 0$

$$\phi(y+h)-\phi(y) = o\,(h^{1/p}).$$

[For $$\phi(y+h)-\phi(y) = \int_0^\infty \frac{f(x)}{x}\,[\sin\{x(y+h)\} - \sin xy]\,dx$$

$$= 2\int_0^\infty \frac{f(x)}{x}\,\sin\tfrac{1}{2}xh\cos x(y+\tfrac{1}{2}h)\,dx.$$

Hence

$$|\phi(y+h)-\phi(y)| \leqslant 2\int_0^\infty \left|f(x)\,\frac{\sin\frac{1}{2}xh}{x}\right|dx$$

$$\leqslant 2\left\{\int_0^\infty |f(x)|^p\,dx\right\}^{1/p}\left\{\int_0^\infty \left|\frac{\sin\frac{1}{2}xh}{x}\right|^{p/(p-1)}dx\right\}^{1-1/p}$$

The first factor is a constant, and the second factor, on putting $x = \xi/h$, is seen to be a multiple of $h^{1/p}$. This gives the required result with O instead of o. If, however, we apply the above argument to the integrals over $(0, \delta)$ and (Δ, ∞), where δ is arbitrarily small and Δ arbitrarily large, and notice that

$$\left|\int_\delta^\Delta \frac{f(x)}{x}\,\sin\tfrac{1}{2}xh\cos x(y+\tfrac{1}{2}h)\,dx\right| \leqslant \int_\delta^\Delta \frac{|f(x)|}{x}\,|\sin\tfrac{1}{2}xh|\,dx = O(h)$$

for fixed δ and Δ, the required result follows.]

12. Show that the integral

$$\phi(y) = \int_0^\infty f(x) \frac{\sin xy}{\sqrt{x}}\, dx$$

is absolutely convergent if $f(x)$ belongs to L^p, where $1 < p < 2$; and that, as $y \to 0$,

$$\phi(y) = o\,(y^{1/p-\frac{1}{2}}).$$

13. If $f(x)$ is uniformly continuous over $(0, \infty)$, and belongs to a class $L^p(0, \infty)$, then $f(x) \to 0$ as $x \to \infty$.

14. If $f(x)$ belongs to $L^p(0, \infty)$, where $p > 1$, so do the functions

$$\phi(x) = \frac{1}{x} \int_0^x f(t)\, dt, \qquad \psi(x) = \int_x^\infty \frac{f(t)}{t}\, dt.$$

[Hardy (17) and (19). Consider $\phi(x)$, for example. It is bounded except as $x \to 0$ or $x \to \infty$. Hence

$$\int_a^b |\phi(x)|^p\, dx$$

exists for $0 < a < b < \infty$; and it is sufficient to prove that this integral remains bounded as $a \to 0$ and $b \to \infty$.

We may suppose without loss of generality that $f(t) \geqslant 0$. Let

$$f_1(x) = \int_0^x f(t)\, dt.$$

Then $x^{1-p}\{f_1(x)\}^p$ tends to zero, both as $x \to 0$ and as $x \to \infty$; for

$$\{f_1(x)\}^p \leqslant \int_0^x \{f(t)\}^p\, dt \left(\int_0^x dt\right)^{p-1} = x^{p-1} \int_0^x \{f(t)\}^p\, dt,$$

whence the result for $x = 0$ follows. Again, if $x > \xi$, a similar argument shows that

$$f_1(x) \leqslant \int_0^\xi f(t)\, dt + \left\{x^{p-1} \int_\xi^x \{f(t)\}^p\, dt\right\}^{1/p},$$

and we can choose ξ so large that the last factor is arbitrarily small, for all $x > \xi$. This gives the result for $x \to \infty$.

We write
$$\int_a^b \{\phi(x)\}^p\, dx = \int_a^b \{f_1(x)\}^p x^{-p}\, dx,$$

and integrate by parts, obtaining

$$\left[\frac{\{f_1(x)\}^p x^{1-p}}{1-p}\right]_a^b + \frac{p}{p-1} \int_a^b \{f_1(x)\}^{p-1} f(x) x^{1-p}\, dx$$

$$= o\,(1) + \frac{p}{p-1} \int_a^b \{\phi(x)\}^{p-1} f(x)\, dx.$$

Hence

$$\int_a^b \{\phi(x)\}^p\,dx \leqslant o(1) + \frac{p}{p-1}\left\{\int_a^b \{\phi(x)\}^p\,dx\right\}^{1-1/p}\left\{\int_a^b \{f(x)\}^p\,dx\right\}^{1/p},$$

and dividing by the factor

$$\left\{\int_a^b \{\phi(x)\}^p\,dx\right\}^{1-1/p},$$

and making $a \to 0$, $b \to \infty$, we obtain

$$\left\{\int_0^\infty \{\phi(x)\}^p\,dx\right\}^{1/p} \leqslant \frac{p}{p-1}\left\{\int_0^\infty \{f(x)\}^p\,dx\right\}^{1/p}.$$

We leave the corresponding process for $\psi(x)$ to the reader.]

15. Prove that, with the hypotheses of the previous example, the integrals

$$\int_0^\infty |\phi(x)|^q x^{q/p-1}\,dx, \qquad \int_0^\infty |\psi(x)|^q x^{q/p-1}\,dx$$

are convergent for $q \geqslant p$.

16. If $f(x)$ belongs to $L^p(0, \infty)$, where $p > 1$, and

$$\phi(x) = \int_0^\infty e^{-xy} f(y)\,dy,$$

then $y^{1-2/p}\phi(y)$ belongs to L^p.

[For $$|\phi(x)| \leqslant \int_0^{1/x} |f(y)|\,dy + \int_{1/x}^\infty \frac{|f(y)|}{xy}\,dy,$$

and the result follows from ex. 14.]

17. If $f(x)$ belongs to $L^p(a, b)$, there is a continuous function $g(x)$ such that

$$\int_a^b |f(x)-g(x)|^p\,dx < \epsilon.$$

[The result for bounded $f(x)$ follows at once from § 12.2, and the general result may then be deduced from this.]

18. If $f(x)$ belongs to L^p over an interval including (a, b), then

$$\lim_{h\to 0}\int_a^b |f(x+h)-f(x)|^p\,dx = 0.$$

[The result is immediate for continuous functions. For the general result use the previous example.]

19. If $f(x)$ belongs to $L^p(-\infty, \infty)$, then

$$\lim_{h\to 0}\int_{-\infty}^\infty |f(x+h)-f(x)|^p\,dx = 0.$$

20. If $f(x)$ belongs to $L^p(-\infty, \infty)$, and $g(x)$ to $L^{p/(p-1)}(-\infty, \infty)$, then

$$F(t) = \int_{-\infty}^{\infty} f(x+t)g(x)\,dx$$

is a continuous function of t.

[For $|F(t+h)-F(t)|$

$$\leqslant \left\{\int_{-\infty}^{\infty} |f(x+t+h)-f(x+t)|^p\,dx\right\}^{1/p} \left\{\int_{-\infty}^{\infty} |g(x)|^{p/(p-1)}\,dx\right\}^{1-1/p}\Bigg]$$

21. The function $F(t)$ of the previous example tends to zero at infinity.

$$\left[\text{Write } \int_{-\infty}^{\infty} = \int_{-\infty}^{-\frac{1}{2}t} + \int_{-\frac{1}{2}t}^{\infty}\right]$$

22. If $f(x)$ belongs to $L^p(-\infty, \infty)$, then

$$F(x) = \int_{-\infty}^{\infty} \frac{f(t)}{1+(x-t)^2}\,dt$$

is continuous, and belongs to $L^p(-\infty, \infty)$.

[Use the inequality

$$|F(x)|^p \leqslant \int_{-\infty}^{\infty} \frac{|f(t)|^p\,dt}{\{1+(x-t)^2\}^{\frac{1}{2}p}} \left\{\int_{-\infty}^{\infty} \frac{dt}{\{1+(x-t)^2\}^{\frac{1}{2}p/(p-1)}}\right\}^{p-1}\Bigg]$$

23. If $f(x)$ is integrable, the integral

$$f_\alpha(x) = \frac{1}{\Gamma(\alpha)} \int_0^x (x-t)^{\alpha-1} f(t)\,dt \qquad (\alpha > 0)$$

exists almost everywhere, and $f_\alpha(x)$ is integrable.

[The function $f_\alpha(x)$ is the integral of $f(x)$ of order α; for some properties of such integrals see Hardy (12) and Hardy and Littlewood (5).]

24. If $f(x)$ belongs to $L^p(p > 1)$, show by the method of ex. 20 that $f_\alpha(x)$ is continuous if $\alpha > 1/p$.

25. If $f_{\alpha,\beta}(x)$ denotes the integral of order β of $f_\alpha(x)$, then

$$f_{\alpha,\beta}(x) = f_{\alpha+\beta}(x) \qquad (\alpha > 0,\ \beta > 0),$$

wherever the right-hand side exists.

[We have to invert the repeated integral

$$\int_0^x (x-t)^{\beta-1}\,dt \int_0^t (t-u)^{\alpha-1} f(u)\,du$$

and use Ch. I, ex. 18. The integral $\int_\delta^x \int_0^{t-\delta}$ may be inverted by § 12.6. We can then make $\delta \to 0$ and use the theorem of § 10.82.]

CHAPTER XIII

FOURIER SERIES

13.1. Trigonometrical series and Fourier series. A trigonometrical series is a series of the form

$$\tfrac{1}{2}a_0 + \sum_{n=1}^{\infty} (a_n \cos nx + b_n \sin nx), \tag{1}$$

where the coefficients a_0, a_1, b_1,... are independent of x. The problem of representing a given function $f(x)$ by a series of this form was first encountered by Fourier in a problem of the conduction of heat. Subsequently it was found that these series play an important part in the theory of functions of a real variable, and it is from this point of view that we shall consider them here.

We naturally begin by trying to find formulae for the coefficients a_n, b_n, in terms of the given function $f(x)$. Suppose that the series converges uniformly, or even boundedly, to $f(x)$; we may then multiply by $\cos mx$, where m is a positive integer, and integrate term by term over the interval $(0, 2\pi)$. Since

$$\int_0^{2\pi} \cos mx \cos nx \, dx = \pi \ (n = m), \qquad = 0 \ (n \neq m)$$

and

$$\int_0^{2\pi} \cos mx \sin nx \, dx = 0$$

for all values of n, we obtain the result

$$a_m = \frac{1}{\pi} \int_0^{2\pi} f(x) \cos mx \, dx. \tag{2}$$

The same formula also gives a_0; and similarly, multiplying by $\sin mx$ and integrating term by term,

$$b_m = \frac{1}{\pi} \int_0^{2\pi} f(x) \sin mx \, dx. \tag{3}$$

The formulae (2) and (3) are known as the Euler-Fourier formulae for the coefficients.

There is, however, no *a priori* reason for supposing that a given function can be expanded in a boundedly convergent trigonometrical series. The above process is therefore not a proof that the coefficients necessarily have the above form. What it really suggests is that we should adopt a different point of view. Instead of starting with the series, and assuming that it has a certain property, we start from the function, and define the coefficients by the above formulae. We then consider the properties of the series so formed.

Suppose, then, that we are given a function $f(x)$, integrable in the sense of Lebesgue over the interval $(0, 2\pi)$. Then the integrals (2) and (3) exist, and the numbers a_m, b_m defined by them are called the *Fourier coefficients of* $f(x)$. The trigonometrical series of the form (1), with these coefficients, is called the Fourier series of $f(x)$.

The scheme of the chapter is as follows. We first try to determine conditions under which the Fourier series converges to $f(x)$. A number of these conditions are found, but they are all rather special ones (§§ 13.11–13.25). We next consider a generalized kind of convergence (summability $(C, 1)$), and find that it enables us to put the theory into a more systematic form (§§ 13.3–13.35). In the following sections we consider some problems of term-by-term integration; and this leads us to consider properties of the Fourier coefficients themselves, apart from the Fourier series. In §§ 13.8–13.86 we return to the question of the relation between Fourier series in particular and trigonometrical series in general. Lastly we give some of the corresponding theory of Fourier integrals.

13.11. The convergence problem. The first problem which we have to consider is whether the series formed in the above manner converges, and, if it does, whether its sum is $f(x)$.

At the time when Fourier series first came into use, there seemed to many mathematicians to be something paradoxical in saying that an 'arbitrary' function could be represented by a series of functions, each of which is continuous and periodic. The reader who has examined the peculiarities of some of the series in Chapter I is perhaps prepared to believe that even this is possible; and we shall show that the series does, substantially, do what is required of it. We must not, however, expect too much.

In the first place, every term of the series has the period 2π; hence the sum of the series, if there is one, also has the period 2π. We therefore define the function $f(x)$ first in the interval

$0 \leqslant x < 2\pi$; outside this interval we define it by periodicity, i.e. by the equation

$$f(x+2\pi) = f(x).$$

Secondly, it is impossible that, whatever $f(x)$ is, the series should converge to the sum $f(x)$ for every value of x. Consider, for example, two functions $f(x)$ and $g(x)$ which differ at one point only. They have the same Fourier series, so that it cannot represent both functions at every point. More generally, two 'equivalent' functions, i.e. functions which are equal almost everywhere, have the same Fourier series, which therefore cannot represent them both if they differ anywhere.

Actually we shall see that the series does represent the function, provided that the function is not too complicated; and even in the most complicated cases, the series still represents in some sense the main features of the function.

13.12. Fourier series and Laurent series. There is a close formal connexion between a Fourier series and a Laurent series. Let $F(z)$ be a one-valued analytic function, regular for $R' < |z| < R$. Then

$$F(z) = \sum_{n=-\infty}^{\infty} c_n z^n,$$

where
$$c_n = \frac{1}{2\pi i} \int\limits_{|z|=r} \frac{F(z)}{z^{n+1}}\, dz \qquad (R' < r < R).$$

Putting $z = re^{i\theta}$, we have

$$F(re^{i\theta}) = \sum_{n=-\infty}^{\infty} A_n e^{in\theta},$$

where
$$A_n = \frac{1}{2\pi} \int\limits_0^{2\pi} F(re^{i\phi}) e^{-in\phi}\, d\phi.$$

The expansion may also be written

$$F(re^{i\theta}) = A_0 + \sum_{n=1}^{\infty} \{(A_n + A_{-n})\cos n\theta + i(A_n - A_{-n})\sin n\theta\},$$

where

$$A_0 = \frac{1}{2\pi} \int\limits_0^{2\pi} F(re^{i\phi})\, d\phi, \qquad A_n + A_{-n} = \frac{1}{\pi} \int\limits_0^{2\pi} F(re^{i\phi}) \cos n\phi\, d\phi,$$

$$i(A_n - A_{-n}) = \frac{1}{\pi} \int\limits_0^{2\pi} F(re^{i\phi}) \sin n\phi\, d\phi.$$

We have thus expressed the Laurent series in the form of a Fourier series. The fact that in this case the series represents the function, and indeed converges uniformly to it, follows from the theory of analytic functions. In general we assume much less about the function than that it is analytic, and the problem requires quite different methods.

13.2. Dirichlet's integral. Let $0 \leqslant x < 2\pi$, and let

$$s_n = s_n(x) = \tfrac{1}{2}a_0 + \sum_{m=1}^{n} (a_m \cos mx + b_m \sin mx). \tag{1}$$

This partial sum can be represented as a definite integral. We have

$$s_n = \frac{1}{2\pi}\int_0^{2\pi} f(t)\,dt +$$

$$+\frac{1}{\pi}\sum_{m=1}^{n}\left\{\cos mx \int_0^{2\pi} f(t)\cos mt\,dt + \sin mx \int_0^{2\pi} f(t)\sin mt\,dt\right\}$$

$$= \frac{1}{\pi}\int_0^{2\pi}\left\{\tfrac{1}{2} + \sum_{m=1}^{n}\cos m(x-t)\right\} f(t)\,dt = \frac{1}{2\pi}\int_0^{2\pi}\frac{\sin(n+\frac{1}{2})(x-t)}{\sin\frac{1}{2}(x-t)} f(t)\,dt.$$

Putting $t = x+u$, this becomes

$$s_n = \frac{1}{2\pi}\int_{-x}^{2\pi-x}\frac{\sin(n+\frac{1}{2})u}{\sin\frac{1}{2}u} f(x+u)\,du,$$

or, since the integrand has the period 2π, and so takes the same values in $(2\pi - x, 2\pi)$ as in $(-x, 0)$,

$$s_n = \frac{1}{2\pi}\int_0^{2\pi}\frac{\sin(n+\frac{1}{2})u}{\sin\frac{1}{2}u} f(x+u)\,du. \tag{2}$$

This formula is known as *Dirichlet's integral*. It may also be written in the form

$$s_n = \frac{1}{2\pi}\int_0^{\pi}\frac{\sin(n+\frac{1}{2})u}{\sin\frac{1}{2}u}\{f(x+u)+f(x-u)\}\,du. \tag{3}$$

This is obtained by writing $u = -v$ in the range $(\pi, 2\pi)$, so that this part of (2) becomes

$$\int_{-2\pi}^{-\pi} \frac{\sin(n+\frac{1}{2})v}{\sin\frac{1}{2}v} f(x-v)\,dv = \int_{0}^{\pi} \frac{\sin(n+\frac{1}{2})u}{\sin\frac{1}{2}u} f(x-u)\,du$$

by periodicity.

Suppose, in particular, that $f(x) = 1$ for all values of x. Then $a_0 = 2$, and all the rest of the Fourier coefficients are zero, so that $s_n = 1$ for $n > 0$. In this case the above formula becomes

$$1 = \frac{1}{2\pi} \int_{0}^{\pi} \frac{\sin(n+\frac{1}{2})u}{\sin\frac{1}{2}u}\, 2\,du.$$

Multiplying this by s, and subtracting from (3), we have

$$s_n - s = \frac{1}{2\pi} \int_{0}^{\pi} \frac{\sin(n+\frac{1}{2})u}{\sin\frac{1}{2}u} \{f(x+u)+f(x-u)-2s\}\,du. \tag{4}$$

A necessary and sufficient condition that the series should converge to the sum s is, therefore, that this integral should tend to zero. The 'convergence problem' is the problem of determining under what conditions the integral tends to zero, and, when it does so, whether $s = f(x)$. We may consider the convergence problem for one particular value of x, for all values of x, or for almost all values of x; or for some other set of values of x. We begin by considering one particular value of x.

13.21. The Riemann-Lebesgue theorem. The following theorem is fundamental in the theory.

If $f(x)$ is integrable over (a, b), then as $\lambda \to \infty$

$$\int_a^b f(x)\cos\lambda x\,dx \to 0, \qquad \int_a^b f(x)\sin\lambda x\,dx \to 0.$$

Consider, for example, the cosine integral. If $f(x)$ is an integral, we may integrate by parts, and obtain

$$\int_a^b f(x)\cos\lambda x\,dx = \left[f(x)\frac{\sin\lambda x}{\lambda}\right]_a^b - \frac{1}{\lambda}\int_a^b f'(x)\sin\lambda x\,dx.$$

The last integral is bounded, so that the whole is $O(1/\lambda)$.

In the general case, given ϵ, we can (§ 12.2) define an absolutely continuous function $\phi(x)$ such that

$$\int_a^b |f(x)-\phi(x)|\,dx < \epsilon.$$

Then $$\left|\int_a^b \{f(x)-\phi(x)\}\cos\lambda x\,dx\right| \leqslant \int_a^b |f(x)-\phi(x)|\,dx < \epsilon$$

for all values of λ; and, by the first part,

$$\left|\int_a^b \phi(x)\cos\lambda x\,dx\right| < \epsilon \qquad (\lambda > \lambda_0).$$

Hence $$\left|\int_a^b f(x)\cos\lambda x\,dx\right| < 2\epsilon \qquad (\lambda > \lambda_0),$$

the required result. A similar proof applies to the sine integral.

There is an alternative proof on the lines of § 13.72, using the example of § 12.2.

13.22. The Riemann-Lebesgue theorem has the following important consequences:

The Fourier coefficients of any integrable function tend to zero.

This is the particular case of the theorem where $\lambda = n$, and the limits are 0 and 2π.

The behaviour of the Fourier series for a particular value of x depends on the behaviour of the function in the immediate neighbourhood of this point only.

Let δ be a positive number less than π, and let $g(t) = f(t)$ in the interval $x-\delta < t < x+\delta$, and $g(t) = 0$ in the rest of the interval $(x-\pi, x+\pi)$. Let the partial sums of the Fourier series of $g(t)$ be denoted by S_n. Then

$$S_n = \frac{1}{2\pi}\int_0^\pi \frac{\sin(n+\frac{1}{2})u}{\sin\frac{1}{2}u}\{g(x+u)+g(x-u)\}\,du$$

$$= \frac{1}{2\pi}\int_0^\delta \frac{\sin(n+\frac{1}{2})u}{\sin\frac{1}{2}u}\{f(x+u)+f(x-u)\}\,du.$$

Hence

$$s_n - S_n = \frac{1}{2\pi}\int_\delta^\pi \frac{\sin(n+\frac{1}{2})u}{\sin\frac{1}{2}u}\{f(x+u)+f(x-u)\}\,du.$$

Now the function

$$\operatorname{cosec} \tfrac{1}{2}u\{f(x+u)+f(x-u)\}$$

is integrable over (δ, π) if $\delta > 0$; and hence, by the Riemann-Lebesgue theorem,

$$s_n - S_n \to 0.$$

Hence, however small δ may be, the behaviour of s_n depends on the nature of $f(t)$ in the interval $(x-\delta, x+\delta)$ only, and is not affected by the values which it takes outside this interval.

It is this property which makes it possible for the series to represent an arbitrary function; but the series only represents the function at the point x as a sort of limit of its average value over the interval $(x-\delta, x+\delta)$, and this will be equal to $f(x)$ only if the behaviour of the function is sufficiently simple. As we have already remarked in § 13.11, the value of $f(t)$ at the point $t = x$ itself does not determine or affect in any way the sum of the series.

13.23. Convergence tests. We first put the 'necessary and sufficient condition for convergence to the sum s' into a more convenient form. Let

$$\phi(u) = f(x+u)+f(x-u)-2s.$$

Then the condition, by § 13.2 (4), is

$$\lim_{n\to\infty} \int_0^\pi \frac{\sin(n+\frac{1}{2})u}{\sin \frac{1}{2}u}\, \phi(u)\, du = 0. \tag{1}$$

We may replace this by

$$\lim_{n\to\infty} \int_0^\delta \frac{\sin(n+\frac{1}{2})u}{\sin \frac{1}{2}u}\, \phi(u)\, du = 0, \tag{2}$$

where $0 < \delta \leqslant \pi$; for, by the Riemann-Lebesgue theorem, the difference between the integrals in (1) and (2) tends to zero. Next we may replace (2) by

$$\lim_{n\to\infty} \int_0^\delta \frac{\sin(n+\frac{1}{2})u}{u}\, \phi(u)\, du = 0; \tag{3}$$

for $(\operatorname{cosec}\frac{1}{2}u - 2/u)\phi(u)$ is integrable over $(0, \delta)$, and so, by the Riemann-Lebesgue theorem,

$$\lim_{n\to\infty} \int_0^\delta \sin(n+\tfrac{1}{2})u \left\{\frac{1}{\sin\frac{1}{2}u} - \frac{2}{u}\right\} \phi(u)\, du = 0.$$

We are now in a position to state some tests for convergence.

13.231. Dini's test. *If $\phi(u)/u$ is integrable over $(0, \delta)$, then the series converges to the sum s.*

This theorem is at once obvious from the above formula (3) and the Riemann-Lebesgue theorem. It should of course be remembered that the integrability of $\phi(u)/u$ in the Lebesgue sense implies 'absolute integrability'. The existence of

$$\lim_{\epsilon\to 0} \int_\epsilon^\delta \frac{\phi(u)}{u}\, du$$

is not a sufficient condition for convergence.

Examples. (i) At any point where $f(x)$ is differentiable, the series converges to the sum $f(x)$.

[At such a point, $\phi(u)/u$ is bounded.]

(ii) More generally, if $f(x)$ satisfies the 'Lipschitz condition' of order α, i.e.

$$f(x+h) - f(x) = O(|h|^\alpha) \qquad (0 < \alpha < 1),$$

then the series converges to the sum $f(x)$.

13.232. Jordan's test. *If $f(t)$ is of bounded variation in the neighbourhood of $t = x$, then the series converges to the sum*

$$\tfrac{1}{2}\{f(x+0) + f(x-0)\}.$$

Since 'bounded variation' means 'bounded variation over an interval', this condition is really one for convergence over an interval.

We know that, if $f(x)$ is of bounded variation, the limits $f(x+0)$ and $f(x-0)$ exist. Hence

$$\phi(u) = f(x+u) + f(x-u) - f(x+0) - f(x-0)$$

is of bounded variation in an interval to the right of $u = 0$, and $\phi(u) \to 0$ as $u \to 0$. Hence we may write

$$\phi(u) = \phi_1(u) - \phi_2(u),$$

where ϕ_1 and ϕ_2 are positive increasing functions of u; each of these functions tends to the same limit as $u \to 0$; and we may, by subtracting a constant from each function, arrange that this limit shall be zero.

Suppose that δ is so small that $\phi(u)$ is of bounded variation in the interval $(0,\delta)$. Then

$$\int_0^\delta \frac{\sin(n+\frac{1}{2})u}{u}\,\phi(u)\,du$$
$$= \int_0^\delta \frac{\sin(n+\frac{1}{2})u}{u}\,\phi_1(u)\,du - \int_0^\delta \frac{\sin(n+\frac{1}{2})u}{u}\,\phi_2(u)\,du$$
$$= J_1 - J_2,$$

say. Consider the integral J_1. Given ϵ, choose η so small that $\phi_1(\eta) < \epsilon$. Then, by the second mean-value theorem,

$$\int_0^\eta \frac{\sin(n+\frac{1}{2})u}{u}\,\phi_1(u)\,du = \phi_1(\eta-0)\int_\xi^\eta \frac{\sin(n+\frac{1}{2})u}{u}\,du \quad (0 < \xi < \eta)$$
$$= \phi_1(\eta-0)\int_{(n+\frac{1}{2})\xi}^{(n+\frac{1}{2})\eta} \frac{\sin v}{v}\,dv.$$

The last integral is bounded for all values of n, ξ, and η, so that

$$\left|\int_0^\eta \frac{\sin(n+\frac{1}{2})u}{u}\,\phi_1(u)\,du\right| < A\epsilon.$$

Having fixed η, by the Riemann-Lebesgue theorem

$$\left|\int_\eta^\delta \frac{\sin(n+\frac{1}{2})u}{u}\,\phi_1(u)\,du\right| < \epsilon \qquad (n > n_0).$$

Hence $J_1 \to 0$; and similarly $J_2 \to 0$. This proves the theorem.

In particular *if $f(x)$ has only a finite number of maxima and minima and a finite number of discontinuities in the interval $(0, 2\pi)$, its Fourier series is convergent for all values of x to the sum* $\frac{1}{2}\{f(x+0)+f(x-0)\}$. For such a function is of bounded variation in the whole interval. These conditions are known as *Dirichlet's conditions*. They are, of course, satisfied in many cases; but they have the disadvantage that the sum of two functions which satisfy them does not itself necessarily satisfy them.

In connexion with Jordan's test, it is interesting to note that *if $f(x)$ is a function of bounded variation over* $(0, 2\pi)$, *its Fourier series is boundedly convergent.*

For let $0 \leqslant x < \pi$, and write Dirichlet's integral for $s_n(x)$ in the form

$$\frac{1}{2\pi}\int\limits_{-\frac{1}{2}\pi}^{\frac{3}{2}\pi}\frac{\sin(n+\frac{1}{2})(x-t)}{\sin\frac{1}{2}(x-t)}f(t)\,dt.$$

Since
$$\frac{1}{\sin\frac{1}{2}(x-t)}-\frac{1}{\frac{1}{2}(x-t)}$$

is bounded for $-\frac{3}{2}\pi < x-t < \frac{3}{2}\pi$, this differs by a bounded function from

$$\frac{1}{\pi}\int\limits_{-\frac{1}{2}\pi}^{\frac{3}{2}\pi}\frac{\sin(n+\frac{1}{2})(x-t)}{x-t}f(t)\,dt.$$

Let $f(t) = f_1(t)-f_2(t)$, where f_1 and f_2 are positive non-decreasing in $(-\frac{1}{2}\pi, \frac{3}{2}\pi)$. Then, by the second mean-value theorem,

$$\int\limits_{-\frac{1}{2}\pi}^{\frac{3}{2}\pi}\frac{\sin(n+\frac{1}{2})(x-t)}{x-t}f_1(t)\,dt = f_1(\tfrac{3}{2}\pi)\int\limits_{\xi}^{\frac{3}{2}\pi}\frac{\sin(n+\frac{1}{2})(x-t)}{x-t}\,dt$$
$$(-\tfrac{1}{2}\pi < \xi < \tfrac{3}{2}\pi),$$

which is bounded for all n and x, as in the above proof. A similar result holds for f_2. Hence the series is boundedly convergent over $(0, \pi)$, and similarly over $(\pi, 2\pi)$.

13.233. de la Vallée-Poussin's test.* *If the function*

$$\psi(t) = \frac{1}{t}\int\limits_0^t \phi(u)\,du$$

is of bounded variation in an interval to the right of $t = 0$, then the series is convergent. If s is so chosen that $\psi(t) \to 0$ as $t \to 0$, the sum of the series is s.

For
$$\phi(t) = \frac{d}{dt}\{t\psi(t)\} = \psi(t)+t\psi'(t).$$

Since $\psi(t)$ is of bounded variation and tends to zero, the part of the integral § 13.23 (3) involving it tends to zero, as in Jordan's test; and since $\psi'(t)$ is integrable (§ 11.54), the part involving $t\psi'(t)$ tends to zero, as in Dini's test.

* The same test was given previously by du Bois-Reymond, but of course with Riemann integrals.

13.24. Relations between the above tests.* Consider the function

$$f(x)=\frac{1}{\log 1/x} \quad (0<x<\pi), \qquad =0 \quad (\pi\leqslant x\leqslant 2\pi).$$

This function is bounded and monotonic in the neighbourhood of $x=0$, so that Jordan's condition is satisfied, and the series converges. But Dini's condition is not satisfied, since the integral

$$\int_0^\delta \frac{dt}{t\log 1/t}$$

is divergent. Thus *Dini's condition does not include Jordan's.*

On the other hand, *Jordan's condition does not include Dini's.* For consider the function

$$f(x)=x^\alpha \sin 1/x \quad (0<x<\pi), \qquad =0 \quad (\pi\leqslant x\leqslant 2\pi),$$

where $0<\alpha<1$. Then Dini's condition for convergence at $x=0$ is obviously satisfied. But the function is not of bounded variation (Ch. XI, ex. 5), i.e. Jordan's condition is not satisfied.

Lastly, *de la Vallée-Poussin's test includes both Dini's and Jordan's*, i.e. if either Dini's or Jordan's condition is fulfilled, then so is de la Vallée-Poussin's.

We first remark that *if $g(x)$ is of bounded variation in* $(0,\delta)$, *then so is*

$$G(x)=\frac{1}{x}\int_0^x g(t)\,dt.$$

For $g(x)=g_1(x)-g_2(x)$, where $g_1(x)$ and $g_2(x)$ are positive, non-decreasing, and bounded; and

$$G(x)=\frac{1}{x}\int_0^x g_1(x)\,dx-\frac{1}{x}\int_0^x g_2(x)\,dx=G_1(x)-G_2(x),$$

say, and it is easily seen that $G_1(x)$ and $G_2(x)$ are both positive, non-decreasing, and bounded. Hence $G(x)$ is of bounded variation.

The relation between Jordan's test and de la Vallée-Poussin's test follows at once; if $\phi(t)$ is of bounded variation, so is $\psi(t)$.

* For a detailed discussion of this question see Hardy (13).

Now consider Dini's test. If $\phi(u)/u$ is integrable,

$$\chi(t) = \int_0^t \frac{\phi(u)}{u}\,du$$

is a function of bounded variation; and

$$\psi(t) = \frac{1}{t}\int_0^t u\frac{d}{du}\{\chi(u)\}\,du = \chi(t) - \frac{1}{t}\int_0^t \chi(u)\,du,$$

which is also of bounded variation, by the above remark. Hence de la Vallée-Poussin's condition is satisfied.

13.25. Convergence throughout an interval. If one of the above conditions is fulfilled at all points of an interval, of course the series converges throughout the interval; and if the condition is fulfilled uniformly, the convergence is uniform. The simplest case is as follows.

The Fourier series of $f(x)$ converges uniformly to $f(x)$ in any interval interior to an interval where $f(x)$ is continuous and of bounded variation.

For in such an interval we can write $f(x) = f_1(x) - f_2(x)$, where $f_1(x)$ and $f_2(x)$ are continuous and non-decreasing. Then, by the property of uniform continuity, we can find η so that

$$|f_1(x+h) - f_1(x)| < \epsilon \qquad (|h| < \eta),$$

the choice of η depending only on ϵ and not on the value of x in the interval. It will be seen on referring to the proof of Jordan's test that this implies the uniform convergence of the integral dealt with in proving the test. We have also to show that the parts of Dirichlet's integral which have been shown to tend to 0, actually tend uniformly to 0; the reader should have no difficulty in verifying this.

The property of uniform convergence is, however, not so important as might be expected in the case of Fourier series, because questions of term-by-term integration can be dealt with under much more general conditions (§ 13.5).

No simple restriction on $f(x)$ which ensures that the Fourier series shall be convergent *almost everywhere*, without obviously proving more than this, appears to be known. It might, for example, be conjectured that continuity would be such a condition; but no result of the kind suggested has been proved. On the other hand, a condition bearing not on the function itself, but on the Fourier coefficients, has been given: *the Fourier series is convergent almost everywhere if the series*

$$\sum (a_n^2+b_n^2)\log n$$

*is convergent.**

13.3. Summation of series by arithmetic means. If a series $u_1+u_2+...$ is not convergent, i.e. if $s_n = u_1+...+u_n$ does not tend to a limit, it is sometimes possible to associate with the series a 'sum' in a less direct way. The simplest such method is 'summation by arithmetic means'. We take the arithmetic mean

$$\sigma_n = \frac{s_1+s_2+...+s_n}{n}$$

of the partial sums of the given series. If $s_n \to s$, then also $\sigma_n \to s$; for if $s_n = s+\delta_n$, then

$$\sigma_n = s+\frac{\delta_1+\delta_2+...+\delta_n}{n},$$

and the last term tends to zero if $\delta_n \to 0$, by the lemma of § 1.23.

But σ_n may tend to a limit even though s_n does not. Consider, for example, the series

$$1-1+1-1+....$$

Here the partial sums s_n are alternately 1 and 0, and it is easily seen that $\sigma_n \to \frac{1}{2}$.

A series for which σ_n tends to a limit is said to be summable by arithmetic means, or by Cesàro's means of the first order, or $(C, 1)$.

Examples. (i) The series $1+0-1+1+0-1+...$ is summable $(C, 1)$ to the sum $\frac{2}{3}$.

(ii) The series $\sin x + \sin 2x + \sin 3x + ...$ is summable $(C, 1)$ for all values of x; the sum is $\frac{1}{2}\cot\frac{1}{2}x$ if x is not an even multiple of π, and otherwise is 0.

(iii) The series $\frac{1}{2} + \cos x + \cos 2x + \cos 3x + ...$ is summable $(C, 1)$ to the sum zero if x is not an even multiple of π.

(iv) If $\sum u_n$ is summable $(C, 1)$, $s_n = o(n)$.

[For $s_n = n\sigma_n-(n-1)\sigma_{n-1}$.]

(v) Let $t_n = u_1+2u_2+...+nu_n$. If $\sum u_n$ is summable $(C, 1)$, a necessary and sufficient condition that it should be convergent is $t_n = o(n)$.

[For $t_n = (n+1)s_n-n\sigma_n$.]

* Plessner (2).

(vi) If $\sum u_n$ is summable $(C, 1)$, and $u_n = o(1/n)$, then $\sum u_n$ is convergent.

[For $t_n = o(n)$, by the lemma of § 1.23. The result is analogous to Tauber's theorem.]

(vii) A necessary and sufficient condition that $\sum u_n$ should be summable $(C, 1)$ is that

$$\sum \frac{t_n}{n(n+1)}$$

should be convergent.

$$\left[\text{For} \qquad \sum_{n=1}^{N} \frac{t_n}{n(n+1)} = \frac{N}{N+1}\sigma_N.\right]$$

(viii) If $\sum u_n$ is summable $(C, 1)$, and $u_n = O(1/n)$, then $\sum u_n$ is convergent.

[Hardy; this is analogous to Littlewood's extension of Tauber's theorem. If $\sum u_n$ is not convergent, then $t_N > A_1 N$, or $t_N < -A_1 N$, for an infinity of values of N—say e.g. the former. Since

$$t_{n+1} = t_n + (n+1)u_n > t_n - A_2,$$

we have $$t_{N+\nu} > \tfrac{1}{2}A_1 N \qquad (0 \leqslant \nu < \tfrac{1}{2}NA_1/A_2).$$

Hence $$\sum_{n=N}^{N+\frac{1}{2}NA_1/A_2} \frac{t_n}{n(n+1)} > A,$$

and by (vii) the series is not summable $(C, 1)$.]

(ix) A series of positive terms is summable $(C, 1)$ only if it is convergent.

[If $s_n \to \infty$, then $\sigma_n \to \infty$.]

13.31. Summability of Fourier series. It was discovered by Fejér* that the method of summation by arithmetic means applies particularly well to Fourier series. We write

$$\sigma_n = \frac{s_0 + s_1 + \ldots + s_{n-1}}{n},$$

where s_n is given by § 13.21 (3). Hence

$$\sigma_n = \frac{1}{2n\pi}\int_0^{\pi} \frac{\sin\frac{1}{2}u + \sin\frac{3}{2}u + \ldots + \sin(n-\frac{1}{2})u}{\sin\frac{1}{2}u}\{f(x+u)+f(x-u)\}\,du$$

$$= \frac{1}{2n\pi}\int_0^{\pi} \frac{\sin^2\frac{1}{2}nu}{\sin^2\frac{1}{2}u}\{f(x+u)+f(x-u)\}\,du. \tag{1}$$

This formula is known as *Fejér's integral*. Its importance is due to the fact that the factor $\sin^2\frac{1}{2}nu/\sin^2\frac{1}{2}u$ is positive. This makes

* Fejér (1).

it much easier to deal with Fejér's integral than with Dirichlet's, in which the corresponding factor, $\sin(n+\frac{1}{2})u/\sin\frac{1}{2}u$, oscillates between positive and negative values.

In the particular case where $f(x) = 1$, the formula becomes

$$1 = \frac{1}{2n\pi}\int\limits_0^\pi \frac{\sin^2\frac{1}{2}nu}{\sin^2\frac{1}{2}u}\, 2\, du,$$

since now $\sigma_n = 1$ for $n > 0$. Hence, multiplying by s and subtracting,

$$\sigma_n - s = \frac{1}{2n\pi}\int\limits_0^\pi \frac{\sin^2\frac{1}{2}nu}{\sin^2\frac{1}{2}u}\{f(x+u)+f(x-u)-2s\}\, du. \qquad (2)$$

A necessary and sufficient condition that the series should be summable $(C, 1)$ to the sum s is, therefore, that the integral (2) should tend to zero.

As in the convergence problem, we can simplify the condition. We write

$$\phi(u) = f(x+u)+f(x-u)-2s$$

as before. Then, if δ is any positive number less than π, a necessary and sufficient condition that the series should be summable $(C, 1)$ to s is

$$\lim_{n\to\infty}\frac{1}{n}\int\limits_0^\delta \frac{\sin^2\frac{1}{2}nu}{\sin^2\frac{1}{2}u}\,\phi(u)\, du = 0; \qquad (3)$$

for

$$\left|\frac{1}{n}\int\limits_\delta^\pi \frac{\sin^2\frac{1}{2}nu}{\sin^2\frac{1}{2}u}\,\phi(u)\, du\right| \leqslant \frac{1}{n}\int\limits_\delta^\pi \frac{|\phi(u)|}{\sin^2\frac{1}{2}u}\, du,$$

which plainly tends to zero. Finally, the condition may be put in the form

$$\lim_{n\to\infty}\frac{1}{n}\int\limits_0^\delta \frac{\sin^2\frac{1}{2}nu}{u^2}\,\phi(u)\, du = 0; \qquad (4)$$

for

$$\left|\frac{1}{n}\int\limits_0^\delta \sin^2\tfrac{1}{2}nu\left\{\frac{1}{\sin^2\frac{1}{2}u} - \frac{1}{(\frac{1}{2}u)^2}\right\}\phi(u)\, du\right|$$

$$\leqslant \frac{1}{n}\int\limits_0^\delta \left\{\frac{1}{\sin^2\frac{1}{2}u} - \frac{1}{(\frac{1}{2}u)^2}\right\}|\phi(u)|\, du,$$

which tends to zero.

13.32. Fejér's theorem. *The Fourier series of $f(x)$ is summable $(C, 1)$ to the sum*

$$\tfrac{1}{2}\{f(x+0)+f(x-0)\}$$

for every value of x for which this expression has a meaning. In particular, the series is summable $(C, 1)$ to the sum $f(x)$ at every point where $f(x)$ is continuous.

We now put $s = \tfrac{1}{2}\{f(x+0)+f(x-0)\}$ in the above formulae. Then $\phi(u) \to 0$ with u, and we have to prove that 13.31 (4) is true. Suppose that $|\phi(u)| \leqslant \epsilon$ for $u \leqslant \eta$. Then

$$\left|\frac{1}{n}\int_0^\delta \frac{\sin^2\frac{1}{2}nu}{u^2}\phi(u)\,du\right| \leqslant \frac{1}{n}\int_0^\eta \frac{\sin^2\frac{1}{2}nu}{u^2}\epsilon\,du + \frac{1}{n}\int_\eta^\delta \frac{\sin^2\frac{1}{2}nu}{u^2}|\phi(u)|\,du$$

$$\leqslant \frac{\epsilon}{n}\int_0^\eta \frac{\sin^2\frac{1}{2}nu}{u^2}\,du + \frac{1}{n}\int_\eta^\delta \frac{|\phi(u)|}{u^2}\,du$$

$$= I_1 + I_2,$$

say. Now

$$\frac{1}{n}\int_0^\eta \frac{\sin^2\frac{1}{2}nu}{u^2}\,du = \frac{1}{2}\int_0^{\frac{1}{2}n\eta} \frac{\sin^2 v}{v^2}\,dv < \frac{1}{2}\int_0^\infty \frac{\sin^2 v}{v^2}\,dv,$$

which is a constant. Hence $I_1 < A\epsilon$. Having fixed η, it is clear that $I_2 \to 0$ as $n \to \infty$. This proves the theorem.

13.33. Summability throughout an interval. The following theorem is an almost immediate consequence of Fejér's theorem.

The Fourier series of $f(x)$ is uniformly summable in any interval included in an interval where $f(x)$ is continuous.

For $f(x)$ is uniformly continuous in any such interval, and so, in the above proof, the choice of η depends only on ϵ and not on x. The result follows at once from this.

Weierstrass's approximation theorem. *If $f(x)$ is continuous in (a, b), and ϵ is a given positive number, there is a polynomial $p(x)$ such that*

$$|f(x)-p(x)| < \epsilon \qquad (a \leqslant x \leqslant b).$$

We can make a preliminary transformation so that the interval considered lies within $(0, 2\pi)$. Then, by the above theorem, there is a 'trigonometrical polynomial' $\sigma_n(x)$ such that

$|f(x)-\sigma_n(x)| < \frac{1}{2}\epsilon$ throughout the interval. If we replace each sine and cosine in $\sigma_n(x)$ by a sufficiently large number of terms in its power series, we obtain a polynomial $p(x)$ such that $|\sigma_n(x)-p(x)| < \frac{1}{2}\epsilon$ throughout the interval. This proves the theorem.

13.34. Almost everywhere summability. As long as we restrict ourselves to ordinary convergence, we cannot show that the Fourier series of a function represents the function in general, without imposing some rather heavy restriction on the function. The theory of summability removes this defect.

The Fejér-Lebesgue theorem. *The Fourier series of $f(x)$ is summable $(C,1)$ to the sum $f(x)$, for every value of x for which*

$$\int_0^t |f(x+u)-f(x)|\, du = o(t). \tag{1}$$

In particular, it is summable $(C,1)$ to $f(x)$ almost everywhere.

We have shown in § 11.6 that the condition (1) is satisfied for almost all values of x, for any integrable function. The second part of the theorem therefore follows at once from the first.

Let x be a point where (1) is satisfied, and take $s = f(x)$ in the formulae of § 13.31. Then

$$\int_0^t |\phi(u)|\, du = \int_0^t |f(x+u)+f(x-u)-2f(x)|\, du$$

$$\leqslant \int_0^t |f(x+u)-f(x)|\, du + \int_0^t |f(x-u)-f(x)|\, du = o(t).$$

Let
$$\Phi(t) = \int_0^t |\phi(u)|\, du,$$

and, given ϵ, choose η so that $\Phi(t) < \epsilon t$ for $t \leqslant \eta$. We suppose that $n > 1/\eta$, and write

$$\int_0^\delta \frac{\sin^2 \frac{1}{2}nu}{u^2}\phi(u)\, du = \int_0^{1/n} + \int_{1/n}^{\eta} + \int_{\eta}^{\delta} = J_1+J_2+J_3.$$

Then, since $\sin^2\theta \leqslant \theta^2$,

$$|J_1| \leqslant (\tfrac{1}{2}n)^2 \int_0^{1/n} |\phi(u)|\, du < \tfrac{1}{4}\epsilon n,$$

$$|J_2| \leqslant \int_{1/n}^{\eta} \frac{|\phi(u)|}{u^2}\,du = \frac{\Phi(\eta)}{\eta^2} - n^2\Phi\left(\frac{1}{n}\right) + 2\int_{1/n}^{\eta} \frac{\Phi(u)}{u^3}\,du$$

$$< \frac{\epsilon}{\eta} + 2\epsilon \int_{1/n}^{\eta} \frac{du}{u^2} < \epsilon/\eta + 2\epsilon n < 3\epsilon n,$$

and obviously
$$|J_3| < \frac{A}{\eta^2}.$$

Hence
$$\left|\frac{1}{n}\int_0^{\delta} \frac{\sin^2\frac{1}{2}nu}{u^2}\phi(u)\,du\right| < \tfrac{1}{4}\epsilon + 3\epsilon + \frac{A}{n\eta^2},$$

and the required result follows on choosing first ϵ, then η, and then n.

13.35. An immediate corollary is that *a trigonometrical series cannot be the Fourier series of two functions which differ in a set of positive measure.* For if it is the Fourier series of $f(x)$ and of $g(x)$, it is summable $(C, 1)$ both to $f(x)$ and to $g(x)$ almost everywhere. Hence $f(x) = g(x)$ almost everywhere.

13.4. A continuous function with a divergent Fourier series. While we have seen that the continuity of a function is a sufficient condition for its Fourier series to be summable $(C, 1)$, for convergence we have had to assume other conditions. That this is really in accordance with the facts is shown by the following example, due to Fejér,* of a Fourier series which is divergent at a point, although the function which gives rise to it is continuous.

13.41. We first require a lemma.

The sum

$$\phi(n,r,x) = \frac{\cos(r+1)x}{2n-1} + \frac{\cos(r+2)x}{2n-3} + \dots + \frac{\cos(r+n)x}{1} -$$
$$- \frac{\cos(r+n+1)x}{1} - \frac{\cos(r+n+2)x}{3} - \dots - \frac{\cos(r+2n)x}{2n-1}$$

is bounded for all values of n, r, and x.

* Fejér (2), (3), (4).

We have

$$\phi(n, r, x) = \sum_{\nu=1}^{n} \frac{\cos(r+n-\nu+1)x}{2\nu-1} - \sum_{\nu=1}^{n} \frac{\cos(r+n+\nu)x}{2\nu-1}$$

$$= 2\sin(r+n+\tfrac{1}{2})x \sum_{\nu=1}^{n} \frac{\sin(\nu-\frac{1}{2})x}{2\nu-1}$$

$$= 2\sin(r+n+\tfrac{1}{2})x\left\{\sum_{\lambda=1}^{2n} \frac{\sin\frac{1}{2}\lambda x}{\lambda} - \frac{1}{2}\sum_{\mu=1}^{n} \frac{\sin\mu x}{\mu}\right\},$$

and each of the sums in the bracket is bounded (§ 1.76).

13.42. Let G_n denote the group of $2n$ numbers

$$\frac{1}{2n-1}, \frac{1}{2n-3}, \ldots, \frac{1}{3}, 1, -1, -\frac{1}{3}, \ldots, -\frac{1}{2n-1}.$$

Let $\lambda_1, \lambda_2, \ldots$ denote an increasing sequence of integers. Take the numbers of the groups $G_{\lambda_1}, G_{\lambda_2}, \ldots$ in order, and multiply each of the numbers of the group G_{λ_ν} by ν^{-2}. We obtain the sequence

$$\frac{1}{1^2}\frac{1}{2\lambda_1-1}, \ldots, -\frac{1}{1^2}\frac{1}{2\lambda_1-1}, \frac{1}{2^2(2\lambda_2-1)}, \frac{1}{2^2(2\lambda_2-3)}, \ldots,$$

say $\alpha_1, \alpha_2, \ldots$.

Now consider the series

$$\sum_{n=1}^{\infty} \alpha_n \cos nx. \tag{1}$$

Suppose first that the terms corresponding to each group G_{λ_ν} are bracketed together. The bracketed series is

$$\sum_{n=1}^{\infty} \frac{\phi(\lambda_n, 2\lambda_1+2\lambda_2+\ldots+2\lambda_{n-1}, x)}{n^2}, \tag{2}$$

which is absolutely and uniformly convergent, by the lemma. The sum of the series (2), say $f(x)$, is therefore a continuous function.

We next observe that the series (1) is the Fourier series of $f(x)$. For since (2) is uniformly convergent, we may multiply it by $\cos mx$ or $\sin mx$ and integrate term by term. The integral of each term is zero, except that of the one containing the term $\alpha_m \cos mx$; and from this we obtain

$$\int_0^{2\pi} f(x)\cos mx\, dx = \pi\alpha_m.$$

The numbers α_m are therefore the Fourier cosine coefficients of $f(x)$.

We show finally that the numbers λ_ν can be chosen so that the series (1) is divergent at the point $x = 0$, i.e. that the series $\alpha_1 + \alpha_2 + \ldots$ is divergent. Let s_n be its nth partial sum. Then

$$s_{2\lambda_1+2\lambda_2+\ldots+2\lambda_{\nu-1}+\lambda_\nu} = \frac{1}{\nu^2}\left(\frac{1}{2\lambda_\nu - 1} + \frac{1}{2\lambda_\nu - 3} + \ldots + \frac{1}{3} + 1\right) \sim \frac{\log \lambda_\nu}{2\nu^2}.$$

If the numbers λ_ν tend to infinity sufficiently rapidly, e.g. if $\lambda_\nu = \nu^{\nu^2}$, it follows that $s_n \to \infty$ as $n \to \infty$ through a certain sequence of values. Hence the series is divergent.

13.43. Fejér's example, together with a simple argument depending on Dirichlet's integral, enables us to say how large the partial sums s_n of a Fourier series of a continuous function can be.

If $f(x)$ is continuous, then

$$s_n = o(\log n);$$

and no more is true, since, if $\psi(n)$ is a function which decreases steadily to zero, however slowly, there is a Fourier series of a continuous function for which

$$s_n > \psi(n) \log n$$

for arbitrarily large values of n.

For the first part, we have to prove that

$$\int_0^\delta \frac{\sin(n+\frac{1}{2})u}{u} \phi(u)\, du = o(\log n)$$

if $\phi(u) \to 0$ as $u \to 0$. Suppose that $|\phi(u)| < \epsilon$ for $u \leqslant \eta$; and, if $n + \frac{1}{2} > 1/\eta$, put

$$\int_0^\delta \frac{\sin(n+\frac{1}{2})u}{u} \phi(u)\, du = \int_0^{1/(n+\frac{1}{2})} + \int_{1/(n+\frac{1}{2})}^{\eta} + \int_\eta^\delta = I_1 + I_2 + I_3.$$

Then
$$|I_1| \leqslant (n+\tfrac{1}{2}) \int_0^{1/(n+\frac{1}{2})} |\phi(u)|\, du < \epsilon,$$

$$|I_2| \leqslant \int_{1/(n+\frac{1}{2})}^{\eta} \frac{\epsilon}{u}\, du < \epsilon \log(n+\tfrac{1}{2}),$$

and $$|I_3| \leqslant \frac{1}{\eta} \int_{\eta}^{\delta} |\phi(u)|\, du.$$

The result clearly follows from these inequalities.

The second part is obtained by taking λ_ν sufficiently large in Fejér's example. Suppose that $\lambda_\nu > 2\nu$, and let

$$n = 2\lambda_1 + 2\lambda_2 + \ldots + 2\lambda_{\nu-1} + \lambda_\nu.$$

Then $$\lambda_\nu < n < 2\nu\lambda_\nu < \lambda_\nu^2.$$

Now $s_n > \psi(n)\log n$ for sufficiently large values of ν, if

$$\frac{\log \lambda_\nu}{2\nu^2} > \psi(n)\log n;$$

and since $\psi(n)\log n < \psi(\lambda_\nu)\log \lambda_\nu^2$, this is true if

$$\psi(\lambda_\nu) < \frac{1}{4\nu^2};$$

and this will be so if the numbers λ_ν tend to infinity rapidly enough.

13.5. Integration of Fourier series. *Any Fourier series, whether convergent or not, may be integrated term by term between any limits; that is, the sum of the integrals of the separate terms is the integral of the function of which the series is the Fourier series.*

Let $f(x)$ have the Fourier coefficients a_n, b_n, and let

$$F(x) = \int_0^x \{f(t) - \tfrac{1}{2}a_0\}\, dt.$$

Then $F(x)$ is periodic, continuous, and of bounded variation. Hence it can be expanded in a Fourier series, say

$$F(x) = \tfrac{1}{2}A_0 + \sum_{n=1}^{\infty} (A_n \cos nx + B_n \sin nx),$$

convergent for all values of x. Here

$$A_n = \frac{1}{\pi} \int_0^{2\pi} F(x) \cos nx\, dx$$

$$= \frac{1}{\pi} \left[F(x) \frac{\sin nx}{n}\right]_0^{2\pi} - \frac{1}{n\pi} \int_0^{2\pi} \{f(x) - \tfrac{1}{2}a_0\} \sin nx\, dx$$

$$= -\frac{1}{n\pi}\int_0^{2\pi} f(x)\sin nx\,dx = -\frac{b_n}{n};$$

and $$B_n = \frac{1}{\pi}\int_0^{2\pi} F(x)\sin nx\,dx$$

$$= \frac{1}{\pi}\left[-F(x)\frac{\cos nx}{n}\right]_0^{2\pi} + \frac{1}{n\pi}\int_0^{2\pi}\{f(x)-\tfrac{1}{2}a_0\}\cos nx\,dx$$

$$= \frac{1}{n\pi}\int_0^{2\pi} f(x)\cos nx\,dx = \frac{a_n}{n},$$

the integrated terms vanishing since $F(2\pi) = F(0) = 0$. Hence

$$F(x) = \tfrac{1}{2}A_0 + \sum_{n=1}^{\infty}\frac{a_n\sin nx - b_n\cos nx}{n}$$

Putting $x = 0$, we obtain

$$\tfrac{1}{2}A_0 = \sum_{n=1}^{\infty}\frac{b_n}{n},$$

and, adding,

$$F(x) = \sum_{n=1}^{\infty}\frac{a_n\sin nx + b_n(1-\cos nx)}{n}.$$

This proves the theorem.

13.51. An interesting particular case is that *the series*

$$\sum_{n=1}^{\infty}\frac{b_n}{n}$$

is convergent. This remark enables us to write down convergent trigonometrical series which are not Fourier series. A simple example is

$$\sum_{n=2}^{\infty}\frac{\sin nx}{\log n}.$$

This is convergent for all values of x, but it cannot be the Fourier series of its sum, since the series

$$\sum\frac{1}{n\log n}$$

is divergent. Actually the sum of this trigonometrical series is not integrable in the sense of Lebesgue, and it is easy to prove directly that the sum of the integrated series $\sum \cos nx/n \log n$ tends to infinity as $x \to 0$.

13.52. The following alternative proof of the above integration theorem is also interesting. We know that the series

$$\frac{\sin(x-t)}{1}+\frac{\sin 2(x-t)}{2}+\ldots = \phi(t)$$

is boundedly convergent. Hence we may multiply by $f(t)/\pi$ and integrate term by term over $(0, 2\pi)$. On the left we obtain

$$\sum_{n=1}^{\infty} \frac{1}{n\pi} \int_0^{2\pi} \sin n(x-t) f(t)\, dt = \sum_{n=1}^{\infty} \frac{a_n \sin nx - b_n \cos nx}{n},$$

the integrated series. On the right we get

$$\frac{1}{\pi}\int_0^{2\pi} \phi(t) f(t)\, dt = \frac{1}{\pi}\int_{x-2\pi}^{x} \phi(t) f(t)\, dt = \frac{1}{\pi}\int_{x-2\pi}^{x} \tfrac{1}{2}(\pi - x + t) f(t)\, dt$$

$$= \frac{1}{2\pi}\big[(\pi-x+t)F(t)\big]_{x-2\pi}^{x} - \frac{1}{2\pi}\int_{x-2\pi}^{x} F(t)\, dt = F(x) - \frac{1}{2\pi}\int_0^{2\pi} F(t)\, dt,$$

since $F(x-2\pi) = F(x)$. The result now follows as before.

13.53. A similar method leads to the following more general integration theorem.

A Fourier series may be multiplied by any function of bounded variation and integrated term by term between any finite limits.

Let $$g(x) = \tfrac{1}{2}\alpha_0 + \sum_{n=1}^{\infty} (\alpha_n \cos nx + \beta_n \sin nx)$$

be a function of bounded variation. The series being boundedly convergent (§ 13.232), we may multiply by any integrable function $f(x)$ and integrate term by term over $(0, 2\pi)$. We obtain

$$\frac{1}{\pi}\int_0^{2\pi} f(x) g(x)\, dx = \tfrac{1}{2} a_0 \alpha_0 + \sum_{n=1}^{\infty} (a_n \alpha_n + b_n \beta_n), \qquad (1)$$

where a_n, b_n, are the Fourier coefficients of $f(x)$. This is the same result as we should have obtained by multiplying the Fourier series for $f(x)$ by $g(x)$ and integrating term by term over $(0, 2\pi)$.

A similar result may be obtained for other ranges of integration by replacing $g(x)$ by 0 outside the required range.

13.54. Parseval's theorem. If $f(x)$ is of bounded variation, we may put $g(x) = f(x)$ in 13.53 (1), and obtain

$$\frac{1}{\pi}\int_0^{2\pi} \{f(x)\}^2\,dx = \tfrac{1}{2}a_0^2 + \sum_{n=1}^{\infty} (a_n^2 + b_n^2).$$

This is known as Parseval's theorem. We shall show in § 13.63 that it is true under much more general conditions than those we have so far assumed.

13.6. Functions of the class L^2: Bessel's inequality. Let $f(x)$ be a function of the class $L^2(0, 2\pi)$, with Fourier coefficients a_n, b_n. Then

$$\phi(x) = f(x) - \tfrac{1}{2}a_0 - \sum_{m=1}^{n} (a_m \cos mx + b_m \sin mx)$$

also belongs to L^2; and

$$\frac{1}{\pi}\int_0^{2\pi} \{\phi(x)\}^2\,dx = \frac{1}{\pi}\int_0^{2\pi} \{f(x)\}^2\,dx + \tfrac{1}{2}a_0^2 + \sum_{m=1}^{n} (a_m^2 + b_m^2) -$$

$$-\frac{a_0}{\pi}\int_0^{2\pi} f(x)\,dx - \frac{2}{\pi}\sum_{m=1}^{n}\int_0^{2\pi} (a_m \cos mx + b_m \sin mx) f(x)\,dx$$

$$= \frac{1}{\pi}\int_0^{2\pi} \{f(x)\}^2\,dx - \tfrac{1}{2}a_0^2 - \sum_{m=1}^{n} (a_m^2 + b_m^2),$$

by the Euler-Fourier formulae. Since the left-hand side is not negative, it follows that

$$\tfrac{1}{2}a_0^2 + \sum_{m=1}^{n} (a_m^2 + b_m^2) \leqslant \frac{1}{\pi}\int_0^{2\pi} \{f(x)\}^2\,dx \tag{1}$$

for all values of n. This result is known as Bessel's inequality.

Since the right-hand side of (1) is independent of n, it follows that *the series*

$$\tfrac{1}{2}a_0^2 + \sum_{m=1}^{\infty} (a_m^2 + b_m^2) \tag{2}$$

is convergent. Also

$$\tfrac{1}{2}a_0^2 + \sum_{m=1}^{\infty} (a_m^2 + b_m^2) \leqslant \frac{1}{\pi}\int_0^{2\pi} \{f(x)\}^2\,dx. \tag{3}$$

13.61. Parseval's theorem for continuous functions. We have seen that, for functions of bounded variation, the above inequality becomes an equality, viz. Parseval's theorem. The same result for continuous functions may be proved as follows. If $f(x)$ is continuous, $\sigma_n(x)$ tends uniformly to $f(x)$, and hence.

$$\lim_{n\to\infty}\frac{1}{\pi}\int_0^{2\pi}\{f(x)-\sigma_n(x)\}f(x)\,dx = 0.$$

Now $$\sigma_n(x) = \tfrac{1}{2}a_0+\sum_{m=1}^{n-1}(a_m\cos mx+b_m\sin mx)\left(1-\frac{m}{n}\right),$$

and, evaluating the integral as in the previous section, we obtain

$$\frac{1}{\pi}\int_0^{2\pi}\{f(x)\}^2\,dx-\tfrac{1}{2}a_0^2-\sum_{m=1}^{n-1}(a_m^2+b_m^2)\left(1-\frac{m}{n}\right)\to 0.$$

Parseval's formula therefore holds if the series is summed $(C, 1)$. Since by § 13.6 the series is convergent, it follows from § 13.3 that it holds in the ordinary sense.*

There is no difficulty in extending this proof to functions which have simple discontinuities. Actually Parseval's theorem holds for all functions of the class L^2. We shall prove this as a corollary of the theorem of the next section.

13.62. The Riesz-Fischer theorem. Let

$$\tfrac{1}{2}a_0+\sum_{n=1}^{\infty}(a_n\cos nx+b_n\sin nx) \tag{1}$$

be any trigonometrical series with coefficients such that the series 13.6 (2) is convergent. Nothing that we have proved so far in this chapter enables us to decide whether such a series is a Fourier series. The problem is solved by means of the theory of mean convergence (§ 12.5). This theory was in fact originally constructed to deal with this very problem.

The following theorem was proved almost simultaneously by F. Riesz and Fischer.†

If the numbers a_n, b_n are such that the series 13.6 (2) *is convergent, then the series* (1) *is the Fourier series of a function $f(x)$*

* A number of different proofs under various conditions are given by Julia, *Exercices d'analyse*, 180–6.

† F. Riesz (1), Fischer (1).

of the class L^2. The partial sums of the series converge in mean to $f(x)$.

Denoting the nth partial sum of (1) by $s_n(x)$, we have

$$\int_0^{2\pi} \{s_n(x)-s_m(x)\}^2\,dx = \int_0^{2\pi} \Big\{ \sum_{\nu=m+1}^{n} (a_\nu \cos \nu x + b_\nu \sin \nu x)\Big\}^2\,dx$$

$$= \pi \sum_{\nu=m+1}^{n} (a_\nu^2+b_\nu^2),$$

all the product terms disappearing on integration. The right-hand side tends to zero when m and n tend independently to infinity. Hence $s_n(x)$ converges in mean to a function, $f(x)$ say, of the class L^2.

Also, by § 12.53,

$$\lim_{n\to\infty} \int_0^{2\pi} s_n(x)\cos \nu x\,dx = \int_0^{2\pi} f(x)\cos \nu x\,dx.$$

But the integral on the left is equal to πa_ν, if $n \geqslant \nu$. Hence

$$a_\nu = \frac{1}{\pi} \int_0^{2\pi} f(x)\cos \nu x\,dx,$$

i.e. a_ν is the νth Fourier cosine coefficient of $f(x)$. Similarly b_ν is the νth sine coefficient. Hence the given trigonometrical series is the Fourier series of the function $f(x)$.

It is important to observe that it is here that the Lebesgue integral first plays an indispensable part in the theory. Most of the previous analysis is true for Riemann integrals and elementary generalized absolutely convergent integrals. Here the result shows that the extension to Lebesgue integrals is really necessary.

13.63. Parseval's theorem for functions of the class L^2. Let $f(x)$ be any function belonging to $L^2(0, 2\pi)$, and let its Fourier series have the usual form. Then the series 13.6 (2) is convergent. Hence, by the Riesz-Fischer theorem, the partial sums $s_n(x)$ converge in mean to a function $g(x)$, of which the given series is the Fourier series. Hence, by § 13.35, $g(x) = f(x)$ almost everywhere. Also, by § 12.52,

$$\lim_{n\to\infty} \int_0^{2\pi} \{s_n(x)\}^2\,dx = \int_0^{2\pi} \{f(x)\}^2\,dx,$$

and on evaluating the integral on the left-hand side we obtain Parseval's formula.

The more general formula 13.53 (1) *also holds if* $f(x)$ *and* $g(x)$ *are any two functions of the class* L^2. For Parseval's formula holds for the functions $f(x)+g(x)$ and $f(x)-g(x)$, and the result stated follows on subtraction.

13.7. Properties of Fourier coefficients. Originally the Fourier coefficients were merely the material out of which the Fourier series was constructed. But the coefficients have some interesting properties of their own. In fact Bessel's inequality, and the theorems of Parseval and Riesz-Fischer call attention to the problem of the behaviour of the Fourier coefficients of given classes of functions and give some important information about it.

The first theorem of this kind (§ 13.22) is that *the Fourier coefficients of any integrable function tend to zero.* On the other hand, they do not tend to zero in any definite order; that is, any theorem such as '$a_n = O(1/\log n)$ for all integrable functions' is certainly false. For consider the series

$$\sum_{n=1}^{\infty} \frac{\cos k_n x}{n^2},$$

where k_n denotes a sequence of positive integers which tends to infinity rapidly as $n \to \infty$. The series is uniformly convergent, and so is the Fourier series of its sum; and

$$a_{k_n} = \frac{1}{n^2},$$

which falsifies any theorem of the kind suggested, if k_n tends to infinity rapidly enough.

13.71. Suppose next that $f(x)$ belongs to the class L^2. This does not enable us to prove any more about the order of the coefficients; in fact the function defined by the above series is clearly continuous, and so belongs to L^2. But we do obtain a definite result about the *average* order, viz. that

$$\sum (a_n^2+b_n^2)$$

is convergent (§ 13.6).

This result has been generalized so as to apply to other

Lebesgue classes; *if $f(x)$ belongs to L^p, where $1 < p \leqslant 2$, then the series*

$$\sum (|a_n|^{p/(p-1)} + |b_n|^{p/(p-1)})$$

is convergent.

The proof of this theorem is, however, too long to be given here.*

There is also a corresponding extension of the Riesz-Fischer theorem: *if the series*

$$\sum (|a_n|^p + |b_n|^p),$$

where $1 < p \leqslant 2$, is convergent, then the numbers a_n, b_n are the Fourier coefficients of a function of the class $L^{p/(p-1)}$.

Both these theorems cease to be true if $p > 2$, so that they are not converses of each other unless $p = 2$.

13.72. If we make still more special assumptions about the function, we obtain new results about the coefficients. *Suppose that $f(x)$ satisfies a Lipschitz condition of order α, i.e. as $h \to 0$*

$$f(x+h) - f(x) = O(|h|^\alpha) \qquad (0 < \alpha \leqslant 1)$$

uniformly with respect to x. Then

$$a_n = O(n^{-\alpha}), \qquad b_n = O(n^{-\alpha}).$$

For

$$a_n = \frac{1}{\pi}\int_0^{2\pi} f(x)\cos nx\,dx = -\frac{1}{\pi}\int_{-\pi/n}^{2\pi-\pi/n} f\left(\frac{\pi}{n}+t\right)\cos nt\,dt$$

$$= -\frac{1}{\pi}\int_0^{2\pi} f\left(\frac{\pi}{n}+t\right)\cos nt\,dt,$$

and hence also

$$a_n = \frac{1}{2\pi}\int_0^{2\pi} \left\{f(x) - f\left(\frac{\pi}{n}+x\right)\right\}\cos nx\,dx$$

$$= \int_0^{2\pi} O\left(\frac{1}{n^\alpha}\right)dx = O\left(\frac{1}{n^\alpha}\right);$$

and similarly for b_n.

13.73. The next result of this kind is that *if $f(x)$ is of bounded variation, then*

$$a_n = O(1/n), \qquad b_n = O(1/n).$$

* W. H. Young (2), (3), (5), (6), Hausdorff (1), F. Riesz (4).

For $f(x) = f_1(x) - f_2(x)$, where $f_1(x)$ and $f_2(x)$ are positive and non-decreasing. Hence, by the second mean-value theorem,

$$\int_0^{2\pi} f_1(x)\cos nx\,dx = f_1(2\pi)\int_\xi^{2\pi}\cos nx\,dx \qquad (0 < \xi < 2\pi)$$

$$= -f_1(2\pi)\frac{\sin n\xi}{n} = O\left(\frac{1}{n}\right),$$

and a similar result holds for the other integral.

An alternative proof of Jordan's theorem (§ 13.232) can be deduced from this result. If $f(x)$ is of bounded variation, its Fourier series is summable $(C, 1)$ to the sum $\frac{1}{2}\{f(x+0)+f(x-0)\}$, by § 13.32. Since $a_n = O(1/n)$, $b_n = O(1/n)$, the series actually converges to this sum (§ 13.3, ex. (viii)).

If $f(x)$ is an integral, and has the period 2π, then

$$a_n = o(1/n), \qquad b_n = o(1/n).$$

For if
$$f(x) = f(0) + \int_0^x \phi(t)\,dt \qquad (x \geqslant 0),$$

then
$$\pi a_n = \left[f(x)\frac{\sin nx}{n}\right]_0^{2\pi} - \frac{1}{n}\int_0^{2\pi}\phi(x)\sin nx\,dx,$$

$$\pi b_n = \left[-f(x)\frac{\cos nx}{n}\right]_0^{2\pi} + \frac{1}{n}\int_0^{2\pi}\phi(x)\cos nx\,dx.$$

The integrated terms are zero, since $f(2\pi) = f(0)$ and the integrals on the right tend to zero, by the Riemann-Lebesgue theorem. This proves the theorem.

If $f'(x)$ satisfies special conditions, such as a Lipschitz condition, still further results of the same kind can, of course, be obtained.

13.8. Uniqueness of trigonometrical series. At the beginning of the chapter we associated with an integrable function a particular trigonometrical series, viz. the Fourier series of the function; and we have shown that the Fourier series does, in various ways, represent the function. The reader might, however, still contend that we had attached undue importance to Fourier series, and that there might be other types of trigonometrical series in which a given function could be expanded.

It is difficult to give a complete solution of this problem. If, however, we assume enough about the set of points where the series converge, we can show that, if a trigonometrical series converges to a given function, it is the only such series which does so; and therefore that, if the function can be expanded in a convergent Fourier series, it cannot be expanded in a convergent trigonometrical series of any other form.

The theory is due to Riemann, du Bois-Reymond, and Cantor. The theorem which we shall prove is as follows.

If two trigonometrical series converge to the same sum in the interval $(0, 2\pi)$, *with the possible exception of a finite number of points, then corresponding coefficients in the two series are equal, i.e. the series are identical.*

This is not all that is known, and more general theorems will be found, e.g., in Hobson's *Theory of Functions*, §§ 420–50. But some extensions which might naturally be suggested are not true; if we say 'are summable $(C, 1)$' instead of 'converge', the theorem becomes false, as is shown by § 13.3, ex. (iii).

The question whether a given trigonometrical series is a Fourier series is really a problem of integral equations. We are given numbers a_0, a_1, b_1,..., and it is required to determine whether there is an integrable function $f(x)$ such that the Euler-Fourier formulae § 13.1 (2), (3), are true. The question is not settled by mere convergence, since a trigonometrical series may be everywhere convergent without being a Fourier series (§ 13.51). But if it converges uniformly, or boundedly, or in mean with index p $(p \geqslant 1)$, then it is a Fourier series; and the theorems of §§ 13.62–13.71 enable us to state conditions for mean convergence, with $p \geqslant 2$.

Another theorem which would naturally suggest itself is that if a trigonometrical series converges almost everywhere to an integrable function, then it is the Fourier series of the function; but this is not necessarily true, and the state of affairs is rather complicated.

The proof of the theorem stated above depends on a number of lemmas.

13.81. Cantor's lemma. *If* $a_n \cos nx + b_n \sin nx$ *tends to* 0 *for all values of* x *in an interval, then* a_n *and* b_n *tend to* 0.

Suppose that $a_n \cos nx + b_n \sin nx \to 0$ in the interval (α, β).

If the lemma is false, there is a constant A and a sequence of values of n for which $a_n^2+b_n^2>A$. Hence, as $n\to\infty$ through this sequence, the function

$$f_n(x)=\frac{(a_n\cos nx+b_n\sin nx)^2}{a_n^2+b_n^2}$$

converges boundedly to 0 in (α,β). Hence, by the theorem of bounded convergence,

$$\int_\alpha^\beta f_n(x)\,dx\to 0.$$

But, evaluating the integral, we find that

$$\int_\alpha^\beta f_n(x)\,dx=\tfrac{1}{2}(\beta-\alpha)+O\left(\frac{1}{n}\right),$$

and this gives a contradiction. This proves the lemma.

13.82. Suppose now that the series

$$\tfrac{1}{2}a_0+\sum_{n=1}^{\infty}(a_n\cos nx+b_n\sin nx) \tag{1}$$

converges to the sum $f(x)$ in $(0,2\pi)$, except possibly at a finite number of points. Let

$$F(x)=\tfrac{1}{4}a_0x^2-\sum_{n=1}^{\infty}\frac{a_n\cos nx+b_n\sin nx}{n^2}. \tag{2}$$

Since by Cantor's lemma a_n and b_n tend to zero, this series is uniformly convergent, and $F(x)$ is continuous, for all values of x. If we could differentiate twice term by term, we should have $F''(x)=f(x)$. We cannot necessarily do this, and instead have to proceed as follows.

Riemann's First Theorem. *If*

$$G(x,h)=\frac{F(x+2h)+F(x-2h)-2F(x)}{4h^2}, \tag{3}$$

then $G(x,h)\to f(x)$ as $h\to 0$, for all values of x for which the series (1) *converges to $f(x)$.*

We have

$$\cos n(x+2h)+\cos n(x-2h)-2\cos nx=-4\cos nx\sin^2 nh,$$
$$\sin n(x+2h)+\sin n(x-2h)-2\sin nx=-4\sin nx\sin^2 nh.$$

and hence

$$G(x,h) = \tfrac{1}{2}a_0 + \sum_{n=1}^{\infty} (a_n \cos nx + b_n \sin nx)\frac{\sin^2 nh}{n^2h^2}. \tag{4}$$

The nth term of (4) tends to the nth term of (1) as $h \to 0$. Hence it is sufficient to prove that the series (4) converges uniformly with respect to h. Let r_n denote the remainder of the series (1) after the term in $\sin nx$. Then $r_n \to 0$, say $|r_n| < \epsilon$ for $n \geqslant N$. Hence

$$\sum_{n=N}^{\infty} (a_n \cos nx + b_n \sin nx)\frac{\sin^2 nh}{n^2h^2} = \sum_{n=N}^{\infty} (r_n - r_{n+1})\left(\frac{\sin nh}{nh}\right)^2$$

$$= r_N\left(\frac{\sin Nh}{Nh}\right)^2 - \sum_{N+1}^{\infty} r_n\left[\left(\frac{\sin nh}{nh}\right)^2 - \left\{\frac{\sin(n+1)h}{(n+1)h}\right\}^2\right],$$

and the modulus of this does not exceed

$$\epsilon + \epsilon \sum_{N+1}^{\infty} \int_{nh}^{(n+1)h} \left|\frac{d}{dt}\left(\frac{\sin^2 t}{t^2}\right)\right| dt < \epsilon + \epsilon \int_0^{\infty} \left|\frac{d}{dt}\left(\frac{\sin^2 t}{t^2}\right)\right| dt,$$

the last integral being convergent. Hence (4) is uniformly convergent, and the result follows.

13.83. Riemann's Second Theorem. *If a_n and b_n tend to zero, then*

$$\lim_{h \to 0} \frac{F(x+2h) + F(x-2h) - 2F(x)}{2h} = 0.$$

for all values of x.

We have to prove that

$$a_0 h + 2\sum_{n=1}^{\infty} (a_n \cos nx + b_n \sin nx)\frac{\sin^2 nh}{n^2 h}$$

tends to zero. Given ϵ, we have

$$|a_n \cos nx + b_n \sin nx| < \epsilon \qquad (n > N).$$

Since $\sin^2 nh \leqslant n^2h^2$ for $n \leqslant 1/h$, the modulus of the sum does not exceed

$$A|h| + 2\sum_{n=1}^{N} A|h| + 2\sum_{N<n\leqslant 1/h} \epsilon h + 2\sum_{n>1/h} \frac{\epsilon}{n^2 h}$$

$$< AN|h| + 2\epsilon + \frac{2\epsilon}{h}\int_{1/h}^{\infty} \frac{du}{(u-1)^2} < AN|h| + A\epsilon,$$

and the result follows by choosing first ϵ and then h sufficiently small.

13.84. Schwarz's theorem. *If $F(x)$ is continuous in an interval (a, b), and*

$$\lim_{h\to 0}\frac{F(x+h)+F(x-h)-2F(x)}{h^2}=0$$

for all values of x in the interval, then $F(x)$ is a linear function.

The expression on the left is called the generalized second derivative of $F(x)$. If $F(x)$ has an ordinary second derivative, the generalized second derivative is equal to it, and the result follows at once.

To prove the theorem, consider the function

$$\phi(x)=F(x)-F(a)-\frac{x-a}{b-a}\{F(b)-F(a)\}.$$

We have $\phi(a)=0$ and $\phi(b)=0$. If $\phi(x)=0$ for all values of x, the result follows. Otherwise it takes values different from zero, say, for example, positive values. Suppose that $\phi(c)>0$. Let

$$\psi(x)=\phi(x)-\tfrac{1}{2}\epsilon(x-a)(b-x),$$

where ϵ is positive and so small that $\psi(c)>0$. Then $\psi(x)$ has a positive upper bound, say at $x=\xi$, which it attains, since it is continuous. Hence

$$\psi(\xi+h)+\psi(\xi-h)-2\psi(\xi)\leqslant 0.$$

But

$$\frac{\psi(\xi+h)+\psi(\xi-h)-2\psi(\xi)}{h^2}=\frac{F(\xi+h)+F(\xi-h)-2F(\xi)}{h^2}+\epsilon,$$

and the right-hand side tends to ϵ as $h\to 0$. This gives a contradiction. Similarly the supposition that $\phi(x)$ takes negative values leads to a contradiction. Hence $\phi(x)=0$ for all values of x, which is the desired result.

13.85. The proof of the main theorem now follows from Schwarz's theorem. It is sufficient to prove that, if a trigonometrical series converges to zero except at a finite number of points, then it must vanish identically. If the series 13.82 (1) has this property, the function $F(x)$ is continuous, and its generalized second derivative is zero except at a finite number of points. Hence $F(x)$ is linear in the interval between any two exceptional points, and the straight lines which form the graph join at the exceptional points. Now, taking x in Riemann's second theorem to be an exceptional point, it follows from the lemma that the slopes of the lines on the two sides of the exceptional point must be the same. Hence $F(x)$ is linear throughout the whole interval $(0, 2\pi)$, say

$$F(x) = ax+b.$$

Hence
$$\sum_{n=1}^{\infty} \frac{a_n \cos nx + b_n \sin nx}{n^2} = \tfrac{1}{4}a_0 x^2 - ax - b.$$

Since the sum of the series is periodic, a_0 and a must be zero. Then, the series being uniformly convergent, we may multiply by $\cos mx$ or $\sin mx$ and integrate term by term; and we obtain

$$\frac{\pi a_m}{m^2} = -b \int_0^{2\pi} \cos mx\, dx = 0, \qquad \frac{\pi b_m}{m^2} = -b \int_0^{2\pi} \sin mx\, dx = 0,$$

for $m > 0$. This completes the proof.

13.9. Fourier series for any range. All our series so far have represented functions with the period 2π. A series of the form

$$f(x) = \tfrac{1}{2}a_0 + \sum_{n=1}^{\infty} \left(a_n \cos\frac{nx}{\lambda} + b_n \sin\frac{nx}{\lambda}\right)$$

represents a function with the period $2\pi\lambda$. Formulae for the coefficients may be calculated as before; we obtain

$$a_n = \frac{1}{\pi\lambda} \int_{-\pi\lambda}^{\pi\lambda} f(t)\cos\frac{nt}{\lambda}\, dt, \qquad b_n = \frac{1}{\pi\lambda} \int_{-\pi\lambda}^{\pi\lambda} f(t)\sin\frac{nt}{\lambda}\, dt.$$

Naturally the whole theory can be applied to series of this kind.

13.91. Fourier's integral formula. The above expansion may be written

$$f(x) = \frac{1}{2\pi\lambda} \int_{-\pi\lambda}^{\pi\lambda} f(t)\, dt + \sum_{n=1}^{\infty} \frac{1}{\pi\lambda} \int_{-\pi\lambda}^{\pi\lambda} f(t)\cos\frac{n(x-t)}{\lambda}\, dt.$$

Suppose now that $\lambda \to \infty$. Then the series on the right behaves very much like one of the sums by which a Riemann integral is defined. In fact, if we write $u_n = n/\lambda$, it is

$$\sum_{n=1}^{\infty} (u_{n+1} - u_n)\phi(u_n),$$

where $$\phi(u) = \frac{1}{\pi}\int_{-\pi\lambda}^{\pi\lambda} f(t)\cos u(x-t)\,dt.$$

If, therefore, we make $\lambda \to \infty$, and ignore such difficulties as the fact that $\phi(u)$ depends on λ, and that the approximating sum is an infinite series, we obtain

$$f(x) = \frac{1}{\pi}\int_0^\infty du \int_{-\infty}^\infty \cos u(x-t)f(t)\,dt.$$

This is Fourier's integral formula. It represents a function defined over $(-\infty, \infty)$ in the same way that a Fourier series represents a function with a finite period.

The difficulty of justifying a proof on these lines would be considerable. A direct consideration of the formula suggested is comparatively easy.

13.92. Suppose that $f(x)$ is integrable in the Lebesgue sense over $(-\infty, \infty)$. Then the integral

$$\int_{-\infty}^\infty \cos u(x-t)f(t)\,dt$$

converges uniformly with respect to u over any finite range. We may therefore integrate with respect to u over $(0, U)$, and invert the order of integration. Thus

$$\int_0^U du \int_{-\infty}^\infty \cos u(x-t)f(t)\,dt = \int_{-\infty}^\infty \frac{\sin U(x-t)}{x-t}f(t)\,dt.$$

Given ϵ, we can choose T so large that

$$\int_{-\infty}^{-T} |f(t)|\,dt < \epsilon, \qquad \int_T^\infty |f(t)|\,dt < \epsilon,$$

and we may suppose that $T > |x|+1$, x being supposed fixed. Then

$$\left|\int_{-\infty}^{-T} \frac{\sin U(x-t)}{x-t}f(t)\,dt\right| < \epsilon, \qquad \left|\int_T^\infty \frac{\sin U(x-t)}{x-t}f(t)\,dt\right| < \epsilon,$$

for all values of U. Having fixed T, the integrals

$$\int_{-T}^{x-\delta} \frac{\sin U(x-t)}{x-t}f(t)\,dt, \qquad \int_{x+\delta}^{T} \frac{\sin U(x-t)}{x-t}f(t)\,dt$$

tend to zero as $U \to \infty$, by the Riemann-Lebesgue lemma. Hence

$$\frac{1}{\pi}\int_0^U du \int_{-\infty}^{\infty} \cos u(x-t) f(t)\, dt = \frac{1}{\pi}\int_{x-\delta}^{x+\delta} \frac{\sin U(x-t)}{x-t} f(t)\, dt + o(1)$$

$$= \frac{1}{\pi}\int_0^{\delta} \frac{\sin Ut}{t} \{f(x+t)+f(x-t)\}\, dt + o(1).$$

The value of the limit, as $U \to \infty$, therefore depends only on the behaviour of $f(t)$ in the immediate neighbourhood of $t = x$; and the problem has been reduced to the discussion of an integral similar to Dirichlet's. Any of the convergence criteria of §§ 13.231–3 apply equally well to this problem. *In particular*

$$\lim_{U\to\infty} \frac{1}{\pi}\int_0^U du \int_{-\infty}^{\infty} \cos u(x-t) f(t)\, dt = \tfrac{1}{2}\{f(x+0)+f(x-0)\}$$

if $f(t)$ is integrable over $(-\infty, \infty)$, and of bounded variation in an interval including $t = x$.

13.93. Fourier transforms. If $f(x)$ is an even function, Fourier's integral becomes

$$f(x) = \frac{2}{\pi}\int_0^{\infty} \cos xu\, du \int_0^{\infty} \cos ut\, f(t)\, dt, \tag{1}$$

the term involving $\sin ut$ vanishing identically. This is Fourier's cosine formula. Similarly for an odd function we obtain Fourier's sine formula

$$f(x) = \frac{2}{\pi}\int_0^{\infty} \sin xu\, du \int_0^{\infty} \sin ut\, f(t)\, dt. \tag{2}$$

If we write $$g(x) = \sqrt{\left(\frac{2}{\pi}\right)}\int_0^{\infty} \cos xt\, f(t)\, dt, \tag{3}$$

then (1) gives $$f(x) = \sqrt{\left(\frac{2}{\pi}\right)}\int_0^{\infty} \cos xt\, g(t)\, dt. \tag{4}$$

There is therefore a reciprocal relation between the functions $f(x)$ and $g(x)$; a pair of functions connected in one sense or another by these formulae are known as *Fourier cosine transforms* of one another. Thus, for example, if $f(x)$ belongs to $L(0,\infty)$, and is of bounded variation in any finite interval, then (3) is absolutely convergent, and (4) holds in the sense that the integral converges (not necessarily absolutely) to

$$\tfrac{1}{2}\{f(x+0)+f(x-0)\}.$$

Similarly from (2) we obtain the reciprocal formulae

$$h(x)=\sqrt{\left(\frac{2}{\pi}\right)}\int\limits_0^\infty \sin xt\, f(t)\,dt, \qquad f(x)=\sqrt{\left(\frac{2}{\pi}\right)}\int\limits_0^\infty \sin xt\, h(t)\,dt, \tag{5}$$

and $f(x)$ and $h(x)$ are *Fourier sine transforms*.

13.94. Integration of Fourier integrals. It is convenient to notice at this point a theorem similar to that of § 13.5: *the formula obtained by integrating* 13.93 (1),

$$\int\limits_0^\xi f(x)\,dx=\frac{2}{\pi}\int\limits_0^\infty \frac{\sin \xi u}{u}\,du\int\limits_0^\infty \cos ut\, f(t)\,dt,$$

holds for any function $f(t)$ integrable over $(0,\infty)$.

For

$$\int\limits_0^U \frac{\sin \xi u}{u}\,du\int\limits_0^\infty \cos ut\, f(t)\,dt=\int\limits_0^\infty f(t)\,dt\int\limits_0^U \frac{\sin \xi u \cos ut}{u}\,du$$

by uniform convergence; and the inner integral on the right is bounded for all U and t; for it is equal to

$$\frac{1}{2}\int\limits_0^U \frac{\sin(\xi+t)u}{u}\,du+\frac{1}{2}\int\limits_0^U \frac{\sin(\xi-t)u}{u}\,du$$

$$=\frac{1}{2}\int\limits_0^{U(\xi+t)} \frac{\sin v}{v}\,dv\pm\frac{1}{2}\int\limits_0^{U|\xi-t|} \frac{\sin v}{v}\,dv,$$

the sign being that of $\xi-t$, and

$$\int\limits_0^V \frac{\sin v}{v}\,dv$$

is a bounded function of V. Hence, by Lebesgue's convergence theorem, we may make $U \to \infty$ under the integral sign. Since

$$\int_0^\infty \frac{\sin \xi u \cos ut}{u}\, du = \tfrac{1}{2}\pi \quad (t<\xi), \qquad =0 \quad (t>\xi),$$

the result now follows.

A similar result may be obtained from Fourier's sine formula.

13.95. Fourier transforms of the class L^2. The analysis of § 13.93 gives conditions under which the reciprocal formulae connecting Fourier transforms hold; but they suffer from the defect that, while the formulae are symmetrical in $f(x)$ and $g(x)$, the conditions which these functions satisfy are quite different. An alternative set of conditions, which has perfect symmetry, can be obtained by considering functions of the class L^2, and using the theory of mean convergence.*

Let $f(x)$ belong to the class $L^2(0,\infty)$. Then the formulae for cosine transforms hold in the sense that, as $a \to \infty$, the integral

$$g_a(x) = \sqrt{\left(\frac{2}{\pi}\right)} \int_0^a \cos xt\, f(t)\, dt \tag{1}$$

converges in mean to a function $g(x)$ of the class $L^2(0,\infty)$; and

$$f_a(x) = \sqrt{\left(\frac{2}{\pi}\right)} \int_0^a \cos xt\, g(t)\, dt \tag{2}$$

converges in mean to $f(x)$.

We prove this by a method suggested by the formal process of § 13.92. Let

$$a_n = \int_{n/\lambda}^{(n+1)/\lambda} f(x)\, dx \qquad (n=1,2,\ldots).$$

Then, as $\lambda \to \infty$, the sum

$$\Phi_{m,n} = \sum_{\nu=m+1}^{n} a_\nu \cos \frac{\nu x}{\lambda}$$

tends to the integral

$$\int_a^b \cos ux\, f(u)\, du,$$

if $0 \leqslant a < b$, and $m=[\lambda a]$, $n=[\lambda b]-1$; for the difference is

* Plancherel (1), (2), (3); Titchmarsh (1), (3); Hardy (12); Pollard (1).

$$\sum_{\nu=m+1}^{n} \int_{\nu/\lambda}^{(\nu+1)/\lambda} \left(\cos ux - \cos\frac{\nu x}{\lambda}\right) f(u)\,du + \\ + \int_{a}^{(m+1)/\lambda} \cos ux f(u)\,du + \int_{(n+1)/\lambda}^{b} \cos ux f(u)\,du,$$

and
$$\left|\cos ux - \cos\frac{\nu x}{\lambda}\right| \leqslant \frac{|x|}{\lambda},$$

so that the sum is $O(1/\lambda)$, while the last two integrals plainly tend to zero. Further, the convergence is clearly uniform with respect to x for $0 \leqslant x \leqslant X$.

Now we can apply to $\Phi_{m,n}$ an argument similar to that used in proving the Riesz-Fischer theorem. We have

$$a_n^2 \leqslant \int_{n/\lambda}^{(n+1)/\lambda} \{f(x)\}^2\,dx \int_{n/\lambda}^{(n+1)/\lambda} dx = \frac{1}{\lambda} \int_{n/\lambda}^{(n+1)/\lambda} \{f(x)\}^2\,dx;$$

and hence

$$\int_0^{\pi\lambda} \Phi_{m,n}^2\,dx = \tfrac{1}{2}\pi\lambda \sum_{\nu=m+1}^{n} a_\nu^2 \leqslant \tfrac{1}{2}\pi \int_{(m+1)/\lambda}^{(n+1)/\lambda} \{f(x)\}^2\,dx \leqslant \tfrac{1}{2}\pi \int_a^b \{f(x)\}^2\,dx,$$

and *a fortiori*
$$\int_0^X \Phi_{m,n}^2\,dx \leqslant \tfrac{1}{2}\pi \int_a^b \{f(x)\}^2\,dx$$

if $\pi\lambda > X$. Keeping X fixed and making $\lambda \to \infty$, we obtain

$$\int_0^X \{g_b(x) - g_a(x)\}^2\,dx \leqslant \int_a^b \{f(x)\}^2\,dx,$$

and then, making $X \to \infty$,

$$\int_0^\infty \{g_b(x) - g_a(x)\}^2\,dx \leqslant \int_a^b \{f(x)\}^2\,dx. \tag{3}$$

Since the right-hand side tends to zero as $a \to \infty$, $b \to \infty$, so does the left-hand side; that is, $g_a(x)$ converges in mean to a function, $g(x)$ say, of the class $L^2(0, \infty)$.

The same argument now shows that the integral (2) converges in mean, to a function $\phi(x)$, say. We have to prove that $\phi(x) = f(x)$ almost everywhere, and for this it is sufficient to show that

$$\int_0^\xi \phi(x)\,dx = \int_0^\xi f(x)\,dx \tag{4}$$

for all values of ξ. Now

$$\int_0^\xi \phi(x)\,dx = \lim_{a\to\infty}\int_0^\xi f_a(x)\,dx = \lim_{a\to\infty}\sqrt{\left(\frac{2}{\pi}\right)}\int_0^\xi dx\int_0^a \cos xt\,g(t)\,dt$$

$$= \lim_{a\to\infty}\sqrt{\left(\frac{2}{\pi}\right)}\int_0^a \frac{\sin\xi t}{t}\,g(t)\,dt = \sqrt{\left(\frac{2}{\pi}\right)}\int_0^\infty \frac{\sin\xi t}{t}\,g(t)\,dt.$$

On the other hand, for $0<\xi<a$,

$$\int_0^\xi f(x)\,dx = \frac{2}{\pi}\int_0^\infty \frac{\sin\xi u}{u}\,du\int_0^a \cos ut\,f(t)\,dt = \sqrt{\left(\frac{2}{\pi}\right)}\int_0^\infty \frac{\sin\xi u}{u}\,g_a(u)\,du,$$

by § 13.94, $f(x)$ being integrable over $(0,a)$. Making $a\to\infty$, and observing that $\sin\xi u/u$ belongs to L^2, we obtain, by § 12.53,

$$\int_0^\xi f(x)\,dx = \sqrt{\left(\frac{2}{\pi}\right)}\int_0^\infty \frac{\sin\xi u}{u}\,g(u)\,du.$$

This proves (4), and completes the proof of the theorem.

There is, of course, a similar theorem for Fourier sine transforms.

13.96. We can also obtain a formula corresponding to Parseval's theorem. Putting $a=0$ in 13.95 (3), we have

$$\int_0^\infty \{g_b(x)\}^2\,dx \leqslant \int_0^b \{f(x)\}^2\,dx \leqslant \int_0^\infty \{f(x)\}^2\,dx,$$

and making $b\to\infty$, by § 12.51

$$\int_0^\infty \{g(x)\}^2\,dx \leqslant \int_0^\infty \{f(x)\}^2\,dx.$$

But since the relation between $f(x)$ and $g(x)$ is reciprocal, the opposite inequality also holds. Hence in fact

$$\int_0^\infty \{g(x)\}^2\,dx = \int_0^\infty \{f(x)\}^2\,dx. \tag{1}$$

Finally, if $\phi(x)$ also belongs to L^2, and $\psi(x)$ is its transform, then $g(x)+\psi(x)$ is the transform of $f(x)+\phi(x)$. Hence

$$\int_0^\infty \{g(x)+\psi(x)\}^2\,dx = \int_0^\infty \{f(x)+\phi(x)\}^2\,dx,$$

and, subtracting (1) and the corresponding formula for ϕ and ψ, we obtain

$$\int_0^\infty g(x)\psi(x)\,dx = \int_0^\infty f(x)\phi(x)\,dx. \tag{2}$$

MISCELLANEOUS EXAMPLES

1. If $f(x)$ is first defined in $(0, \pi)$, then in $(-\pi, 0)$ by the equation $f(-x) = f(x)$, and elsewhere by periodicity, show that $f(x)$ has the Fourier cosine series

$$\tfrac{1}{2}a_0 + \sum_{n=1}^\infty a_n \cos nx,$$

where
$$a_n = \frac{2}{\pi}\int_0^\pi f(t)\cos nt\,dt.$$

Similarly, if $f(-x) = -f(x)$, then $f(x)$ has the Fourier sine series

$$\sum_{n=1}^\infty b_n \sin nx,$$

where
$$b_n = \frac{2}{\pi}\int_0^\pi f(t)\sin nt\,dt.$$

2. Show that

$$e^{ax} = \frac{e^{2\pi a}-1}{\pi}\left\{\frac{1}{2a} + \sum_{n=1}^\infty \frac{a\cos nx - n\sin nx}{a^2+n^2}\right\} \qquad (0 < x < 2\pi),$$

$$e^{ax} = \frac{e^{a\pi}-1}{a\pi} + \frac{2}{\pi}\sum_{n=1}^\infty \{(-1)^n e^{a\pi}-1\}\frac{a\cos nx}{a^2+n^2} \qquad (0 < x < \pi),$$

$$e^{ax} = \frac{2}{\pi}\sum_{n=1}^\infty \{1-(-1)^n e^{a\pi}\}\frac{n\sin nx}{a^2+n^2} \qquad (0 < x < \pi).$$

Find the sums of the series when $x = 0$.

3. Sum the series

$$\sum_{n=1}^\infty \frac{a\cos nx}{a^2+n^2}, \qquad \sum_{n=1}^\infty \frac{n\sin nx}{a^2+n^2} \qquad (0 < x < 2\pi).$$

4. Expand in Fourier series valid over $(0, 2\pi)$, and also in Fourier cosine and sine series valid over $(0, \pi)$, the functions

$$1, \quad x, \quad x^2, \quad x^3, \quad \cos ax, \quad \sin ax, \quad \cosh ax, \quad \sinh ax,$$
$$e^{ax}\cos bx, \quad e^{ax}\sin bx, \quad [x/\pi], \quad [2x/\pi].$$

Consider the values of x for which the series converge to a value different from the value of the function expanded.

5. Prove that, if $-1 < r < 1$,

$$\frac{1-r^2}{1-2r\cos\theta+r^2} = 1+2\sum_{n=1}^{\infty} r^n \cos n\theta$$

for all values of θ.

6. If a_n, b_n denote the Fourier coefficients of $f(x)$, then for $-1 < r < 1$

$$\tfrac{1}{2}a_0+\sum_{n=1}^{\infty}(a_n\cos nx+b_n\sin nx)r^n = \frac{1}{2\pi}\int_0^{2\pi}\frac{1-r^2}{1-2r\cos(x-t)+r^2}f(t)\,dt.$$

7. Prove that

$$\lim_{r\to 1}\frac{1}{2\pi}\int_0^{2\pi}\frac{1-r^2}{1-2r\cos(x-t)+r^2}f(t)\,dt = \tfrac{1}{2}\{f(x+0)+f(x-0)\}$$

for all values of x for which the right-hand side exists.

[The discussion is similar to that of Fejér's integral.]

8. Show that, if $f(x)$ is bounded, then

$$s_n = O(\log n).$$

9. Show that, if $m \leqslant f(x) \leqslant M$, then

$$m \leqslant \sigma_n(x) \leqslant M$$

for all values of n and x.

10. Show that, if $m \leqslant f(x) \leqslant M$, and

$$|a_n| \leqslant \frac{A_1}{n}, \qquad |b_n| \leqslant \frac{A_2}{n},$$

then
$$m-A_1-A_2 \leqslant s_n \leqslant M+A_1+A_2.$$

[Use the formula

$$s_n = \sigma_{n+1}-\frac{1}{n+1}\sum_{\nu=1}^{n}\nu(a_\nu\cos\nu x+b_\nu\sin\nu x).]$$

11. Show that

$$\frac{\pi-x}{2} = \frac{\sin x}{1}+\frac{\sin 2x}{2}+\frac{\sin 3x}{3}+\ldots \qquad (0 < x < 2\pi),$$

and deduce that

$$\left|\frac{\sin x}{1}+\frac{\sin 2x}{2}+\ldots+\frac{\sin nx}{n}\right| \leqslant \tfrac{1}{2}\pi+1$$

for all values of n and x.

[Compare § 1.76. The actual upper bound of the partial sums is $\int_0^{\pi}\frac{\sin x}{x}\,dx = 1.85\ldots$; see Gronwall (1).]

12. Use Parseval's theorem to sum the series

$$\sum_{n=1}^{\infty}\frac{1}{n^2}, \qquad \sum_{n=1}^{\infty}\frac{1}{(a^2+n^2)^2}, \qquad \sum_{n=1}^{\infty}\frac{n^2}{(a^2+n^2)^2}.$$

13. A necessary and sufficient condition that

$$a_n = O(e^{-(k-\epsilon)n}), \qquad b_n = O(e^{-(k-\epsilon)n}),$$

where $k > 0$, for all positive values of ϵ, is that $f(x)$ should be almost everywhere equal to the value on the real axis of an analytic function $f(z)$, which is regular for $-k < y < k$, and has the period 2π.

14. Construct a Fourier series for which

$$s_n(0) > \frac{\log n}{\log\log n}$$

for arbitrarily large values of n.

15. Show that if, in the Fourier series of § 13.42, we substitute $\nu! x$ for x in the terms corresponding to the group of numbers G_{λ_ν}, we obtain a series which is also the Fourier series of a continuous function, and which diverges for all values of x such that x/π is a rational number.

16. Show that, if the series

$$\sum_{m=0}^{\infty} \alpha_m \cos(2^m x)$$

is a Fourier series, it is convergent for almost all values of x.

[In this case the formula used in example 10 becomes

$$\sigma_{2^k} - s_{2^k} = \frac{1}{2^k} \sum_{m=1}^{k-1} 2^m \alpha_m \cos(2^m x),$$

and since $\alpha_m \to 0$ the right-hand side tends to 0 for all values of x. Hence s_{2^k} tends to a limit wherever σ_{2^k} does, i.e. almost everywhere. See Kolmogoroff (1).]

17. If $f(x) = x^{-\alpha}$, where $0 < \alpha < 1$, for $0 < x \leqslant 2\pi$, show that, as $n \to \infty$,

$$a_n \sim \frac{n^{\alpha-1}}{2\Gamma(\alpha)\cos\frac{1}{2}\pi\alpha}, \qquad b_n \sim \frac{n^{\alpha-1}}{2\Gamma(\alpha)\sin\frac{1}{2}\pi\alpha}.$$

Show that $f(x)$ belongs to L^p if $p < 1/\alpha$, and that $\sum(|a_n|^q + |b_n|^q)$ is divergent if $q < 1/(1-\alpha)$.

[See Bromwich, *Infinite Series* (2nd ed.), § 174, Ex. 5, and Haslam-Jones (1). The result should be compared with the extended Riesz-Fischer theorem referred to in § 13.71. It shows that the exponent of convergence of the series of coefficients is the 'best possible'.]

18. A function $f(x)$ is equal to $\nu^\alpha \cos(\nu^2 x)$ in the intervals

$$\frac{\pi}{(\nu+1)^\beta} < x \leqslant \frac{\pi}{\nu^\beta}, \qquad \nu = 1, 2, \ldots,$$

where $0 < \alpha < \beta < 1$, and is defined in $(-\pi, 0)$ by the relation

$$f(-x) = -f(x).$$

Show that $f(x)$ is integrable in the Lebesgue sense, and that its Fourier sine coefficients satisfy

$$b_n = O(n^{\frac{1}{2}\alpha-\frac{1}{2}}\log n).$$

By taking α small enough and β/α near enough to 1, show that the convergence of $\sum|b_n|^q$, where $q > 2$, is not sufficient to ensure that $f(x)$ shall belong to L^p, where $p = p(q) > 1$.

[The point of the example is that if $q = 2$ the convergence of $\sum |b_n|^q$ *does* imply that $f(x)$ belongs to L^2, and there is an abrupt change in the state of affairs when q becomes greater than 2.

We have

$$b_n = \frac{1}{\pi}\sum_{\nu=1}^{\infty} \nu^{\alpha} \int\limits_{\pi/(\nu+1)^{\beta}}^{\pi/\nu^{\beta}} \{\sin(n+\nu^2)x + \sin(n-\nu^2)x\}\, dx.$$

The terms for which $\sqrt{n}-2 \leqslant \nu \leqslant \sqrt{n}+2$ are

$$O(\nu^{\alpha-\beta-1}) = O(n^{\frac{1}{2}(\alpha-\beta-1)});$$

the terms for which $\nu \leqslant \sqrt{n}-2$ are

$$O\left(\sum_{\nu\leqslant\sqrt{n}-2} \frac{\nu^{\alpha}}{n-\nu^2}\right) = O\left(n^{\frac{1}{2}\alpha}\int\limits_0^{\sqrt{n}-1} \frac{du}{n-u^2}\right)$$

$$= O\left(n^{\frac{1}{2}\alpha-\frac{1}{2}}\int\limits_0^{1-n^{-\frac{1}{2}}} \frac{dv}{1-v^2}\right) = O(n^{\frac{1}{2}\alpha-\frac{1}{2}}\log n),$$

and a similar result holds for the remaining terms. See also Titchmarsh (2).]

19. Show that the function

$$f(x) = -x + \lim_{m\to\infty} \int\limits_0^x (1+\cos t)(1+\cos 4t)...(1+\cos 4^{m-1}t)\, dt$$

is continuous and of bounded variation, and has the period 2π; but that, if b_n is its nth Fourier sine coefficient, nb_n does not tend to zero so that $f(x)$ is not an integral.

[This example is due to F. Riesz (3). Let $\tau_m(x)$ denote the integrand. It is a cosine polynomial of order

$$1+4+...+4^{m-1} = \tfrac{1}{3}(4^m-1).$$

On multiplying by $1+\cos 4^m x$, the first new term involves

$$\cos\{4^m - \tfrac{1}{3}(4^m-1)\}x = \cos\tfrac{1}{3}(2.4^m+1)x,$$

which is of higher order than any of the terms in $\tau_m(x)$. Hence $\tau_{m+1}(x)$ is obtained by adding new terms to $\tau_m(x)$ without altering the existing ones. Also it is easily seen that all the coefficients lie between 0 and 1.

Let α_m be the number of non-vanishing terms in τ_m. The recurrence relation $\alpha_{m+1} = 3\alpha_m - 1$ is easily verified. Hence $\alpha_{m+1}-\alpha_m = 3(\alpha_m-\alpha_{m-1})$, $\alpha_{m+1}-\alpha_m = 3^m$. Hence, if $0 < x \leqslant 2\pi$,

$$\left|\int\limits_0^x \{\tau_{m+1}(t)-\tau_m(t)\}\, dt\right| \leqslant 2\pi\frac{3^m}{\frac{1}{3}(2.4^m+1)}.$$

Hence $\int_0^x \tau_m(t)\, dt$ tends uniformly to a limit, i.e. $f(x)$ is continuous. Also

$\tau_m(x)$ is non-decreasing, and so is its limit. Hence $f(x)$ is of bounded variation. Finally $b_{4^m} = 1/4^m$.]

20. Show that, if $a_n \cos nx + b_n \sin nx \to 0$ in a set of positive measure, then $a_n \to 0$ and $b_n \to 0$.

21. Show that the reciprocal formulae

$$F(x) = \int_{-\infty}^{\infty} e^{ixt} f(t)\, dt, \qquad f(x) = \frac{1}{2\pi} \int_{-\infty}^{\infty} e^{-ixt} F(t)\, dt,$$

hold under the same conditions as Fourier's integral.

22. Show that Mellin's inversion formulae

$$\phi(s) = \int_{0}^{\infty} x^{s-1} \psi(x)\, dx, \qquad \psi(x) = \frac{1}{2\pi i} \int_{c-i\infty}^{c+i\infty} \phi(s) x^{-s}\, ds,$$

may, with suitable conditions, be deduced from the formulae of the previous example.

23. Show that the functions

$$x^{-\frac{1}{2}}, \qquad e^{-\frac{1}{2}x^2}, \qquad \operatorname{sech} x\sqrt{(\tfrac{1}{2}\pi)}$$

are their own Fourier cosine transforms, and that

$$x^{-\frac{1}{2}}, \qquad xe^{-\frac{1}{2}x^2}, \qquad \frac{1}{e^{x\sqrt{(2\pi)}}-1} - \frac{1}{x\sqrt{(2\pi)}}$$

are their own sine transforms.

24. Express $e^{-a|x|}$, where $a > 0$, as a Fourier integral. Verify the formula 13.96 (2) in the case where $f(x) = e^{-ax}$, $\phi(x) = e^{-bx}$.

25. Evaluate the integral

$$\int_{0}^{\infty} \frac{\sin ax \sin bx}{x^2}\, dx$$

by means of the formula 13.96 (2).

26. Let $f(x)$ belong to $L(0, \infty)$, and be continuous and steadily decreasing to zero as $x \to \infty$ (or be the difference between two functions of this type). Let $\alpha > 0$, $\alpha\beta = 2\pi$, and let $g(x)$ be the Fourier cosine transform of $f(x)$. Then

$$\sqrt{\alpha}\left\{\tfrac{1}{2}f(0) + \sum_{n=1}^{\infty} f(n\alpha)\right\} = \sqrt{\beta}\left\{\tfrac{1}{2}g(0) + \sum_{n=1}^{\infty} g(n\beta)\right\}.$$

[This is known as Poisson's formula. It is easily verified that

$$\sqrt{\beta}\left\{\tfrac{1}{2}g(0) + \sum_{m=1}^{n} g(m\beta)\right\}$$

$$= \frac{\sqrt{\alpha}}{2\pi} \int_{0}^{\pi} f\left(\frac{t}{\beta}\right) \frac{\sin(n+\frac{1}{2})t}{\sin \frac{1}{2}t}\, dt + \frac{\sqrt{\alpha}}{2\pi} \sum_{m=1}^{\infty} \int_{(2m-1)\pi}^{(2m+1)\pi} f\left(\frac{t}{\beta}\right) \frac{\sin(n+\frac{1}{2})t}{\sin \frac{1}{2}t}\, dt.$$

This differs from the left-hand side of Poisson's formula by

$$\frac{\sqrt{\alpha}}{2\pi}\int_{0}^{\pi}\left\{f\left(\frac{t}{\beta}\right)-f(0)\right\}\frac{\sin(n+\frac{1}{2})t}{\sin\frac{1}{2}t}\,dt+$$

$$+\frac{\sqrt{\alpha}}{2\pi}\sum_{m=1}^{\infty}\int_{(2m-1)\pi}^{(2m+1)\pi}\left\{f\left(\frac{t}{\beta}\right)-f\left(\frac{2m\pi}{\beta}\right)\right\}\frac{\sin(n+\frac{1}{2})t}{\sin\frac{1}{2}t}\,dt.$$

The given conditions ensure that this series converges uniformly with respect to n; in fact, it is easily seen from the second mean-value theorem that the general term is $O[f\{(2m-1)\pi/\beta\}]$ independently of n; and each term tends to zero as $n \to \infty$ (as in the proof of Jordan's test), and the result follows.

For other conditions for the formula see Linfoot (1), Mordell (2).]

27. Verify Poisson's formula for the function $f(x) = 1/(1+x^2)$. [The result is equivalent to that of § 3.22, ex. (iii).]

28. Deduce from Poisson's formula that if $x > 0$

$$\sum_{n=-\infty}^{\infty} e^{-n^2x^2} = \frac{\sqrt{\pi}}{x}\sum_{n=-\infty}^{\infty} e^{-n^2\pi^2/x^2}.$$

29. Sum the series $\sum_{n=1}^{\infty} n^{-\nu}J_\nu(n\beta)$, where $\beta > 0$, $\nu > \frac{1}{2}$, by means of Poisson's formula and the first result of Ch. 1, ex. 5.

BIBLIOGRAPHY

This is a list of books (so far as possible in English) where the subjects touched on here may be consulted further.

Chapter 1. The standard works in English are—

BROMWICH, T. J. I'A. *Theory of Infinite Series.* London, ed. 2, 1926.

KNOPP, K. *Theory and Application of Infinite Series.* English translation by Miss R. C. Young, London, 1928.

And, on the calculus side,

CHAUNDY, T. W. *The Differential Calculus.* Oxford, 1935.

Chs. 2–5.

COPSON, E. T. *Theory of Functions of a Complex Variable.* Oxford, 1935.

DIENES, P. *The Taylor Series.* Oxford, 1931.

WHITTAKER, E. T., and WATSON, G. N. *Modern Analysis.* Cambridge, ed. 4, 1927.

On Cauchy's theorem specially, see

WATSON, G. N. *Complex Integration and Cauchy's Theorem.* Cambridge tracts, No. 15, 1914.

Ch. 6. See DIENES, and also

CARATHÉODORY, C. *Conformal Representation.* Cambridge tracts, No. 28, 1932.

Ch. 7. See DIENES, and also

LANDAU, E. *Darstellung und Begründung einiger neuerer Ergebnisse der Funktionentheorie.* Berlin, ed. 2, 1929.

Ch. 8.

VALIRON, G. *Lectures on the General Theory of Integral Functions.* Toulouse, 1923.

NEVANLINNA, R. *Le théorème de Picard-Borel et la théorie des fonctions méromorphes.* Paris, 1929.

NEVANLINNA, R. *Eindeutige analytische Funktionen.* Berlin, 1936.

Ch. 9.

HARDY, G. H., and RIESZ, M. *The General Theory of Dirichlet's Series.* Cambridge tracts, No. 18, 1915.

BESICOVITCH, A. S. *Almost Periodic Functions.* Cambridge, 1932.

Chs. 10–12.

HOBSON, E. W. *The Theory of Functions of a Real Variable.* Cambridge, ed. 2, 1921–6.

KESTELMAN, H. *Modern Theories of Integration.* Oxford, 1937.

LITTLEWOOD, J. E. *The Elements of the Theory of Real Functions.* Cambridge, ed. 2, 1926.

SAKS, S. *Theory of the Integral.* English translation by L. C. Young. Warsaw, 1937.

YOUNG, L. C. *The Theory of Integration.* Cambridge tracts, No. 21, 1927.

Ch. 13.

CARSLAW, H. S. *Introduction to the Theory of Fourier's Series and Integrals.* London, ed. 3, 1930.

PALEY, R., and WIENER, N. *Fourier Transforms in the Complex Domain.* New York, 1934.

TITCHMARSH, E. C. *Introduction to the Theory of Fourier Integrals.* Oxford, 1937.

WIENER, N. *The Fourier Integral and certain of its Applications.* Cambridge, 1933.

ZYGMUND, A. *Trigonometrical Series.* Warsaw, 1935,

and Hobson's treatise referred to above.

ORIGINAL MEMOIRS REFERRED TO IN THE TEXT

BACKLUND, R. J. (**1**) Über die Nullstellen der Riemannschen Zetafunktion, *Acta Math.* 41 (1918), 345–75.

BESICOVITCH, A. (**1**) Ueber die Beziehung zwischen dem Maximum und Minimum des Moduls einer ganzen Funktion von der Ordnung < 1, *Bull. Acad. Sc. Russ.* (1924), 17–28.

BOHNENBLUST, H. F. (**1**) Note on singularities of power series, *Proc. Nat. Acad. Science U.S.A.* 16 (1930), 752–4.

BOHR, H. (**1**), (**2**), (**3**) Zur Theorie der Fastperiodischen Funktionen, *Acta Math.* 45 (1924), 29–127; 46 (1925), 101–214; 47 (1925), 237–81.

(**4**) On the limit values of analytic functions, *Journal London Math. Soc.* 2 (1927), 180–1.

CARLEMAN, T. (**1**) Über die Approximation analytischer Funktionen durch lineare Aggregate von vorgegebenen Potenzen, *Arkiv för Mat. Astr. o. Fys.* 17 (1922), No. 9.

CARLSON, F. (**1**) Sur une classe de séries de Taylor, thesis, Upsala, 1914.

(**2**), (**3**) Contributions à la théorie des séries de Dirichlet, *Arkiv för Mat. Astr. o. Fys.*, 16 (1922), No. 18, and 19 (1926), No. 25.

Chaundy, T. W., and Jolliffe, A. E. (**1**) The uniform convergence of a certain class of trigonometrical series, *Proc. London Math. Soc.* (2), 15 (1916), 214–16.

Egoroff, D. T. (**1**) Sur les suites de fonctions mesurables, *Comptes Rendus* 152 (1911), 244–6.

Estermann, T. (**1**) On certain functions represented by Dirichlet series, *Proc. London Math. Soc.* (2), 27 (1928), 435–48.

(**2**) On Ostrowski's gap theorem, *Journal London Math. Soc.* 7 (1932), 19–20.

Fatou, P. (**1**) Séries trigonométriques et séries de Taylor, *Acta Math.* 30 (1906), 335–400.

Fejér, L. (**1**) Untersuchungen über Fouriersche Reihen, *Math. Annalen*, 58 (1904), 51–69.

(**2**) Beispiele stetiger Funktionen mit divergenter Fourierreihe, *Journal für Math.* 137 (1910), 1–5.

(**3**) Eine stetige Funktion, deren Fouriersche Reihe divergiert, *Rendiconti di Palermo* 18 (1910), 402–4.

(**4**) Über gewisse Potenzreihen an der Konvergenzgrenze, *Münchener Bericht.* 40 (1910), No. 3.

(**5**) Über die Positivität von Summen, die nach trigonometrischen oder Legendreschen Funktionen fortschreiten, *Acta Reg. Univ. Hungaricae Francisco-Josephinae*, 2 (1925), 75–86.

Fischer, E. Sur la convergence en moyenne, *Comptes Rendus*, 144 (1907), 1022–4.

Gronwall, T. H. (**1**) Über die Gibbssche Erscheinung und die trigonometrischen Summen $\sin x + \frac{1}{2}\sin 2x + \ldots + \frac{1}{n}\sin nx$. *Math. Annalen*, 72 (1912), 228–43.

Hadamard, J. (**1**) Essai sur l'étude des fonctions données par leur developpement de Taylor, *Journal de Math.* (4), 8 (1892), 101–86.

(**2**) Étude sur les propriétés des fonctions entières et en particulier d'une fonction considérée par Riemann, *Journal de Math.* (4), 9 (1893), 171–215.

(**3**) Théorème sur les séries entières, *Acta Math.* 22 (1899), 55–64.

Hardy, G. H. (**1**) On differentiation and integration of divergent series, *Trans. Camb. Phil. Soc.* 19 (1904), 297–321.

(**2**) On the zeros of certain classes of integral Taylor series, *Proc. London Math. Soc.* (2), 2 (1905), 332–9 and 401–31.

(**3**) On double Fourier series, *Quart. J. of Math.* 37 (1905), 53–79.

(**4**) On the function $P_\rho(x)$, *Quart J. of Math.* 37 (1905), 146–72.

(**5**) A note on the continuity or discontinuity of a function defined by an infinite product, *Proc. London Math. Soc.* (2), 7 (1908), 40–8.

(**6**) Further researches in the theory of divergent series and integrals, *Trans. Camb. Phil. Soc.* 21 (1908), 1–48.

HARDY, G. H. (*cont.*)

(**7**) Theorems connected with Maclaurin's test for the convergence of series, *Proc. London Math. Soc.* (2), 9 (1909), 126–44.

(**8**) The mean value of the modulus of an analytic function, *Proc. London Math. Soc.* (2), 14 (1914), 269–77.

(**9**) Weierstrass's non-differentiable function, *Trans Amer. Math. Soc.* 17 (1916), 301–25.

(**10**) The application of Abel's method of summation to Dirichlet series, *Quart. J. of Math.* 47 (1916), 176–92.

(**11**) Sir George Stokes and the concept of uniform convergence, *Proc. Camb. Phil. Soc.* 19 (1918), 148–56.

(**12**) On some properties of integrals of fractional order, *Messenger of Math.* 47 (1918), 145–50.

(**13**) On certain criteria for the convergence of the Fourier series of a continuous function, *Messenger of Math.* 49 (1920), 149–55.

(**14**) On two theorems of F. Carlson and S. Wigert, *Acta Math.* 42 (1920), 327–39.

(**15**) On the integration of Fourier series, *Messenger of Math.* 51 (1922), 186–92.

(**16**) On Fourier transforms, *Messenger of Math.* 53 (1924), 135–42.

(**17**) An inequality between integrals, *Messenger of Math.* 54 (1925), 150–6.

(**18**) A theorem concerning harmonic functions, *Journal London Math. Soc.* 1 (1926), 130–1.

(**19**) Further inequalities between integrals, *Messenger of Math.* 57 (1927), 12–16.

(**20**) Prolegomena to a chapter on inequalities, *Journal London Math. Soc.* 4 (1929), 61–78.

HARDY, G. H., and LITTLEWOOD, J. E. (**1**) Contributions to the arithmetic theory of series, *Proc. London Math. Soc.* (2), 11 (1911) 411–78.

(**2**) Tauberian theorems concerning power series and Dirichlet's series whose coefficients are positive, *Proc. London Math. Soc.* (2), 13 (1914), 174–91.

(**3**) Abel's theorem and its converse, *Proc. London Math. Soc.* (2), 18 (1918), 205–35.

(**4**) Abel's theorem and its converse II, *Proc. London Math. Soc.* (2), 22 (1923), 254–69.

(**5**) Some properties of fractional integrals, *Math. Zeitschrift*, 27 (1928), 565–606.

(**6**) A convergence criterion for Fourier series, *Math. Zeitschrift*, 28 (1928), 612–34.

HASLAM-JONES, U. S. (**1**) A note on the Fourier coefficients of unbounded functions, *Journal London Math Soc.* 2 (1927), 151–4.

HAUSDORFF, F. (1) Eine Ausdehnung des Parsevalschen Satzes über Fourierreihen, *Math. Zeitschrift*, 16 (1923), 163–9.

HILLE, E., and TAMARKIN, J. D. (1) Remarks on a known example of a monotone continuous function, *American Math. Monthly*, 36 (1929), 255–64.

HOBSON, E. W. (1) Generalization of a theorem of F. Riesz, *Journal London Math. Soc.* 1 (1926), 211–18.

HURWITZ, A. (1) Ueber die Nullstellen der Bessel'schen Function, *Math. Annalen*, 33 (1889), 246–66.

IZUMI, S. (1) On the distribution of the zero points of sections of a power series, *Japanese Journal of Math.* 4 (1927), 29–32.

JENTZSCH, R. (1) Untersuchungen zur Theorie der Folgen analytischer Funktionen, *Acta Math.* 41 (1917), 219–70.

KARAMATA, J. (1) Über die Hardy-Littlewoodschen Umkehrungen des Abelschen Stetigkeitssatzes, *Math. Zeitschrift*, 32 (1930), 319–20.

KNOPP, K. (1) Über Lambertsche Reihen, *Journal für Math.* 142 (1912), 283–315.

(2) Ein einfaches Verfahren zur Bildung stetiger, nirgends differenzierbarer Funktionen, *Math. Zeitschrift*, 2 (1918), 1–26.

KOLMOGOROFF, A. (1) Une contribution à l'étude de la convergence des séries de Fourier, *Fundamenta Math.* 5 (1924), 96–7.

LANDAU, E. (1) Über eine Verallgemeinerung des Picardschen Satzes, *Sitzungsber. Preuss. Akad. Wissens.* (1904), 1118–33.

(2), (3), (4) Abschätzung der Koeffizientensumme einer Potenzreihe, *Archiv der Math. und Phys.* (3), 21 (1913), 42–50 and 250–5; and (3), 24 (1916), 250–60.

(5) Über die Zetafunktion und die Hadamardsche Theorie der ganzen Funktionen, *Math. Zeitschrift*, 26 (1927), 170–5.

LANDAU, E., and WALFISZ, A. (1) Über die Nichtfortsetzbarkeit einiger durch Dirichletsche Reihen definierter Funktionen, *Rendiconti di Palermo*, 44 (1919), 82–6.

LINFOOT, E. H. (1) A sufficiency condition for Poisson's formula. *Journal London Math. Soc.* 4 (1928), 54–61.

LITTLEWOOD, J. E. (1) A general theorem on integral functions of finite order, *Proc. London Math. Soc.* (2), 6 (1908), 189–204.

(2) On a class of conditionally convergent infinite products, *Proc. London Math. Soc.* (2), 8 (1910), 195–9.

(3) The converse of Abel's theorem on power series, *Proc. London Math. Soc.* (2), 9 (1911), 434–48.

(4) On the zeros of the Riemann zeta-function, *Proc. Camb. Phil. Soc.* 22 (1924), 295–318.

MONTEL, P. (1) Sur les familles de fonctions analytiques qui admettent des valeurs exceptionnelles dans un domaine, *Annales de l'École Normale*, (3) 23 (1912), 487–535.

MORDELL, L. J. (1) On power series with the circle of convergence as a line of essential singularities, *Journal London Math. Soc.* 2 (1927), 146–8.

(2) Poisson's summation formula and the Riemann zeta-function. *Journal London Math. Soc.* 4, (1928), 285–91.

OSTROWSKI, A. (1) On representation of analytical functions by power series, *Journal London Math. Soc.* 1 (1926), 251–63.

PHRAGMÉN, E., and LINDELÖF, E. (1) Sur une extension d'un principe classique de l'analyse, *Acta Math.* 31 (1908), 381–406.

PLANCHEREL, M. (1) Contribution à l'étude de la représentation d'une fonction arbitraire par des intégrales définies, *Rend. di Palermo*, 30 (1910), 289–335.

(2) Sur la convergence et sur la sommation par les moyennes de Cesàro de $\lim_{z=\infty} \int_a^z f(x) \cos xy\, dx$, *Math. Annalen*, 76 (1915), 315–26.

(3) Sur les formules d'inversion de Fourier et de Hankel, *Proc. London Math. Soc.* (2), 24 (1925), 62–70.

PLESSNER, A. (1) Zur Theorie der konjugierten trigonometrischen Reihen, *Mitteilungen des Math. Seminars der Univ. Giessen*, 1923.

(2) Über Konvergenz von trigonometrischen Reihen, *Journal für Math.* 155 (1926), 15–25.

POLLARD, S. (1) On Fourier's integral, *Proc. London Math. Soc.* (2), 26 (1927), 12–24.

POLYA, G. (1) On the zeros of an integral function represented by Fourier's integral, *Messenger of Math.* 52 (1923), 185–8.

(2) On an integral function of an integral function, *Journal London Math. Soc.* 1 (1926), 12–15.

(3) On the minimum modulus of integral functions, *Journal London Math. Soc.* 1 (1926), 78–86.

(4) Untersuchungen über Lücken und Singularitäten von Potenzreihen, *Math. Zeitschrift*, 29 (1929), 549–640.

RAJCHMAN, A., and SAKS, S. (1) Sur la dérivabilité des fonctions monotones, *Fundamenta Math.* 4 (1923), 204–13.

RAMANUJAN, S. (1) Some formulae in the analytic theory of numbers, *Messenger of Math.* 45 (1915), 81–4.

RIESZ, F. (1) Über orthogonale Funktionensysteme, *Göttinger Nachrichten* (1907), 116–22.

(2) Untersuchungen über Systeme integrierbarer Funktionen, *Math. Annalen*, 69 (1910), 449–97.

(3) Über die Fourierkoeffizienten einer stetigen Funktion von beschränkter Schwankung, *Math. Zeitschrift*, 2 (1918), 312–15.

(4) Über eine Verallgemeinerung der Parsevalschen Formel, *Math. Zeitschrift*, 18 (1923), 117–24.

RIESZ, M. (1) Sur le principe de Phragmén-Lindelöf, *Proc. Camb. Phil. Soc.* 20 (1920), 205–7; and correction, *ibid.* 21 (1921), 6.

RITT, J. F. (1) Representation of analytic functions as infinite products, *Math. Zeitschrift*, 32 (1930), 1–3.

SIERPINSKI, W. (1) Un lemme métrique, *Fundamenta Math.* 4 (1923), 201–3.

TITCHMARSH, E. C. (1) Hankel transforms, *Proc. Camb. Phil. Soc.* 21 (1923), 463–73.

(2) A note on the Riesz-Fischer theorem in the theory of trigonometrical series, *Proc. London Math. Soc.* (2), 22 (1923), Records for February.

(3) A contribution to the theory of Fourier transforms, *Proc. London Math. Soc.* (2), 23 (1924), 279–89.

(4) Conjugate trigonometrical series, *Proc. London Math. Soc.* (2), 24 (1925), 109–30.

(5) A theorem on infinite products, *Journal London Math. Soc.* 1 (1926), 35–7.

(6) On integral functions with real negative zeros, *Proc. London Math. Soc.* (2), 26 (1927), 185–200.

(7) A theorem on Lebesgue integrals, *Journal London Math. Soc.* 2 (1927), 36–7.

(8) On an inequality satisfied by the zeta-function of Riemann, *Proc. London Math. Soc.* (2) 28 (1929), 70–80.

VALIRON, G. (1) Sur les fonctions entières d'ordre nul et d'ordre fini, *Annales de Toulouse*, (3), 5 (1913), 117–257.

WAERDEN, B. L. VAN DER. (1) Ein einfaches Beispiel einer nicht-differenzierbaren stetigen Funktion, *Math. Zeitschrift*, 32 (1930), 474–5.

WATSON, G. N. (1) Theorems stated by Ramanujan (II): Theorems on summation of series. *Journal London Math. Soc.* 3 (1928), 216–25.

WIGERT, S. (1) Sur un théorème concernant les fonctions entières, *Arkiv för Mat. Astr. o. Fys.* 11 (1916), No. 22.

WILSON, B. M. (1) Proofs of some formulae enunciated by Ramanujan, *Proc. London Math. Soc.* (2), 21 (1922), 235–55.

WIMAN, A. (1) Über eine Eigenschaft der ganzen Funktionen von der Höhe Null, *Math. Annalen*, 76 (1915), 197–211.

YOUNG, W. H. (1) On the integration of Fourier series, *Proc. London Math. Soc.* (2), 9 (1910), 449–62.

(2) Sur la généralisation du theorème de Parseval, *Comptes Rendus*, 155 (1912), 30–3.

(3) Sur la sommabilité d'une fonction dont la série de Fourier est donnée, *Comptes Rendus*, 155 (1912), 472–5.

(4) On classes of summable functions and their Fourier series, *Proc. Royal Soc.* (A), 87 (1912), 225–9.

YOUNG, W. H. (*cont.*)

(**5**) On the multiplication of successions of Fourier constants, *Proc. Royal Soc.* (A), 87 (1912), 331–9.

(**6**) On the determination of the summability of a function by means of its Fourier constants, *Proc. London Math. Soc.* (2), 12 (1913), 71–88.

(**7**) On restricted Fourier series and the convergence of power series, *Proc. London Math. Soc.* 17 (1918), 353–66.

YOUNG, W. H. and YOUNG, G. C. (**1**) On the existence of a differential coefficient, *Proc. London Math. Soc.* (2), 9 (1910), 325–35.

(**2**) On the theorem of Riesz-Fischer, *Quart. J. of Math.* 44 (1913), 49–88.

ZYGMUND, A. (**1**) On a theorem of Ostrowski, *Journal London Math. Soc.* 6 (1931), 162–3.

GENERAL INDEX